Analysis and Applications of Mathematical Fluid Dynamics

Analysis and Applications of Mathematical Fluid Dynamics

Editor

Ramoshweu Solomon Lebelo

MDPI

Basel • Beijing • Wuhan • Barcelona • Belgrade • Novi Sad • Cluj • Manchester

Editor
Ramoshweu Solomon Lebelo
Vaal University of Technology
Vanderbijlpark
South Africa

Editorial Office
MDPI
St. Alban-Anlage 66
4052 Basel, Switzerland

This is a reprint of articles from the Special Issue published online in the open access journal *Mathematics* (ISSN 2227-7390) (available at: https://www.mdpi.com/journal/mathematics/special_issues/J4QN4HK2X1).

For citation purposes, cite each article independently as indicated on the article page online and as indicated below:

Lastname, A.A.; Lastname, B.B. Article Title. *Journal Name* **Year**, *Volume Number*, Page Range.

ISBN 978-3-7258-0753-6 (Hbk)
ISBN 978-3-7258-0754-3 (PDF)
doi.org/10.3390/books978-3-7258-0754-3

Contents

About the Editor

Ramoshweu Solomon Lebelo

RS Lebelo is an Associate Professor in the Education Department at the Vaal University of Technology. His research areas are in Computational Fluid Dynamics (CFDs) and Mathematics Education. He has published extensively in topics relating to CFDs in international peer-reviewed journals indexed in Scopus and ISI and to date has presented his work in more than 31 conferences. He is a reviewer for over 20 peer-reviewed journals and sits on the Editorial Board of several international peer-reviewed journals. He has guest-edited seven Special Issues in reputable journals since 2015.

He co-edited two research books as follows: "International Journal of Engineering Research in Africa Vol. 55" (published by Scientific.Net (2020–2021)) and "Modelling and Numerical Simulations with Differential Equations in Mathematical Biology, Medicine and the Environment" (published by Frontiers in Applied Mathematics and Statistics,). He co-authored the textbook "Mathematics for Engineering Students" (published by Juta & Company (Pty) Ltd.) and authored the research book "Mathematical Approach to Transient Heat Analysis" (published by Lambert Academic Publishing). Furthermore, he authored two spiritual books as follows: "Suffer with Christ" (published by Xlibris Publishing) and "The Real You" (published by Blessed Hope Publishing).

mathematics

MDPI

Article

Periodic Flows in a Viscous Stratified Fluid in a Homogeneous Gravitational Field

Yuli D. Chashechkin * and Artem A. Ochirov

Ishlinsky Institute for Problems in Mechanics RAS, 119526 Moscow, Russia; ochir@ipmnet.ru
* Correspondence: chakin@ipmnet.ru; Tel.: +7-(495)-434-01-92

Abstract: The density of a fluid or gas, which depends on the temperature, pressure and concentration of dissolved substances or suspended particles, changes under the influence of a large number of physical factors. We assume that an undisturbed liquid is heterogeneous. The propagation of periodic flows in viscous, uniformly stratified fluids is considered. The analysis is based on a system of fundamental equations for the transfer of energy, momentum and matter in periodic flows. Taking into account the compatibility condition, dispersion relations are constructed for two-dimensional internal, acoustic and surface linear periodic flows with a positive definite frequency and complex wave number in a compressible viscous fluid exponentially stratified by density. The temperature conductivity and diffusion effects are neglected. The obtained regularly perturbed solutions of the dispersion equations describe the conventional weakly damped waves. The families of singular solutions, specific for every kind of periodic flow, characterize the before unknown thin ligaments that accompany each type of wave. In limited cases, the constructed regular solutions transform into well-known expressions for a viscous homogeneous and an ideal fluid. Singular solutions are degenerated in a viscous homogeneous fluid or disappear in an ideal fluid. The developing method of the fundamental equation system analysis is directed to describe the dynamics and spatial structure of periodic flows in heterogeneous fluids in linear and non-linear approximations.

Keywords: heterogeneous fluid; stratification; viscosity; compressibility; linear models; complete description; dispersion relations

MSC: 76A02; 76Q05; 76M45

Citation: Chashechkin, Y.D.; Ochirov, A.A. Periodic Flows in a Viscous Stratified Fluid in a Homogeneous Gravitational Field. *Mathematics* **2023**, *11*, 4443. https://doi.org/10.3390/math11214443

Academic Editor: Ramoshweu Solomon Lebelo

Received: 28 September 2023
Revised: 23 October 2023
Accepted: 24 October 2023
Published: 26 October 2023

1. Introduction

In natural, laboratory and industrial conditions, the density of a liquid or gas depends on the temperature, pressure, concentration of dissolved substances or suspended particles. It is not constant and changes under the influence of a large number of physical factors. An oscillating source forms waves that propagate over long distances in a medium with a weak dissipation. Historically, it is customary to distinguish acoustic waves, the existence of which is provided by the compressibility of the medium and gravitational waves associated with the action of the gravity field. Inertial waves propagate in a globally rotating medium and capillary waves run at the interface between the media. The existence of a large group of hybrid waves is provided by the combined action of a number of factors [1,2].

In the mass forces (gravity and inertia) field, the fluid medium is separated. Heavy particles sink, light particles float up, and the medium is naturally stratified. Compressibility under the action of hydrostatic pressure has an additional impact on density. The choice of the coordinate system depends on the overall geometry of the problem. The consideration of flows with scales much smaller than the Earth's radius is carried out in a Cartesian coordinate system with an axis z pointing vertically upwards. The acceleration of gravity g is directed downwards.

The density distribution in the direction of gravity $\rho(z)$ is characterized by the scale $\Lambda = |d \ln \rho(z)/dz|^{-1}$, frequency $N = \sqrt{g/\Lambda}$ and buoyancy period $T_b = 2\pi/N$. In the atmosphere and ocean, the average buoyancy period lies in the range of $3 < T_b < 10$ min [3,4]. In the "instantaneous" density profiles of the atmosphere and ocean, thin, highly gradient interfaces are expressed. They separate thick, more homogeneous layers, thereby forming a "fine structure" of the medium [4,5].

In practice, several characteristic types of average density distributions have been identified. Further, the models of continuous (linear or exponential), two-layer or multi-layer stratification (the last two with a persistent density gap) will be used. In a large group of flows, density variations are much less than the average value.

At the end of the 18th century, B. Franklin observed sea fluctuations in the free surface and the interface between water and olive oil in a ship lighting lamp, which was later mounted on a swing. He noted the need to analyze the influence of fluid density heterogeneity in mathematical research [6]. Initially, the effects of stratification began to be taken into account in calculations of the internal wave propagation in the atmosphere and ocean, which were carried out by famous English scientists G.G. Stokes [7], Lord Rayleigh [8], H. Lamb [9] and others.

A systematic study of the influence of stratification on the pattern of flows in the atmosphere and ocean, including navigation (the "dead water" effect), which was noticed in ancient times, gained interest after the publication of the scientific results from F. Nansen's polar expeditions [10]. V. Ekman developed the methodology and planned the experiments. In order to conduct laboratory studies of the phenomenon of the "dead water", he used a review of the first publications on the theory of internal waves in the treatise [11]. In a series of thorough experiments, V. Ekman determined the conditions for the generation of large waves using a moving model of a ship at a smoothed interface between fresh and salt (sea) water and determined the influence of the movement mode on the position of the model's hull and drag [10]. However, in general, the work on the consideration of the equations of internal waves and the "exotic" phenomenon of "dead water" fell out of scientific circulation for more than half a century and did not affect the development of the general theory of fluid flows.

At least two of the reasons have to be noted: the smallness of the density variations compared to its average value, limiting the effect on inertial properties, and the insufficient development of the mathematical apparatus. G.G. Stokes noted in a fundamental article [12], written several years before a thorough study of wave propagation in homogeneous and layer-by-layer stratified media [13], "As it is quite useless to consider cases of the utmost degree of generality, I shall suppose the fluid to be homogeneous..." However, a few years later, he also emphasized the limitations of the approximation used: "The three equations of which (1) is the type are not the general equations of motion which apply to a heterogeneous fluid when internal friction is taken into account, which are those numbered (10) in my former paper, but are applicable to a homogeneous incompressible fluid, or to a homogeneous elastic fluid subject to small variations of density, such as those which accompany sonorous vibrations" [7].

Accordingly, when studying the waves of other types—acoustic [14] or gravitational-capillary—at the interface between the atmosphere and the hydrosphere [1,15], the unperturbed density was assumed to be homogeneous. Here and further, general rotation effects and associated inertial waves [1,16] are not considered.

The interest in the mathematical study of the stratification influence started to form in the middle of the last century. During this period, the precision instruments identified the thin, highly gradient structure of the Baltic Seawaters [5]. Next, the flows induced by diffusion on an inclined wall in a continuously stratified atmosphere were discovered [17]. The development of interest in studying the influence of stratification was facilitated by the papers [18,19], which showed the important role of diffusion-induced flows on topography not only in the atmosphere, where they manifested themselves not only in the form of mountain and valley winds but also in the ocean. At the same time, experimental [20] and theoretical studies of internal waves in continuously stratified media [21] were developing. Numerous expeditions have shown the existence of fine structure and its influence on the dynamics of the atmosphere and ocean in various regions of the Earth.

The number of original articles and reviews describing the influence of stratification on individual phenomena (internal waves, currents and vortices) was increasing rapidly. The propagation of acoustic vibrations in a continuously stratified medium was considered [22]. The influence of viscosity was initially taken into account only in terms of the exponential attenuation of wave amplitudes [1,21]. It was analyzed in more detail when describing the propagation of gravitational surface [23–29], internal [30] and acoustic waves [31], considering the boundary layers formed simultaneously with the waves.

From the general content of papers and monographs [1,2,11,15,16], it follows that the basis of a rational mathematical description of inhomogeneous fluid flows is a system of fundamental equations-differential analogues of the momentum, energy and matter conservation laws with physically justified initial and boundary conditions. All the equations that were originally presented in the first edition of the treatise [1], published in 1944, were quite complex for general analysis. In practice, the reduced forms of the general system of equations are usually used, which makes it possible to study the properties of individual flow components, such as waves, vortices, jets, and wakes with the required degree of completeness. In this work, the main attention is paid to the analysis of periodic flows, the temporal variability of which is proportional to a function of the form $f \propto \exp(-i\omega t)$.

In the experiment, as in the early stage of the development of the analytical theory of waves [11], it was emphasized that the measured physical quantities-parameters of periodic flows, such as the period T_w (frequency ω,), length λ, group c_g and phase c_{ph} velocity of the wave, are characterized by real numbers. From the very beginning of the theoretical study, periodic flows began to be described using complex numbers, introduced to reduce the notation and convenience of calculations. The immersion of problems in the algebra of complex numbers leads to the expansion of the dimension of the problem space and the emergence of additional "physically unrealizable" solutions. Accordingly, there is a need to select a part of the solutions corresponding to the initial formulation, with the introduction of criteria explaining the procedure.

The physical interpretation of the solutions depends on the choice of the algorithm for the rules for immersing the problem in the algebra of complex numbers. Traditionally, starting with the works of scientists in the 19th century, the frequency ω of a waveform $f \propto \exp(ikx - i\omega t)$ has been chosen as a complex value. Its real part determines the dispersion relation, the functional relationship between frequency ω and wave vector k, and the imaginary part determines the stability condition and the wave attenuation coefficient [1]. An innumerable number of works, including popular monographs, are devoted to the study of flow and wave stability [32,33]. The history of the development of flow stability studies is traced in detail in [34]. Researchers investigated the problem of finding the liquid surface shape and the criteria for the development of instability under the action of various destabilizing factors, such as surface electric charge (Tonks–Frenkel instability) [35,36], Rayleigh-Taylor and Marangoni thermal convective instabilities [37], etc.

However, the amplitude and wavelength change within the distance of the source, but the frequency of periodic motion remains constant, as it follows from the consideration of the experimental patterns of non-dispersive waves, propagation in a medium at rest. In this regard, it is natural to maintain the frequency, which is a measure of the wave energy, as a positive definite real quantity in calculations, and take the wave number to be complex [38]. Substituting expansions of this type into a linearized system of fundamental equations, the solution of which is found using methods of singular perturbation theory [39], allows for a new classification of the structural components of periodic flows based on the properties of complete solutions.

This part of the solutions of the fundamental equation system, which includes regularly perturbed functions, characterizes waves slowly decaying in the direction of propagation in weakly dissipative media. Singularly perturbed components of the solution describe ligaments—thin flows that determine the structure of the medium in both linear and weakly nonlinear approximations [40–42].

In the hydrosphere and atmosphere, there are types of waves that differ significantly in frequency (in particular, acoustic and internal waves in the thickness of a stratified liquid [1,11,21,22,31]) or in the distribution of displacement amplitudes in depth (surface and internal waves) [1,11,21]. This makes it possible to study their properties within the framework of individual specialized equations-acoustics [1,22,31], internal [21] surface gravity or capillary waves [11,43]. Modern researchers often consider the problem of acoustic wave propagation in compressible media with complex structures using numerical and analytical methods [44,45]. At the same time, an important part of the periodic flow, which determines the fine structure of the flow, remains without attention.

The patterns of propagation of a set of two-dimensional periodic disturbances—waves and ligaments—in an incompressible fluid, when the reduced continuity equation allows us to introduce a stream function convenient for analysis, are considered in the thickness [38] and on the surface of a viscous stratified fluid [46]. This paper is the first to consider the problem of propagating a complete set of two-dimensional infinitesimal periodic disturbances in a continuously stratified compressible fluid.

2. System of Fundamental Equations of Periodic Flows in the Atmosphere and Ocean

2.1. The Complete System of Equations Determining the Flow of the Liquid

Periodic wave processes occurring in a viscous liquid are considered. The liquids existing in nature are heterogeneous. The inhomogeneous distribution of density ρ is determined using the equation of state:

$$\rho = \rho(P, S, s_n, T). \tag{1}$$

The symbol P denotes pressure, S stands for entropy, s_n denotes salinity of the n-th impurity and T stands for temperature

Far from the conditions of phase transitions, the values of the temperature gradient and the impurity content are limited, and it is permissible to use a linearized equation of state:

$$\begin{gathered} \rho = \rho_0\left(1 - \alpha_T(T - T_0) + \alpha_P(P - P_0) + \sum_n \alpha_{s_n}(s_n - s_{n0})\right). \\ \alpha_T = -\frac{1}{\rho}\frac{\partial \rho}{\partial T},\ \alpha_P = \frac{1}{\rho}\left(\frac{\partial \rho}{\partial P}\right)_S,\ \alpha_{s_n} = \frac{1}{\rho}\frac{\partial \rho}{\partial s_n} \end{gathered} \tag{2}$$

Here, α_T denotes the coefficient of thermal expansion of the liquid; α_P stands for the coefficient of adiabatic compressibility of the liquid; α_{s_n} denotes the coefficient of contraction of the n-th impurity; and T_0, P_0, s_{n0} refer to the reference level of temperature, pressure and salinity, respectively.

The fundamental system of equations in addition to the equation of state consists of the equations for describing the matter transfer, the concentration of impurity, temperature and momentum. Taking into account that the neglect of thermophores is the Ludwig–Soret effect [47,48] and Dufour effect [49], the system of equations is written as follows [38,46]:

$$G = G(P, S, s_n, T), \rho = \rho(P, S, s_n, T) \tag{3}$$

$$\partial_t \rho + \nabla_j\left(p^j\right) = Q_\rho, \tag{4}$$

$$\partial_t\left(p^i\right) + \nabla_j \Pi^{ij} = \rho g^i + 2\varepsilon^{ijk} p_j \Omega_k + Q^i, \tag{5}$$

$$\partial_t(\rho T) + \nabla_j\left(p^j T\right) = \Delta(\kappa_T \rho T) + Q_T, \tag{6}$$

$$\partial_t(\rho s_n) + \nabla_j\left(p^j s_n\right) = \Delta(\kappa_s \rho s_n) + Q_{s_n}. \tag{7}$$

Here, G is Gibbs potential; $Q_\rho, Q^i, Q_T, Q_{s_n}$ represent the source of mass, momentum, temperature and the salinity concentration, respectively; $\mathbf{p}$ denotes momentum; $\Pi^{ij} = \rho u^i u^j + P\delta^{ij} - \sigma^{ij}$ stands for the momentum flux density tensor; u^i is the component of the fluid velocity $\mathbf{u} = \mathbf{p}/\rho$; δ^{ij} is the Kronecker delta; $\sigma^{ij} = \mu\left(\frac{\partial u^i}{\partial x^j} + \frac{\partial u^j}{\partial x^i} - \frac{2}{3}\delta^{ij}\frac{\partial u^k}{\partial x^k}\right) + \zeta\delta^{ij}\frac{\partial u^k}{\partial x^k}$ denotes the viscous stress tensor; μ, ζ are dynamic and bulk viscosities, respectively; $\mathbf{g}$ is the gravity acceleration; ε^{ijk} is the Levi-Civita symbol; $\boldsymbol{\Omega}$ is the global rotation angular velocity; and κ_T, κ_{s_n} stand for thermal and mass diffusivity, respectively.

Equations (1), (3)–(7) form a fundamental system of equations that determine the fluid flow. The complete solution of the system of Equations (3)–(7) defines all components of flow in liquids—waves: acoustic, gravitational (internal and surface), capillary, hybrid and ligaments—accompanying components that identify the fine flow structure. Usually, researchers ignore the fine structure, limiting themselves to a partial solution of a system of equations. In this work, we construct a theory that takes into account all flow components.

To complete the formulation, it is necessary to add initial and boundary conditions of the problem. The initial conditions depend on the shape and type of the oscillation source. Often, when studying the properties of periodic flows, instead of initial conditions, researchers specify the type of solution and look for steady-state solutions of a given type. No-slip, no-flux boundary and initial conditions on the surface of a solid impermeable body Σ are written as follows:

$$\mathbf{u}\big|_{\Sigma} = 0,\ \mathbf{u}\big|_{t\leq 0} = 0,\ P\big|_{t\leq 0} = P_0,\ s_n\big|_{t\leq 0} = s_{n0},\ T\big|_{t\leq 0} = T_0, \tag{8}$$

If the distance to the boundaries greatly exceeds the characteristic dimensions of the observed phenomena, then a model of an unbounded medium is often used. In this case, the boundary conditions are transformed into the conditions of physical implementation–attenuation with removal:

$$\mathbf{u}\big|_{\mathbf{r}\to\infty} \to 0, \tag{9}$$

If the model under consideration contains a free surface or interface between layers of immiscible liquids, then it is necessary to add standard hydrodynamic boundary conditions: kinematic and dynamic boundary conditions. The kinematic boundary condition is written for both contacting layers (or for one medium in the case of a free surface): the substantial derivative of the function F defining the shape of the free surface is equal to zero at the boundary:

$$\frac{DF}{Dt} \equiv \frac{\partial F}{\partial t} + (\mathbf{u}\cdot\nabla)F = 0, \tag{10}$$

Dynamic boundary conditions are determined using the balance of forces at the interface (free surface of the liquid):

$$n^k_{(1)}\sigma^{ik}_{(1)} + n^k_{(2)}\sigma^{ik}_{(2)} = 0, \tag{11}$$

Here, $\mathbf{n}$ is the unit normal vector, and the subscript (1) and (2) refer to the two contacting media. If the model takes into account the effects of surface tension, then on the right side of (11), it is necessary to take into account Laplace forces as well.

2.2. The Reduced System of Equations

The fundamental system of equations is complete and allows one to determine the patterns of changes in basic physical quantities during the propagation of periodic disturbances in continuous media. Since the complete system of equations is of a high order and very complex to analyze, it is simplified to study the properties of individual processes. An extremely simplified model in which it is possible to track the dynamics and evolution of the structure of periodic flows takes into account the uneven distribution of density, without indicating the physical nature of the heterogeneity formation.

The system of Equations (3)–(7) is noticeably reduced in the constant temperature model in the absence of impurities in a weakly compressible fluid. The consideration is carried out in a Cartesian coordinate system $Oxyz$ in which the Oz axis is directed against the direction of the gravity acceleration $\mathbf{g}$. The Oxy plane determines the position of the reference level. In a weakly compressible viscous fluid, bulk viscosity takes on a zero value. In the absence of mass sources, $Q_\rho = 0$ and under the assumptions made, the reduced system of equations will take the following form:

$$\rho(\partial_t \mathbf{u} + (\mathbf{u} \cdot \nabla)\mathbf{u}) = \rho\nu\Delta\mathbf{u} - \nabla P + \rho\mathbf{g} \tag{12}$$

$$\partial_t \rho + \mathbf{u} \cdot \nabla\rho + \rho \mathrm{div}\mathbf{u} = 0 \tag{13}$$

$$\rho = \rho_0(z)(1 + \widetilde{\rho}(x,y,z,t)) \tag{14}$$

The initial stratification $\rho_0(z)$ when describing models is often defined as a linear $\rho_0(z) = \rho_{00}(1 - z/\Lambda)$ or exponential $\rho_0(z) = \rho_{00}\exp(-z/\Lambda)$ function. The symbol ρ_{00} indicates the density value at the reference level $z = 0$, and the symbol $\Lambda = |d\ln\rho/dz|^{-1}$ characterizes the stratification scale. In nature, liquids are usually weakly stratified and the scale of stratification is on the order of tens or hundreds of kilometers. When considering phenomena with characteristic dimensions much smaller than the scale of stratification, the density value in the linear and exponential stratification models turns out to be practically the same. In this case, the researchers select the more user-friendly mathematical model. Real measurements show that in an atmosphere with good accuracy, stratification can be considered linear or exponential [3].

A stably stratified liquid is characterized by the limiting frequency of its own mechanical vibrations [8]—buoyancy frequency—the square of which is given by

$$N^2 = -\frac{g}{\rho}\frac{d\rho}{dz} \tag{15}$$

The equation of state (2) under the assumptions made is simplified as follows:

$$\rho = \rho_0(z)(1 - \alpha_P(P - P_0)) \tag{16}$$

Fluid pressure is represented as the sum of reference level pressure P, hydrostatic pressure and perturbation pressure $\widetilde{P}$:

$$P = P_0 + \int_z^0 \rho(x,y,\xi,t)g d\xi + \widetilde{P}(x,y,z,t) \tag{17}$$

Taking into account the equation of state (16), the definition of the velocity of sound $c^2 = (\partial P/\partial\rho)_S$ and the definition of pressure (17), the relation (15) for the buoyancy frequency takes the following form:

$$N^2 = \frac{g^2}{c^2}\left(\frac{c_P}{c_V} - 1\right), \tag{18}$$

Here, c_P, c_V is the heat capacity at constant pressure and at constant volume, respectively.

The resulting system of equations, despite significant simplifications, qualitatively completely describes periodic flows in viscous inhomogeneous continuous media. The boundary and initial conditions will not change.

In the model under consideration, there are intrinsic parameters. These parameters determine the characteristic scales of the flow components and the characteristic times of their observation. A set of kinetic coefficients allows one to form their own parameters.

Intrinsic parameters for liquids with the parameters of water and air are presented in Tables 1 and 2, respectively.

Table 1. Intrinsic parameters of hydrosphere.

Parameter	Fluid			
	Stratified		Homogeneous	
	Strongly	Weakly	Potentially	Actually
Buoyancy frequency N, s^{-1}	1	0.01	0.00001	0.0
Buoyancy period T_b	10 s	10 min	10 days	∞
Scale of stratification Λ	10 m	100 km	10^8 km	∞
Viscous wave scale $\delta_N^{g\nu} = (g\nu)^{1/3}N^{-1}$, cm	2	200	$2 \cdot 10^5$	∞
Stokes microscale $\delta_N^{\nu} = \sqrt{\nu/N}$, cm	0.1	1	30	∞

Table 2. Intrinsic parameters of the atmosphere.

Parameter	Fluid			
	Stratified		Homogeneous	
	Strongly	Weakly	Potentially	Actually
Buoyancy frequency N, s^{-1}	1	0.01	0.00001	0.0
Buoyancy period T_b	10 s	10 min	10 days	∞
Scale of stratification Λ	10 m	100 km	10^8 km	∞
Viscous wave scale $\delta_N^{g\nu} = (g\nu)^{1/3}N^{-1}$, cm	5	500	$5 \cdot 10^5$	∞
Stokes microscale $\delta_N^{\nu} = \sqrt{\nu/N}$, cm	0.4	4	120	∞

The natural parameters presented in the table have to be supplemented with temporal and spatial scales that do not depend on the level of fluid stratification. Taking into account compressibility, a time scale $\tau_c^{\nu} = \nu/c^2$ is added. It takes values for water $\tau_c^{\nu} \simeq 4 \cdot 10^{-13}$ s and for air $\tau_c^{\nu} \simeq 10^{-10}$ s. Spatial scale $\delta_c^{\nu} = \nu/c$ is added. It takes values for water $\delta_c^{\nu} \simeq 7 \cdot 10^{-10}$ m, and for air $\delta_c^{\nu} \simeq 5 \cdot 10^{-8}$ m. In viscous liquids (homogeneous and heterogeneous), a capillary-viscous time scale appears $\tau_{\nu g}^{\gamma} = \gamma/\nu g$. The symbol $\gamma = \sigma/\rho_{00}$ denotes the surface tension coefficient of the liquid σ normalized to the equilibrium density value ρ_{00}. For water, the capillary-viscous time scale takes on values $\tau_{\nu g}^{\gamma} \simeq 7$ s, and for air $\tau_{\nu g}^{\gamma} \simeq 400$ s. The spatial scale in viscous liquids $\delta_g^{\nu} = \sqrt[3]{\nu^2/g}$ has the value $\delta_g^{\nu} \simeq 5 \cdot 10^{-5}$ m in water and $\delta_g^{\nu} \simeq 3 \cdot 10^{-4}$ m in air. The capillary length $\delta_g^{\gamma} = \sqrt{\gamma/g}$ is in both the viscous liquid model and in the inviscid liquid one. For water capillary length, it takes the value $\delta_g^{\gamma} \simeq 3 \cdot 10^{-3}$ m, and for air it is $\delta_g^{\gamma} \simeq 8 \cdot 10^{-2}$ m.

Small disturbances of physical quantities (pressure, density, velocity) often occur in nature. Let us solve the problem using the decomposition method for a small parameter that plays the role of the amplitude of periodic movements.

3. Periodic Flows in the Thickness of a Uniformly Stratified Liquid

3.1. Linearization of the Equation System

The perturbations of the target values (velocity, density and pressure) are considered small. To obtain dispersion relations, we linearize the system of Equations (12)–(14), (16). If we assume that the fluid is exponentially stratified, then in a linear approximation in terms of the amplitude of periodic motion, the reduced system of fundamental equations is written as follows:

$$\begin{cases} \partial_t\widetilde{\rho} - \frac{w}{\Lambda} + \partial_x u + \partial_y v + \partial_z w = 0 \\ \partial_t u - \nu\Delta u + \frac{1}{\rho_{00}}\partial_x\widetilde{P} = 0 \\ \partial_t v - \nu\Delta v + \frac{1}{\rho_{00}}\partial_y\widetilde{P} = 0 \\ \partial_t w - \nu\Delta w + \frac{1}{\rho_{00}}\partial_z\widetilde{P} + g\widetilde{\rho} = 0 \\ \frac{1}{\rho_{00}c^2}\partial_t\widetilde{P} - \frac{wg}{c^2} + \partial_x u + \partial_y v + \partial_z w = 0 \end{cases} \tag{19}$$

Here u, v, w are the components of the velocity field $\mathbf{u} = (u, v, w)$. We look for the solution of the equation system (19) in the form of periodic flows $\propto \exp(i\omega t)$:

$$\begin{pmatrix} u \\ v \\ w \\ \widetilde{P} \\ \widetilde{\rho} \end{pmatrix} = \begin{pmatrix} U_m \\ V_m \\ W_m \\ P_m \\ P_m \end{pmatrix} \exp(i\mathbf{kr} - i\omega t) = \begin{pmatrix} U_m \\ V_m \\ W_m \\ P_m \\ P_m \end{pmatrix} \exp(ik_x x + ik_y y + ik_z z - i\omega t) \tag{20}$$

Here, U_m, V_m, W_m, P_m, P_m are the amplitudes of the corresponding quantities; $\mathbf{k}$ is the wave vector, the components of which have the right to be complex values k_x, k_y, k_z; and the frequency of periodic motion ω is considered positive definite.

3.2. Dispersion Relation: Classification of Flow Components

By substituting the type of solution (20) into the system of Equation (19), we obtain a system of algebraic equations. The compatibility condition of the algebraic equations system determines the dispersion relations between the components of the wave vector and the frequency of periodic motion:

$$\begin{gathered} D_\nu(k)\left(\omega^2 D_\nu^2(k) - \omega N^2 D_\nu(k) + c^2 k_\perp^2 N_c^2 - c^2\omega k^2 D_\nu(k)\right) = 0, \\ D_\nu(k) = \omega + i\nu k^2,\ k^2 = k_x^2 + k_y^2 + k_z^2,\ k_\perp^2 = k_x^2 + k_y^2,\ N^2 = \frac{g}{\Lambda},\ N_c^2 = N^2 - \frac{g^2}{c^2} \end{gathered} \tag{21}$$

Dispersion relation (21) coincides with the relation obtained earlier [50] in which the limiting transition to a non-rotating weakly compressible fluid (second viscosity $\zeta \to 0$) has been made. It is convenient to find and analyze the regular and singular components of the solution to the dispersion relation (21) in dimensionless variables if one chooses the scales of the problem as non-dimensional parameters. Intrinsic scales characterize the spatial and temporal dimensions of the observed phenomena (see Tables 1 and 2). We choose the inverse buoyancy frequency $\tau_b = N^{-1}$ as the time scale, and the viscous wave scale $\delta_N^{g\nu} = (g\nu)^{1/3}N^{-1}$ as the spatial scale. With the selected non-dimensional parameters, the dispersion relation (21) is written as follows:

$$\begin{gathered} (ik_*^2\varepsilon + \omega_*)\left(k_{\perp *}^2\left(\frac{\varepsilon}{\eta} - \frac{1}{\varepsilon^2}\right) + \omega_*^2(ik_*^2\varepsilon + \omega_*)^2 - \omega_*(ik_*^2\varepsilon + \omega_*) - k_*^2\omega_*\frac{\varepsilon}{\eta}(ik_*^2\varepsilon + \omega_*)\right) = 0, \\ \varepsilon = \frac{\delta_g^\nu}{\delta_N^{g\nu}} = \frac{\sqrt{\nu/N}}{(g\nu)^{1/3}N^{-1}} = \frac{N\nu^{1/3}}{g^{2/3}},\ \eta = \frac{\tau_c^\nu}{\tau_b} = \frac{N\nu}{c^2} \end{gathered} \tag{22}$$

The ratio of the natural parameters of the medium-viscous scale δ_g^ν and viscous wave scale $\delta_N^{g\nu}$ and the time scale ratio η characterize the small parameters of the problem. The dimensionless components of the wave vector and the dimensionless frequency are indicated by the subscript «*». Since at the highest degree of Equation (22), there is a small parameter, the equation is singularly perturbed with respect to k_{*z}. Consequently, the solution of the form $k_{*z} = k_{*z}(k_{*x}, k_{*y}, \omega_*)$ contains regular and singular components. The solutions of Equation (22) are written as follows:

$$k_{*z} = \pm\sqrt{-k_{*\perp}^2 + \frac{i\omega_*}{\varepsilon}} \tag{23}$$

$$k_{*z} = \pm\sqrt{\frac{-\varepsilon\omega_*\left(\omega_* + 2k_{*\perp}^2\varepsilon(i+\eta\omega_*) - i\eta(2\omega_*^2 - 1)\right) - \sqrt{-\varepsilon^2\omega_*^2(\eta + i\omega_*) + 4\omega_* k_{*\perp}^2(\varepsilon^3 - \eta)(i + \eta\omega_*)}}{2\varepsilon^2\omega_*(i+\eta\omega_*)}} \tag{24}$$

$$k_{*z} = \pm\sqrt{\frac{-\varepsilon\omega_*\left(\omega_* + 2k_{*\perp}^2\varepsilon(i+\eta\omega_*) - i\eta(2\omega_*^2 - 1)\right) + \sqrt{-\varepsilon^2\omega_*^2(\eta + i\omega_*) + 4\omega_* k_{*\perp}^2(\varepsilon^3 - \eta)(i + \eta\omega_*)}}{2\varepsilon^2\omega_*(i+\eta\omega_*)}} \tag{25}$$

In dimensional form, the roots (23)–(25) are written as follows:

$$k_z = \pm\sqrt{-k_\perp^2 + \frac{i\omega}{\nu}} \tag{26}$$

$$k_{*z} = \pm\sqrt{\frac{-\left(i\nu N^2 + 2\nu\omega(\nu k_\perp^2 - i\omega) + c^2(2i\nu k_\perp^2 + \omega)\right) - \sqrt{-\nu^2 N^4 + 2\nu c^2(2\nu k_\perp^2 N_c^2 - i\omega N^2) + c^4\left(\omega^2 + \frac{4i\nu N_c^2 k_\perp^2}{\omega}\right)}}{2\nu(ic^2 + \nu\omega)}} \tag{27}$$

$$k_{*z} = \pm\sqrt{\frac{-\left(i\nu N^2 + 2\nu\omega(\nu k_\perp^2 - i\omega) + c^2(2i\nu k_\perp^2 + \omega)\right) + \sqrt{-\nu^2 N^4 + 2\nu c^2(2\nu k_\perp^2 N_c^2 - i\omega N^2) + c^4\left(\omega^2 + \frac{4i\nu N_c^2 k_\perp^2}{\omega}\right)}}{2\nu(ic^2 + \nu\omega)}} \tag{28}$$

The choice of a sign in solutions (23)–(25) or (26)–(28) is determined by the boundary conditions for the decay of periodic motion with distance from the source of disturbances. The solutions (25) and (28) describe the regular component and determine the wave motion, the solutions (23)–(24) and (26)–(27) determine the singular component of the solution and define two types of ligaments. Wave roots (25) or (28) can be obtained approximately using regular decomposition. Ligament roots (23)–(24) or (26)–(27) can be obtained approximately using singular value decomposition [39]. The presented expressions are exact solutions of the dispersion relation. By substituting numerical values of kinetic coefficients, the components of the wave vector corresponding to both ligament and wave solutions can be calculated.

To verify the solutions obtained, we consider some limiting cases.

4. High-Frequency Acoustic Waves

Let us consider the limit of high-frequency oscillations corresponding to acoustic oscillations if their oscillation frequency significantly exceeds the buoyancy frequency of the medium $\omega \gg N$ [22,50]. In this approximation, the dispersion Equation (21) is rewritten as follows:

$$D_\nu(k)\left(D_\nu(k)\omega\left(D_\nu(k)\omega - c^2k^2\right) - g^2k_\perp^2\right) = 0, \tag{29}$$

The solution of dispersion relation (29) is written as follows:

$$\begin{aligned}
k_z &= \pm\sqrt{-k_\perp^2 + \tfrac{i\omega}{\nu}};\\
k_z &= \pm\sqrt{-k_\perp^2 - \frac{c^2\omega - 2i\nu\omega^2 + \sqrt{c^4\omega^2 - \frac{4g^2\nu k_\perp^2}{\omega}(ic^2+\nu\omega)}}{2\nu(ic^2+\nu\omega)}};\\
k_z &= \pm\sqrt{-k_\perp^2 - \frac{c^2\omega - 2i\nu\omega^2 - \sqrt{c^4\omega^2 - \frac{4g^2\nu k_\perp^2}{\omega}(ic^2+\nu\omega)}}{2\nu(ic^2+\nu\omega)}};
\end{aligned} \tag{30}$$

The roots (30) describe the wave motion and two attached ligaments. The sign-in solution (30) is chosen based on the need for attenuation of periodic motion $\mathrm{Im}(k_z) > 0$ when moving in the positive direction of the axis Oz. For oppositely directed motion, the solutions are symmetrical.

When moving to a 2D formulation (if we consider the movement to be independent of the horizontal coordinate y), one of the ligaments degenerates and the solution contains one wave and one ligament component:

$$\begin{aligned} k_z &= \pm\sqrt{-k_x^2 - \frac{c^2\omega - 2i\nu\omega^2 + \sqrt{c^4\omega^2 - \frac{4g^2\nu k_x^2}{\omega}(ic^2+\nu\omega)}}{2\nu(ic^2+\nu\omega)}}; \\ k_z &= \pm\sqrt{-k_x^2 - \frac{c^2\omega - 2i\nu\omega^2 - \sqrt{c^4\omega^2 - \frac{4g^2\nu k_x^2}{\omega}(ic^2+\nu\omega)}}{2\nu(ic^2+\nu\omega)}}; \end{aligned} \tag{31}$$

In the limit of an inviscid fluid, the dispersion relation (29) is simplified even further and written in the following form:

$$\omega^2\left(\omega^2 - k^2c^2\right) - g^2k_\perp^2 = 0, \tag{32}$$

The ligament components of the solution to relation (32) degenerate and only the wave component remains:

$$k_z = \pm\sqrt{-k_\perp^2 - \frac{g^2k_\perp^2}{c^2\omega^2} + \frac{\omega^2}{c^2}} \tag{33}$$

5. Low-Frequency Gravity Waves

In the limit of low-frequency oscillations $\omega \ll N$, the dispersion relation (21) takes the following form:

$$D_\nu(k)\left(c^2\omega i k^4\nu - c^2N^2k_\perp^2 + c^2k^2\omega^2 + N^2\omega D_\nu(k) + g^2k_\perp^2\right) = 0, \tag{34}$$

The relation (34) also contains a solution in the form of a wave disturbance and two attached ligaments:

$$\begin{aligned} k_z &= \pm\sqrt{-k_\perp^2 + \frac{i\omega}{\nu}}; \\ k_z &= \pm\sqrt{-k_\perp^2 - \frac{ic^2\omega^2 - N^2\nu\omega + \sqrt{N^4\nu^2\omega^2 + 4ic^2k_\perp^2\omega\nu(g^2 - c^2N^2) - c^4\omega^4}}{2c^2\nu\omega}}; \\ k_z &= \pm\sqrt{-k_\perp^2 - \frac{ic^2\omega^2 - N^2\nu\omega - \sqrt{N^4\nu^2\omega^2 + 4ic^2k_\perp^2\omega\nu(g^2 - c^2N^2) - c^4\omega^4}}{2c^2\nu\omega}}; \end{aligned} \tag{35}$$

When transitioning to a flat formulation, one of the ligaments degenerates.

Relation (34) also contains a solution in the form of a wave disturbance and two attached ligaments:

$$\begin{aligned} k_z &= \pm\sqrt{-k_x^2 - \frac{ic^2\omega^2 - N^2\nu\omega + \sqrt{N^4\nu^2\omega^2 + 4ic^2k_x^2\omega\nu(g^2 - c^2N^2) - c^4\omega^4}}{2c^2\nu\omega}}; \\ k_z &= \pm\sqrt{-k_x^2 - \frac{ic^2\omega^2 - N^2\nu\omega - \sqrt{N^4\nu^2\omega^2 + 4ic^2k_x^2\omega\nu(g^2 - c^2N^2) - c^4\omega^4}}{2c^2\nu\omega}}; \end{aligned} \tag{36}$$

In an ideal liquid, the dispersion relation (34) is simplified as follows:

$$c^2k^2\omega^2 - c^2k_\perp^2N^2 + N^2\omega^2 + g^2k_\perp^2 = 0, \tag{37}$$

The solution (37), which is represented only by a wave component, in an ideal liquid transforms into a well-known expression that does not include the wavelength $\omega^2 = N^2\sin^2\theta$, which describes the geometry of the wave packet in the shape of a "St. Andrew's cross" (θ is the angle of inclination of the wave vector to the horizontal) [20,21]. Ligaments in the ideal fluid model degenerate:

$$k_z = \pm\sqrt{-k_\perp^2 + k_\perp^2\frac{N^2c^2 - g^2}{c^2\omega^2} - \frac{N^2}{c^2}}; \tag{38}$$

The limiting cases discussed in paragraphs 4 and 5 show that ligaments are observed in the entire frequency range from infra-low-frequency mechanical vibrations to high-frequency sound vibrations. The fine structure of the flow accompanies wave motion and requires attention when analyzing phenomena.

6. Periodic Flows in a Two-Layer System of Stratified Liquids

In a two-layer system, which consists of a stratified weakly compressible ocean and a stratified compressible atmosphere, it is necessary to write down the boundary conditions at the interface. In a two-layer system, the pressure in both media is written in the form of the sum of hydrostatic pressure and perturbation pressure, so in a 2D formulation (if we consider the movement independent of the horizontal coordinate y) it is written as follows:

$$P^{o,a} = \int_z^{\zeta} \rho^{o,a}(x,\xi,t) g d\xi + \widetilde{P}^{o,a}(x,z,t) \tag{39}$$

Here and further, the superscripts "o" and "a" denote quantities related to the ocean (the lower denser liquid) and the atmosphere (the upper less dense liquid), respectively. The symbol $\zeta = \zeta(x,t)$ denotes the function that determines the deviation of the interface between media from the equilibrium position $z = 0$. The system of equations of motion, taking into account Expression (39), is written as follows:

$$z < \zeta : \ \partial_t \mathbf{u}^o - \nu^o \Delta \mathbf{u}^o + \frac{1}{\rho_{00}^o} \nabla P^o - \rho^o \mathbf{g} = 0 \tag{40}$$

$$\partial_t \rho^o + \mathbf{u}^o \cdot \nabla \rho^o + \rho^o \mathrm{div} \mathbf{u}^o = 0 \tag{41}$$

$$\rho^o = \rho_0^o(z)(1 - \alpha_P^o(P^o - P_0^o)) \tag{42}$$

$$z > \zeta : \ \partial_t \mathbf{u}^a - \nu^a \Delta \mathbf{u}^a + \frac{1}{\rho_{00}^a} \nabla P^a - \rho^a \mathbf{g} = 0 \tag{43}$$

$$\partial_t \rho^a + \mathbf{u}^a \cdot \nabla \rho^a + \rho^a \mathrm{div} \mathbf{u}^a = 0 \tag{44}$$

$$\rho^a = \rho_0^a(z)(1 - \alpha_P^a(P^a - P_0^a)) \tag{45}$$

The system of Equations (40)–(45) is supplemented with the boundary conditions at the interface: $z = \xi$

$$z = \zeta : \ \partial_t \zeta + u^o \partial_x \zeta = w^o \tag{46}$$

$$\partial_t \zeta + u^a \partial_x \zeta = w^a \tag{47}$$

$$P^o - 2\rho^o \nu^o \mathbf{n} \cdot ((\mathbf{n} \cdot \nabla)\mathbf{u}^o) = P^a - 2\rho^a \nu^a \mathbf{n} \cdot ((\mathbf{n} \cdot \nabla)\mathbf{u}^a) - \sigma \mathrm{div} \mathbf{n} \tag{48}$$

$$\mathbf{u}^o \cdot \boldsymbol{\tau} = \mathbf{u}^a \cdot \boldsymbol{\tau} \tag{49}$$

$$\rho^o \nu^o (\boldsymbol{\tau} \cdot ((\mathbf{n} \cdot \nabla)\mathbf{u}^o) + \mathbf{n} \cdot ((\boldsymbol{\tau} \cdot \nabla)\mathbf{u}^o)) = \rho^a \nu^a (\boldsymbol{\tau} \cdot ((\mathbf{n} \cdot \nabla)\mathbf{u}^a) + \mathbf{n} \cdot ((\boldsymbol{\tau} \cdot \nabla)\mathbf{u}^a)) \tag{50}$$

$$\mathbf{n} = \frac{\nabla(z-\zeta)}{|\nabla(z-\zeta)|} = \left(\frac{-\partial_x \zeta}{\sqrt{1+(\partial_x \zeta)^2}}, \frac{1}{\sqrt{1+(\partial_x \zeta)^2}} \right), \ \boldsymbol{\tau} = \left(\frac{1}{\sqrt{1+(\partial_x \zeta)^2}}, \frac{\partial_x \zeta}{\sqrt{1+(\partial_x \zeta)^2}} \right)$$

Here, σ is the coefficient of surface tension at the interface between contacting media, and $\mathbf{n}, \boldsymbol{\tau}$ are the normal and tangent vectors to the interface, respectively. After carrying out the linearization procedure and transferring the boundary conditions to the equilibrium surface $z = 0$ [51], the mathematical formulation in a linear approximation takes the following form:

$$\begin{array}{c} z<0: \int_z^\zeta \mathrm{e}^{-\frac{\xi}{\Lambda^o}} g\partial_x\widetilde{\rho}^o(x,\xi,t) + \mathrm{e}^{-\frac{\zeta}{\Lambda^o}} g\partial_x\zeta + \partial_t u^o - \nu^o\Delta u^o + \partial_x\widetilde{P}^o = 0 \\ \mathrm{e}^{-\frac{z}{\Lambda^o}}\partial_t w^o - \nu^o \mathrm{e}^{-\frac{z}{\Lambda^o}}\Delta w^o + \frac{\partial_z\widetilde{P}^o}{\rho_{00}^o} = 0 \\ \partial_t\widetilde{\rho}^o - \frac{w^o}{\Lambda^o} + \partial_x u^o + \partial_z w^o = 0 \\ \frac{1}{\rho_{00}^o c^{o2}}\partial_t\widetilde{P}^o - \frac{w^o g}{c^{o2}} + \partial_x u^o + \partial_z w^o = 0 \end{array} \tag{51}$$

$$\begin{array}{c} z>0: \int_z^\zeta \mathrm{e}^{-\frac{\xi}{\Lambda^a}} g\partial_x\widetilde{\rho}^a(x,\xi,t) + \mathrm{e}^{-\frac{\zeta}{\Lambda^a}} g\partial_x\zeta + \partial_t u^a - \nu^a\Delta u^a + \partial_x\widetilde{P}^a = 0 \\ \mathrm{e}^{-\frac{z}{\Lambda^a}}\partial_t w^a - \nu^a \mathrm{e}^{-\frac{z}{\Lambda^a}}\Delta w^a + \frac{\partial_z\widetilde{P}^a}{\rho_{00}^a} = 0 \\ \partial_t\widetilde{\rho}^a - \frac{w^a}{\Lambda^a} + \partial_x u^a + \partial_z w^a = 0 \\ \frac{1}{\rho_{00}^a c^{a2}}\partial_t\widetilde{P}^a - \frac{w^a g}{c^{a2}} + \partial_x u^a + \partial_z w^a = 0 \end{array} \tag{52}$$

$$\begin{array}{c} z=0: \partial_t\zeta - w^o = 0,\ \partial_t\zeta - w^a = 0,\ u^o - u^a = 0, \\ \widetilde{P}^o - \widetilde{P}^a + 2\rho_{00}^a\nu^a\partial_z w^a - 2\rho_{00}^o\nu^o\partial_z w^o + \sigma\partial_{xx}\zeta = 0, \\ \rho^o\nu^o(\partial_z u^o + \partial_x w^o) - \rho^a\nu^a(\partial_z u^a + \partial_x w^a) = 0 \end{array} \tag{53}$$

We look for a solution to the system of Equations (51)–(53) in the form of periodic flows $\propto \exp(i\omega t)$:

$$\begin{pmatrix} u^{o,a} \\ w^{o,a} \\ \widetilde{P}^{o,a} \\ \widetilde{\rho}^{o,a} \\ \zeta \end{pmatrix} = \begin{pmatrix} U_m^{o,a} \\ W_m^{o,a} \\ P_m^{o,a} \\ P_m^{o,a} \\ A_m \end{pmatrix} \exp(i\mathbf{k}^{o,a}\mathbf{r} - i\omega t) = \begin{pmatrix} U_m^{o,a}\exp(ik_z^{o,a}z) \\ W_m^{o,a}\exp(ik_z^{o,a}z) \\ P_m^{o,a}\exp(ik_z^{o,a}z) \\ P_m^{o,a}\exp(ik_z^{o,a}z) \\ A_m \end{pmatrix} \exp(ik_x x - i\omega t) \tag{54}$$

Substituting the type of solution (54) into the main Equations (51) and (52) leads to a system of algebraic equations connecting the components of wave vectors k_x, $k_z^{o,a}$ and the frequency of periodic disturbances ω:

$$\begin{pmatrix} -gk_x^2 + \frac{\omega(i+k_z^o\Lambda^o)\left(\nu^o\left(k_x^2+k_z^{o2}\right)-i\omega\right)}{\Lambda^o} & -k_x\left(N^{o2}(i+k_z^o\Lambda^o) + \omega\left(\nu^o\left(k_x^2+k_z^{o2}\right)-i\omega\right)\right) & 0 & 0 \\ 0 & \nu^o\left(k_x^2+k_z^{o2}\right)-i\omega & \frac{ik_z^o\mathrm{e}^{\frac{z}{\Lambda^o}}}{\rho_{00}^o} & 0 \\ ik_x & ik_z^o - \frac{1}{\Lambda^o} & 0 & -i\omega \\ ik_x & -\frac{g}{c^{o2}} + ik_z^o & -\frac{i\omega}{\rho_{00}^o c^{o2}} & 0 \\ -gk_x^2 + \frac{\omega(i+k_z^a\Lambda^a)\left(\nu^a\left(k_x^2+k_z^{a2}\right)-i\omega\right)}{\Lambda^a} & -k_x\left(N^{a2}(i+k_z^a\Lambda^a) + \omega\left(\nu^a\left(k_x^2+k_z^{a2}\right)-i\omega\right)\right) & 0 & 0 \\ 0 & \nu^a\left(k_x^2+k_z^{a2}\right)-i\omega & \frac{ik_z^a\mathrm{e}^{\frac{z}{\Lambda^a}}}{\rho_{00}^a} & 0 \\ ik_x & ik_z^a - \frac{1}{\Lambda^a} & 0 & -i\omega \\ ik_x & -\frac{g}{c^{a2}} + ik_z^a & -\frac{i\omega}{\rho_{00}^a c^{a2}} & 0 \end{pmatrix} = \begin{pmatrix} 0 \\ 0 \\ 0 \\ 0 \\ 0 \\ 0 \\ 0 \\ 0 \end{pmatrix} \tag{55}$$

The resulting system is divided into two independent systems of equations that describe the relationships between the upper and lower media. The compatibility condition for each of the systems leads to dispersion relations for the lower one:

$$\begin{array}{c} \frac{\omega}{c^{o2}\Lambda^{o2}}\left[\omega\left(\nu^o\left(k_x^2+k_z^{o2}\right)-i\omega\right)\left(-gk_x^2\Lambda^o + \omega\left(i+k_z^0\Lambda^o\right)\left(\nu^o\left(k_x^2+k_z^{o2}\right)-i\omega\right)\right)+\right. \\ +\mathrm{e}^{\frac{z}{\Lambda^o}}k_z^o\left(\left(g+ic^{o2}k_z^{o2}\right)\left(-gk_x^2\Lambda^o + \omega\left(i+k_z^0\Lambda^o\right)\left(\nu^o\left(k_x^2+k_z^{o2}\right)-i\omega\right)\right)+\right. \\ \left.\left.+\,c^{o2}k_x^2\Lambda^o\left(N^{o2}\left(ik_z^o\Lambda^o-1\right)+\omega\left(i\nu\left(k_x^2+k_z^{o2}\right)+\omega\right)\right)\right)\right]=0 \end{array} \tag{56}$$

and top liquid:

$$\begin{aligned}&\tfrac{\omega}{c^{a2}\Lambda^{a2}}\left[\omega\left(\nu^a\left(k_x^2+k_z^{a2}\right)-i\omega\right)\left(-gk_x^2\Lambda^a+\omega\left(i+k_z^a\Lambda^a\right)\left(\nu^a\left(k_x^2+k_z^{a2}\right)-i\omega\right)\right)+\right.\\&+e^{\frac{z}{\Lambda^a}}k_z^a\left(\left(g+ic^{a2}k_z^{a2}\right)\left(-gk_x^2\Lambda^a+\omega\left(i+k_z^a\Lambda^a\right)\left(\nu^a\left(k_x^2+k_z^{a2}\right)-i\omega\right)\right)+\right.\\&\left.\left.+\,c^{a2}k_x^2\Lambda^a\left(N^{a2}\left(ik_z^a\Lambda^a-1\right)+\omega\left(i\nu\left(k_x^2+k_z^{a2}\right)+\omega\right)\right)\right)\right]=0\end{aligned} \tag{57}$$

Let us consider expressions (56) and (57) in a dimensionless form. We choose the natural parameters of each medium as non-dimensional scales: as the time scale, we take the inverse buoyancy frequency $\tau_b = N^{-1}$, and as the spatial scale, we select the viscous wave scale $\delta_N^{g\nu} = (g\nu)^{1/3}N^{-1}$:

$$\begin{aligned}&\tfrac{\omega_*}{\varepsilon^{o2}}\left[\varepsilon^o\eta^o\omega_*\left(\varepsilon^o\omega_*\left(k_{z*}^o+i\varepsilon^o\right)\left(\varepsilon^o k_{z*}^{o2}-i\omega_*\right)+\varepsilon^o k_{x*}^4\left(-1+\varepsilon^{o2}\omega_* k_{z*}^o+i\varepsilon^{o3}\omega_*\right)+\right.\right.\\&\left.+k_{x*}^2\left(2\varepsilon^{o3}\omega_* k_{z*}^{o3}-2i\varepsilon^{o2}\omega_*^2 k_{z*}^o+\omega_*\left(i+2\varepsilon^{o3}\omega_*\right)+k_{z*}^{o2}\left(-\varepsilon^o+2i\varepsilon^{o4}\omega_*\right)\right)\right)+\\&+e^{\frac{z}{\Lambda^o}}k_{z*}^o\left(i\varepsilon^{o4}\omega_* k_{x*}^4+\varepsilon^o\omega_*\left(k_{z*}^o+i\varepsilon^o\right)\left(\varepsilon^{o2}k_{z*}^o-i\eta^o\right)\left(i\varepsilon^o k_{z*}^{o2}+\omega_*\right)-\right.\\&\left.\left.-\,k_{x*}^2\left(\eta^o-2i\varepsilon^{o4}\omega_* k_{z*}^{o2}+\varepsilon^{o5}\omega_* k_{z*}^o-\varepsilon^{o2}\eta^o\omega_* k_{z*}^o+\varepsilon^{o3}\left(1-i\eta^o\omega_*-\omega_*^2\right)\right)\right)\right]=0\end{aligned} \tag{58}$$

$$\begin{aligned}&\tfrac{\omega_*}{\varepsilon^{a2}}\left[\varepsilon^a\eta^a\omega_*\left(\varepsilon^a\omega_*\left(k_{z*}^a+i\varepsilon^a\right)\left(\varepsilon^a k_{z*}^{a2}-i\omega_*\right)+\varepsilon^a k_{x*}^4\left(-1+\varepsilon^{a2}\omega_* k_{z*}^a+i\varepsilon^{a3}\omega_*\right)+\right.\right.\\&\left.+k_{x*}^2\left(2\varepsilon^{a3}\omega_* k_{z*}^{a3}-2i\varepsilon^{a2}\omega_*^2 k_{z*}^a+\omega_*\left(i+2\varepsilon^{a3}\omega_*\right)+k_{z*}^{a2}\left(-\varepsilon^a+2i\varepsilon^{a4}\omega_*\right)\right)\right)+\\&+e^{\frac{z}{\Lambda^a}}k_{z*}^a\left(i\varepsilon^{a4}\omega_* k_{x*}^4+\varepsilon^a\omega_*\left(k_{z*}^a+i\varepsilon^a\right)\left(\varepsilon^{a2}k_{z*}^a-i\eta^a\right)\left(i\varepsilon^a k_{z*}^{a2}+\omega_*\right)-\right.\\&\left.\left.-\,k_{x*}^2\left(\eta^a-2i\varepsilon^{a4}\omega_* k_{z*}^{a2}+\varepsilon^{a5}\omega_* k_{z*}^a-\varepsilon^{a2}\eta^a\omega_* k_{z*}^a+\varepsilon^{a3}\left(1-i\eta^a\omega_*-\omega_*^2\right)\right)\right)\right]=0\end{aligned} \tag{59}$$

$$\varepsilon^a = N^a\sqrt[3]{\frac{\nu^a}{g^2}},\ \varepsilon^o = N^o\sqrt[3]{\frac{\nu^o}{g^2}},\ \eta^a = \frac{N^a\nu^a}{c^{a2}},\ \eta^o = \frac{N^o\nu^o}{c^{o2}}.$$

Expressions (58) and (59) are reduced to the dispersion relations in an incompressible fluid when passing to the limit $c^{o,a} \to \infty\,(\eta^{o,a} \to 0)$:

$$\omega_*\left(i\varepsilon^o\left(k_{x*}^2+k_{z*}^{o2}\right)+\omega_*\right)\left(\varepsilon^o\omega_*\left(k_{z*}^o+i\varepsilon^o\right)\left(\varepsilon^o k_{z*}^{o2}-i\omega_*\right)+k_{x*}^2\left(-1+\varepsilon^{o2}\omega_* k_{z*}^o+i\varepsilon^{o3}\omega_*\right)\right)=0 \tag{60}$$

$$\omega_*\left(i\varepsilon^a\left(k_{x*}^2+k_{z*}^{a2}\right)+\omega_*\right)\left(\varepsilon^a\omega_*\left(k_{z*}^a+i\varepsilon^a\right)\left(\varepsilon^a k_{z*}^{a2}-i\omega_*\right)+k_{x*}^2\left(-1+\varepsilon^{a2}\omega_* k_{z*}^a+i\varepsilon^{a3}\omega_*\right)\right)=0 \tag{61}$$

The small parameter $\eta^{o,a}$ for liquids with the parameters of water and air turns out to be significantly smaller than the small parameter $\varepsilon^{o,a}$. The approximate solutions of dispersion relations (58) and (59) $k_{*z}^{o,a}$ have the following form:

$$k_{*z}^{o,a} = k_{0*z}^{o,a} + \eta k_{1*z}^{o,a} \tag{62}$$

In solution $k_{0*z}^{o,a}$ (62) takes one of the following values:

$$k_{0*z}^{o,a} = 0; \tag{63}$$

$$k_{0*z}^{o,a} = -\frac{i\varepsilon^{o,a}}{4} - \frac{1}{2}\sqrt{-\frac{\varepsilon^{o,a}}{4} - \frac{2\varepsilon^{o,a}k_{x*}^2 - i\omega_*}{\varepsilon^{o,a}} + \theta} \pm \frac{1}{2}\sqrt{-\frac{\varepsilon^{o,a2}}{2} - \frac{2\varepsilon^{o,a}k_{x*}^2 - i\omega_*}{\varepsilon^{o,a}} - \theta - \frac{i\varepsilon^{o,a3} - 8i\left(\varepsilon^{o,a}k_{x*}^2 - i\omega_*\right) + 4i\left(2\varepsilon^{o,a}k_{x*}^2 - i\omega_*\right)}{4\sqrt{-\frac{\varepsilon^{o,a2}}{4} + \theta - \frac{2\varepsilon^{o,a}k_{x*}^2 - i\omega_*}{\varepsilon^{o,a}}}}} \tag{64}$$

$$k_{0*z}^{o,a} = -\frac{i\varepsilon^{o,a}}{4} + \frac{1}{2}\sqrt{-\frac{\varepsilon^{o,a}}{4} - \frac{2\varepsilon^{o,a}k_{x*}^2 - i\omega_*}{\varepsilon^{o,a}} + \theta} \pm \frac{1}{2}\sqrt{-\frac{\varepsilon^{o,a2}}{2} - \frac{2\varepsilon^{o,a}k_{x*}^2 - i\omega_*}{\varepsilon^{o,a}} - \theta + \frac{i\varepsilon^{o,a3} - 8i\left(\varepsilon^{o,a}k_{x*}^2 - i\omega_*\right) + 4i\left(2\varepsilon^{o,a}k_{x*}^2 - i\omega_*\right)}{4\sqrt{-\frac{\varepsilon^{o,a2}}{4} - \frac{2\varepsilon^{o,a}k_{x*}^2 - i\omega_*}{\varepsilon^{o,a}} + \theta}}}; \tag{65}$$

$$\theta = \frac{\left(i+\sqrt{3}\right)\left(\alpha+\sqrt{\alpha^2-4\beta^3}\right)^{1/3}}{6\cdot 2^{1/3}\varepsilon^{o,a}\omega_*} + \frac{2\varepsilon^{o,a}k_{x*}^2 - i\omega_*}{3\varepsilon^{o,a}} + \frac{\left(i-\sqrt{3}\right)2^{1/3}\beta}{3\varepsilon^{o,a}\left(\alpha+\sqrt{\alpha^2-4\beta^3}\right)^{1/3}}$$

$$\beta = \left(-16\varepsilon^{o,a2}k_{x*}^4\omega_* + \omega_*^2\left(3i\varepsilon^{o,a}+\omega_*\right) + k_{x*}^2\left(-3\varepsilon^{o,a4}\omega_* + 4i\varepsilon^{o,a}\left(-3+4\omega_*^2\right)\right)\right)$$

$$\alpha = \omega_*^2\left(128i\varepsilon^{o,a3}\omega_* k_{x*}^6 + 2\omega_*^3\left(-9i\varepsilon^{o,a3} + \omega_*\right) + 12\varepsilon^{o,a2}k_{x*}^4\left(-12 + 3i\varepsilon^{o,a3}\omega_* + 16\omega_*^2\right) + 3k_{x*}^2\left(9\varepsilon^{o,a4}\left(-1 + 2\omega_*^2\right) - 4i\varepsilon^{o,a}\omega_*\left(-6 + 5\omega_*^2\right)\right)\right)$$

and $k_{1*z}^{o,a}$ takes the corresponding (63)–(65) values:

$$k_{1*z}^{o,a} = \frac{\left(k_{0z*}^{o,a} + \varepsilon^{o,a} e^{-\frac{z}{\Lambda^{o,a}}}\left(\varepsilon^{o,a}\omega_*\left(k_{x*}^2 + k_{0z*}^{o,a2} - i\omega_*\right)\right)\right)\left(\varepsilon^{o,a}\omega_*\left(k_{0z*}^{o,a} + i\varepsilon^{o,a}\right)\left(\varepsilon^{o,a}k_{0z*}^{o,a2} - i\omega_*\right) + k_{x*}^2\left(-1 + \varepsilon^{o,a2}k_{0z*}^{o,a}\omega_* + i\varepsilon^{o,a3}\omega_*\right)\right)}{\varepsilon^{o,a3}\left(-i\varepsilon^{o,a}\omega_* k_{x*}^4 + \omega_* k_{0z*}^{o,a}\left(-5i\varepsilon^{o,a}k_{0z*}^{o,a3} + 4\varepsilon^{o,a2}k_{0z*}^2 - 3\omega_* k_{0z*}^{o,a} - 2i\varepsilon^{o,a}\omega_*\right) + k_{x*}^2\left(1 - 6i\varepsilon^{o,a}\omega_* k_{0z*}^{o,a2} + 2\varepsilon^{o,a2}\omega_* k_{0z*}^{o,a} - \omega_*^2\right)\right)} \quad (66)$$

Additional conditions for physical implementation are imposed on solutions (63)–(66):

$$\mathrm{Im}(k_{*x}) > 0,\ \mathrm{Im}(k_{*z}^o) < 0,\ \mathrm{Im}(k_{*z}^a) > 0 \quad (67)$$

Taking (67) into account, solution (63) turns out to be physically unrealizable in both media. Solution (64) describes a regular solution with respect to a small parameter $\varepsilon^{o,a}$ and the corresponding wave component of a periodic flow. Solution (65) describes a singular solution with respect to a small parameter $\varepsilon^{o,a}$ and corresponds to the ligament component of the periodic flow. To distinguish the roots, we introduce a redesignation for singular solutions $k_{*l}^{o,a}$. Mathematically, the solutions corresponding to the wave component are determined by the following condition:

$$\left|\mathrm{Re}(k_{*z}^{o,a})\right| \gg \left|\mathrm{Im}(k_{*z}^{o,a})\right| \quad (68)$$

and the solutions corresponding to the ligament component are determined by the following mathematical condition:

$$\left|\mathrm{Re}\left(k_{*l}^{o,a}\right)\right| \sim \left|\mathrm{Im}\left(k_{*l}^{o,a}\right)\right| \quad (69)$$

Taking into account the ligament components, the form of the complete solution (54) is rewritten as follows:

$$\begin{pmatrix} u^{o,a} \\ w^{o,a} \\ \widetilde{p}^{o,a} \\ \widetilde{\rho}^{o,a} \\ \zeta \end{pmatrix} = \begin{pmatrix} U_m^{o,a}\left(\exp(ik_z^{o,a}z) + \Theta\exp\left(ik_l^{o,a}z\right)\right) \\ W_m^{o,a}\left(\exp(ik_z^{o,a}z) + \Theta\exp\left(ik_l^{o,a}z\right)\right) \\ P_m^{o,a}\left(\exp(ik_z^{o,a}z) + \Theta\exp\left(ik_l^{o,a}z\right)\right) \\ P_m^{o,a}\left(\exp(ik_z^{o,a}z) + \Theta\exp\left(ik_l^{o,a}z\right)\right) \\ A_m \end{pmatrix} \exp(ik_x x - i\omega t) \quad (70)$$

Substituting the form of solution (70) for the boundary conditions (53), we obtain the dispersion relations connecting the components of the wave vector k_x with the frequency of wave motion ω. Substituting the approximate solutions (64), (66), (65) and (66) into the resulting relation, we obtain a dispersion equation. Restrictions (67) are imposed on the solution, and thus physically realizable roots are selected. The resulting expressions are cumbersome and difficult to analyze. Let us consider some limiting cases.

We consider the behavior of oscillations far from the interface between the media. In this case, we assume that $|z| \gg 1$ for the lower liquid and for the upper liquid. Thus, for the ocean in the dispersion relation (56), we can neglect the second term and use the following:

$$\frac{\omega^2\left(\nu^o\left(k_x^2 + k_z^{o2}\right) - i\omega\right)\left(-gk_x^2\Lambda^o + \omega\left(i + k_z^0\Lambda^o\right)\left(\nu^o\left(k_x^2 + k_z^{o2}\right) - i\omega\right)\right)}{c^{o2}\Lambda^{o2}} = 0 \quad (71)$$

or in a dimensionless form:

$$\omega_*^2\eta^o\left(\omega_*(k_{z*}^o + i\varepsilon^o)\left(\varepsilon^o k_{z*}^{o2} - i\omega_*\right) + k_{x*}^4\left(-1 + \varepsilon^{o2}\omega_* k_{z*}^o + i\varepsilon^{o3}\omega_*\right)\right) = 0 \quad (72)$$

The solutions to expression (71) (or (72)) are found exactly, but due to their cumbersomeness, they are not given here.

For the atmosphere in the dispersion relation (57), based on similar reasoning, we neglect the first term and obtain the dispersion relation, which is far from the interface:

$$\frac{\omega k_z^a\left(\left(g + ic^{a2}k_z^{a2}\right)\left(-gk_x^2\Lambda^a + \omega\left(i + k_z^a\Lambda^a\right)\left(\nu^a\left(k_x^2 + k_z^{a2}\right) - i\omega\right)\right) + c^{a2}k_x^2\Lambda^a\left(N^{a2}\left(ik_z^a\Lambda^a - 1\right) + \omega\left(i\nu\left(k_x^2 + k_z^{a2}\right) + \omega\right)\right)\right)}{c^{a2}\Lambda^{a2}} = 0 \quad (73)$$

or in a dimensionless form:

$$\begin{aligned}&\tfrac{\omega_*^2}{\varepsilon^a}\eta^a\big(\varepsilon^a\omega_*(k^a_{z*}+i\varepsilon^a)\big(\varepsilon^a k^{a2}_{z*}-i\omega_*\big)+\varepsilon^a k^4_{x*}\big(-1+\varepsilon^{a2}\omega_* k^a_{z*}+i\varepsilon^{a3}\omega_*\big)+\\&+k^2_{x*}\big(2\varepsilon^{a3}\omega_* k^{a3}_{z*}-2i\varepsilon^{a2}\omega_*^2k^a_{z*}+\omega_*\big(i+2\varepsilon^{a3}\omega_*\big)+k^{a2}_{z*}\big(-\varepsilon^a+2i\varepsilon^{a4}\omega_*\big)\big)\big)=0\end{aligned} \tag{74}$$

The solutions of expression (73) (or (74)) are not given here due to their cumbersomeness. For waves near the surface, we can assume that the dispersion relations (56) and (57) are simplified:

$$\begin{aligned}&\tfrac{\omega}{c^{o2}\Lambda^{o2}}\big[\omega\big(\nu^o\big(k_x^2+k_z^{o2}\big)-i\omega\big)\big(-gk_x^2\Lambda^o+\omega\big(i+k_z^o\Lambda^o\big)\big(\nu^o\big(k_x^2+k_z^{o2}\big)-i\omega\big)\big)+\\&\quad+k_z^o\big(\big(g+ic^{o2}k_z^{o2}\big)\big(-gk_x^2\Lambda^o+\omega\big(i+k_z^o\Lambda^o\big)\big(\nu^o\big(k_x^2+k_z^{o2}\big)-i\omega\big)\big)+\\&\qquad+c^{o2}k_x^2\Lambda^o\big(N^{o2}\big(ik_z^o\Lambda^o-1\big)+\omega\big(i\nu\big(k_x^2+k_z^{o2}\big)+\omega\big)\big)\big)\big]=0\end{aligned} \tag{75}$$

$$\begin{aligned}&\tfrac{\omega}{c^{a2}\Lambda^{a2}}\big[\omega\big(\nu^a\big(k_x^2+k_z^{a2}\big)-i\omega\big)\big(-gk_x^2\Lambda^a+\omega\big(i+k_z^a\Lambda^a\big)\big(\nu^a\big(k_x^2+k_z^{a2}\big)-i\omega\big)\big)+\\&\quad+k_z^a\big(\big(g+ic^{a2}k_z^{a2}\big)\big(-gk_x^2\Lambda^a+\omega\big(i+k_z^a\Lambda^a\big)\big(\nu^a\big(k_x^2+k_z^{a2}\big)-i\omega\big)\big)+\\&\qquad+c^{a2}k_x^2\Lambda^a\big(N^{a2}\big(ik_z^a\Lambda^a-1\big)+\omega\big(i\nu\big(k_x^2+k_z^{a2}\big)+\omega\big)\big)\big)\big]=0\end{aligned} \tag{76}$$

or in a dimensionless form:

$$\begin{aligned}&\tfrac{\omega_*}{\varepsilon^{o2}}\big[\varepsilon^o\eta^o\omega_*\big(\varepsilon^o\omega_*(k^o_{z*}+i\varepsilon^o)\big(\varepsilon^o k^{o2}_{z*}-i\omega_*\big)+\varepsilon^o k^4_{x*}\big(-1+\varepsilon^{o2}\omega_* k^o_{z*}+i\varepsilon^{o3}\omega_*\big)+\\&\quad+k^2_{x*}\big(2\varepsilon^{o3}\omega_* k^{o3}_{z*}-2i\varepsilon^{o2}\omega_*^2k^o_{z*}+\omega_*\big(i+2\varepsilon^{o3}\omega_*\big)+k^{o2}_{z*}\big(-\varepsilon^o+2i\varepsilon^{o4}\omega_*\big)\big)\big)+\\&\qquad+k^o_{z*}\big(i\varepsilon^{o4}\omega_* k^4_{x*}+\varepsilon^o\omega_*(k^o_{z*}+i\varepsilon^o)\big(\varepsilon^{o2}k^o_{z*}-i\eta^o\big)\big(i\varepsilon^o k^{o2}_{z*}+\omega_*\big)-\\&-k^2_{x*}\big(\eta^o-2i\varepsilon^{o4}\omega_* k^{o2}_{z*}+\varepsilon^{o5}\omega_* k^o_{z*}-\varepsilon^{o2}\eta^o\omega_* k^o_{z*}+\varepsilon^{o3}\big(1-i\eta^o\omega_*-\omega_*^2\big)\big)\big)\big]=0\end{aligned} \tag{77}$$

$$\begin{aligned}&\tfrac{\omega_*}{\varepsilon^{a2}}\big[\varepsilon^a\eta^a\omega_*\big(\varepsilon^a\omega_*(k^a_{z*}+i\varepsilon^a)\big(\varepsilon^a k^{a2}_{z*}-i\omega_*\big)+\varepsilon^a k^4_{x*}\big(-1+\varepsilon^{a2}\omega_* k^a_{z*}+i\varepsilon^{a3}\omega_*\big)+\\&\quad+k^2_{x*}\big(2\varepsilon^{a3}\omega_* k^{a3}_{z*}-2i\varepsilon^{a2}\omega_*^2k^a_{z*}+\omega_*\big(i+2\varepsilon^{a3}\omega_*\big)+k^{a2}_{z*}\big(-\varepsilon^a+2i\varepsilon^{a4}\omega_*\big)\big)\big)+\\&\qquad+k^a_{z*}\big(i\varepsilon^{a4}\omega_* k^4_{x*}+\varepsilon^a\omega_*(k^a_{z*}+i\varepsilon^a)\big(\varepsilon^{a2}k^a_{z*}-i\eta^a\big)\big(i\varepsilon^a k^{a2}_{z*}+\omega_*\big)-\\&-k^2_{x*}\big(\eta^a-2i\varepsilon^{a4}\omega_* k^{a2}_{z*}+\varepsilon^{a5}\omega_* k^a_{z*}-\varepsilon^{a2}\eta^a\omega_* k^a_{z*}+\varepsilon^{a3}\big(1-i\eta^a\omega_*-\omega_*^2\big)\big)\big)\big]=0\end{aligned} \tag{78}$$

Nevertheless, despite their simpler appearance, the roots of expressions (77) and (78), as well as complete expressions, can only be found asymptotically or numerically.

7. Discussion

The expressive properties of periodic flows in fluids—the regularity of wave displacements of the liquid-free surface, the high speed of sound vibrations propagation and the clarity of the pattern of periodic internal waves beams—formed the basis for the generally accepted classification of waves and predetermined the rules for constructing mathematical models of the phenomenon. To describe each wave process in a linear [1,2,16,21] or nonlinear approximation [41], its own system of equations was developed based on the system of fundamental equations of mechanics of fluids and gases [1,2,4,16], and general physical considerations [31,52].

Under natural conditions, sharp disturbances lead to the formation of several types of waves, which propagate with their own phase and group velocities and differ in attenuation laws. The parameters of wave processes—periods, wavelength, group and phase propagation velocities—are described by real numbers. The mathematical description of periodic flows is carried out in the algebra of complex numbers. The use of wave representations by exponential functions of complex frequency and complex wave vector allows us to construct the dispersion relations [1,2] and evaluate the stability of the flows under study [32,33].

Taking into account the special physical properties of the wave frequency—the measure of the energy of periodic motion—in this work, as in [38,40,42,46], the wave frequency ω is assumed to be real, and the wave number $\mathbf{k}$ is taken to be complex. In this approximation, the degree of the dispersion relation corresponds to the order of the system of differential equations. The solutions of the system of governing equations, constructed using methods of singular perturbation theory and by taking into account the type of small parameter of the process under study, contain two types of solutions. The real part of some wave numbers is large, and the imaginary part is small. The other types have real and imaginary parts of the same order. Accordingly, some of the solutions, including solutions with small values of the imaginary part of the wave vectors, contain functions that are regular in the small parameter and describe waves. For each type of wave, its own dispersion equation is constructed.

Another part of the solutions with large values of the wave vector imaginary parts determines the ligaments, which correspond to thin high-gradient fibers and interfaces in the thickness of a stratified liquid [30,38]. From the given analysis, it follows that specific ligaments accompany all types of waves—surface, internal and acoustic ones. The consideration of the ligaments' influence made it possible to pre-calculate the parameters of reflected and leaking waves, which occur when the reflecting beams of the internal waves of the critical level separate the medium with a high buoyancy frequency from a low-frequency layer not exceeding the wave frequency [5]. It is consistent with the data of later experiments [53].

From the theoretical point of view, the number of ligaments accompanying the wave is determined by the completeness degree taking into account the factors influencing the density and the dimension of the problem space. The minimum number—two ligaments—accompany two-dimensional waves in a medium with one dissipative parameter (kinematic viscosity). Their thickness is determined by the scale of the periodic Stokes flow $\delta_\omega^\nu = \sqrt{\nu/\omega}$ [13]. Considering the three-dimensionality of space, the effects of thermal diffusivity and diffusion lead to an increase in the number of ligaments with different properties [40]. The effects of nonlinear interaction between ligaments can increase the mutual influence of waves of different types [42].

The developed methodology for constructing complete solutions makes it possible to describe not only the wave component of a periodic flow but also the fine structure, manifested in the form of ligaments—thin jets accompanying the wave motion. The parameters of the observed phenomena in the process of propagation of periodic disturbances in liquids and gases, which are determined using the properties of the medium, define the requirements for the experimental methodology and the resolution (spatial and temporal) of the equipment for observing the complete picture of flows.

8. Conclusions

For the first time in a unified formulation, the propagation of infinitesimal periodic disturbances in the thickness and on the surface of a viscous compressible exponentially stratified fluid has been studied based on a system of fundamental equations. The analysis of linearized equations has been carried out using the methods of singular perturbation theory, taking into account the compatibility condition. The dispersion relations for periodic flows with a real positive definite frequency and complex wave number are calculated and analyzed. Complete solutions of the dispersion relations containing regular and singular roots are found. Regular roots, which determine the wave components of periodic flows, are regularly reduced to known dispersion relations for waves in a homogeneous viscous or ideal fluid. Singular roots define the ligament components of periodic flows. Ligaments describe the fine structure of periodic flows and characterize thin high-gradient jets and interfaces.

The general properties of solutions are that acoustic or internal waves propagating in the thickness, as well as gravitational waves at the interface of infinitely deep media, are accompanied by ligaments forming a fine structure of the medium. In extreme cases, the obtained relationships transform into known expressions for waves a viscous incompressible and an ideal homogeneous fluid.

The further application of the obtained expressions in studying the physical properties of periodic flows in configuration space and in comparison with experimental data using high-resolution instruments is of scientific and practical interest.

Author Contributions: Conceptualization, Y.D.C. and A.A.O.; methodology, Y.D.C. and A.A.O.; validation, Y.D.C. and A.A.O.; formal analysis, Y.D.C.; investigation, Y.D.C.; resources, Y.D.C.; data curation, Y.D.C.; writing—original draft preparation, Y.D.C. and A.A.O.; writing—review and editing, Y.D.C.; visualization, A.A.O.; supervision, Y.D.C.; project administration, Y.D.C.; funding acquisition, Y.D.C. All authors have read and agreed to the published version of the manuscript.

Funding: The work was supported by the Russian Science Foundation (project 19-19-00598-P "Hydrodynamics and energetics of drops and droplet jets: formation, motion, break-up, interaction with the contact surface" https://rscf.ru/en/project/19-19-00598/ (accessed on 23 October 2023)).

Data Availability Statement: Not applicable.

Conflicts of Interest: The authors declare no conflict of interest.

References

1. Landau, L.D.; Lifshitz, E.M. *Fluid Mechanics;* Course of Theoretical Physics; Pergamon Press: Oxford, UK, 1987; Volume 6, 560p.
2. Müller, P. *The Equations of Oceanic Motions;* CUP: Cambridge, UK, 2006; 302p.
3. NASA. *US Standard Atmosphere 1976—NOAA-S/T-76-1562. NASA-TM-X-74335;* Accession Number 77N16482; NASA: Washington, DC, USA. Available online: https://ntrs.nasa.gov/citations/19770009539 (accessed on 23 October 2023).
4. Matveev, L.T. *Fizika Atmosfery;* Gidrometeoizdat: St. Petersburg, Russia, 2000; 777p. (In Russian)
5. Fedorov, K.N. *The Thermohaline Finestructure of the Ocean;* Pergamon Marine Series; Elsevier: Amsterdam, The Netherlands, 2013; 180p.
6. Franklin, B. Behavior of oil on water. Letter to J. Pringle. In *Experiments and Observations on Electricity;* R. Cole: London, UK, 1769; pp. 142–144.
7. Stokes, G.G. On the theory of oscillatory waves. In *Mathematical and Physical Papers (Cambridge Library Collection—Mathematics);* Cambridge University Press: Cambridge, UK, 2010; pp. 197–229. [CrossRef]
8. Rayleigh, L. Investigation of the character of the equilibrium of an incompressible heavy fluid of variable density. *Proc. Lond. Math. Soc.* **1882**, *s114*, 170–177. [CrossRef]
9. Lamb, H. On atmospheric oscillations. *Proc. Roy. Soc.* **1911**, *84*, 551–574. [CrossRef]
10. Ekman, V.W. On Dead Water. In *The Norwegian North Polar Expedition 1893–1896. Scientific Results;* Nansen, F., Ed.; Yakov Dyewad: Christiania, Norway, 1906; Volume 15, 152p.
11. Lamb, H. *A Treatise on the Mathematical Theory of the Motion of Fluids;* Cambridge University Press: Cambridge, UK, 1879; 258p.
12. Stokes, G.G. On the theories of the internal friction of fluids in motion and of the equilibrium and motion of elastic solids. In *Mathematical and Physical Papers (Cambridge Library Collection—Mathematics);* Cambridge University Press: Cambridge, UK, 2010; pp. 75–129. [CrossRef]
13. Stokes, G.G. On the Effect of Internal Friction of Fluids on the Motion of Pendulums. In *Mathematical and Physical Papers (Cambridge Library Collection—Mathematics);* Cambridge University Press: Cambridge, UK, 2010; pp. 1–10. [CrossRef]
14. Rayleigh, J.W.S. *Theory of Sound,* 2nd ed.; Dover: New York, NY, USA, 1945.
15. Kochin, N.E.; Kibel, I.A.; Roze, N.V. *Theoretical Hydromechanics;* John Wiley & Sons Ltd.: Chichester, NH, USA, 1964; 560p.
16. Vallis, G.K. *Atmospheric and Oceanic Fluid Dynamics;* CUP: Cambridge, UK, 2017; 745p.
17. Prandtl, L. *Führer Durch die Strömungslehre;* Vieweg, Vieweg und Sohn: Braunschweig, Germany, 1942.
18. Phillips, O. On flows induced by diffusion in a stably stratified fluid. *Deep Sea Res. Oceanogr. Abstr.* **1970**, *17*, 435–443. [CrossRef]
19. Wunsch, C. On oceanic boundary mixing. *Deep Sea Res. Oceanogr. Abstr.* **1970**, *17*, 293–301. [CrossRef]
20. Turner, J.S. *Buoyancy Effects in Fluids;* Cambridge Monographs on Mechanics; Cambridge University Press: Cambridge, UK, 1980; 412p.
21. Lighthill, J.M. *Waves in Fluids;* Cambridge Mathematical Library; Cambridge University Press: Cambridge, UK, 1978; 594p.
22. Tolstoy, I.; Clay, C.S. *Ocean Acoustics: Theory and Experiment in Underwater Sound;* McGraw-Hill: New York, NY, USA, 1966; 293p.
23. Longuet-Higgins, M.S. Mass transport in water waves. *Phil. Trans. R. Soc. London Ser. A Math. Phys. Sci.* **1953**, *245*, 535–581. [CrossRef]
24. Longuet-Higgins, M.S. Mass transport in the boundary layer at a free oscillating surface. *J. Fluid Mech.* **1960**, *8*, 293–306. [CrossRef]
25. Liu, A.; Davis, S. Viscous attenuation of mean drift in water waves. *J. Fluid Mech.* **1977**, *81*, 63–84. [CrossRef]
26. Robertson, S.; Rousseaux, G. Viscous dissipation of surface waves and its relevance to analogue gravity experiments. In *Fluid Dynamics;* Cornell University Press: New York, NY, USA, 2018. [CrossRef]
27. Dore, B. Mass transport in layered fluid systems. *J. Fluid Mech.* **1970**, *40*, 113–126. [CrossRef]
28. Zhang, W.; Jin, H. Nonlinear Stability of the Monotone Traveling Wave for the Isothermal Fluid Equations with Viscous and Capillary Terms. *Mathematics* **2023**, *11*, 1734. [CrossRef]
29. Pei, F.; Wu, G.; Guo, Y. Construction of Infinite Series Exact Solitary Wave Solution of the KPI Equation via an Auxiliary Equation Method. *Mathematics* **2023**, *11*, 1560. [CrossRef]
30. Kistovich, Y.V.; Chashechkin, Y.D. Linear theory of beams of internal wave propagation an arbitrarily stratified liquid. *J. Appl. Mech. Tech. Phys.* **1998**, *39*, 302–309. [CrossRef]

31. Krasil'nikov, V.A.; Krylov, V.V. *Vvedeniye v Fizicheskuyu Akustiku*; Nauka: Moskva, Russia, 1984; 400p. (In Russian)
32. Lin, C.C. *The Theory of Hydrodynamic Stability*; Cambridge University Press: Cambridge, UK, 1955; 155p.
33. Chandrasekhar, S. *Hydrodynamic and Hydromagnetic Stability, International Series of Monographs on Physics*; Clarendon Press: Oxford, UK, 1961; 654p.
34. Darrigol, O. *Worlds of Flow: A History of Hydrodynamics from the Bernoullis to Prandt*; Oxford University Press: Oxford, UK, 2005; 356p.
35. Suvorov, V.G.; Zubarev, N.M. Formation of the Taylor cone on the surface of liquid metal in the presence of an electric field. *J. Phys. D Appl. Phys.* **2003**, *37*, 289. [CrossRef]
36. Zubarev, N.M. Exact solutions of the equations of motion of liquid helium with a charged free surface. *J. Exp. Theor. Phys.* **2002**, *94*, 534–544. [CrossRef]
37. Zeytounian, R.K. The Benard–Marangoni thermocapillary-instability problem. *Phys. Uspekhi* **1998**, *41*, 241. [CrossRef]
38. Chashechkin, Y.D. Foundations of engineering mathematics applied for fluid flows. *Axioms* **2021**, *10*, 286. [CrossRef]
39. Nayfeh, A.H. *Introduction to Perturbation Technique*; John Wiley & Sons: New York, NY, USA, 1993; 536p.
40. Chashechkin, Y.D. Singularly perturbed components of flows—Linear precursors of shock waves. *Math. Model. Nat. Phenom.* **2018**, *13*, 17. [CrossRef]
41. Whitham, G.B. *Linear and Nonlinear Waves*; Wiley Interscience: New York, NY, USA, 1999; 660p.
42. Chashechkin, Y.D. Conventional partial and new complete solutions of the fundamental equations of fluid mechanics in the problem of periodic internal waves with accompanying ligaments generation. *Mathematics* **2021**, *9*, 586. [CrossRef]
43. Thomson, W. Hydrokinetic solutions and observations. *Lond. Edinb. Dublin Philos. Mag. J. Sci.* **1871**, *42*, 362–377. [CrossRef]
44. Shishlenin, M.; Savchenko, N.; Novikov, N.; Klyuchinskiy, D. Modeling of 2D Acoustic Radiation Patterns as a Control Problem. *Mathematics* **2022**, *10*, 1116. [CrossRef]
45. Fellah, Z.E.A.; Fellah, M.; Ongwen, N.O.; Ogam, E.; Depollier, C. Acoustics of Fractal Porous Material and Fractional Calculus. *Mathematics* **2021**, *9*, 1774. [CrossRef]
46. Chashechkin, Y.D.; Ochirov, A.A. Periodic waves and ligaments on the surface of a viscous exponentially stratified fluid in a uniform gravity field. *Axioms* **2022**, *11*, 402. [CrossRef]
47. Soret, C. Sur l'etat d'équilibre que prend au point de vue de sa concentration une dissolution saline primitivement homogène dont deux parties sont portées a des températures difféntes. *Arch. Sci. Phys. Nat.* **1879**, *2*, 48.
48. Mortimer, R.G.; Eyring, H. Elementary transition state theory of the Soret and Dufour effects. *Proc. Natl. Acad. Sci. USA* **1980**, *77*, 1728–1731. [CrossRef] [PubMed]
49. Dufour, L. The Diffusion Thermoeffect. *Arch. Sci. Phys. Nat.* **1872**, *45*, 9–12.
50. Kistovich, A.V.; Chashechkin, Y.D. Regular and singular components of periodic flows in the fluid interior. *J. Appl. Math. Mech.* **2007**, *71*, 762. [CrossRef]
51. Joseph, D.D. Domain perturbations: The higher order theory of infinitesimal water waves. *Arch. Ration. Mech. Anal.* **1973**, *51*, 295. [CrossRef]
52. Rudenko, O.V. Giant nonlinearities in structurally inhomogeneous media and the fundamentals of nonlinear acoustic diagnostic techniques. *Phys. Uspekhi* **2006**, *176*, 1011–1036. [CrossRef]
53. Paoletti, M.S.; Swinney, H.L.; Paoletti, M.S.; Swinney, H.L. Propagating and evanescent internal waves in a deep ocean model. *J. Fluid Mech.* **2012**, *706*, 571–583. [CrossRef]

mathematics
MDPI

Article

Mathematical Modeling of Pseudoplastic Nanofluid Natural Convection in a Cavity with a Heat-Generating Unit and Solid Finned Heat Sink

Daria S. Loenko and Mikhail A. Sheremet *

Laboratory on Convective Heat and Mass Transfer, Tomsk State University, 634050 Tomsk, Russia; d.s.loenko@mail.tsu.ru
* Correspondence: sheremet@math.tsu.ru; Tel.: +7-3822-529740

Abstract: The power-law nanofluid natural convection in a chamber with a thermally generating unit and a solid ribbed structure has been studied in this work. A mixture of carboxymethylcellulose with water and copper nanoparticles is a working fluid illustrating pseudoplastic properties. The effective properties of the nanoliquid have been described by experimental correlations reflecting the temperature effect. The governing equations have been formulated on the basis of the conservation laws of mass, momentum and energy employing non-primitive parameters such as stream function and vorticity. The defined boundary value problem has been worked out by the finite difference technique using an independently developed calculation system. The Rayleigh number is fixed for analysis ($Ra = 10^5$). The paper analyzes the influence of the nanoparticles volume fraction, an increase in which reduces the temperature in the case of the one edge presence. An analysis of the rib height has shown that its growth leads to a weakening of the convective heat transfer, but at the same time, the source temperature also decreases. Increasing the number of fins from 1 to 3 also helps to reduce the average temperature of the heat-generated element by 15%.

Keywords: natural convection; pseudoplastic nanofluid; square enclosure; mathematical modeling; finned structure; radiator

MSC: 76A05; 76M20; 76R10; 80A20

Citation: Loenko, D.S.; Sheremet, M.A. Mathematical Modeling of Pseudoplastic Nanofluid Natural Convection in a Cavity with a Heat-Generating Unit and Solid Finned Heat Sink. *Mathematics* **2023**, *11*, 3868. https://doi.org/10.3390/math11183868

Academic Editor: Ramoshweu Solomon Lebelo

Received: 3 August 2023
Revised: 4 September 2023
Accepted: 8 September 2023
Published: 11 September 2023

1. Introduction

Cooling or heating due to convective heat transfer is one of the most attractive mechanisms of heat transfer in thermal engineering systems, as it has a number of advantages in simplicity and economy. Natural convection in various geometric configurations has a wide range of applications in many fields of technology, such as aerospace engineering, textile engineering, the automotive industry, heat storage systems, microelectronic devices design, etc. But researchers are constantly looking for more efficient heat transfer modes that can be obtained by changing the geometry of the system or working fluid [1–3]. One way to improve the heat transfer is to use expanded surfaces or fins. In many supplements in mechanical engineering, chemical engineering, power engineering, heat recovery, surface research, etc., fins have a wide range of applications. Based on this, many researchers investigate the efficiency of their use through analytical, numerical, or experimental studies [4]. Another way to intensify the heat transfer is to use highly thermally conductive liquids as working media. These media include nanofluids, which consist of a base medium and highly thermally conductive solid additives, which increase the thermal characteristics of the entire liquid.

The combination of these mentioned improvements can lead to more efficient cooling/heating of various elements. For example, Hidki et al. [5] have studied the thermal convection of a $Cu+Al_2O_3$/water hybrid nanoliquid in a closed chamber with two heat-generating blocks. Simulations have shown that the presence of solid particles reduces the temperature of the blocks by up to 18%. Mounting the fins on the blocks surface reduces their temperature by up to 12%. The influence of the heated fin presence on the hybrid nanofluid natural convection has been studied by Iftikhar et al. [6]. The working medium is a mixture of ethylene glycol with copper and silver nanoparticles. The outcomes have demonstrated that the presence of nanoadditives enhances the fluid flow rate by 47.4%, in contrast to the case of a pure base medium. The work of M. Hatami [4] is similar to the study of Iftikhar et al. [6], but the side walls and the upper one are cooled. Additionally, two types of nanoparticles, namely, TiO_2 and Al_2O_3, are studied. The authors have found that the use of titanium oxide nanoparticles leads to an intensification of the convective energy transport in the chamber. The influence of the ribs number, nanoparticles volume fraction and Rayleigh number on the thermal convective energy transport of the Cu/H_2O nanofluid in the annular space has been investigated by Waqas et al. [7]. The analysis has demonstrated that a growth in the solid particles concentration and Rayleigh number improves the energy transport and flow velocity. Shahsavar et al. [8] have studied the Ag/water nanofluid natural convection in a concentric channel. Additionally, entropy generation has been analyzed. The authors have found that a growth in the number of edges results in a rise in the mean Nusselt number to 35.50% and entropy production rate to 39.07%. A similar geometry was used by Tayebi et al. in [9] where the authors also investigated the influence of the Lorentz force on the Al_2O_3/H_2O nanoliquid thermogravitational convection. It has been shown that the strength of the convective energy transport decreases with a growth of the fins size and the Hartmann number, but increases with Ra and ϕ. Yasmin et al. [10] used a hexagonal nanofluid chamber under the impact of a ribbed circular cylinder. The horizontal borders of the cavity are hot, the rest are cold. As a result of the study, it has been ascertained that a growth in the fins' height and the nanoadditives' concentration improves the energy transport. Hejri and Malekshah [11] have analyzed the best geometric characteristics for a cooling system consisting of a finned structure filled with CuO/H_2O nanofluid. The authors have also studied the influence of the shape of nanoparticles on the process. The analysis has shown that thin ribs and plate-shaped particles are more efficient for the considered phenomenon. The presence of two heat-conducting ribs in a tilted porous chamber filled with ferrofluid Fe_3O_4/H_2O has been studied by Al-Farhany et al. [12]. The authors found that the fins' length and the distance between them have a significant effect on the intensity of the convective heat transfer. We should note that the addition of Fe_3O_4 nanoparticles intensifies the convective heat transfer regardless of the ribs' length and the distance between them. A similar geometry was investigated by Yan et al. [13], where the main difference is that the edges can have different inclination angles. The influence of the Al_2O_3 nanoparticles' shape has been also studied in this work. The main conclusion is that an increase in the concentration of nanoparticles does not always lead to an intensification of the convective heat transfer. The highest energy transport coefficient was obtained in the case of inclined fins. An analogous system has been studied by Siavashi et al. [14] but with porous fins. The authors revealed that a growth in Da enhances the heat transfer in the cavity. An increase in the number of ribs or their length does not have an essential impact on the mean Nusselt number.

Esfe et al. [15] modeled a 3D free convection in a differentially heated chamber under the porous ribs; influence on a hot wall. CuO/H_2O nanofluid has been used as a working medium. The simulation results have demonstrated that an addition of porous fins decreases the fluid flow rate, and a growth in the nanoadditives' volume fraction leads to an intensification of convective energy transport. Gireesha et al. [16] considered the free-convective energy transport between a hybrid $Ag+MgO/H_2O$ nanofluid and a hot moving porous fin. It has been found that the presence of nanoadditives intensifies the energy transport, and an increase in the Biot number increases the rate of the heat removal from the fin surface. A ribbed square body with a rotating hot rectangular block inside has been examined by Aly et al. [17]. The chamber was filled with Cu/H_2O nanofluid, and the hot block also had two hot ribs on the sides. The results demonstrated that the presence of ribs reduces the distribution of isotherms and lines of constant concentration inside the cavity and changing the block angle from 0° to 90° can increase the flow rate by 9.59% and 31.53%, respectively. Ganesh et al. [1] performed the mathematical modeling of the Casson nanofluid MWCNT/sodium alginate flow in a casing with wavy horizontal walls. In the housing center, there is a circular heat-insulated barrier and two differentially heated fins. The simulation has shown that a growth in the Rayleigh number, Casson parameter, nanoadditives concentration, and wall waviness amplitude increases the heat transfer rate. The work of Siva et al. [18] is devoted to an analytical study of the electromagnetic hydrodynamic flow of a non-Newtonian fluid in a microchannel. The authors found that an increase in couple stresses leads to an increase in the flow velocity, but the critical Hartmann number is unchanged. Mehta and Mondal [19] carried out a numerical simulation of a natural convective heat transfer in a cavity with wavy walls, which was filled with MWCNT+Fe_3O_4/water nanofluid. The performed analysis showed that an increase in the volume fraction of nanoparticles and the amplitude of the wall wave leads to an increase in the values of the average Nusselt number. Mondal and Wongwises [20] studied the unsteady electro-MHD nanofluid flow in a rotating channel. The authors showed that the combination of the Coriolis and Lorentz forces with the rheological liquid nature has an interesting result for the flow reversal in the channel. Moreover, for a higher-rotation Reynolds number, the flow reversal owing to the strong Coriolis force influence is impossible due to a growth in the nanoliquid's effective viscosity.

A brief analysis of studies of the convective heat transfer of nanofluids in various cavities with ribs showed that this topic is very popular and useful from a theoretical and practical point of view. However, there are still many unexplored phenomena in this area. Therefore, the purpose of this work is to evaluate the efficiency of using a ribbed solid structure for cooling a local heat source employing laminar free-convective energy transfer of a power law nanofluid. The novelty of this study is an analysis of the combined effect of the pseudoplastic nanofluid and solid finned heat sink on the cooling of the heat-generating element. Moreover, the nanofluid's physical properties have been described using experimentally based correlations with the temperature influence. At the same time, the computational analysis was conducted using non-primitive variables that allow reducing the computational time and improving the physical analysis. Obtained results can be used in the modeling and optimizing of the passive cooling systems for various electronic systems.

2. Formulation of the Problem

The geometric scheme of the solution region (see Figure 1) is a square closed cavity, the upper and part of the side borders of which are maintained at a fixed low temperature T_c. A heating element of the fixed volumetric thermal production Q is placed in the center of the chamber lower wall. Above the source there is a heat-conducting copper profile, along the perimeter of which the outer walls are heat-insulated. The height of the ribs H varies during the study. The cavity is filled with a suspension of carboxymethylcellulose (CMC) with water and copper nanoparticles. The thermal characteristics of the materials used are presented in Table 1. The analyzed finned heat sink with nanoenhanced CMC/water suspension can be considered as a passive cooling system for more effective thermal energy removal from the heat-generating element. Such a combination of two different approaches allows enhancing the heat removal compared to each of these techniques.

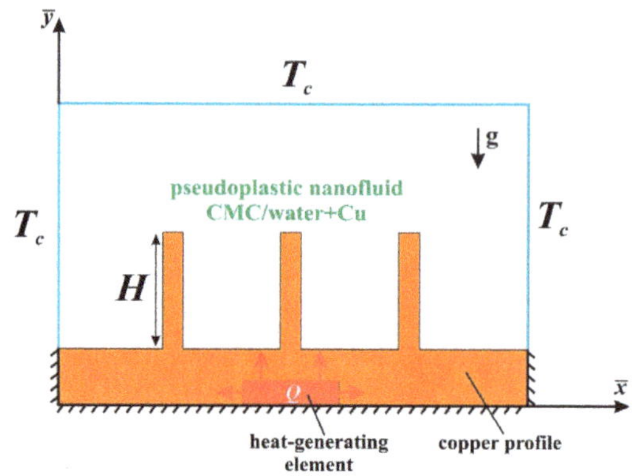

Figure 1. Geometry for the considered problem.

Table 1. Thermal properties of materials [21,22].

Properties	c_p [J/kg·K]	ρ [kg/m^3]	k [W/m·K]
Base fluid CMC (0.1%)/water	4179	997.1	0.613
Nanoparticles (Cu)	385	8933	400
Heat source (silicon)	710	2330	150
Ribbed heat sink (Cu)	385	8920	400

The pseudoplastic behavior of the liquid motion is defined by the Ostwald–deWaele power law [23]:

$$\tau_{ij} = 2\mu_{nf} D_{ij} \tag{1}$$

The effective viscosity coefficient of a nanofluid is simulated employing the relation of Guo et al. [24]:

$$\frac{\mu_{nf}}{\mu_{bf}} = \left(1 + 2.5\phi + 6.5\phi^2\right)\left(1 + 350\frac{\phi}{d_p}\right) \tag{2}$$

The viscosity of the host liquid was defined using the relationship: $\mu_{bf} = K(2D_{kl}D_{kl})^{\frac{n-2}{2}}$. Here, K is the flux density factor; D_{kl} are components of the strain rate tensor; n is the fluid behavior index, which is equal to 0.91. Considering that for $n < 1$, the working fluid has a pseudoplastic nature. Such fluids are described by viscosity that reduces with a growth in the strain rate, which is very efficient for the cooling problems.

The nanoliquid heat conductivity was calculated using the experimental correlation of Jang and Choi [25]:

$$\frac{k_{nf}}{k_{bf}} = (1-\phi) + 0.01\frac{k_p}{k_{bf}}\phi + \left(18 \cdot 10^6\right)\frac{d_{bf}}{d_p}Re^2 Pr\phi \tag{3}$$

The control equations reflecting the process of laminar transient convective energy transport employing the Oberbeck–Boussinesq approach based on the dimensional primitive variables "velocity and pressure" are:

$$\frac{\partial \overline{u}}{\partial \overline{x}} + \frac{\partial \overline{v}}{\partial \overline{y}} = 0 \tag{4}$$

$$\rho_{nf}\left(\frac{\partial \overline{u}}{\partial t} + \overline{u}\frac{\partial \overline{u}}{\partial \overline{x}} + \overline{v}\frac{\partial \overline{u}}{\partial \overline{y}}\right) = -\frac{\partial \overline{p}}{\partial \overline{x}} + 2\frac{\partial}{\partial \overline{x}}\left[\mu_{nf}\frac{\partial \overline{u}}{\partial \overline{x}}\right] + \frac{\partial}{\partial \overline{y}}\left[\mu_{nf}\left(\frac{\partial \overline{u}}{\partial \overline{y}} + \frac{\partial \overline{v}}{\partial \overline{x}}\right)\right] \tag{5}$$

$$\rho_{nf}\left(\frac{\partial \overline{v}}{\partial t}+\overline{u}\frac{\partial \overline{v}}{\partial \overline{x}}+\overline{v}\frac{\partial \overline{v}}{\partial \overline{y}}\right)=-\frac{\partial \overline{p}}{\partial \overline{y}}+\frac{\partial}{\partial \overline{x}}\left[\mu_{nf}\left(\frac{\partial \overline{u}}{\partial \overline{y}}+\frac{\partial \overline{v}}{\partial \overline{x}}\right)\right]+2\frac{\partial}{\partial \overline{y}}\left[\mu_{nf}\frac{\partial \overline{v}}{\partial \overline{y}}\right]+g(\rho\beta)_{nf}(T-T_c) \quad (6)$$

$$(\rho c)_{nf}\left(\frac{\partial T}{\partial t}+\overline{u}\frac{\partial T}{\partial \overline{x}}+\overline{v}\frac{\partial T}{\partial \overline{y}}\right)=\frac{\partial}{\partial \overline{x}}\left(k_{nf}\frac{\partial T}{\partial \overline{x}}\right)+\frac{\partial}{\partial \overline{y}}\left(k_{nf}\frac{\partial T}{\partial \overline{y}}\right) \quad (7)$$

Heat conduction equations for the energy source and the solid finned heat sink are:

$$(\rho c)_{hs}\frac{\partial T}{\partial t}=k_{hs}\left(\frac{\partial^2 T}{\partial \overline{x}^2}+\frac{\partial^2 T}{\partial \overline{y}^2}\right)+Q \quad (8)$$

$$(\rho c)_r\frac{\partial T}{\partial t}=k_r\left(\frac{\partial^2 T}{\partial \overline{x}^2}+\frac{\partial^2 T}{\partial \overline{y}^2}\right) \quad (9)$$

To reduce the mathematical difficulties, the stream function $\overline{\psi}$ ($\overline{u}=\partial\overline{\psi}/\partial\overline{y}$, $\overline{v}=-\partial\overline{\psi}/\partial\overline{x}$) and vorticity ($\overline{\omega}=\partial\overline{v}/\partial\overline{x}-\partial\overline{u}/\partial\overline{y}$), as well as reference parameters presented in Table 2, are introduced into system (4)–(9). L is chosen as the length scale and $\Delta T=QL^2/k_{hs}$ is used for the temperature difference.

Table 2. Reference parameters.

Parameters	Formula
Velocity	$\sqrt{g\beta L\Delta T}$
Time	$\sqrt{L/(g\beta\Delta T)}$
Stream function	$\sqrt{g\beta L^3\Delta T}$
Vorticity	$\sqrt{g\beta\Delta T/L}$
Temperature	$\Delta T=QL^2/k_{hs}$

As a result, the non-dimensional governing equations are:

$$\frac{\partial^2\Psi}{\partial X^2}+\frac{\partial^2\Psi}{\partial Y^2}=-\Omega \quad (10)$$

$$\frac{\partial\Omega}{\partial\tau}+\frac{\partial\Psi}{\partial Y}\frac{\partial\Omega}{\partial X}-\frac{\partial\Psi}{\partial X}\frac{\partial\Omega}{\partial Y}=H_1(\phi)\left(\frac{Ra}{Pr}\right)^{\frac{n-2}{2}}\left[\nabla^2(\overline{M}\Omega)+S_\Omega\right]+H_2(\phi)\frac{\partial\Theta}{\partial X} \quad (11)$$

$$\frac{\partial\Theta}{\partial\tau}+\frac{\partial\Psi}{\partial Y}\frac{\partial\Theta}{\partial X}-\frac{\partial\Psi}{\partial X}\frac{\partial\Theta}{\partial Y}=\frac{H_3(\phi)}{\sqrt{Ra\cdot Pr}}\left[\frac{\partial}{\partial X}\left(\frac{k_{nf}}{k_{bf}}\cdot\frac{\partial\Theta}{\partial X}\right)+\frac{\partial}{\partial Y}\left(\frac{k_{nf}}{k_{bf}}\cdot\frac{\partial\Theta}{\partial Y}\right)\right] \quad (12)$$

The heat conduction equations for the energy source and the radiator are:

$$\frac{\partial\Theta_{hs}}{\partial\tau}=\frac{\alpha_{hs}/\alpha_{bf}}{\sqrt{Ra\cdot Pr}}\left(\frac{\partial^2\Theta_{hs}}{\partial X^2}+\frac{\partial^2\Theta_{hs}}{\partial Y^2}+1\right) \quad (13)$$

$$\frac{\partial\Theta_r}{\partial\tau}=\frac{\alpha_r/\alpha_{bf}}{\sqrt{Ra\cdot Pr}}\left(\frac{\partial^2\Theta_r}{\partial X^2}+\frac{\partial^2\Theta_r}{\partial Y^2}\right) \quad (14)$$

Here, the used additonal functions are $\overline{M}$ (non-dimensional viscosity of the host liquid), S_Ω (source term) and $H_1(\phi)$, $H_2(\phi)$, $H_3(\phi)$ (additional functions reflecting the thermal characteristics of the nanosuspension). These additional functions can be presented as follows:

$$\overline{M}=\left[4\left(\frac{\partial^2\Psi}{\partial X\partial Y}\right)^2+\left(\frac{\partial^2\Psi}{\partial Y^2}-\frac{\partial^2\Psi}{\partial X^2}\right)^2\right]^{\frac{n-1}{2}},\ S_\Omega=2\left[\frac{\partial^2\overline{M}}{\partial X^2}\frac{\partial^2\Psi}{\partial Y^2}+\frac{\partial^2\overline{M}}{\partial Y^2}\frac{\partial^2\Psi}{\partial X^2}-2\frac{\partial^2\overline{M}}{\partial X\partial Y}\frac{\partial^2\Psi}{\partial X\partial Y}\right]$$

$$H_1(\phi)=\frac{\mu_{nf}}{\mu_{bf}}\frac{\rho_{bf}}{\rho_{nf}}=\frac{\mu_{nf}/\mu_{bf}}{(1-\phi+\phi\rho_p/\rho_{bf})},\ H_2(\phi)=\frac{(\rho\beta)_{nf}}{(\rho\beta)_{bf}}\frac{\rho_{bf}}{\rho_{nf}}=\frac{1-\phi+\phi(\rho\beta)_p/(\rho\beta)_{bf}}{1-\phi+\phi\rho_p/\rho_{bf}},$$
$$H_3(\phi)=\frac{(\rho c)_{bf}}{(\rho c)_{nf}}=\frac{1}{1-\phi+\phi(\rho c)_p/(\rho c)_{bf}}$$

The initial and boundary conditions for the system (10)–(14) in a dimensionless form are as follows:

$$\begin{aligned}
&\tau = 0 \rightarrow \Psi = \Omega = 0, \Theta = 0.0;\\
&\tau > 0 \rightarrow\\
&X = 0 \,\&\, X = 1, 0 \le Y \le 0.1, \tfrac{\partial \Theta}{\partial X} = 0;\\
&X = 0 \,\&\, X = 1, 0.1 < Y \le 1, \Psi = 0, \tfrac{\partial \Psi}{\partial X} = 0, \Theta = 0;\\
&Y = 0, 0 \le X \le 1, \tfrac{\partial \Theta}{\partial Y} = 0;\\
&Y = 1, 0 \le X \le 1, \Psi = 0, \tfrac{\partial \Psi}{\partial Y} = 0, \Theta = 0.
\end{aligned}$$

At the heat source surface, one can find: $\left\{ \Theta_{hs} = \Theta_r, \frac{k_{hs}}{k_r}\frac{\partial \Theta_{hs}}{\partial n} = \frac{\partial \Theta_r}{\partial n} \right.$.

At the radiator surface, we have: $\Psi = 0, \Omega = -\frac{\partial^2 \Psi}{\partial n^2}, \left\{ \Theta_{nf} = \Theta_{bf}, \frac{k_{nf}}{k_r}\frac{\partial \Theta_{nf}}{\partial n} = \frac{\partial \Theta_r}{\partial n} \right.$.

For a solution to the formulated system of unsteady differential Equations (10)–(14) with the corresponding restrictions, the finite difference technique was used [26–28]. A successive under-relaxation algorithm was used to work out the approximated Poisson equation. The non-dimensional viscosity was discretized employing a regularization technique [26]. The Samarskii locally one-dimensional difference procedure [27,28] was applied to reduce Equations (11) and (12) to the system of one-dimensional equations. Further on, the convective terms were approximated using the donor cells scheme, and the diffusion terms were discretized by central differences. The obtained systems of linear algebraic equations were solved by the Thomas method.

The developed numerical tehcnique and the prepared computation code were verified employing the model problems. Figure 2 demonstrates the geometry of the model problem. Power-law fluid circulates in a differentially heated enclosure. The horizontal walls are thermally insulated. Comparison of the outcomes was performed for the average Nusselt number calculated at the hot wall $\left(Nu_{avg} = -\int_0^1 \frac{\partial \Theta}{\partial X} dY \right)$, depending on time and the fluid behavior index n as shown in Figure 3. Obtained results are represented by white symbols, while the results of [24] are shown in black. It can be seen that the difference in the results does not exceed 4%; this illustrates a very good agreement.

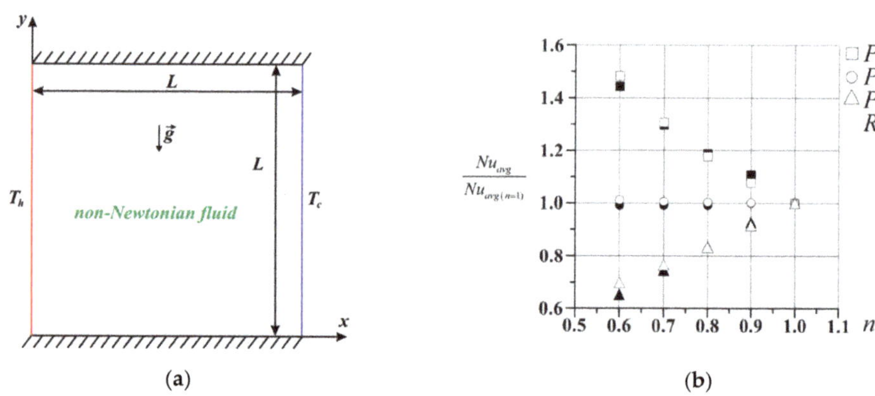

Figure 2. Testing the program code: (**a**) Model domain of interest; (**b**) Average Nusselt number behavior in comparison with [29].

An additional study was performed for the mesh effect on the solution convergence using the mean Nusselt number and mean heater temperature presented in Figure 4 for three ribs at $Ra = 10^5$, $\phi = 0.01$, $\delta = 0.2$. The mean Nusselt number was defined by $Nu_{avg} = -\frac{1}{1.8}\int_0^{1.8} \frac{\partial \Theta}{\partial n} d\varsigma$ at the finned heat sink border, where ς is the natural coordinate along the finned heat sink border. It can be seen that the meshes of 300×300 and 400×400 units did not lead to significant discrepancies; therefore, the further calculations were carried out on a uniform rectangular grid of 300×300 elements to optimize the computation time.

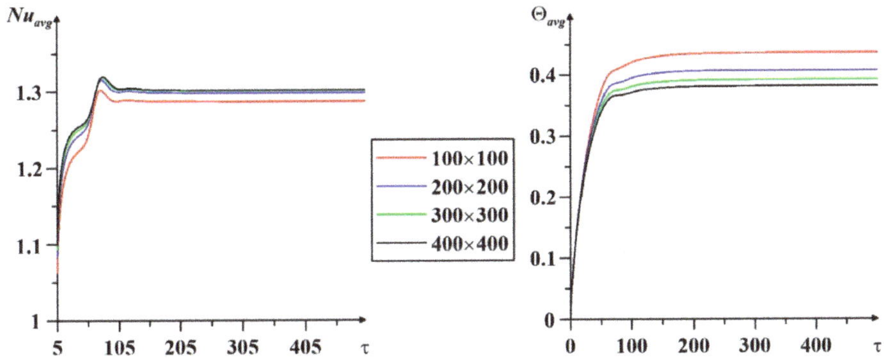

Figure 3. Influence of grid parameters.

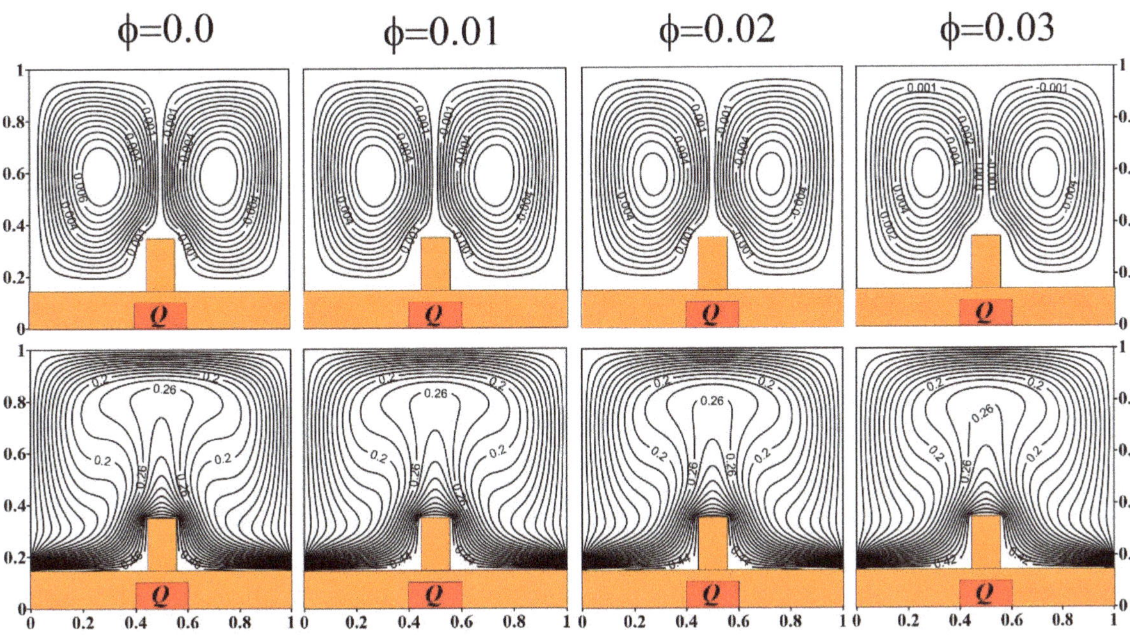

Figure 4. Nanoparticles volume fraction influence on the streamlines and isotherms in the case of one edge at δ = 0.2.

3. Results

This research simulates the process of free convective energy transport of a power-law nanofluid in an enclosure with a heat source and a rib structure. In the course of the study, the impact of the governing characteristics was analyzed, including the volume fraction of nanoparticles (ϕ = 0.0–0.03), the height of the ribs (δ = 0.1–0.3) and the number of ribs (1–3). The Rayleigh number is fixed for all results ($Ra = 10^5$). The outcomes are shown by the distribution of streamlines and isotherms, as well as curves of the average Nusselt number over the radiator surface and the mean heater temperature.

Figure 4 demonstrates the distribution of streamlines and isotherms depending on the nanoparticles' volume fraction in the case of one heat sink fin at δ = 0.2. The streamlines in the first row illustrate the structure of the nanofluid motion within the chamber. It consists of two symmetrical convective cells, the flow in which occurs in opposite directions. It should be noted that their shape practically does not change with a growth in the nanoadditives' concentration. The second row presents isotherms that reflect the temperature stratification in the chamber. If you look at them closely, you can see that a two-dimensional heat plume has formed above the rib, which characterizes an intense convective heat transfer. At the same time, according to the isotherm with a value of 0.26, one can trace a slight reduction in the chamber temperature, which indicates the efficiency of using such a cooling system for a heated element.

The impact of the copper nanoparticles' concentration has also been estimated from the values of the mean Nusselt number and mean heater temperature, which are demonstrated in Figure 5 at $\delta = 0.2$. It can be seen that with an increase in ϕ, the average Nusselt number decreases, which indicates a weakening of the convective energy transport in the enclosure. In this case, the average source temperature changes insignificantly, but still decreases with an increase in ϕ. This phenomenon is explained by a growth in the thermal conductivity of the working fluid due to nanoadditives. It should also be noted that the decrease in the values of Nu_{avg} and Θ_{avg} occurs by the same amount with an increase in ϕ. That is, Nu_{avg} decreases by 0.2, and Θ_{avg} by 0.01 with a rise in the nanoadditives' concentration.

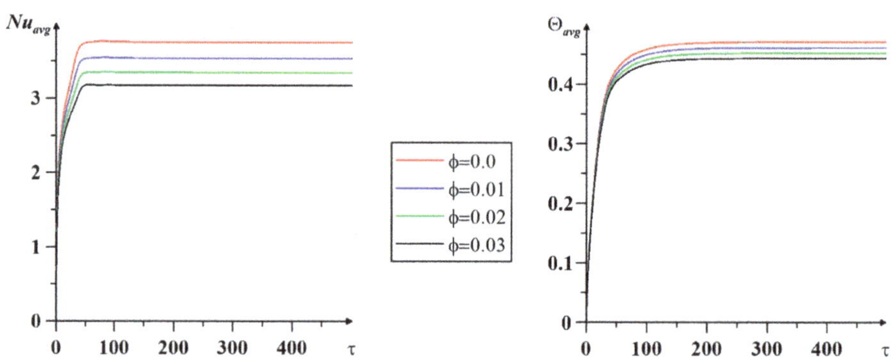

Figure 5. Nanoparticles' volume fraction influence on the average Nusselt number and mean temperature for one rib at $\delta = 0.2$.

An analysis of the nanoadditives' volume fraction influence was carried out for all options for the ribs number. It should be noted that the results in the cases of two and three ribs are similar; therefore, the ϕ effect is presented on the results of modeling the free convection in an enclosure with two solid ribs at $\delta = 0.2$.

It can be seen in Figure 6 that the flow structures were not changed significantly when the second rib was added; the cavity also contains two convective cells. Their size was reduced as the space for the fluid flow was reduced also with the addition of the second fin. In this case, it is noticeable that the streamline density changes with an increase in ϕ. In addition, according to the values of streamlines, it can be noted that the strength of the flow decreases with a growth in the nanoadditives' concentration. If we look closely at the isotherms, we can see that in the cases of $\phi = 0.0$ and 0.01 they practically coincide. Furthermore, the thermal plume degrades to an almost horizontal arrangement of isotherms. This means that in the latter case, the conductive mechanism of heat transfer predominates. At the same time, according to the nearest isotherm to the source, it can be seen that the temperature in the cavity noticeably increases with an increase in ϕ.

Figure 7 reflects the nanoparticles' concentration effect on the integral characteristics of the process at $\delta = 0.2$. Again, we note that with three fins, similar results are obtained; therefore, a variant with two fins is presented. It can be seen from the figure that the mean Nusselt number decreases with increasing]ϕ, which illustrates the attenuation of convective energy transport. It should be noted that in the cases of $\phi = 0.0$, 0.01, and 0.02, Nu_{avg} uniformly decreases by 0.1, but at $\phi = 0.03$, the decrease occurs only by 0.03. In this case, the average source temperature behaves differently. It can be seen that the cases of $\phi = 0.0$ and 0.01 coincided over time and have exactly the same values for the dimensionless time τ from 250 to the end. With further addition of nanoparticles, Θ_{avg} actively increases. In the case of $\phi = 0.03$, it has a maximum value, and also requires more time to reach the stationary regime.

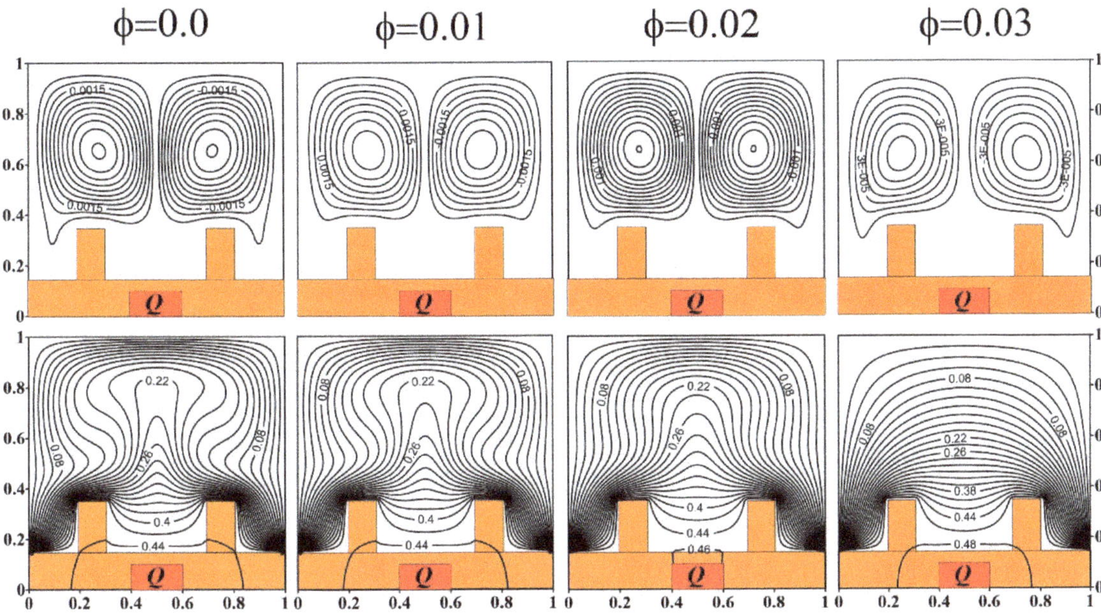

Figure 6. Nanoparticles' concentration influence on the streamlines and isotherms in the case of two fins at δ = 0.2.

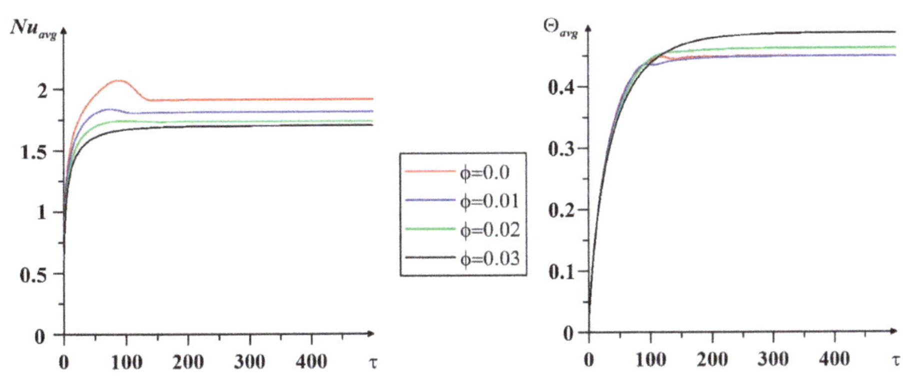

Figure 7. Nanoparticles' volume fraction influence on the average Nusselt number and mean temperature for two fins at δ = 0.2.

Next, an analysis of the rib structure height influence was carried out. Figure 8 shows the distribution of streamlines and isotherms at $\phi = 0.01$ in the case of one rib. The isotherms show that two convective cells are preserved. Changes occurred only in the places where the cells adjoin the edge. Isotherms illustrate a two-dimensional heat plume. A rise in the fin height leads to compaction of the temperature isolines.

Figure 9 reflects the influence of changing the fin height on the mean Nusselt number and the average temperature in the heat-generated element at $\phi = 0.01$. It can be seen that both parameters decrease with an increase in δ. At the same time, Nu_{avg} decreases more actively than Θ_{avg}. It should be noted that at δ = 0.1 and 0.2, the temperature values practically coincide. The case δ = 0.3 corresponds to the lowest temperature of the source, since in this case the enlarged fin removes heat more actively by means of a conductive heat exchange.

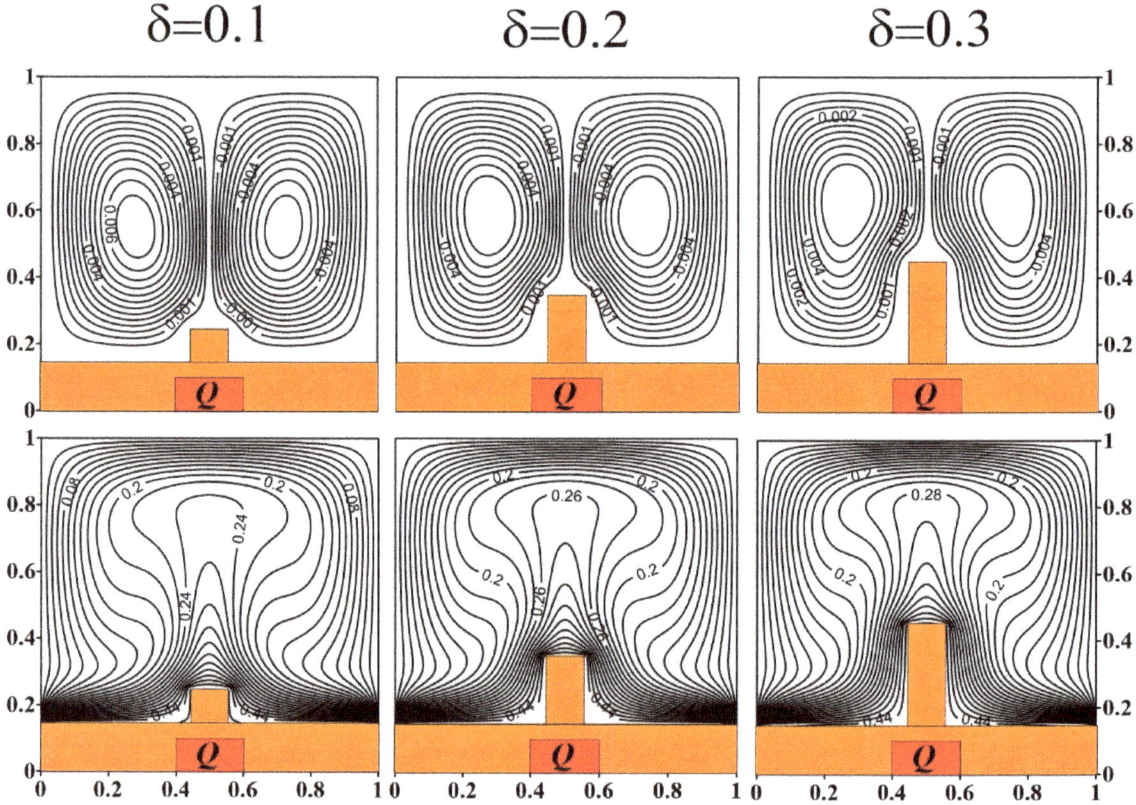

Figure 8. Effect of fin height on streamlines and isotherms at $\phi = 0.01$.

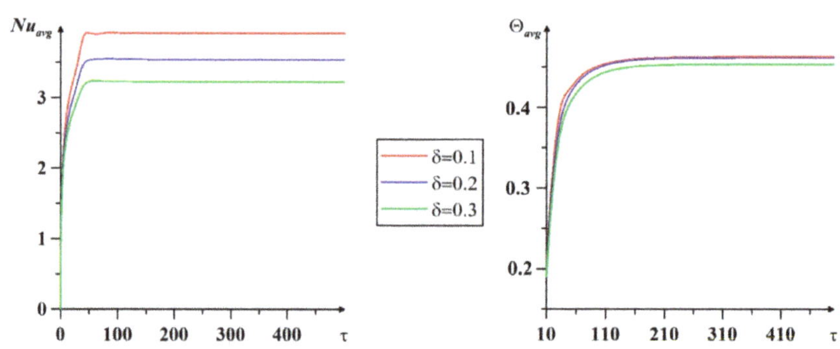

Figure 9. Effect of fin height on the mean Nusselt number and mean temperature at $\phi = 0.01$.

Next, we considered the influence of three ribs height in the cavity at $\phi = 0.01$. It should be noted that the results for two and three ribs are similar, so only three ribs are shown in Figure 10. The streamlines illustrate two convective cells in the chamber. The circulation for the right one is clockwise, while for the left one it is counterclockwise. With a rise in the fins' height, the size of the cells decreases, as well as the density of the streamlines. This is due to the reduced space for the fluid to flow. Note that in the case of $\delta = 0.3$, the cells almost take the form of a square. Isotherms also represent a two-dimensional heat plume. According to the isotherm closest to the source, we can trace the decrease in temperature: for $\delta = 0.1$ it is 0.46, for $\delta = 0.2$ it is 0.42, while for $\delta = 0.3$ it is 0.38. This result indicates an efficient cooling of the heated unit for three fins with a maximum height.

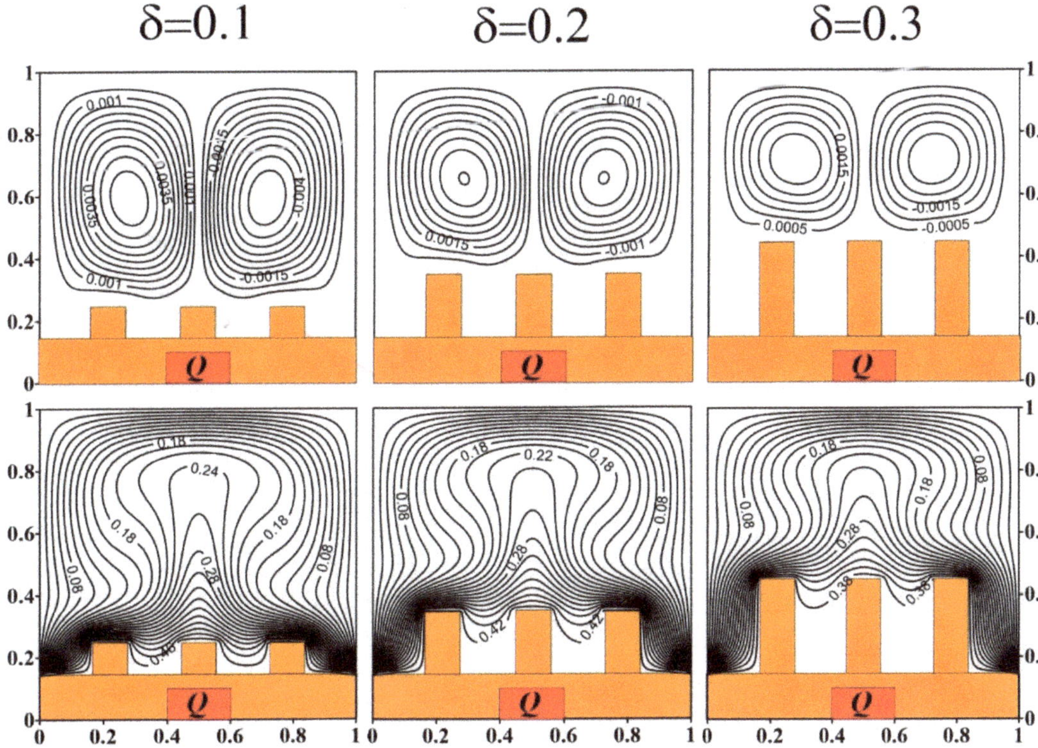

Figure 10. Effect of fin height on streamlines and isotherms in the case of three fins at $\phi = 0.01$.

Parameters Nu_{avg} and Θ_{avg} depending on the ribs' height and time are shown in Figure 11. Both parameters decrease with an increasing δ, as in the case of one rib. Despite the attenuation of convective energy transport in the chamber, the average source temperature also decreases. This phenomenon is also explained by the predominance of conductive energy transport.

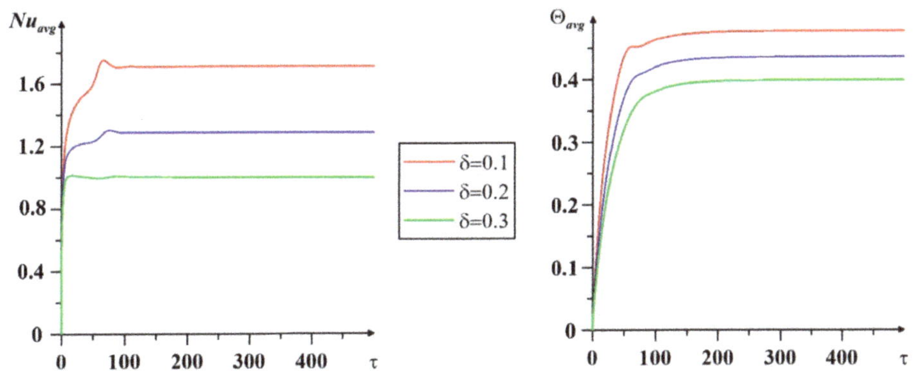

Figure 11. Impact of the fin height on the mean Nusselt number and mean temperature in the case of three fins at $\phi = 0.01$.

An assessment of the solid ribs number influence on the natural convective heat transfer is shown in Figure 12 at $\phi = 0.01$ and $\delta = 0.2$. It can be seen that the most intense convective heat transfer corresponds to calculations with one fin, because in this case Nu_{avg} has the maximum value. But for the same case, Θ_{avg} also has the maximum value. The most efficient cooling, taking into account the selected parameters, occurs in the presence of three fins, despite the fact that the convective heat transfer is the weakest here.

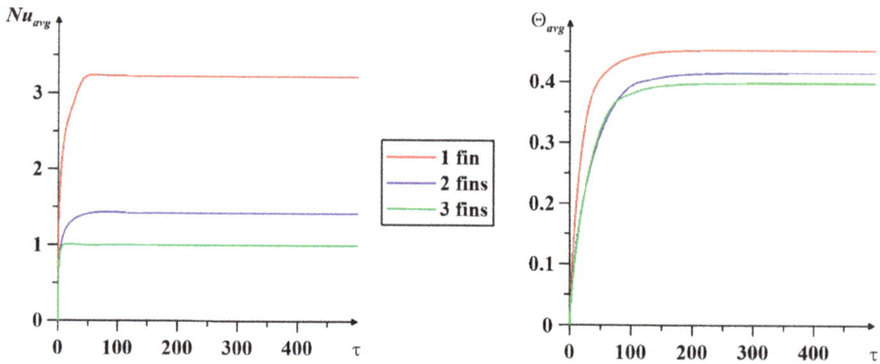

Figure 12. Impact of the fins number on the average Nusselt number and average temperature at $\phi = 0.01$, $\delta = 0.2$.

4. Conclusions

Mathematical modeling of the power-law nanoliquid natural convection in an enclosure with a solid heat sink and a heater of constant internal volumetric heat generation was carried out in this study. In the course of the work, the influence of the nanoadditives' concentration ($\phi = 0.0$–0.03), the ribs' height ($\delta = 0.1$–0.3) and their number on the intensification of the heated element cooling was analyzed. The analysis of the results allowed us to draw the following conclusions:

1. An increase in the concentration of nano-additives in the case of one rib leads to a decrease in the average temperature of the heater, while in the case of two ribs, an increase in the average temperature of the heater is observed with a reduction in the nanoparticles' concentration. Therefore, it is necessary to perform additional analysis for the nanoparticles' concentration influence that depends on the fins number.
2. Increasing the fin height makes it possible to intensify the cooling of the source with any solid fins number.
3. The lowest average source temperature corresponds to the presence of three fins in the cooling system.

The performed analysis showed advantages of the considered cooling system. In future, this formulation can be extended to the following cases—porous finned heat sink, a heat-generating element of variable internal volumetric heat flux, three-dimensional analysis.

Author Contributions: Conceptualization, M.A.S.; methodology, M.A.S.; software, D.S.L.; validation, D.S.L.; investigation, D.S.L.; writing—original draft preparation, D.S.L. and M.A.S.; writing—review and editing, D.S.L. and M.A.S.; visualization, D.S.L.; supervision, M.A.S. All authors have read and agreed to the published version of the manuscript.

Funding: This work was conducted as a government task of the Ministry of Science and Higher Education of the Russian Federation (Project Number FSWM-2020-0036).

Data Availability Statement: Not applicable.

Conflicts of Interest: The authors declare no conflict of interest.

References

1. Ganesh, N.V.; Al-Mdallal, Q.M.; Hirankumar, G.; Kalaivanan, R.; Chamkha, A.J. Buoyancy-driven convection of MWCNT—Casson nanofluid in a wavy enclosure with a circular barrier and parallel hot/cold fins. *Alex. Eng. J.* **2022**, *61*, 3249–3264. [CrossRef]
2. Nemati, H. A general equation based on entropy generation minimization to optimize plate fin heat sink. *Eng. J.* **2018**, *22*, 159–174. [CrossRef]
3. Prabowo, A.D.; Amrizal; Ibrahim, G.A. Geometry optimization of PV/T-TEG collector under different operating conditions using CFD simulation and Taguchi method. *Eng. J.* **2022**, *26*, 1–11. [CrossRef]

4. Hatami, M. Numerical study of nanofluids natural convection in a rectangular cavity including heated fins. *J. Mol. Liq.* **2017**, *233*, 1–8. [CrossRef]
5. Hidki, R.; El Moutaouakil, L.; Boukendil, M.; Charqui, Z.; Zrikem, Z.; Abdelbaki, A. Impact of Cu/Al_2O_3-water hybrid nanofluid on natural convection inside a square cavity with two heat-generating bodies. *Mater. Today Proc.* **2023**, *72*, 3749–3756. [CrossRef]
6. Iftikhar, B.; Javed, T.; Siddiqui, M.A. Natural convective flow of non-isothermal hybrid nanofluid inside the cavity with the influence of a heated fin and non-linear thermal radiation: Second law analysis. *Mater. Today Commun.* **2023**, *34*, 105341. [CrossRef]
7. Waqas, H.; Khan, S.A.; Yasmin, S.; Liu, D.; Imran, M.; Muhammad, T.; Alhushaybari, A.; Farooq, U. Galerkin finite element analysis for buoyancy driven copper-water nanofluid flow and heat transfer through fins enclosed inside a horizontal annulus: Applications to thermal engineering. *Case Stud. Therm. Eng.* **2022**, *40*, 102540. [CrossRef]
8. Shahsavar, A.; Rashidi, M.; Yıldız, Ç.; Arıcı, M. Natural convection and entropy generation of Ag-water nanofluid in a finned horizontal annulus: A particular focus on the impact of fin numbers. *Int. Commun. Heat Mass Transf.* **2021**, *125*, 105349. [CrossRef]
9. Tayebi, T.; Dogonchi, A.S.; Karimi, N.; Ge-JiLe, H.; Chamkha, A.J.; Elmasry, Y. Thermo-economic and entropy generation analyses of magnetic natural convective flow in a nanofluid-filled annular enclosure fitted with fins. *Sustain. Energy Technol. Assess.* **2021**, *46*, 101274. [CrossRef]
10. Yasmin, S.; Khan, S.A.; Fatima, N.; Imran, M.; Tahir, M.; Waqas, H.; Farooq, U.; Xu, Y.-J. Computational analysis of MHD MgO—Water nanofluid flow inside hexagonal enclosure fitted with fins. *Case Stud. Therm. Eng.* **2023**, *43*, 102788. [CrossRef]
11. Hejri, S.; Malekshah, E.H. Cooling of an electronic processor based on numerical analysis on natural convection and entropy production over a dissipating fin equipped with copper oxide/water nanofluid with Koo-Kleinstreuer-Li model. *Therm. Sci. Eng. Prog.* **2021**, *23*, 100916. [CrossRef]
12. Al-Farhany, K.; Al-Chlaihawi, K.K.; Al-dawody, M.F.; Biswas, N.; Chamkha, A.J. Effects of fins on magnetohydrodynamic conjugate natural convection in a nanofluid-saturated porous inclined enclosure. *Int. Commun. Heat Mass Transf.* **2021**, *126*, 105413. [CrossRef]
13. Yan, S.-R.; Pordanjani, A.H.; Aghakhani, S.; Goldanlou, A.S.; Afrand, M. Managment of natural convection of nanofluids inside a square enclosure by different nano powder shapes in presence of fins with different shapes and magnetic field effect. *Adv. Powder Technol.* **2020**, *31*, 2759–2777. [CrossRef]
14. Siavashi, M.; Yousofvand, R.; Rezanejad, S. Nanofluid and porous fins effect on natural convection and entropy generation of flow inside a cavity. *Adv. Powder Technol.* **2018**, *29*, 142–156. [CrossRef]
15. Esfe, M.H.; Barzegarian, R.; Bahiraei, M. A 3D numerical study on natural convection flow of nanofluid inside a cubical cavity equipped with porous fins using two-phase mixture model. *Adv. Powder Technol.* **2020**, *31*, 2480–2492. [CrossRef]
16. Gireesha, B.J.; Sowmya, G.; Khan, M.I.; Öztop, H.F. Flow of hybrid nanofluid across a permeable longitudinal moving fin along with thermal radiation and natural convection. *Comput. Methods Programs Biomed.* **2020**, *185*, 10516. [CrossRef]
17. Aly, A.M.; Mohamed, E.M.; Alsedais, N. Double-diffusive convection from a rotating rectangle in a finned cavity filled by a nanofluid and affected by a magnetic field. *Int. Commun. Heat Mass Transf.* **2021**, *126*, 105363. [CrossRef]
18. Siva, T.; Jangili, S.; Kumbhakar, B.; Mondal, P.K. Unsteady electromagnetohydrodynamic flow of couple stress fluid through a microchannel: A theoretical analysis. *Eur. J. Mech.-B/Fluids* **2022**, *95*, 83–93. [CrossRef]
19. Mehta, S.K.; Mondal, P.K. Free convective heat transfer and entropy generation characteristics of the nanofluid flow inside a wavy solar power plant. *Microsyst. Technol.* **2023**, *29*, 489–500. [CrossRef]
20. Mondal, P.K.; Wongwises, S. Magneto-hydrodynamic (MHD) micropump of nanofluids in a rotating microchannel under electrical double-layer effect. *J. Process Mech. Eng.* **2020**, *234*, 318–330. [CrossRef]
21. Nabwey, H.A.; Rashad, A.M.; Khan, W.A.; Alshber, S.I. Effectiveness of magnetize flow on nanofluid via unsteady natural convection inside an inclined U-shaped cavity with discrete heating. *Alex. Eng. J.* **2022**, *61*, 8653–8666. [CrossRef]
22. Khan, A.U.; Ullah, N.; Al-Zubaidi, A.; Nadeem, S. Finite element analysis for CuO/water nanofluid in a partially adiabatic enclosure: Inclined Lorentz forces and porous medium resistance. *Alex. Eng. J.* **2022**, *61*, 6477–6488. [CrossRef]
23. Khezzar, L.; Siginer, D.; Vinogradov, I. Natural convection of power law fluids in inclined cavities. *Int. J. Therm. Sci.* **2012**, *53*, 8–17. [CrossRef]
24. Guo, S.S.; Luo, Z.Y.; Tao, W.; Zhao, J.F.; Cen, K.F. Viscosity of monodisperse silica nanofluids. *Bull. Chin. Ceram. Soc.* **2006**, *25*, 52–55.
25. Jang, S.P.; Choi, S.U.S. Effects of various parameters on nanofluid thermal conductivity. *J. Heat Transf.* **2007**, *129*, 617–623. [CrossRef]
26. Loenko, D.S.; Sheremet, M.A. Regularization models for natural convection of a pseudoplastic liquid in a closed differentially heated cavity. *Bull. Perm Univ. Phys.* **2021**, *3*, 13–22. [CrossRef]
27. Loenko, D.; Shenoy, A.; Sheremet, M. Influence of the chamber inclination angle and heat-generating element location on thermal convection of power-law medium in a chamber. *Int. J. Numer. Methods Heat Fluid Flow* **2021**, *31*, 134–153. [CrossRef]
28. Shenoy, A.; Sheremet, M.; Pop, I. *Convective Flow and Heat Transfer from Wavy Surfaces: Viscous Fluids, Porous Media and Nanofluids*; CRC Press: Boca Raton, FL, USA, 2016; p. 306.
29. Turan, O.; Sachdeva, A.; Chakraborty, N.; Poole, R.J. Laminar natural convection of power-law fluids in a square enclosure with differentially heated side walls subjected to constant temperatures. *J. Non-Newton. Fluid Mech.* **2011**, *166*, 1049–1063. [CrossRef]

mathematics MDPI

Article

Numerical Investigation of a Combustible Polymer in a Rectangular Stockpile: A Spectral Approach

Adeshina T. Adeosun [1,*], Joel C. Ukaegbu [2] and Ramoshweu S. Lebelo [3]

1 Department of Mathematics, Federal College of Education, Iwo 232102, Nigeria
2 Department of Mathematics, Adeleke University, Ede 232104, Nigeria; ukaegbu.joel@adelekeuniversity.edu.ng
3 Education Department, Vaal University of Technology, Vanderbijlpark 1911, South Africa; sollyl@vut.ac.za
* Correspondence: adeshinata@fceiwo.edu.ng

Abstract: Despite the wide application of combustion in reactive materials, one of the challenges faced globally is the auto-ignition of such materials, resulting in fire and explosion hazards. To avoid this unfortunate occurrence, a mathematical model describing the thermal decomposition of combustible polymer material in a rectangular stockpile is formulated. A nonlinear momentum equation is provided with the assumption that the combustible polymer follows a Carreau constitutive relation. The chemical reaction of the polymer material is assumed to be exothermic; therefore, Arrhenius's kinetic theory is considered in the energy balance equation. The bivariate spectral local linearization scheme (BSLLS) is utilized to provide a numerical solution for the dimensionless equations governing the problem. The obtained results are validated by the collocation weighted residual method (CWRM), and a good agreement is achieved. Dimensionless velocity, temperature, and thermal stability results are presented and explained comprehensively with suitable applications. Some of the obtained results show that thermal criticality increases with increasing power law index (n) and radiation (Ra) values and decreases with increasing variable viscosity (β_1) and material parameter (W_e) values.

Keywords: Carreau fluid; thermal stability; variable thermal conductivity; variable viscosity; BSLLS

MSC: 80A19

Citation: Adeosun, A.T.; Ukaegbu, J.C.; Lebelo, R.S. Numerical Investigation of a Combustible Polymer in a Rectangular Stockpile: A Spectral Approach. *Mathematics* **2023**, *11*, 3510. https://doi.org/10.3390/math11163510

Academic Editor: Lihua Wang

Received: 5 July 2023
Revised: 8 August 2023
Accepted: 10 August 2023
Published: 14 August 2023

1. Introduction

In engineering and industries, burning combustible materials in a slab is important for storing cellulosic materials, solid combustion, refuse cremation, heavy oil recovery, and other processes [1–3]. Polymers can undergo combustion and release a large amount of energy, which can then be used for transportation, generating electric power, and providing heat for various domestic and industrial applications. Compared with other energy sources, like solar cells, wind generators, and turbines, polymers are quite inexpensive. However, there are some challenges with polymer burning, such as fire ignitions due to human negligence and the physical characteristics of hydrocarbon polymers. These have caused extensive property destruction and claimed the lives of an unknown number of people. From this perspective, several authors have been prompted to study the causes of fire ignition in the combustion process and how it can be controlled.

Drysdale [4] described ignition as the process of initiating a fast exothermic reaction, which then propagates and causes changes in the materials involved and as well generates temperatures in excess. Also, ref. [4] distinguished between two kinds of ignition: (1) piloted ignition, in which a flammable mixture is ignited by a pilot such as an electrical spark or an autonomous flame, and (2) auto-ignition, in which flame develops spontaneously within the mixture. Shi and Chew [5] studied polymers' responses to fire under auto-ignition conditions within a cone-shaped calorimeter and concluded that the CO and CO_2 production process for polymers is a two-step reaction. In the first-step reaction, CO

and other flammable substances are generated following Arrhenius's law. The second stage is the oxidation reaction of CO in the presence of air. The whole process is expressed as follows:

$$\begin{cases} Step1 : Polymer + O_2 \rightarrow H_2O + CO + others\ product \\ Step2 : CO + 0.5O_2 \rightarrow CO_2. \end{cases}$$

It was also discovered in [5] that the CO and CO_2 emissions from flaming combustion are greater than those from non-flaming combustion.

Moreover, the combustion of polyethylene and polyvinyl chloride was examined in [6–8]. These investigations used experimental settings, allowing researchers to acquire more information and factors for pyrolysis, ignition, and combustion. Geschwindner et al. [9] incorporated a mix of high-speed planar laser-induced fluorescence of the HO radical (OH-PLIF) and a thermal decomposition analysis to examine the combustion of micrometer-sized polypropylene (PP) particles. They found that the highest density of flame-retardant polymer particles decreases during ignition and the early stage of burning. Lohrer et al. [10,11] investigated the effects of physical factors such as material wetness, atmospheric humidity, and concentrations of oxygen on the combustion of reactive materials and discovered that water in the reactive material increases auto-ignition.

Some mathematical models that are less expensive and faster than experimental approaches have been implemented in the literature to explain the auto-ignition of combustible materials in a stockpile. For instance, a one-step combustion process of heat transfer in a spherical channel was investigated in [12,13]. According to their reports, the system maintains stability as heat escapes into the atmosphere. In addition, enhancement in the chemical reaction rate leads to increased heat generation in the stockpile, resulting in quick auto-ignition. Lebelo et al. [14–16] examined the two-step thermal decomposition of combustible materials in a sphere. They identified that elevation in the two-step kinetic parameter diminishes the rate of heat loss on the sphere's surface, which, in turn, speeds up auto-ignition.

In the literature above reviewed, the authors did not consider flow behavior in their studies. However, an increase or decrease in flow speed contributes to the heat transfer performance of combustible materials. It was discovered in the literature that the Carreau fluid model well describes the flow behavior of polymeric solutions because of its shear rate properties [17]. Several studies on Carreau fluid constitutive relations have been documented. For instance, Siska et al. [18] examined the terminal velocity of non-spherical particles falling through a Carreau fluid and concluded that the Carreau fluid model can well characterize the rheology of various polymeric solutions, such as 1% methylcellulose tylose in glycerol solutions and 3% hydroxyethyl–cellulose Natrosol HHX in glycerol solutions. Ref. [19] reported an intriguing study on the entropy production of Carreau fluid in the presence of infinite shear rate viscosity. Also, the behavior of a Carreau fluid flow past a stretching sheet was extensively analyzed in [20,21]. In addition, the peristaltic movement of a Carreau fluid was also extensively studied in [22–24]. For more on the Carreau fluid model with different configurations, see [25–27].

Motivated by the reviewed literature in [12–14,22,28], this study focuses on an investigation of the thermal decomposition of Carreau fluid in a rectangular stockpile with variable thermophysical properties. It is believed that this present study has not been reported in the literature. However, the outcome of this study could be useful for engineers dealing with the combustion of polymers by determining the conditions necessary for explosions and how to control them. The rest of the article is structured as follows: a mathematical model for the unsteady, fully developed flow and temperature of the polymer is presented in Section 2; Section 3 deals with the application of BSLLM to the dimensionless initial-boundary value problem; in Section 4, an extensive discussion of the obtained findings is provided; and concluding remarks are provided in Section 5.

2. Mathematical Analysis

A transient laminar flow of a reactive incompressible Carreau fluid material in a combustible stockpile positioned at a distance of $2h$ apart is considered (see Figure 1). $\hat{x}-axis$ is parallel to the flow direction, and $\hat{y}-axis$ traversed to it. Initially, the fluid is assumed to be fully developed in the stockpile of the temperature, T_0, and the material's viscosity and thermal conductivity, denoted as $\mu=\mu_0$ and $\kappa=\kappa_0$, respectively, are assumed to be constant. At time $\hat{t}>0$, the combustion process begins, and the material properties become temperature-dependent: $\mu=\mu(\mathrm{T})$ and $\kappa=\kappa(\mathrm{T})$. The chemistry involved in this problem assumes two-step Arrhenius kinetics. The reactant consumption is assumed negligible in this model. We also assume that the means of heat loss into the environment of the temperature, T_a, is mainly via radiation and convection. The influence of density variation with temperature is approximated following Boussinesque approximation. The equations governing the momentum and energy balance under the assumptions above are as follows [14–16,20,26]:

$$\rho\frac{\partial\hat{u}}{\partial\hat{t}}=-\frac{\partial\hat{P}}{\partial\hat{x}}+\frac{\partial}{\partial\hat{y}}\left(\mu(\mathrm{T})\left(1+\Gamma^2\left(\frac{\partial\hat{u}}{\partial\hat{y}}\right)^2\right)^{\frac{n-1}{2}}\frac{\partial\hat{u}}{\partial\hat{y}}\right)+\rho g\beta(T-T_0), \tag{1}$$

$$\begin{gathered}\rho c_p\frac{\partial T}{\partial\hat{t}}=\frac{\partial}{\partial\hat{y}}\left(\kappa(\mathrm{T})\frac{\partial T}{\partial\hat{y}}\right)+\mu(\mathrm{T})\left(\left(1+\Gamma^2\left(\frac{\partial\hat{u}}{\partial\hat{y}}\right)^2\right)^{\frac{n-1}{2}}\left(\frac{\partial\hat{u}}{\partial\hat{y}}\right)^2\right)+\\ Q_1A_1C_1e^{-\frac{E_1}{RT}}+Q_2A_2C_2e^{-\frac{E_2}{RT}}-\varepsilon\sigma(T^4-T_0^4),\end{gathered} \tag{2}$$

with the initial-boundary conditions

$$\left.\begin{array}{l}\hat{u}(\hat{y},0)=0,T(\hat{y},0)=T_0\\ \frac{\partial\hat{u}}{\partial\hat{y}}(0,\hat{t})=0,\ \frac{\partial T}{\partial\hat{y}}(0,\hat{t})=0,\ \text{for }\hat{t}>0\\ \hat{u}(\mathrm{h},\hat{t})=0,\ -\kappa(T)\frac{\partial T}{\partial\hat{y}}(h,\hat{t})=h_t[T(h,\hat{t})-T_a],\ \text{for }\hat{t}>0\end{array}\right\} \tag{3}$$

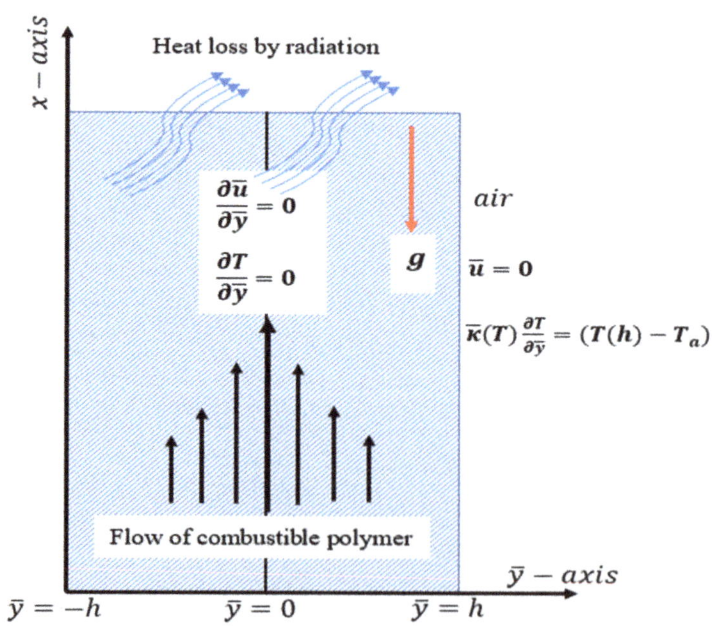

Figure 1. Problem physical geometry.

$\hat{u}$¯axial velocity, $\hat{P}$¯modified pressure, T¯absolute temperature, T_a¯ambient temperature, T_0¯initial temperature of the stockpile, μ_0¯material's dynamic viscosity at temperature T_0, κ_0¯material's thermal conductivity at T_0, c_p¯specific heat at constant pressure, ρ¯material's density, Q_1, Q_2¯first- and second-step heat of reaction, A_1, A_2¯first- and second-step rate constants, C_1, C_2¯first- and second-step reactant's concentration, E_1, E_2¯first and second step activation energies, ε¯stockpile's emissivity $(0 < \varepsilon < 1)$, σ¯Stefan–Boltzmann constant, β¯volumetric coefficient, g¯gravitational acceleration, h_t¯coefficient of heat transfer, Γ¯time constant, n¯dimensionless power law index. $n < 1$ represents shear-thinning fluids, $n = 0$ represents Newtonian fluids, and $n > 1$ represents shear-thickening fluids The variable viscosity and thermal conductivity, $\overline{\mu}(\mathrm{T})$ and $\overline{\kappa}(\mathrm{T})$, are expressed, respectively, as

$$\overline{\mu}(\mathrm{T}) = \mu_0 e^{-b_1(\mathrm{T}-\mathrm{T}_0)} \text{ and } \overline{\kappa}(\mathrm{T}) = \kappa_0 e^{-b_2(\mathrm{T}-\mathrm{T}_0)}, \tag{4}$$

where b_1 and b_2 are dynamic viscosity and thermal conductivity variation parameters. We then introduce the below dimensionless parameters to Equations (1)–(3):

$$\begin{gathered} y = \tfrac{\overline{y}}{h}, \mathrm{x} = \tfrac{\overline{x}}{h}, t = \tfrac{\nu\overline{t}}{h^2}, u = \tfrac{\overline{u}h}{\nu}, \mu = \tfrac{\overline{\mu}}{\mu_0}, \kappa = \tfrac{\overline{\kappa}}{\kappa_0}, \nu = \tfrac{\mu_0}{\rho}, P = \tfrac{\overline{P}h^2}{\nu^2}, \theta = \tfrac{\mathrm{E}_1(T-\mathrm{T}_0)}{RT_0^2}, \theta_a = \tfrac{\mathrm{E}_1(T_a-\mathrm{T}_0)}{RT_0^2} \\ , \beta_1 = \tfrac{b_1 RT_0^2}{E_1}, \beta_2 = \tfrac{b_2 RT_0^2}{E_1}, W_e = \tfrac{\Gamma\nu}{h^2}, \mathrm{Pr} = \tfrac{\mu_0 c_p}{\kappa_0}, \lambda = \tfrac{Q_1 E_1 A_1 h^2 C_1 e^{-\frac{E_1}{RT_0}}}{T_0^2 R\kappa_0}, A = -\tfrac{\partial\hat{P}}{\partial\overline{x}}, Gr = \tfrac{g\beta RT_0^2 h^3\rho^2}{E_1\mu_0^2} \\ \epsilon_1 = \tfrac{RT_0}{E_1}, \epsilon_2 = \tfrac{E_2}{E_1}, V_d = \tfrac{\mu_0^3 e^{\frac{E_1}{RT_0}}}{\rho^2 Q_1 A_1 h^4 C_1}, Ra = \tfrac{\varepsilon\sigma h^2 E_1 T_0^2}{R\kappa_0}, \omega = \tfrac{Q_2 A_2 E_2}{Q_1 A_1 E_1} e^{\frac{(E_1-E_2)}{RT_0}}, Bi = \tfrac{hh_t}{\kappa_0}. \end{gathered} \tag{5}$$

The following dimensionless equations are then obtained:

$$\frac{\partial u}{\partial t} = A + \frac{\partial}{\partial y}\left(e^{-\beta_1\theta}\left(1 + W_e^2\left(\frac{\partial u}{\partial y}\right)^2\right)^{\frac{n-1}{2}}\frac{\partial u}{\partial y}\right) + Gr\theta, \tag{6}$$

$$\begin{aligned} \mathrm{Pr}\tfrac{\partial\theta}{\partial t} = {} & \tfrac{\partial}{\partial y}\left(e^{-\beta_2\theta}\tfrac{\partial\theta}{\partial y}\right) + \lambda\left(e^{\frac{\theta}{1+\epsilon_1\theta}} + \omega e^{\frac{\epsilon_2\theta}{1+\epsilon_1\theta}} + V_d e^{-\beta_1\theta}\left(1 + W_e^2\left(\tfrac{\partial u}{\partial y}\right)^2\right)^{\frac{n-1}{2}}\left(\tfrac{\partial u}{\partial y}\right)^2\right) \\ & -Ra\left((\epsilon_1\theta + 1)^4 - 1\right), \end{aligned} \tag{7}$$

$$\left.\begin{array}{l} u(y,0) = 0,\ \theta(y,0) = 0, \\ \frac{\partial u}{\partial y}(0,t) = 0,\ \frac{\partial\theta}{\partial y}(0,t) = 0,\ \text{for } t > 0 \\ u(1,t) = 0,\ \frac{\partial\theta}{\partial y}(1,t) = -Bie^{\beta_2\theta}[\theta(1,t) - \theta_a],\ \text{for } t > 0 \end{array}\right\}, \tag{8}$$

where $\beta_1, \beta_2, W_e, \mathrm{Pr}, \lambda, \epsilon_1, \epsilon_2, \mathrm{A}, \mathrm{Gr}, \mathrm{V}_d, Ra, \omega, Bi, \theta_a$ are, respectively, the variable viscosity parameter, the variable thermal conductivity parameter, the material parameter, the Prandtl number, the Frank–Kamenetskii parameter, the activation energy parameter, the activation energy ratio parameter, the pressure gradient, the Buoyancy parameter, the viscous heating parameter, the radiation parameter, the two-step exothermic reaction parameter, the Biot number, and the ambient temperature parameter. We also considered the Nusselt number, defined as follows:

$$Nu = \frac{hE_1 q_w}{\kappa(\mathrm{T})RT_0^2} = -\left.\frac{\partial\theta}{\partial y}\right|_{y=1}, \text{ where } q_w = -\kappa(\mathrm{T})\left.\frac{\partial T}{\partial\hat{y}}\right|_{\hat{y}=h}.$$

3. Solution Method

In this section, the BSLLS is implemented to provide a numerical solution for Equations (6)–(8), as outlined in [29,30]. For further studies on the convergence analysis of the BSLLS, see [31]. To adopt the BSLLS, the nonlinear equations, (6) and (7), and the nonlinear convective boundary conditions (8) are, respectively, represented by F, Θ, and Bc.

$$F = A + \frac{\partial}{\partial y}\left(e^{-\beta_1\theta}\left(1+W_e^2\left(\frac{\partial u}{\partial y}\right)^2\right)^{\frac{n-1}{2}}\frac{\partial u}{\partial y}\right) + Gr\theta - \frac{\partial u}{\partial t}, \tag{9}$$

$$\Theta = \frac{\partial}{\partial y}\left(e^{-\beta_2\theta}\frac{\partial\theta}{\partial y}\right) + \lambda\left(e^{\frac{\theta}{1+\epsilon_1\theta}} + \omega e^{\frac{\epsilon_2\theta}{1+\epsilon_1\theta}} + V_d e^{-\beta_1\theta}\left(1+W_e^2\left(\frac{\partial u}{\partial y}\right)^2\right)^{\frac{n-1}{2}}\left(\frac{\partial u}{\partial y}\right)^2\right) \\ -Ra\left((\epsilon_1\theta+1)^4-1\right) - \Pr\frac{\partial\theta}{\partial t}, \tag{10}$$

$$Bc = \frac{\partial\theta}{\partial y}(1,t) + Bie^{\beta_2\theta(1,t)}[\theta(1,t)-\theta_a] \tag{11}$$

The iteration technique (quasi-linearization method) is applied independently to Equations (9)–(11) to arrive at

$$\alpha_{0,r}(y,t)\frac{\partial^2 u_{r+1}}{\partial y^2} + \alpha_{1,r}(y,t)\frac{\partial u_{r+1}}{\partial y} + \alpha_{2,r}(y,t)\frac{\partial u_{r+1}}{\partial t} = \mathbb{R}_{1,r}(y,t), \tag{12}$$

$$\beta_{0,r}(y,t)\frac{\partial^2\theta_{r+1}}{\partial y^2} + \beta_{1,r}(y,t)\frac{\partial\theta_{r+1}}{\partial y} + \beta_{2,r}(y,t)\theta_{r+1} + \beta_{3,r}(y,t)\frac{\partial\theta_{r+1}}{\partial t} = \mathbb{R}_{2,r}(y,t), \tag{13}$$

$$c_{0,r}(1,t)\frac{\partial\theta_{r+1}}{\partial y} + c_{1,r}(1,t)\theta_{r+1} = d_r(1,t), \tag{14}$$

where

$$\alpha_{0,r}(y,t) = \frac{\partial F}{\partial\left(\frac{\partial^2 u}{\partial y^2}\right)} = e^{-\beta_1\theta}\left(1+nW_e^2\left(\frac{\partial u}{\partial y}\right)^2\right)\left(1+W_e^2\left(\frac{\partial u}{\partial y}\right)^2\right)^{\frac{n-3}{2}},$$

$$\alpha_{1,r}(y,t) = \frac{\partial F}{\partial\left(\frac{\partial u}{\partial y}\right)} = e^{-\beta_1\theta}\left(\begin{array}{c}(n-1)W_e^2\frac{\partial u}{\partial y}\left(3+nW_e^2\left(\frac{\partial u}{\partial y}\right)^2\right)\frac{\partial^2 u}{\partial y^2} - \\ \beta_1\left(1+W_e^2\left(\frac{\partial u}{\partial y}\right)^2\right)\left(1+nW_e^2\left(\frac{\partial u}{\partial y}\right)^2\right)\frac{\partial\theta}{\partial y}\end{array}\right)\left(1+W_e^2\left(\frac{\partial u}{\partial y}\right)^2\right)^{\frac{n-5}{2}},$$

$$\alpha_{2,r}(y,t) = \frac{\partial F}{\partial\left(\frac{\partial u}{\partial t}\right)} = -1, \beta_{0,r}(y,t) = \frac{\partial T}{\partial\left(\frac{\partial^2\theta}{\partial y^2}\right)} = e^{-\beta_2\theta}, \beta_{1,r}(y,t) = \frac{\partial T}{\partial\left(\frac{\partial\theta}{\partial y}\right)} = -2\beta_2 e^{-\beta_2\theta}\frac{\partial\theta}{\partial y},$$

$$\beta_{2,r}(y,t) = \frac{\partial T}{\partial\theta} = \frac{\lambda\left(\beta_2 V_d\left(\frac{\partial u}{\partial y}\right)^2(-e^{-\beta_2\theta})(\epsilon_1\theta+1)^2\left(W_e^2\left(\frac{\partial u}{\partial y}\right)^2+1\right)^{\frac{n-1}{2}} + \omega\epsilon_2 e^{\frac{\epsilon_2\theta}{\epsilon_1\theta+1}} + e^{\frac{\theta}{\epsilon_1\theta+1}}\right)}{(\epsilon_1\theta+1)^2} + \\ \beta_2^2 e^{-\beta_2\theta}\left(\frac{\partial\theta}{\partial y}\right)^2 - \beta_2 e^{-\beta_2\theta}\frac{\partial^2\theta}{\partial y^2} - 4\mathrm{Ra}\epsilon_1(\epsilon_1\theta+1)^3,$$

$$\beta_{3,r}(y,t) = \frac{\partial T}{\partial\left(\frac{\partial\theta}{\partial t}\right)} = -\Pr,\ c_{0,r}(1,t) = \frac{\partial Bc}{\partial\left(\frac{\partial\theta}{\partial y}\right)} = 1,$$

$$c_{1,r}(1,t) = \frac{\partial Bc}{\partial\theta} = \beta_2 Bi\theta_r(1,t) + Bie^{\beta_2\theta_r(1,t)} - \beta_2 Bi\theta_a e^{\beta_2\theta_r(1,t)},$$
$$d_r(1,t) = c_{0,r}(1,t)\frac{\partial\theta_r}{\partial y} + c_{1,r}(1,t)\theta_r - Bc_r(1,t),$$
$$\mathbb{R}_{1,r}(y,t) = \alpha_{0,r}(y,t)\frac{\partial^2 u_r}{\partial y^2} + \alpha_{1,r}(y,t)\frac{\partial u_r}{\partial y} + \alpha_{2,r}(y,t)\frac{\partial u_r}{\partial t} - F_r(y,t),$$
$$\mathbb{R}_{2,r}(y,t) = \beta_{0,r}(y,t)\frac{\partial^2\theta_r}{\partial y^2} + \beta_{1,r}(y,t)\frac{\partial\theta_r}{\partial y} + \beta_{2,r}(y,t)\theta_r + \beta_{3,r}(y,t)\frac{\partial\theta_r}{\partial t} - \Theta_r(y,t).$$

The next step is to transform the physical domains, $t \in [0,T]$ and $y \in [0,1]$, respectively, into domains $\tau \in [-1,1]$ and $x \in [-1,1]$ using transformations $t = \frac{T(\tau+1)}{2}$ and $y = \frac{(x+1)}{2}$, with collocation points

$$x_i = \left\{cos(\frac{\pi i}{N_x})\right\}_{i=0}^{N_x} \text{ and } \tau_j = \left\{cos(\frac{\pi j}{N_\tau})\right\}_{j=0}^{N_\tau}.$$

It is assumed that solutions to $u(\tau, x)$ and $\theta(\tau, x)$, in the form of bivariate Lagrange's interpolating polynomials, are defined as:

$$\begin{aligned} u(x,\tau) &\approx \sum_{p=0}^{N_y}\sum_{q=0}^{N_t} u(x_i,\tau_j)L_p(x)L_q(\tau), \\ \theta(x,\tau) &\approx \sum_{p=0}^{N_y}\sum_{q=0}^{N_t} \theta(x_i,\tau_j)L_p(x)L_q(\tau), \end{aligned} \tag{15}$$

where function $L_p(x)$ represents the Lagrange cardinal polynomial of the Chebyshev–Gaus–Lobatto grid points,

$$L_p(x) = \prod_{\substack{p=0\\ p\neq i}}^{N_x} \frac{x - x_i}{x_p - x_i}, \tag{16}$$

with $L_p(x_i) = \delta_{ip} = \begin{cases} 0, i \neq p \\ 1, i = p \end{cases}.$

The $L_q(\tau)$ function is defined similarly. The derivative values at the Chebyshev–Gaus–Lobatto points (x_i, τ_j) are computed as follows:

$$\begin{aligned} \frac{\partial^r u}{\partial x^r}(x_i,\tau_j) &= 2^r \sum_{p=0}^{N_x} D^r_{i,p} u(x_p,\tau_j) = \mathbf{D}^r \mathbf{u}_j, \\ \frac{\partial u}{\partial \tau}(x_i,\tau_j) &= \frac{2}{T}\sum_{q=0}^{N_\tau} d_{j,q} u(x_i,\tau_q) = \sum_{q=0}^{N_\tau} \hat{d}_{j,q}\mathbf{u}_q, \\ \frac{\partial^r \theta}{\partial x^r}(x_i,\tau_j) &= 2^r \sum_{p=0}^{N_x} D^r_{i,p} \theta(x_p,\tau_j) = \mathbf{D}^r \boldsymbol{\theta}_j, \\ \frac{\partial \theta}{\partial \tau}(x_i,\tau_j) &= \frac{2}{T}\sum_{q=0}^{N_\tau} d_{j,q} \theta(x_i,\tau_q) = \sum_{q=0}^{N_\tau} \hat{d}_{j,q}\boldsymbol{\theta}_q, \end{aligned} \tag{17}$$

where r is the order of the derivative, and $\mathbf{D}^r = 2^r D^r_{i,p}$ and $\hat{d}_{j,q} = \frac{2}{T} d_{j,q}$ are Chebyshev differentiation matrices $(N_x+1)\times(N_x+1)$ and $(N_\tau+1)\times(N_\tau+1)$ respectively. $\mathbf{u}_j$ and $\boldsymbol{\theta}_j$ are defined as

$$\begin{aligned} \mathbf{u}_j &= [u(x_0,\tau_j), u(x_1,\tau_j), u(x_2,\tau_j), \ldots, u(x_{N_x},\tau_j)]^T,\ for\ j = 0,1,2,\ldots,N_\tau, \\ \boldsymbol{\theta}_j &= [\theta(x_0,\tau_j), \theta(x_1,\tau_j), \theta(x_2,\tau_j), \ldots, \theta(x_{N_x},\tau_j)]^T, for\ j = 0,1,2,\ldots,N_\tau, \end{aligned} \tag{18}$$

Superscript T denotes a transpose. Substituting Equations (17) and (18) into (12) and (13) yields

$$\begin{aligned} &\alpha_{0,r}(\mathbf{x},\tau_j)\mathbf{D}^2\mathbf{u}_{r+1,j} + \alpha_{1,r}(\mathbf{x},\tau_j)\mathbf{D}\mathbf{u}_{r+1,j} + \alpha_{2,r}(\mathbf{x},\tau_j)\sum_{q=0}^{N_\tau}\hat{d}_{j,q}\mathbf{u}_{r+1,q} = \mathbb{R}_{1,r}(\mathbf{x},\tau_j), \\ &\beta_{0,r}(\mathbf{x},\tau_j)\mathbf{D}^2\boldsymbol{\theta}_{r+1,j} + \beta_{1,r}(\mathbf{x},\tau_j)\mathbf{D}\boldsymbol{\theta}_{r+1,j} + \beta_{2,r}(\mathbf{x},\tau_j)I + \beta_{3,r}(\mathbf{x},\tau_j)\sum_{q=0}^{N_\tau}\hat{d}_{j,q}\boldsymbol{\theta}_{r+1,q} = \mathbb{R}_{2,r}(\mathbf{x},\tau_j), \end{aligned} \tag{19}$$

where

$$\alpha_{k,r}(\mathbf{x},\tau_j)\ (k=0,1,2)=\begin{pmatrix}\alpha_{k,r}(x_0,\tau_j) & & \\ & \ddots & \\ & & \alpha_{k,r}(x_{N_x},\tau_j)\end{pmatrix},$$
$$\beta_{k,r}(\mathbf{x},\tau_j)\ (k=0,1,2,3)=\begin{pmatrix}\beta_{k,r}(x_0,\tau_j) & & \\ & \ddots & \\ & & \beta_{k,r}(x_{N_x},\tau_j)\end{pmatrix},$$

$$\mathbb{R}_{1,r}(\mathbf{x},\tau_j)=\alpha_{0,r}(\mathbf{x},\tau_j)\mathbf{D}^2\mathbf{u}_{r,j}+\alpha_{1,r}(\mathbf{x},\tau_j)\mathbf{D}\mathbf{u}_{r,j}+\alpha_{2,r}(\mathbf{x},\tau_j)\sum_{q=0}^{N_\tau}\hat{d}_{j,q}\mathbf{u}_{r,q}-F_r(\mathbf{x},\tau_j),$$
$$\mathbb{R}_{2,r}(\mathbf{x},\tau_j)=\beta_{0,r}(\mathbf{x},\tau_j)\mathbf{D}^2\boldsymbol{\theta}_{r,j}+\beta_{1,r}(\mathbf{x},\tau_j)\mathbf{D}\boldsymbol{\theta}_{r,j}+\beta_{2,r}(\mathbf{x},\tau_j)\boldsymbol{\theta}_{r,j}+\beta_{3,r}(\mathbf{x},\tau_j)\sum_{q=0}^{N_\tau}\hat{d}_{j,q}\boldsymbol{\theta}_{r,q}-\Theta_r(\mathbf{x},\tau_j),$$

and I is an identity matrix. Applying spectral collocation to the boundary conditions (8) and the convective boundary condition (14), we have

$$\sum_{p=0}^{N_x}2D_{N_x,p}u_{r+1}(x_p,\tau_j)=0,\ u_{r+1}(x_0,\tau_j)=0,$$
$$\sum_{p=0}^{N_x}2D_{N_x,p}\theta_{r+1}(x_p,\tau_j)=0,\ c_{0,r}(x_0,\tau_j)\sum_{p=0}^{N_x}2D_{0,p}\theta_{r+1}(x_p,\tau_j)+c_{1,r}(x_0,\tau_j)\theta_{r+1}(x_0,\tau_j)=d_r(x_0,\tau_j).$$

Imposing boundary conditions on Equation (19) for $j=0,1,\ldots,N_\tau-1$, we obtain the following $N_\tau(N_x+1)\times N_\tau(N_x+1)$ system of matrices:

$$\begin{aligned}&\begin{bmatrix}A_{1(0,0)} & A_{1(0,1)} & \cdots & A_{1(0,N_\tau-1)}\\ A_{1(1,0)} & A_{1(1,1)} & \cdots & A_{1(1,N_\tau-1)}\\ \vdots & \vdots & \vdots & \vdots\\ A_{1(N_\tau-1,0)} & A_{1(N_\tau-1,1)} & \cdots & A_{1(N_\tau-1,N_\tau-1)}\end{bmatrix}\begin{bmatrix}\mathbf{u}_{r+1,0}\\ \mathbf{u}_{r+1,1}\\ \vdots\\ \mathbf{u}_{r+1,N_\tau-1}\end{bmatrix}=\begin{bmatrix}\mathbb{R}_1(\mathbf{x},0)\\ \mathbb{R}_1(\mathbf{x},1)\\ \vdots\\ \mathbb{R}_1(\mathbf{x},N_\tau-1)\end{bmatrix},\\ &\begin{bmatrix}A_{2(0,0)} & A_{2(0,1)} & \cdots & A_{2(0,N_\tau-1)}\\ A_{2(1,0)} & A_{2(1,1)} & \cdots & A_{2(1,N_\tau-1)}\\ \vdots & \vdots & \vdots & \vdots\\ A_{2(N_\tau-1,0)} & A_{2(N_\tau-1,1)} & \cdots & A_{2(N_\tau-1,N_\tau-1)}\end{bmatrix}\begin{bmatrix}\boldsymbol{\theta}_{r+1,0}\\ \boldsymbol{\theta}_{r+1,1}\\ \vdots\\ \boldsymbol{\theta}_{r+1,N_\tau-1}\end{bmatrix}=\begin{bmatrix}\mathbb{R}_2(\mathbf{x},0)\\ \mathbb{R}_2(\mathbf{x},1)\\ \vdots\\ \mathbb{R}_2(\mathbf{x},N_\tau-1)\end{bmatrix},\end{aligned}\tag{20}$$

with

$$\begin{aligned}A_{1(j,j)}&=\alpha_{0,r}(\mathbf{x},\tau_j)\mathbf{D}^2+\alpha_{1,r}(\mathbf{x},\tau_j)\mathbf{D}+\alpha_{2,r}(\mathbf{x},\tau_j)\hat{d}_{j,j}\mathbf{I}=,j=0,1,\cdots,N_\tau-1,\\ A_{1(j,i)}&=\alpha_{2,r}(\mathbf{x},\tau_j)\hat{d}_{j,i}\mathbf{I},\mathit{when}\ j\neq i,\\ \mathbb{R}_1(\mathbf{x},\tau_j)&=\mathbb{R}_{1,r}(\mathbf{x},\tau_j)-\alpha_{2,r}(\mathbf{x},\tau_j)\hat{d}_{j,N_\tau}\mathbf{u}_{r+1,N_\tau},\ \mathit{for}\ j=0,1,\cdots,N_\tau-1,\\ A_{2(j,j)}&=\beta_{0,r}(\mathbf{x},\tau_j)\mathbf{D}^2+\beta_{1,r}(\mathbf{x},\tau_j)\mathbf{D}+\beta_{2,r}(\mathbf{x},\tau_j)\mathbf{I}+\beta_{3,r}(\mathbf{x},\tau_j)\hat{d}_{j,j}\mathbf{I},\ j=0,1,\cdots,N_\tau-1,\\ A_{2(j,i)}&=\beta_{3,r}(\mathbf{x},\tau_j)\hat{d}_{j,i},\ \mathit{when}\ j\neq i,\\ \mathbb{R}_2(\mathbf{x},\tau_j)&=\mathbb{R}_{2,r}(\mathbf{x},\tau_j)-\beta_{3,r}(\mathbf{x},\tau_j)\hat{d}_{j,N_\tau}\boldsymbol{\theta}_{r+1,N_\tau},\ \mathit{for}\ j=0,1,\cdots,N_\tau-1,\end{aligned}$$

The vectors u_{r+1,N_τ} and θ_{r+1,N_τ} correspond to the initial condition given in Equation (8). The matrices (20) are solved iteratively until suitable results are obtained.

Convergence Analysis

The convergence of the BSLLS is evaluated by considering the error norms between two successive iterations. Error norms are defined as

$$\begin{aligned}E_u&=\max_{0<i<N_t}\|u_{r+1,i}-u_{r,i}\|,\\ E_\theta&=\max_{0<i<N_t}\|\theta_{r+1,i}-\theta_{r,i}\|,\end{aligned}\tag{21}$$

E_u and E_θ decrease swiftly as the number of iterations increases (see Figure 2a. This shows that the BSLLS converges within a few iterations. Also, residual error norms are computed to show the accuracy of the BSLLS. Residual error norms are expressed as

$$
\begin{aligned}
R_u &= \max_{0<i<N_t} \|\partial_u(F)\|, \\
R_\theta &= \max_{0<i<N_t} \|\partial_\theta(\Theta)\|,
\end{aligned}
\tag{22}
$$

where ∂_u and ∂_θ are corresponding nonlinear partial differential equations: Equations (9) and (10), respectively. Figure 2b depicts residual errors R_u and R_θ against the number of iterations. Residual errors are found to decrease rapidly with an increasing number of iterations.

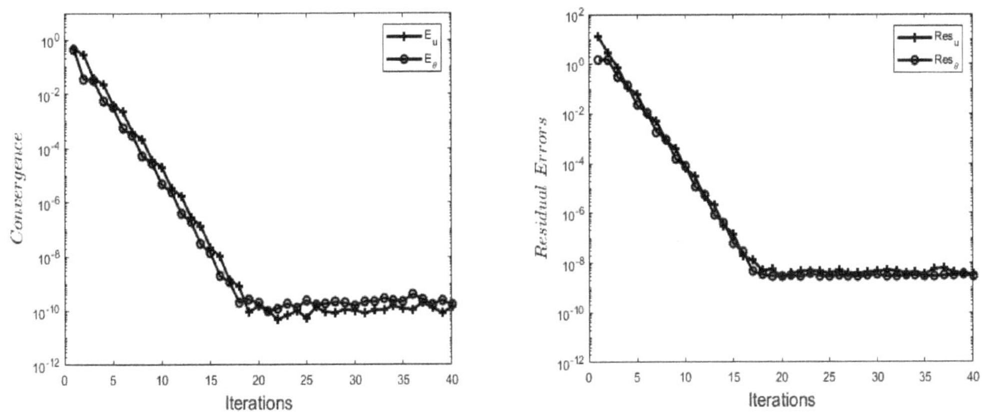

(a) **(b)**

Figure 2. (**a**) Solution convergence against the number of iterations. (**b**) Residual errors against the number of iterations.

4. Results and Discussion

In this section, we employ the parameter values, $n = 0.5$, $W_e = 0.5$, $A = 1$, $\beta_1 = 0.1$, $\beta_2 = 0.1$, $\lambda = 0.1$, $\epsilon_1 = 0.1$, $\epsilon_2 = 0.1$, $\omega = 1$, $Ra = 0.1$, $\Pr = 10$, $Gr = 1$, $V_d = 0.5$, $\theta_a = 0.1$, $A = 1, T = 120$, as default values, unless otherwise stated in graphs and tables. The results obtained by using the BSLLS are validated with the ones obtained using the collocation weighted residual method (see Table 1), and a good agreement is observed.

Table 1. Validation of bivariate spectral local linearization method with Chebyshev collocation method.

y	u_{BSLLM}	u_{CWRM}	θ_{BSLLM}	θ_{CWRM}
0.00	0.8472544166	0.8472544291	0.5231137461	0.5231137500
0.25	0.7969166162	0.7969166280	0.5149631360	0.5149631391
0.50	0.6436905650	0.6436905747	0.4901023867	0.4901023899
0.75	0.3812086637	0.3812086696	0.4472855132	0.4472855163
1.00	-9.59276×10^{-14}	-1.12618×10^{-23}	0.3843857453	0.3843857480

4.1. Transient Profiles for Velocity and Temperature

Figures 3 and 4 show the time development of the velocity and temperature profiles. As time passes, the velocity (Figure 3a) and temperature (Figure 3b) profiles rise until they reach steady-state maximum values. Furthermore, the velocity profile reaches a steady state faster than the temperature profile. This is expected since velocity acts as a source of heat for the combustion process and, hence, increases the temperature profile.

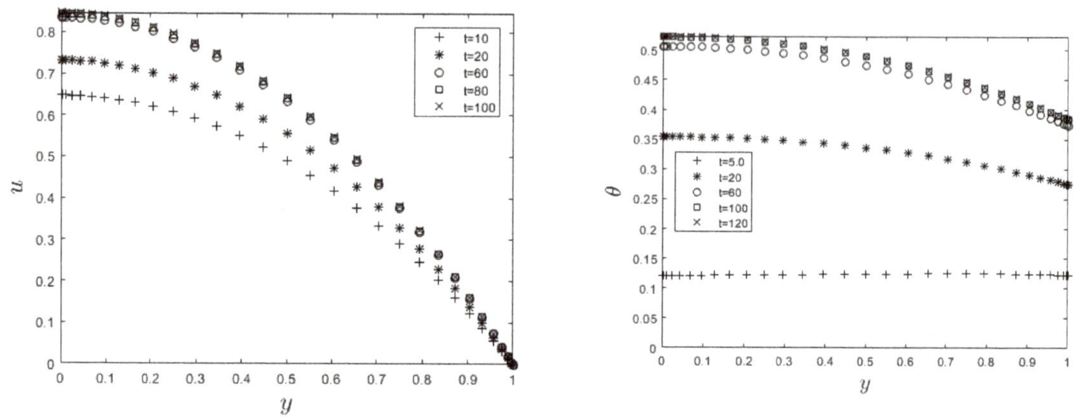

(**a**) (**b**)

Figure 3. (**a**): Velocity steady state condition. (**b**): Temperature steady state condition.

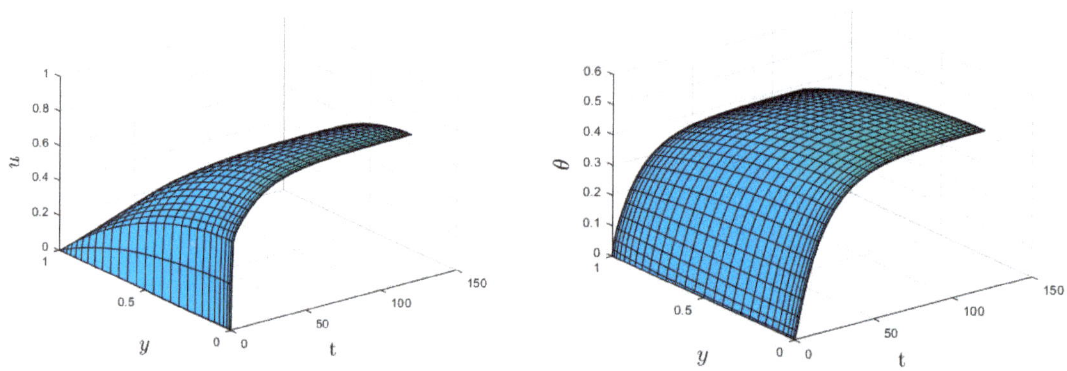

(**a**) (**b**)

Figure 4. (**a**) Three-dimensional velocity profile. (**b**) Three-dimensional temperature profile.

4.2. Solution Blow-Up Profile

The bifurcation plot of the maximum temperature, $(\theta(0))$, versus the Frank–Kamenetskii parameter, λ, is provided to examine the thermal criticality condition of the system. The critical value, λ_c, is computed at a steady state (when combustion is independent of time) to explain auto-ignition during the combustion process. Figure 5 displays a bifurcation diagram explaining the thermal behavior of the system as λ increases. The bifurcation diagram illustrates that the solution to Equation (7), at a steady state, is finite for λ between interval 0 and λ_c. Auto-ignition is observed at the upper limit of the interval, λ_c, and a real solution does not exist when $\lambda > \lambda_c$.

To prevent or control spontaneous ignition, the impact of the thermophysical parameters on the thermal criticality is examined (see Table 2). Thermal criticality increases with increasing values of the power law index (n), radiation (Ra), and the Biot number (Bi) and decreases with increasing values of variable viscosity (β_1), the variable thermal conductivity parameter (β_2), and the material parameter (W_e). This implies that these factors are important in minimizing auto-ignition. In other words, for thermal stability to be maintained during combustion processes, these parameters should be made significant.

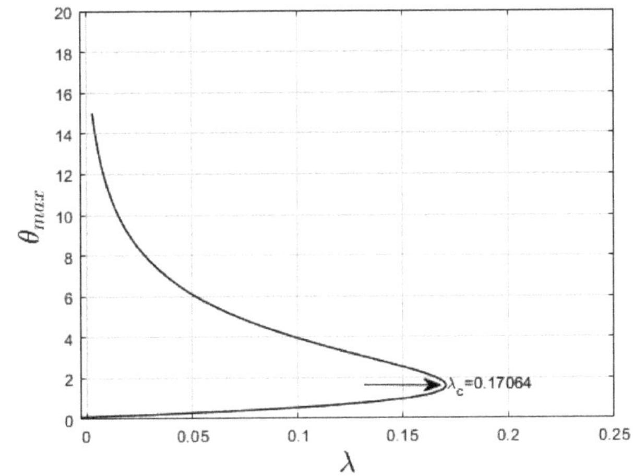

Figure 5. Bifurcation curve.

Table 2. Variations in criticality values for $A = 1$, $\omega = 1$, $Gr = 1$, $\Pr = 10$, $\epsilon_1 = 0.1$, $\epsilon_2 = 0.1$, $V_d = 0.5, T = 120$.

β_1	β_2	$\mathbf{W}_e$	n	Ra	Bi	λ_c
0.1	0.1	0.5	0.5	0.1	1.0	0.17064
0.3	0.1	0.5	0.5	0.1	1.0	0.15603
0.5	0.1	0.5	0.5	0.1	1.0	0.14231
0.1	0.1	0.5	0.5	0.1	1.0	0.17064
0.1	0.3	0.5	0.5	0.1	1.0	0.15942
0.1	0.5	0.5	0.5	0.1	1.0	0.14860
0.1	0.1	0.1	0.5	0.1	1.0	0.18038
0.1	0.1	0.3	0.5	0.1	1.0	0.17695
0.1	0.1	0.5	0.5	0.1	1.0	0.17064
0.1	0.1	0.5	0.5	0.1	1.0	0.17064
0.1	0.1	0.5	1.0	0.1	1.0	0.18081
0.1	0.1	0.5	1.5	0.1	1.0	0.18591
0.1	0.1	0.5	0.5	0.1	1.0	0.17064
0.1	0.1	0.5	0.5	0.3	1.0	0.19447
0.1	0.1	0.5	0.5	0.5	1.0	0.21836
0.1	0.1	0.5	0.5	0.1	0.5	0.10362
0.1	0.1	0.5	0.5	0.1	1.0	0.17064
0.1	0.1	0.5	0.5	0.1	1.5	0.22120

4.3. Dependence of Velocity and Temperature Profiles on Flow Parameters

Figures 6 and 7 depict the impacts of W_e on the velocity and temperature profiles, respectively. Both profiles are elevated as the W_e values increase. This is ascribed to the fact that an increase in W_e makes the fluid thinner, and the resistance force to the flow decreases; hence, the velocity profile increases. Furthermore, internal heat generation, as a result of the viscous term, is high when W_e increases. This leads to an enhancement of the temperature profile. Figures 8 and 9 present the effect of the power law index, n, on the velocity and temperature distributions, respectively. As such, the velocity distribution decreases with increasing n values. When $n > 1$, the resistance force to the flow for shear-thickening fluids becomes maximal, and the velocity profile decreases (see Figure 8). A decrease in fluid speed results in a decrease in viscous heating, and the temperature profile reduces significantly (see Figure 9).

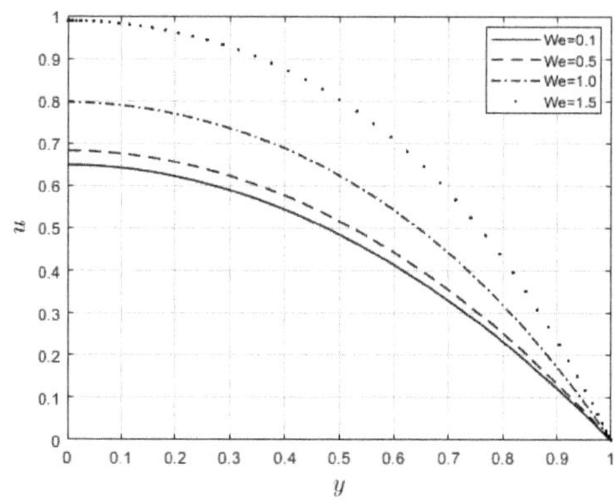

Figure 6. Velocity distributions for W_e.

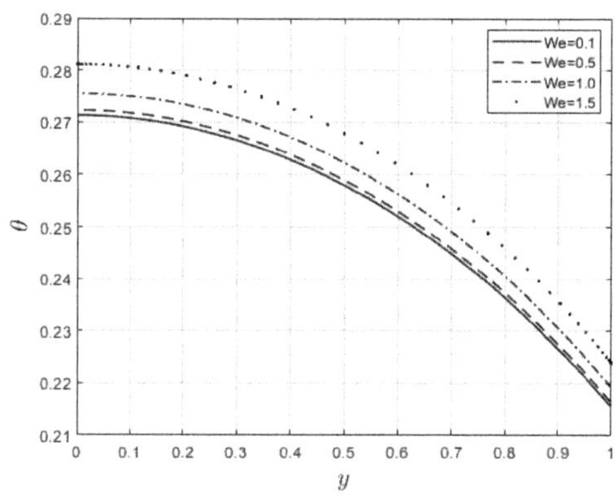

Figure 7. Temperature distribution for W_e.

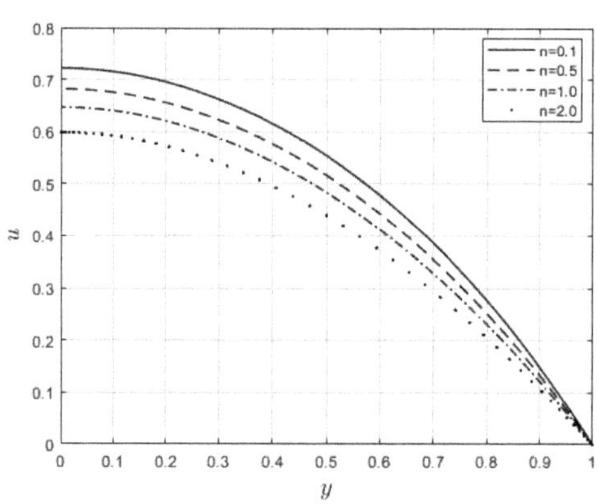

Figure 8. Velocity profiles for n.

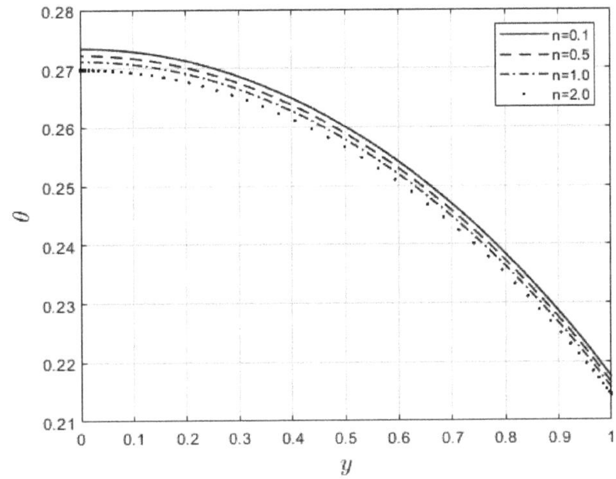

Figure 9. Temperature profiles for n.

Figures 10 and 11 examine the behavior of velocity and temperature distributions subjected to a variable viscosity parameter, β_1. The velocity profile is elevated with increasing values of β_1, as observed in Figure 10. The reason for this is that fluid viscosity reduces as β_1 increases, and the kinetic energy of the fluid molecules increases, leading to velocity distribution enhancement. An increase in the velocity profile naturally enhances the heat source term in the energy equation, resulting in an elevation in the temperature field (see Figure 11). The impact of variable thermal conductivity, β_2, on the velocity and temperature distributions is provided in Figures 12 and 13, respectively. An increase in the values of β_2 increases the temperature profile (see Figure 13). This is attributed to the fact that a rise in β_2 leads to a reduction in the thermal conductivity term, $e^{-\beta_2\theta}$. This leads to the slow, random movement of fluid molecules and thus facilitates heat transfer through the fluid, which consequently enhances the temperature profile. A significant rise in temperature as β_2 increases results in a reduction in the fluid viscosity and thus enhances the flow speed (see Figure 12). Figure 14 depicts the impact of the radiation parameter, Ra, on the fluid temperature. It can be observed that the temperature profile reduces as the Ra values rise. This indicates that more heat exits the stockpile through radiation and, thus, reduces the fluid temperature profile. The same scenario is seen in Figure 15 as the temperature profile reduces with an increasing Biot number, Bi, because more heat escapes the stockpile through the walls. Enhancement in the temperature profile is observed in Figure 16 as the two-step parameter, ω, values increase. This is due to extra heating created by higher values of ω. Figures 17 and 18 illustrate the variation in the Nusselt number, Nu (heat transfer rate at the wall) for different combinations of thermophysical parameters. Nu increases with higher values of the Frank–Kamenetskii parameter, λ, and the two-step reaction parameter, ω, and decreases with higher values of the radiation parameter, Ra, and the Biot number, Bi. These results agree with the work in [32].

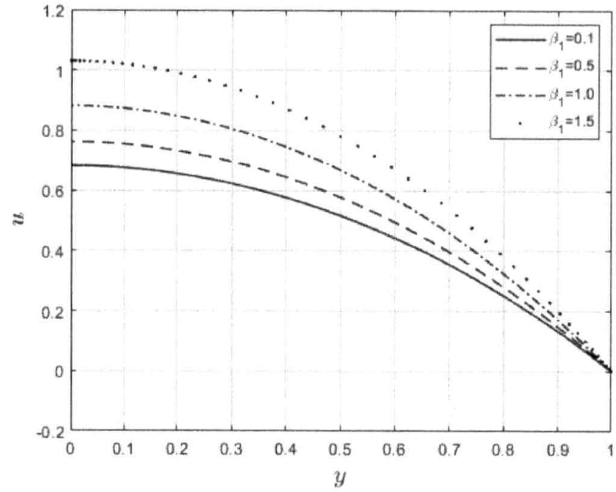

Figure 10. Velocity graph for variation in β_1.

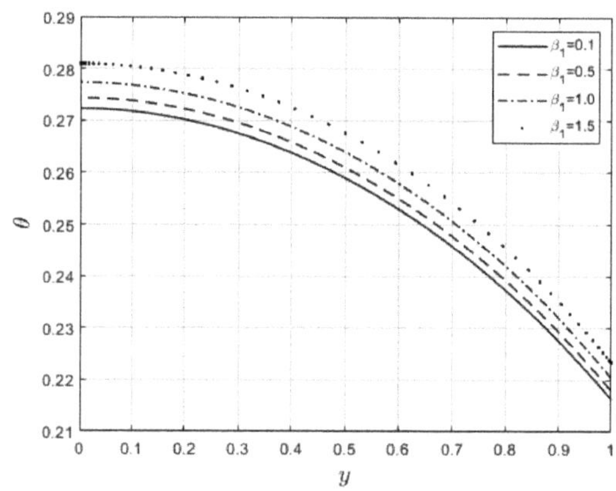

Figure 11. Temperature graph for variation in β_1.

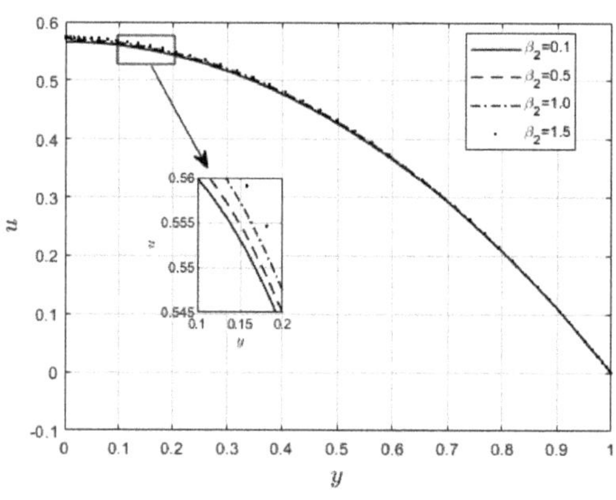

Figure 12. β_2 effect on velocity profiles.

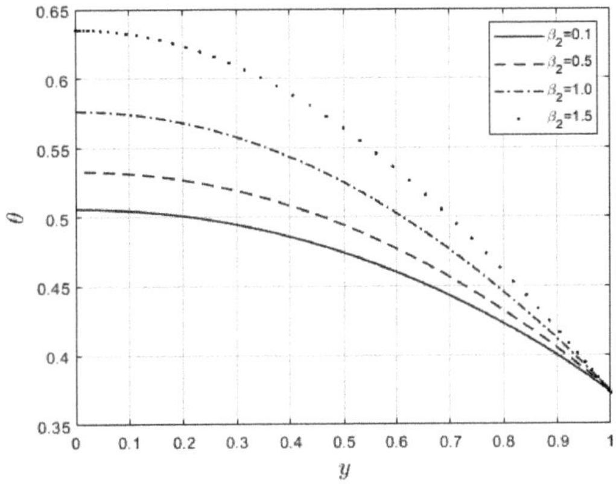

Figure 13. β_2 effect on temperature profiles.

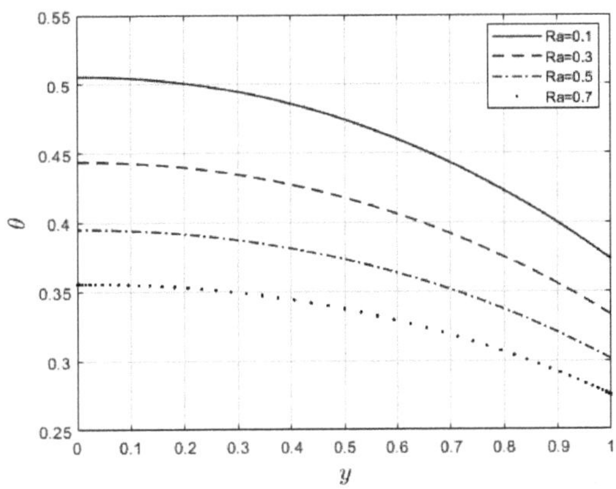

Figure 14. Impact of Ra on temperature distribution.

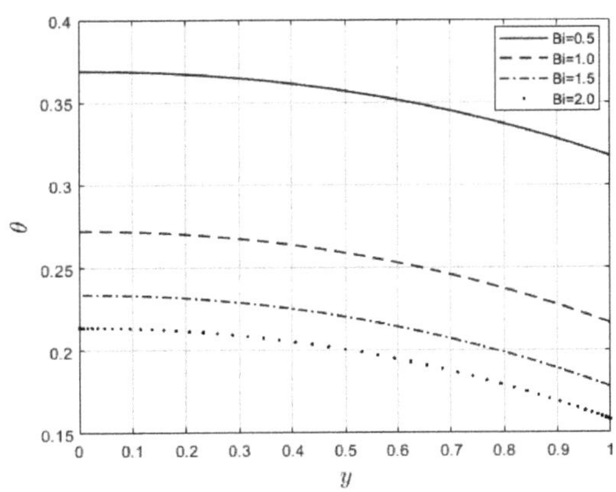

Figure 15. Influence of Bi on temperature profiles.

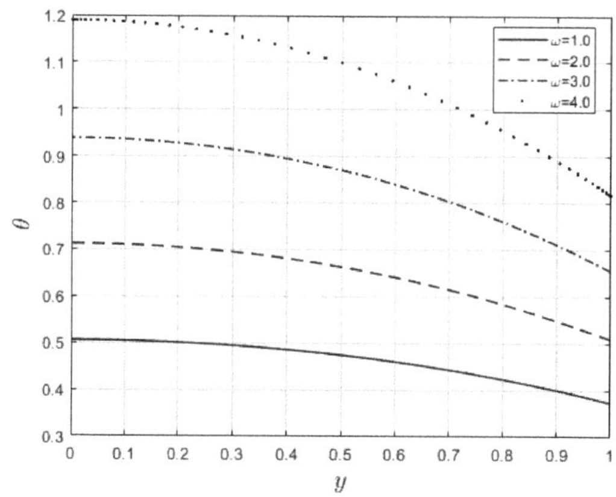

Figure 16. Temperature profiles with change in ω.

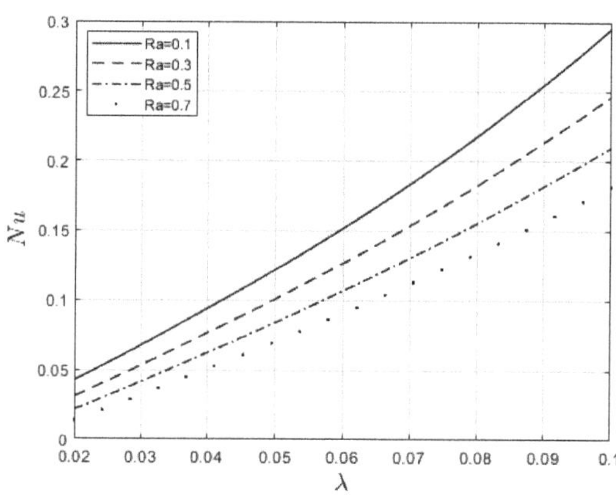

Figure 17. Influence of Ra and λ on the Nusselt number.

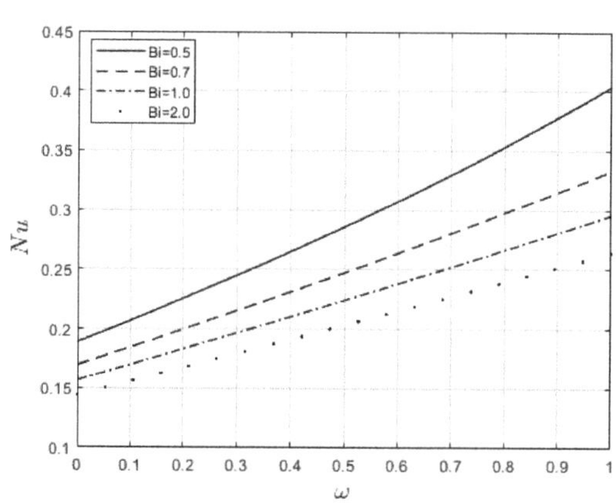

Figure 18. Influence of Bi and ω on the Nusselt number.

5. Concluding Remarks

This article considered the combustion of polymer material in a rectangular stockpile. The rheology of the polymer was assumed to follow a Carreau fluid constitutive relation. The spectral local linearization method was implemented to provide a numerical solution to the problem. The impacts of various thermokinetics factors on the flow and thermal behaviors were examined. From the obtained results, it was found that some parameters (W_e, β_2, *and* ω) improve the combustion process since the temperature profiles increase as the values of these parameters increase. This may speed up thermal ignition and, thus, lead to an explosion. The opposite case can be observed for an increase in the values of n, Ra, *and* Bi. This slows down the chemical reaction and, thus, minimizes the combustion process.

Author Contributions: Conceptualization, A.T.A.; methodology, A.T.A.; software, J.C.U.; validation, J.C.U.; formal analysis, R.S.L.; investigation, J.C.U.; writing—original draft preparation, A.T.A.; writing—review and editing, A.T.A.; supervision, R.S.L.; project administration, J.C.U.; funding acquisition, R.S.L. All authors have read and agreed to the published version of the manuscript.

Funding: This research received no external funding.

Data Availability Statement: No data are associated with this manuscript.

Conflicts of Interest: The authors declare no conflict of interest.

References

1. Xiong, Q.; Kong, S. Modelling effect of interphase transport coefficients on biomass pyrolysis in fluidized beds. *Powder Technol.* **2014**, *262*, 96–105.
2. Xiong, Q.; Kong, S.; Passalacqua, A. Development of a generalized numerical frame work for simulating biomass fast pyrolysis in fluidized-bed reactors. *Chem. Eng. Sci.* **2013**, *99*, 305–313.
3. Tan, Z.; Su, G.; Su, J. Improved lumped models for combined convective and radiative cooling of a wall. *Appl. Therm. Eng.* **2009**, *29*, 2439–2443.
4. Drysdale, D. Ignition: The initiation of flaming combustion. In *An Introduction to Fire Dynamics*, 3rd ed.; Wiley: Hoboken, NJ, USA, 2011; Chapter 6; pp. 225–275.
5. Shi, L.; Chew, M.Y.L. A review of fire processes modeling of combustible materials under external heat flux. *Fuel* **2013**, *106*, 30–50.
6. Sener, A.A.; Demirhan, E. The investigation of using magnesium hydroxide as a flame retardant in the cable insulation material by cross-linked polyethylene. *Mater. Des.* **2008**, *29*, 1376–1379.
7. Gong, T.; Xie, Q.; Huang, X. Fire behaviors of flame-retardant cable part decomposition, swelling and spontaneous ignition. *Fire Saf. J.* **2018**, *95*, 113–121.
8. Xie, Q.; Gong, T.; Huang, X. Fire Zone Diagram of Flame-Retardant Cables: Ignition and Upward Flame Spread. *Fire Technol.* **2021**, *57*, 2643–2659.
9. Geschwindner, C.; Goedderz, D.; Li, T.; Köser, J.; Fasel, C.; Riedel, R.; Altstädt, V.; Bethke, C.; Puchtler, F.; Breu, J.; et al. Investigation of flame retarded polypropylene by high-speed planar laser-induced fluorescence of OH radicals combined with a thermal decomposition analysis. *Exp. Fluids* **2020**, *61*, 30.
10. Lohrer, C.; Krause, U.; Steinbach, J. Self-Ignition of Combustible Bulk Materials Under Various Ambient Conditions. *Process. Saf. Environ. Prot.* **2000**, *83*, 145–150.
11. Lohrer, C.; Schmidt, M.; Krause, U. A study on the influence of liquid water and water vapor on the self-ignition of lignite coal-experiments and numerical simulations. *J. Prev. Process. Ind.* **2005**, *18*, 167–177.
12. Lebelo, R.S. Thermal stability investigation in a reactive sphere of combustible material. *Adv. Math. Phys.* **2016**, *2016*, 9384541.
13. Lebelo, R.S.; Makinde, O.D.; Chinyoka, T. Thermal decomposition analysis in a sphere of combustible materials. *Adv. Mech. Eng.* **2017**, *9*, 1–14.
14. Lebelo, R.S.; Adeosun, A.T.; Gbadeyan, J.A.; Akindeinde, S.O. On the heat transfer stability for a convective reactive material of variable thermal conductivity in a sphere. *Gorteria J.* **2021**, *34*, 62–74.
15. Lebelo, R.S.; Moloi, K.C.; Okosun, K.O.; Mukamuri, M.; Adesanya, S.O.; Muthuvalu, M.S. Two-step low-temperature oxidation for thermal stability analysis of a combustible sphere. *Alex. Eng. J.* **2018**, *57*, 2829–2835.
16. Lebelo, R.S.; Waetzel, M.; Mahlobo, R.K.; Moloi, K.C.; Adesanya, S.O. On transient heat analysis of a two-step convective reactive cylinder. *J. Phys. Conf. Ser.* **2021**, *1730*, 012141.
17. Khan, M.; Hashim. Boundary layer flow and heat transfer to Carreau fluid over a nonlinear stretching sheet. *AIP Adv.* **2015**, *5*, 107203.
18. Siska, B.; Bendova, H.; MacHak, I. Terminal velocity of non-spherical particles falling through a Carreau model fluid. *Chem. Eng. Process* **2005**, *44*, 1312–1319.
19. Reedy, S.; Srihari, P.; Ali, F.; Naikoti, K. Numerical analysis of Carreau fluid flow over a vertical porous microchannel with entropy generation. *Partial. Differ. Equ. Appl. Math.* **2022**, *5*, 100304.
20. Abbas, T.; Rehman, S.; Shah, R.A.; Idrees, M.; Qayyum, M. Analysis of MHD Carreau fluid flow over a stretching permeable sheet with variable viscosity and thermal conductivity. *Physical A* **2020**, *551*, 124225.

21. Megahed, M. Carreau fluid flow due to nonlinearly stretching sheet with thermal radiation, heat flux, and variable thermal conductivity. *Appl. Math. Mech.* **2019**, *40*, 1615–1624.
22. Hayat, T.; Yasmin, H.; Alsaedi, A. Peristatic motion of Carreau fluid in a channel with convective boundary conditions. *Appl. Bionics Biomech.* **2014**, *11*, 157–168.
23. Alqarni, M.M.; Riaz, A.; Firdous, M.; Lali, I.U.; El Sayed, M.; El-Din, T.; ur Rahman, S. Hall currents and EDL effects on multiphase wavy flow of Carreau fluid in a microchannel having oscillating walls: A numerical study. *Front. Phys.* **2022**, *10*, 984277. [CrossRef]
24. Noreen, S.; Kausar, T.; Tripathi, D.; Ain, Q.U.; Lu, D. Heat transfer analysis on creeping flow Carreau fluid driven by peristaltic pumping in an inclined asymmetric channel. *Therm. Sci. Eng. Prog.* **2020**, *17*, 100486. [CrossRef]
25. Asha, S.K.; Beleri, J. Peristatic flow of Carreau nanofluid in presence of joule heat effect in an inclined asymmetric channel by multi-step differential transformation method. *World Sci. News Int. J.* **2022**, *164*, 44–63.
26. Qayyum, M.; Abbas, T.; Afzal, S.; Saeed, S.T.; Akgül, A.; Inc, M.; Mahmoud, K.H.; Alsubaie, A.S. Heat transfer analysis of unsteady MHD Carreau fluid flow over a stretching/shrinking sheet. *Coatings* **2022**, *12*, 1661.
27. Shao, Y.; Wu, A.; Abbas, S.; Khan, W.; Ashraf, I. Thermal management for the shear-rate driven flow of Carreau fluid in a ciliated channel. *Int. Commun. Heat Mass Transf.* **2022**, *139*, 106473.
28. Tshehla, M.S. The flow of a Carreau fluid down an inclined with a free surface. *Int. J. Phys. Sci.* **2011**, *6*, 3896–3910.
29. Tijani, Y.O.; Oloniiju, S.D.; Kasali, K.B.; Akolade, M.T. Nonsimilar solution of a boundary layer flow of a Reiner–Philippoff fluid with nonlinear thermal convection. *Heat Transf.* **2022**, *51*, 5659–5678.
30. Agbaje, T.; Mondal, S.; Motsa, S.; Sibanda, P. A numerical study of unsteady non-Newtonian Powell-Eyring nanofluid flow over a shrinking sheet with heat generation and thermal radiation. *Alex. Eng. J.* **2017**, *56*, 81–91.
31. Magagula, V.M. On the multidomain bivariate spectral local linearization method for solving systems of nonlinear boundary layer partial differential equations. *Int. J. Math. Math. Sci.* **2019**, *2019*, 6423294.
32. Tairu, S.; Makinde, O.D. Analysis of nonlinear heat transfer in a cylindrical solid with two-step exothermic kinetic and radiative heat loss. *Defect Diffus. Forum* **2017**, *377*, 17–28.

mathematics

MDPI

Article

Convective Heat Transfer and Entropy Generation for Nano-Jet Impingement Cooling of a Moving Hot Surface under the Effects of Multiple Rotating Cylinders and Magnetic Field

Lioua Kolsi [1], Fatih Selimefendigil [2,3,*], Samia Larguech [4], Kaouther Ghachem [5], Hind Albalawi [6], Badr M. Alshammari [7] and Taher Labidi [8]

1 Department of Mechanical Engineering, College of Engineering, University of Ha'il, Ha'il 81451, Saudi Arabia
2 Department of Mechanical Engineering, College of Engineering, King Faisal University, Al Ahsa 31982, Saudi Arabia
3 Department of Mechanical Engineering, Celal Bayar University, Manisa 45140, Turkey
4 Department of Electrical Engineering, College of Engineering, Princess Nourah bint Abdulrahman University, P.O. Box 84428, Riyadh 11671, Saudi Arabia
5 Department of Industrial Engineering and Systems, College of Engineering, Princess Nourah Bint Abdulrahman University, P.O. Box 84428, Riyadh 11671, Saudi Arabia
6 Department of Physics, College of Sciences, Princess Nourah bint Abdulrahman University, P.O. Box 84428, Riyadh 11671, Saudi Arabia
7 Department of Electrical Engineering, College of Engineering, University of Ha'il, Ha'il 81451, Saudi Arabia
8 Department of Software Engineering, College of Computer Engineering and Sciences, Prince Sattam bin Abdulaziz University, P.O. Box 151, Al-Kharj 11942, Saudi Arabia
* Correspondence: fgil@kfu.edu.sa

Citation: Kolsi, L.; Selimefendigil, F.; Larguech, S.; Ghachem, K.; Albalawi, H.; Alshammari, B.M.; Labidi, T. Convective Heat Transfer and Entropy Generation for Nano-Jet Impingement Cooling of a Moving Hot Surface under the Effects of Multiple Rotating Cylinders and Magnetic Field. *Mathematics* **2023**, *11*, 1891. https://doi.org/10.3390/math11081891

Academic Editor: Ramoshweu Solomon Lebelo

Received: 30 March 2023
Revised: 12 April 2023
Accepted: 13 April 2023
Published: 17 April 2023

Abstract: In this study, confined slot nano-jet impingement cooling of a hot moving surface is investigated under the combined utilization multiple rotating cylinders and magnetic field. Both convective heat transfer and entropy generation analysis are conducted using a finite element method. Parametric variation of the rotational Reynolds number (Rew between −500 and 500), velocity ratio (VR between 0 and 0.25), Hartmann number (Ha between 0 and 20) and the horizontal location of cylinders (Mx between −8 and 8) are considered. Rotation of the cylinders generally resulted in the degradation of cooling performance while increasing the wall velocity, and the horizontal location of the cylinder was found to positively contribute to this. Heat transfer rate reductions of 20% and 12.5% are obtained using rotations at the highest Rew for the case of stationary (VR = 0) and moving wall (VR = 0.25). When magnetic field at the highest strength is imposed in the rotating cylinder case, the cooling performance is increased by about 18.6%, while it is reduced by about 28% for the non-rotating cylinder case. The hot wall movement contributes, by about 14%, to the overall cooling performance enhancement. Away from the inlet location of the rotating cylinders, thermal performance improvement of 12% is obtained. The entropy generation rises with higher hot wall velocity and higher horizontal distances of the rotating cylinders, while it is reduced with a higher magnetic field for non-rotating cylinders. The best configurations in terms of cooling performance provide 8.7% and 34.2% enhancements for non-rotating and rotating cylinders compared with the reference case of (Rew, VR, Ha, Mx) = (0, 0, 0, 0), while entropy generation becomes 1% and 15% higher.

Keywords: magnetic field; slot jet impingement; finite element method; multiple rotating cylinders; entropy generation; moving wall

MSC: 76D25; 76D55; 80M10; 80M50; 76S05

1. Introduction

The use of impinging jets has been considered for effective cooling in many thermal applications, including solar power, drying, electronic cooling and materials processing such as glass annealing, among others [1–3]. Higher localized heat and mass transfer coefficients can be obtained by using impinging jets. The flow features and basic heat transfer (HT) mechanism by using impinging jets have been provided with important design and operating parameters [4–6]. Due to the complicated nature of the flow field and its interaction with the thermal field, complex geometry and pressure gradients, theoretical treatment of the convective HT with impinging jets is challenging. Magnetic field (MGF) in jet impingement has been considered in several studies. The applications in jet flow with MGF may be relevant in electromagnetic braking. The utilization of MGF in convective HT has been considered in many studies as HT and flow control method. An external MGF, which may be active, partial, non-uniform or time dependent, can be imposed for thermal system and several studies used MGF in slot jet impingement (SJ-I) HT applications. The performance of the SJ-I cooling can be improved by using nanofluids (NFs). This technology of NFs has already been implemented in diverse engineering systems, including energy storage, refrigeration, renewable energy and many more [7–9]. For electrically conducting fluids, nanoparticles can be used in base fluid to alter the thermophysical properties, including thermal and electrical conductivity, while application of MGF will be more efficient. Sheikholeslami and Rokni [10] performed a review for application of NFs in the existence of MGF effects by analyzing many numerical and analytical studies. They also considered the Brownian motion and thermophoresis effects of NFs. Improvements in the thermal performance were reported with NFs while discrepancies between single and two-phase models were noted. M'hamed et al. [11] performed an extensive review of the application of MGF considering NFs. They noted successful application of the flow control with MGF, while some challenges were also mentioned, such as the stability and cost of using NFs. In jet impingement, NFs have been used. In their review work, [12] analyzed the existing studies for NF applications in impinging jet HT. Many aspects of the NF were covered, including non-Newtonian fluid behavior and single- and two-phase modeling approaches. They noted that single-phase non-Newtonian models required higher pumping power even if they had higher HT rates. They recommended considering the application of hybrid NFs, multiphase model and erosion impacts of NFs in jet impingement applications. Even though diverse studies have been considered for the application of NF in jet impingement HT [13–15], there are few works that have considered the MGF and NF impacts with jets together [16–18].

Many different methods of HT enhancement have been considered in J-I cooling. MGF and NF applications have been discussed above. The utilization of rotating circular cylinders (CCs) in HT has been considered before in many studies. The size, rotational speed, conductivity ratio and location of the CCs are some of the most influencing factors for the overall thermal performance [19–21]. In J-I systems, cooling of rotating hot surfaces has been considered in several studies [22–24], but rotating CCs as a control method for cooling performance of SJ-I onto a hot flat surface has only been considered in [25].

Entropy generation analysis (EG-A) can be used in thermal system for performance evaluations and design optimizations. The basics and some applications of EG minimization can be found in several sources [26–28]. The HT and fluid friction irreversibility analysis can be performed by using EG studies in convective HT. EG-A have been considered in NF studies [29–31]. A review of EG in NF systems has been performed by Mahian et al. [32]. They considered different NF and various geometry with different boundary conditions. They noted that the EG can be reduced by using NF depending upon the geometry and flow regime while in viscous dominated micro-channels, and using NFs resulted in increased EG. In another review, Huminic and Huminic [33] presented an overview of the NF and hybrid NF applications for EG of different thermal systems. Microchannel flow and cavities were considered while both experimental and numerical studies were covered. The potential of using NFs in EG minimization was explored. In MGF with NF, EG-A has been been performed in many different studies [34–36].

In this study, slot J-I (SJ-I) cooling of moving hot surface under the combined effects of using MGF and multiple rotating CCs are considered. The combined method can be considered either as a novel HT enhancement technique for SJ-I or rotating CCs or MGF can be present in the thermal system. In the literature, utilization of MGF and NF together has been considered in several studies, while only one study exists for a single rotating CC in SJ-I cooling. As novel contributions to the exiting literate, multiple rotating CCs are used, and movement of the hot surface to be cooled is considered along with the MGF and NF. The study also includes EG analysis for the evaluation of system performance. Figure 1 shows the schematic view of the available HT enhancement methods that can be used for J-I cooling system with the jet and wall features considered in this study. The coupled impacts of moving wall, rotation of multiple CCs and MGF will be explored for cooling of an isothermal hot surface by using confined SJ-I. Performance of the system for varying parameters of interest is indicated in terms of EG, which results in higher possibility to assess the system performance. The outcomes of the current work are useful in initial design and optimization studies for development of cooling systems with SJ-I. The applications can be encountered in microfluidic devices, photovoltaic (PV) /thermal systems, material processing, drying and diverse HT equipment.

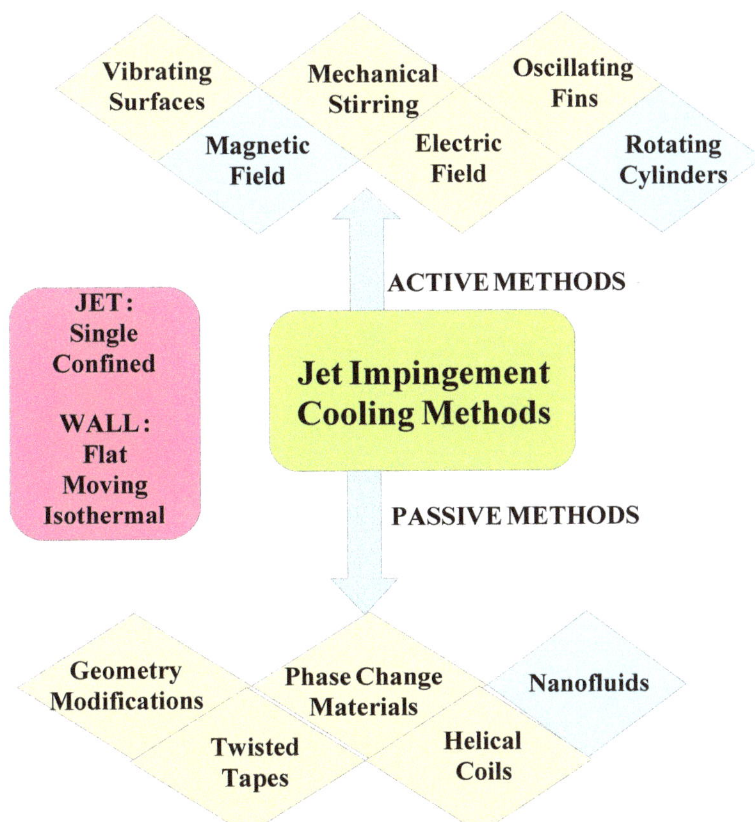

Figure 1. Different available methods that can be used for flow control and thermal performance improvement of jet impingement cooling.

2. Numerical Model

We consider SJ-I cooling of a hot moving wall under the combined utilization of three identical rotating CCs and inclined MGF effects. A schematic view is given in Figure 2. Cold fluid enters with velocity of uc and temperature of Tc, while a hot wall is moving with velocity uw and is kept at a temperature of Th. Velocity ratio (VR) is defined as VR = uw/uc. The slot size is wj, and the hot plate length is L = 150wj and the vertical distance between them is H = 7wj. The rotating CCs are identical and have radius of r = 0.1H and are rotating with same speed of ω. Their center locations are (xc1, yc1) = (0.5L + Mx, 0.35H), (xc2, yc2) = (0.5L + Mx + 0.6H, 0.35H) and (xc3, yc3) = (0.5L + Mx + 0.3H, 0.65H). An external uniform MGF is imposed with inclination of γ with horizontal. The hybrid NF, which contains Ag–MgO nanoparticles in water, is utilized while the solid volume fraction of 2% is considered. The Pr of the base fluid is 6.9. The flow is 2D and laminar. The MGF is uniform, and induced MGF effects along with the electric currents are not taken into account. The impacts of natural convection, radiation and viscous dissipation are ignored.

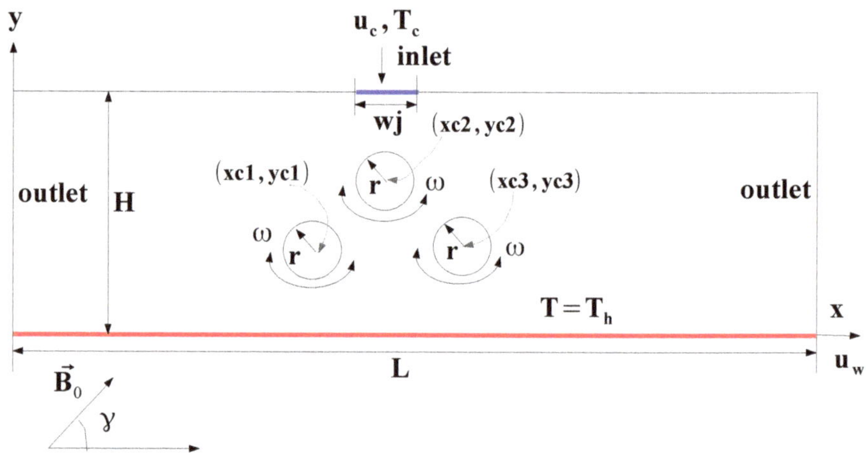

Figure 2. Schematic description of the SJ-I cooling system for a moving hot wall under combined effects of MGF and multiple rotating CCs.

The conservation equations under the above assumptions are:

$$\frac{\partial u}{\partial x} + \frac{\partial v}{\partial y} = 0 \tag{1}$$

$$\begin{aligned} u\frac{\partial u}{\partial x} + v\frac{\partial u}{\partial y} &= -\frac{1}{\rho_{nf}}\frac{\partial p}{\partial x} + \nu_{nf}\left(\nabla^2 u\right) \\ &+ \frac{\sigma_{nf}B_0^2}{\rho_{nf}}\left(v\sin(\gamma)\cos(\gamma) - u\sin^2\gamma\right) \end{aligned} \tag{2}$$

$$\begin{aligned} u\frac{\partial v}{\partial x} + v\frac{\partial v}{\partial y} &= -\frac{1}{\rho_{nf}}\frac{\partial p}{\partial y} + \nu_{nf}\left(\nabla^2 v\right) \\ &+ \frac{\sigma_{nf}B_0^2}{\rho_{nf}}\left(u\sin(\gamma)\cos(\gamma) - v\cos^2\gamma\right) \end{aligned} \tag{3}$$

$$u\frac{\partial T}{\partial x} + v\frac{\partial T}{\partial y} = \alpha_{nf}\nabla^2 T. \tag{4}$$

The related non-dimensional parameters and numbers are:

$$X = \frac{x}{D_h},\ Y = \frac{y}{D_h},\ U = \frac{u}{u_c},\ V = \frac{v}{u_c},\ P = \frac{p}{\rho_{nf} u_c^2}\ \theta = \frac{T - T_c}{T_h - T_c}$$
$$\mathrm{Pr} = \frac{\nu_f}{\alpha_f},\ \ \mathrm{Re} = \frac{u_c D_h}{\nu_{nf}},\ \ \mathrm{Ha} = B_0 D_h \sqrt{\frac{\sigma_{nf}}{\rho_{nf} \nu_f}},\ \ \mathrm{Re}_w = \frac{\omega D_h D_h}{\nu_{nf}} \quad (5)$$

where $D_h = 2w$ is the characteristic length.

$$\frac{\partial U}{\partial X} + \frac{\partial V}{\partial Y} = 0 \quad (6)$$

$$U\frac{\partial U}{\partial X} + V\frac{\partial U}{\partial Y} = -\frac{\partial P}{\partial X} + b_1 \frac{1}{\mathrm{Re}}\left(\nabla^2 U\right) + b_2 \frac{\mathrm{Ha}^2}{\mathrm{Re}}\left(V \sin(\gamma)\cos(\gamma) - U \sin^2 \gamma\right) \quad (7)$$

$$U\frac{\partial V}{\partial X} + V\frac{\partial V}{\partial Y} = -\frac{\partial P}{\partial Y} + b_1 \frac{1}{\mathrm{Re}}\left(\nabla^2 V\right) + b_2 \frac{\mathrm{Ha}^2}{\mathrm{Re}}\left(U \sin(\gamma)\cos(\gamma) - V \cos^2 \gamma\right) \quad (8)$$

$$U\frac{\partial \theta}{\partial X} + V\frac{\partial \theta}{\partial Y} = b_3 \frac{1}{\mathrm{RePr}} \nabla^2 \theta. \quad (9)$$

In the above representation, $b_1 = \frac{\nu_{nf}}{\nu_f}$, $b_2 = \frac{\rho_f}{\rho_{nf}} \frac{\sigma_{nf}}{\sigma_f}$ and $b_3 = \frac{\alpha_{nf}}{\alpha_f}$.

In dimensional form, the boundary conditions are given as:

- Jet inlet, $u = 0, v = -u_c, T = T_c$
- Outlet: $\frac{\partial u}{\partial x} = \frac{\partial T}{\partial x} = 0,\ v = 0$
- At the target plate: $u = u_0, v = 0,\ T = T_h$
- Upper plate walls: $u = v = 0,\ \frac{\partial T}{\partial x} = 0$
- At the rotating cylinder walls: $u = -\omega(y - yci),\ v = \omega(x - xci),\ \frac{\partial T}{\partial n} = 0$

Local and average Nusselt numbers (Nu) are used for thermal performance evaluations, which are given as:

$$\mathrm{Nu}_s = -\frac{k_{nf}}{k_f}\left(\frac{\partial \theta}{\partial n}\right),\ \ \mathrm{Nu}_m = \frac{1}{L}\int_0^L \mathrm{Nu}_s ds. \quad (10)$$

The EG equation is stated as in the following:

$$S = \frac{k_{nf}}{T_0^2}\left[\left(\frac{\partial T}{\partial x}\right)^2 + \left(\frac{\partial T}{\partial y}\right)^2\right] + \frac{\mu_{nf}}{T_0}\left[2\left(\left(\frac{\partial u}{\partial x}\right)^2 + \left(\frac{\partial v}{\partial y}\right)^2\right) + \left(\frac{\partial u}{\partial x} + \frac{\partial v}{\partial y}\right)^2\right]$$
$$+ \frac{\sigma_{nf} B_0^2}{T_0}(u \sin\gamma - v \cos\gamma)^2 \quad (11)$$

where impacts of HT, viscous dissipation and MGF effects are represented by various terms in the above equation.

The Galerkin weighted (GW) residual finite element method (R-FEM) is employed as the solution method. The application of FEM and basic steps in modeling for HT and flow problems can be found in many sources [37,38]. The FEM-based solution of convective HT problems including SJ-I cooling has been considered in many studies [39–41]. In the method, the field variables (f) are approximated by using different ordered Lagrange finite elements as in the following:

$$f = \sum_{k=1}^{N^f} \Psi_k^f F_k, \quad (12)$$

where Ψ^f denotes the shape function and F is the nodal value. The weighted average of the resulting residual (R) is set to be zero as:

$$\int_V WRdV = 0. \tag{13}$$

SUPG method is used to handle the numerical instability while BICGStab is considered for the flow and HT modules of the code. The convergence of the solution is assumed when the relative error for each of the variables satisfy convergence criteria of 10^{-6}.

Grid independence test results are shown in Figure 3a considering average Nu variations with different grid sizes at two different rotations of CCs (Rew = 0 and Rew = -500). A grid with 470,870 number of triangular elements is selected, and its variation in the domain near the CCs is given in Figure 3b. Near-wall refinement is performed.

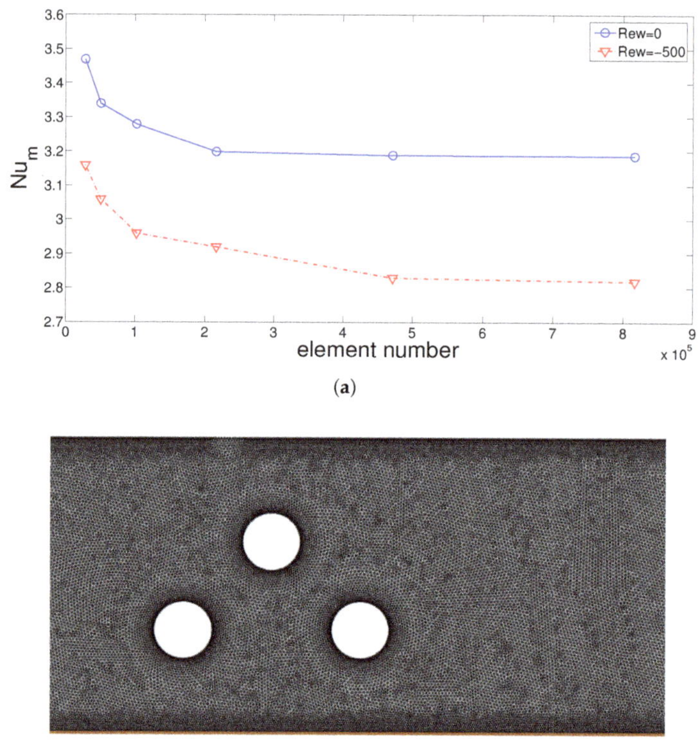

(**a**)

(**b**)

Figure 3. Test results for grid independence: Average Nu for different grid sizes considering rotating and non-rotating CCs (**a**) (Ha = 5, Mx = -1) and grid distribution near the CCs (**b**).

Validation of the code is performed by using different available studies. In the first work, convection in a differentially cavity under MGF effects is analyzed. Comparison results of average Nu at three different MGF strength are shown in Figure 4a by using the available results in [42]. The value of the Grashof number is fixed to Gr $= 2 \times 10^5$ while Ha values of 0, 50 and 100 are taken. The highest difference is seen at Ha = 0, and it is below 5%. For SJ-I cooling, the results available in [43,44] are used. An isothermal plate is considered while Re values of 100 and 300 are taken. The maximum difference between the available results is below 4% (Figure 4b). The MGF effects in convection and SJ-I cooling effects are accurately captured by using the current solver.

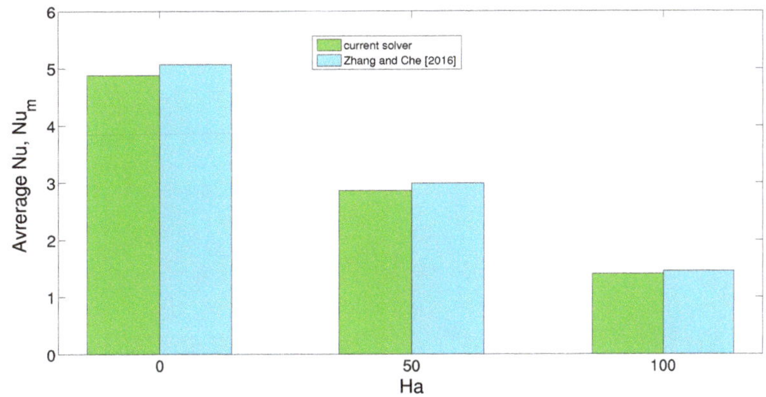

(**a**)

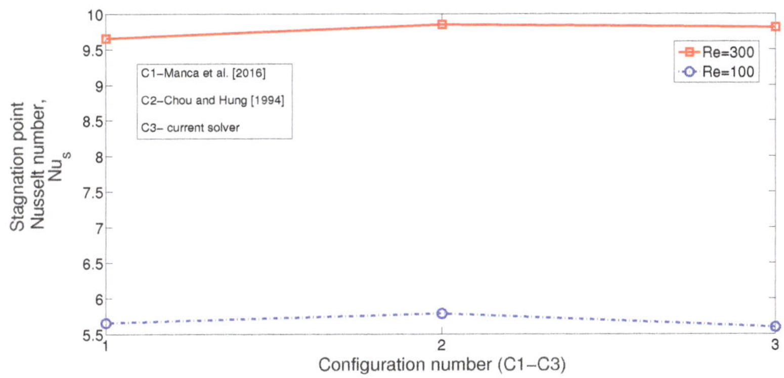

(**b**)

Figure 4. Code validation 1: Average Nu comparisons at three different Ha considering the convection in a differentially heated enclosure. Results in [42] are used, and Grashof number is fixed to Gr = 2×10^5 (**a**). Code validation 2: Stagnation point Nu comparison for SJ-I cooling of an isothermal surface by using the reference results in [43,44] at Reynolds number 100 and 300 (**b**).

3. Results and Discussion

Convective HT and EG analysis for confined SJ-I cooling of a moving hot surface under the combined effects of using MGF and triple rotating CCs is conducted. The CCs are identical and rotating at the same speed, while the hot wall is moving with constant speed. The numerical study is conducted for various values of rotational speed of the CCs (Rew between −500 and 500), velocity ratio (VR between 0 and 0.25), MGF strength (Ha between 0 and 20) and horizontal distance of the rotating CCs to the jet entrance (Mx between −8 and 8). EG-A is also conducted for the varying parameters of interest. Hybrid nanoparticles volume fraction is considered as 2%.

Figure 5 shows the impacts of rotational speeds of the CCs on the streamline variation at two different velocity of the hot wall. In the case of non-moving wall and without the activation of the CCs, vortices form near the inlet due to confinement and entrainment, while the right vortex is smaller in size due to the existence of the CC. A large elongated vortex near the right CC is formed, which extends toward the upper plate. When the CCs start to rotate with higher speeds, the sizes of the inlet region vortices are reduced, while vortices near the left and right CC on the bottom wall are established, and their sizes increase with higher speeds. When the hot bottom wall moves in +x direction without rotation of the CCs, a large vortex on the left bottom wall is formed. When the value of Rew is increased, this vortex zone expands in vertical direction while suppression of the inlet vortices with rotations of the CCs is obtained with higher Rew. Due to the wall movement, no vortex is formed on the right part of the inlet near the bottom wall which exists at VR = 0.

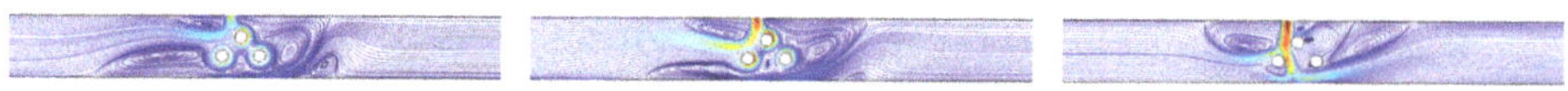

(**a**) Rew = −500, VR = 0 (**b**) Rew = −250, VR = 0 (**c**) Rew = 0, VR = 0

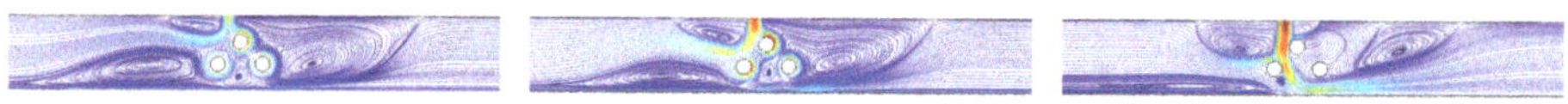

(**d**) Rew = −500, VR = 0.25 (**e**) Rew = −250, VR = 0.25 (**f**) Rew = 0, VR = 0.25

Figure 5. Effects of Rew on streamline variations considering two different velocities of hot wall (Ha = 5, Mx = −1).

Impacts of hot bottom wall velocity on the flow pattern distribution are shown in Figure 6 for two cases of CCs. In the case of non-rotating CCs, on the left part of the bottom wall, a large elongated vortex is formed with higher values of VR. This is attributed to the higher velocity of the hot wall, which induces drag force on the fluid while the flow field is distorted. A vortex is also formed below the left CC, and this region becomes larger with higher VR. When rotations become active, the existence of this elongated vortex on the left part is also seen due to the hot wall movement. Some slight modifications of the vortices near the upper CCs are observed. At highest speed wall velocity (VR = 0.25), rotation of the cylinders results in formation of the large vortices near the right CC adjacent to top wall, while the vortex near the left CC extends in size. The inlet vortex on the left part elongates toward the left due to the rotations of the CCs. Characteristics of the HT are seen in Figure 7 considering the variation of average Nu with respect to changes in Rew and VR. The rotation of the CCs results in reduction of HT performance for CW rotations of the CCs for both stationary (VR = 0) or moving wall (VR = 0.25) cases. In the case of VR = 0, the reduction amount becomes 20%, while it is 12.5% at VR = 0.25 when configurations with rotations CCs at Rew = −500 are compared with the case of non-rotating CCs. The impact of rotation on the HT reduction becomes lower with movement of the hot wall due to the increased HT. When CCW rotations are considered, there is 10% reduction at VR = 0.25, while it becomes 2% increment at VR = 0 when configuration of non-rotating (Rew = 0) and rotating CCs at Rew = 500 are compared. The wall velocity is small compared with the jet velocity, whereas the shear-driven impacts due to wall movement are not significant when the rotations of the CCs are the determining factors. When VR is increased, the average Nu becomes higher for the case of rotating or non-rotating CCs. In both cases, the movement of the wall at the highest velocity contributes 14% increment in the average Nu compared with the stationary wall configuration.

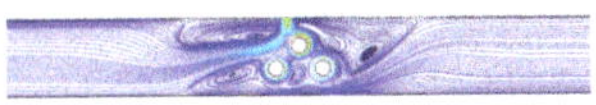

(**a**) Rew = −500, VR = 0

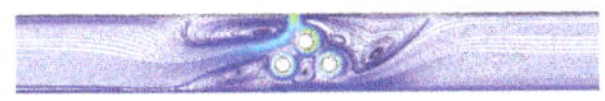

(**b**) Rew = −500, VR = 0.15

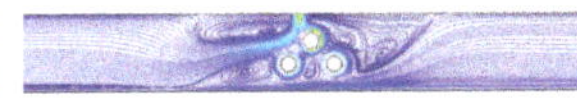

(**c**) Rew = −500, VR = 0.25

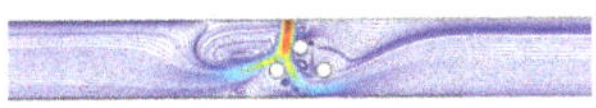

(**d**) Rew = 0, VR = 0

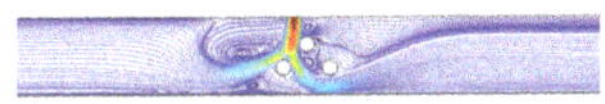

(**e**) Rew = 0, VR = 0.15

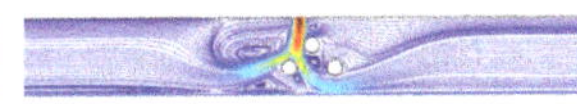

(**f**) Rew = 0, VR = 0.25

Figure 6. Effects of hot wall velocity on streamline variations for rotating and non-rotating CC cases (Ha = 5, Mx = −1).

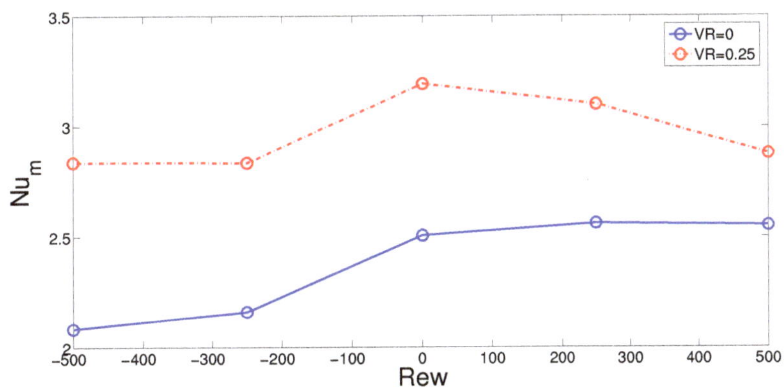

(**a**)

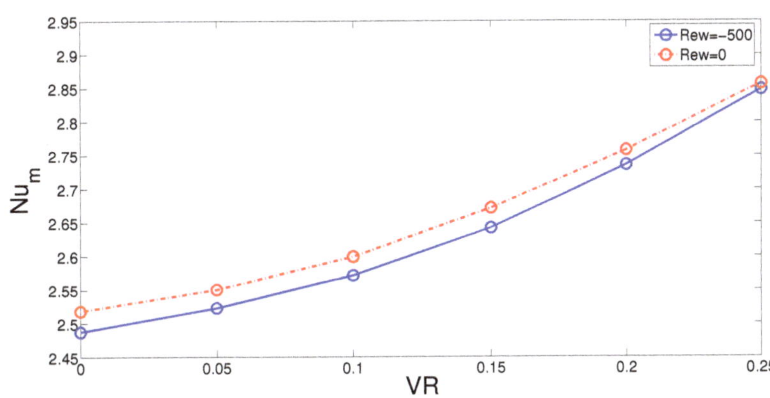

(**b**)

Figure 7. Average Nu versus Rew (**a**) and versus VR (**b**) (Ha = 5, Mx = −1).

The impacts of MGF strength on streamline variations are shown in Figure 8a–h considering both rotating and non-rotating CC cases. When CCs are not rotating (Rew = 0) at Ha = 0, large vortices near the bottom wall (left part), near the right CC on the top wall and near the inlet are established. When MGF strength is increased to Ha = 5, the top and bottom wall vortices are significantly suppressed while at the highest MGF strength (Ha = 20), and most have disappeared. Similar observations can be made with higher MGF strength when the rotations of the CCs become active. In the case of no MGF, two large elongated vortices are formed in the left and right part of the bottom wall that result from the combined utilization of rotation and movement of the hot wall. As the Ha value is increased, the extent of this vortex is largely suppressed. The suppression of vortices due to the complex interaction of the SJ-I, shear effects due to the wall movement and rotation of the CCs are seen by imposing the MGF and increasing its strength. The HT behavior shows different characteristics when imposing the MGF effects depending upon if the rotations are active or not. When the rotations of the CCs are not considered, reduction of the HT is seen until Ha = 10, and after this Ha, it is slightly varying. The amount of HT reduction becomes 28% until Ha = 10 at Rew = 0 (Figure 8i). However, when rotations become active at Rew = −200, the average Nu rises by about 18.6% by increasing the MGF strength at Ha = 5. Further increment of MF strength has little influence on the variation of average Nu. As the rotations of the CCs are activated, the average Nu values are higher until Ha = 10 compared with the non-rotating cylinder case and then become almost identical after this value until Ha = 20. This is attributed to the rotations of the cylinders, which deflect the fluid flow toward the upper wall in the left region of the inlet while effective impinging to the hot wall is hindered. When MGF is used for rotating case, the amount of deflection of this fluid flow toward the upper wall becomes reduced.

The horizontal location of the multiple rotating CCs influence on the flow patterns, as shown in Figure 9a–f considering rotating and non-rotating CCs. When the rotating CCs are away from the inlet jet, large recirculations are formed near the inlet while effective impinging of the jet onto the hot moving plate is obtained. When the rotating CCs are near the inlet jet, due to the rotations of the CCs, inlet vortices are damped out and mechanism of the HT is mainly due to the rotating effects of CCs in contrast to jet impingement HT. When rotations are not used at location Mx = 0, the deflection of the impinging jet on the hot surface is seen due to the presence of stationary bottom cylinder, while below it, a small vortex is established. Average Nu versus location (Mx) of the CCs is shown in Figure 9g. The horizontal location closer to the inlet results in lower HT rates when rotations are active at Rew = −500. Higher Nu values are obtained when the rotating CCs become way from the inlet in horizontal direction. Compared with the case of Mx = 0, average Nu enhancements of 12% and 7.5% are obtained for cases with locations of Mx = −8 and 8. For both rotating and non-rotating CC case, the location of Mx = −2 provides the lowest value of average Nu, where the upper CC is just located below the inlet jet and horizontal distance between the cylinder center and jet center becomes 0. A sudden increment in average Nu is seen when the rotations are not active if Mx is increased from Mx = −2 to Mx = 0, which is not the case with rotations of the CCs due to the rotational effects that hinder the effective impingement of cold fluid onto the hot surface.

EG-A is also considered for the SJ-I problem under combined effects of rotational CCs and MGF. Variations of normalized EG with respect to changes rotational speed of the CCs (Rew) and hot wall velocity (VR) are given in Figure 10. As the velocity of the hot wall is not significant, the shear-driven effects due to wall movement are not profound. The SG values show the dominance of HT irreversibility, while EG generation achieves the minimum value when non-rotating CC case is considered. Higher velocities of the moving wall result in higher entropy production due to the higher HT and viscous irreversibility. At the highest VR, the discrepancy between the rotating and non-rotating increases. The EG rises by about 6.25%/5.25% for non-rotating/rotating cases when comparing the moving wall case (VR = 0.25) with the stationary wall case (VR = 0). The reductions of the EG by activation of rotations in CW direction of the CCs at the highest speed become 18.5% and 16.5% for cases of VR = 0 and VR = 0.25. Depending upon whether the rotations are active or not, the EG shows different behavior with higher MGF strength. For stationary CCs, EG is reduced by about 23% at the highest MGF strength, while it is increased by about 66% for non-stationary CCs at Rew = −200. The behavior shows similar trends as in the average Nu variations due to the higher irreversibility in the HT, as the wall velocity is not significant.

The location of the rotating CCs closer to the inlet position in horizontal direction produces less entropy, while for both cases the EG rises with higher Mx (Figure 11).

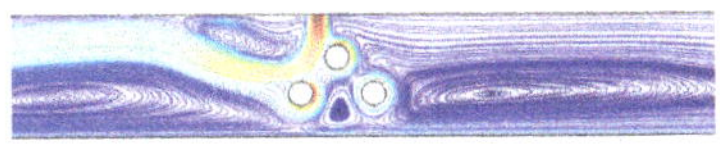

(**a**) Ha = 0, Rew = −200

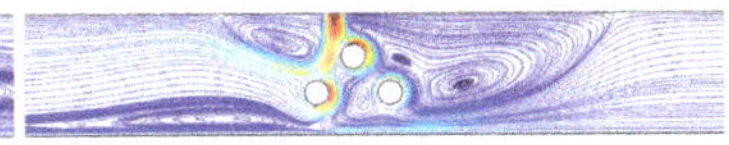

(**b**) Ha = 5, Rew = −200

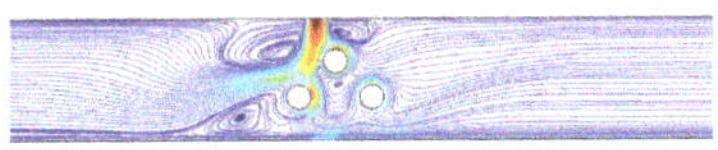

(**c**) Ha = 10, Rew = −200

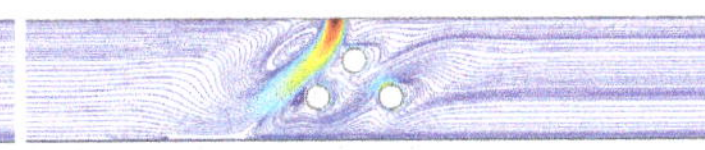

(**d**) Ha = 20, Rew = −200

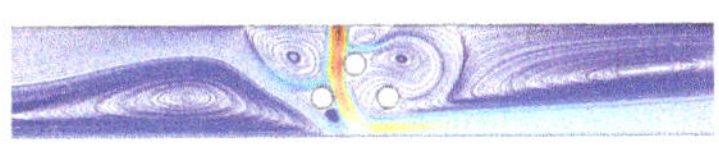

(**e**) Ha = 0, Rew = 0

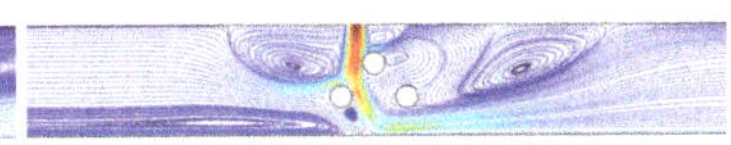

(**f**) Ha = 5, Rew = 0

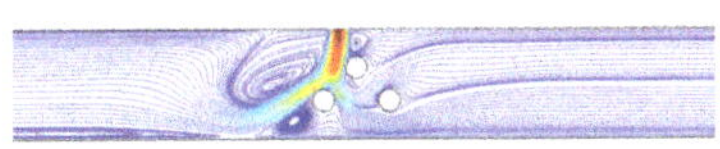

(**g**) Ha = 10, Rew = 0

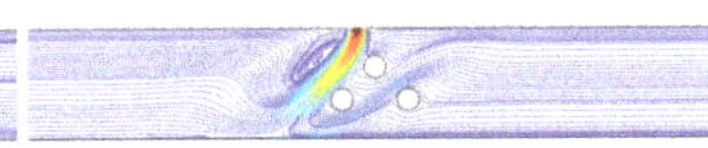

(**h**) Ha = 20, Rew = 0

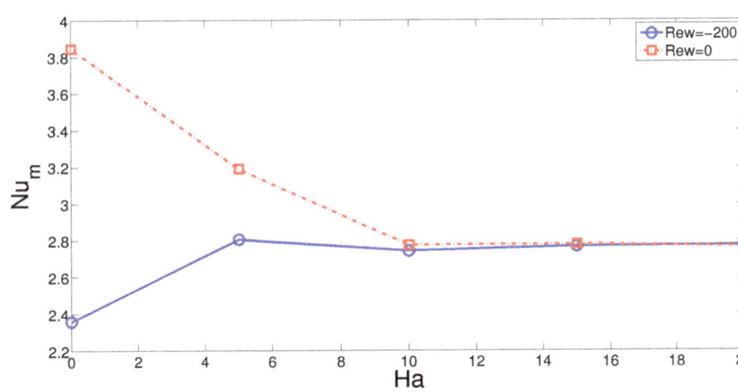

(**i**) Ha = 20, Rew = 0

Figure 8. Impacts of MGF strength on the streamline distributions considering rotating (**a**–**d**), non-rotating CC cases (**e**–**h**) and on the average Nu variations (**i**) (VR = 0.25, Mx = −1).

A summary of different cases in terms of thermal performance and production of entropy is shown in Figure 12. Case 1 (C1) shows the reference configuration for the set of parameters (Rew = 0, VR = 0, Ha = 0, Mx = 0) where non-rotating CCs with stationary hot wall is considered without MGF effects. When rotations are active at Rew = −500, the best case is obtained for (Ha, VR, Mx) = (0, 0.25, −4wj), while thermal performance improvement becomes 8.7%, and without rotation, improvement of up to 34.2% is obtained at (Ha, VR, Mx) = (20, 0.25, −8wj). When no cylinders are installed in the SJ-I system, thermal performance improvements of 14.2% and 5.9% are obtained compared with the reference case at Ha = 0 and Ha = 20. When EG performances are compared, the highest

EG is obtained for case 3, which is 15% higher than the reference case. Case 2 has only 1% higher EG than the reference case. When movement of the wall is not considered, even without installation of CCs, the EG becomes higher either with (Ha = 20) or without MGF effects (Ha = 0).

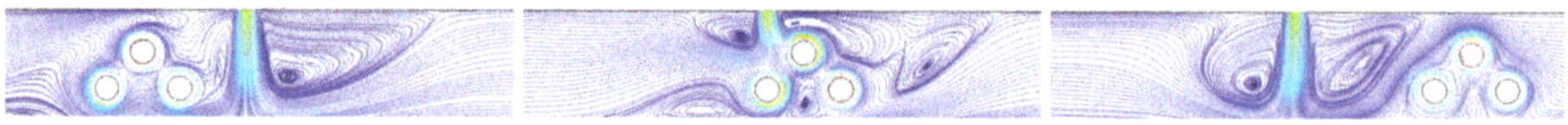

(**a**) Mx = −8wj, Rew = −500 (**b**) Mx = 0, Rew = −500 (**c**) Mx = 8wj, Rew = −500

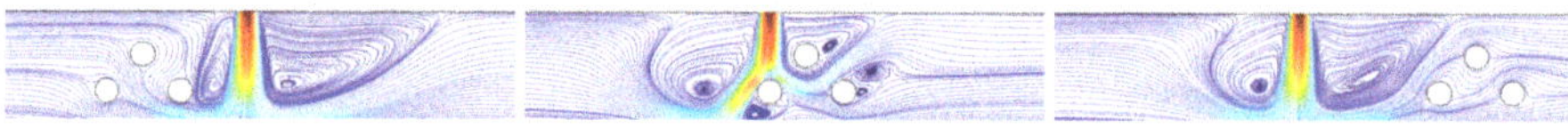

(**d**) Mx = −8wj, Rew = 0 (**e**) Mx = 0, Rew = 0 (**f**) Mx = 8wj, Rew = 0

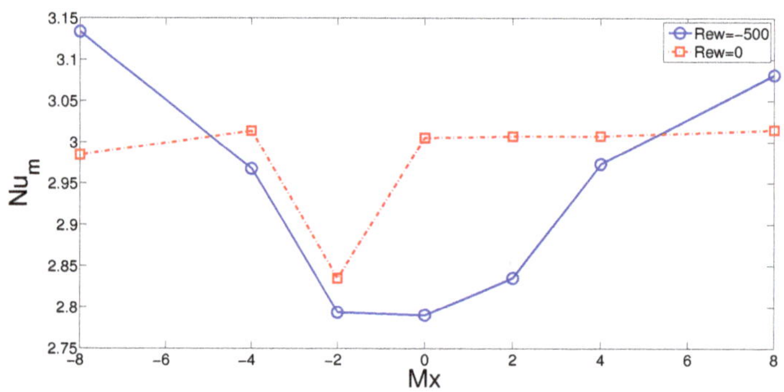

(**g**)

Figure 9. Variations of streamlines with changes in the horizontal location of the multiple CCs considering rotational (**a–c**) and stationary (**d–f**) cases of CCs and impacts of horizontal location of CCs on the variation of average Nu considering rotating and non-rotating CC cases (**g**) (Ha = 5, Mx = −1).

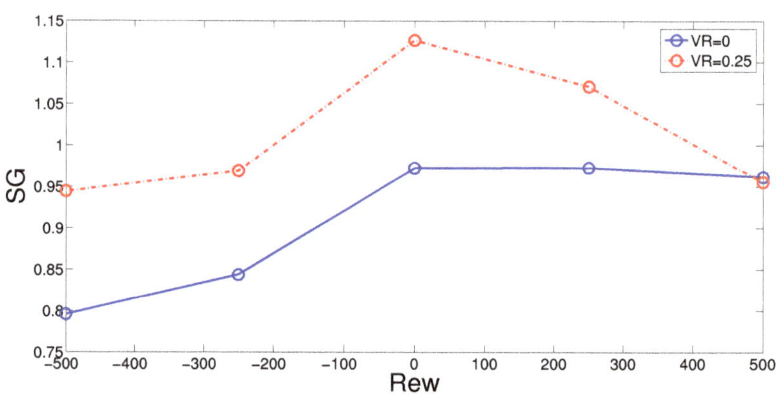

(**a**)

Figure 10. *Cont.*

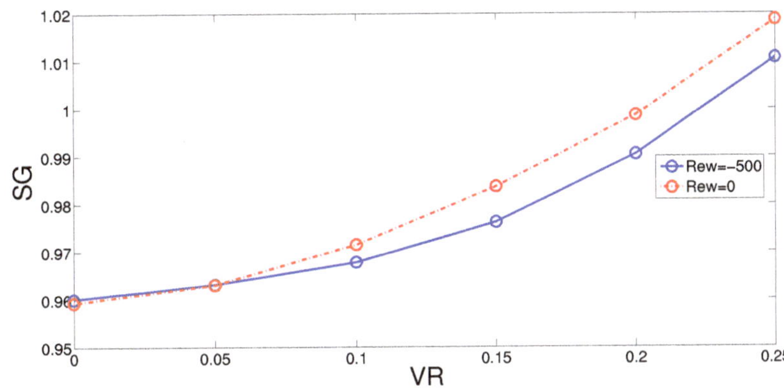

(**b**)

Figure 10. Impacts of rotational Re on the variation of EG for stationary (VR = 0) and moving wall (VR = 0.25) cases (**a**) and impacts of VR on EG for rotating (Rew = −500) and non-rotating CC cases (Rew = 0) CC cases (**b**) (Ha = 5, Mx = −1).

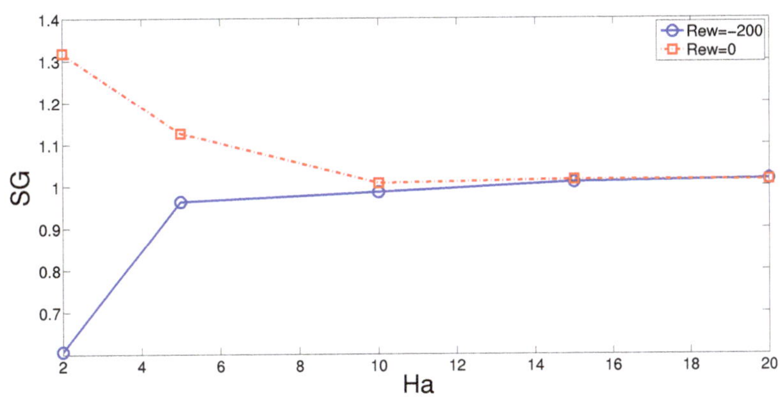

(**a**)

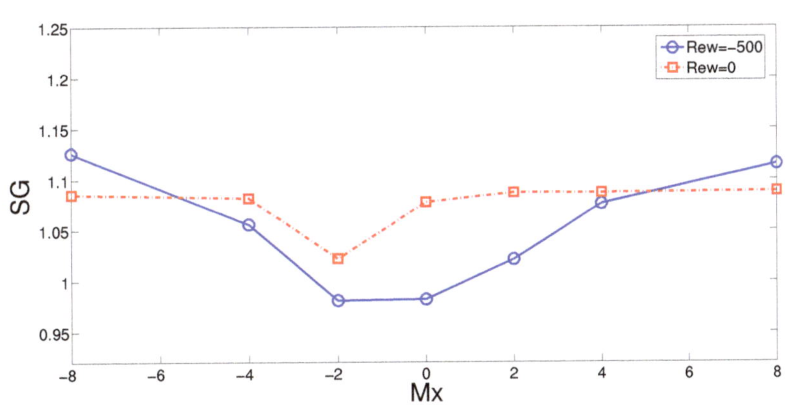

(**b**)

Figure 11. Effects of MGF strength (**a**) horizontal location of the CCs (**b**) on the variation of EG considering rotating and non-rotating CC cases (VR = 0.25, Mx = −1).

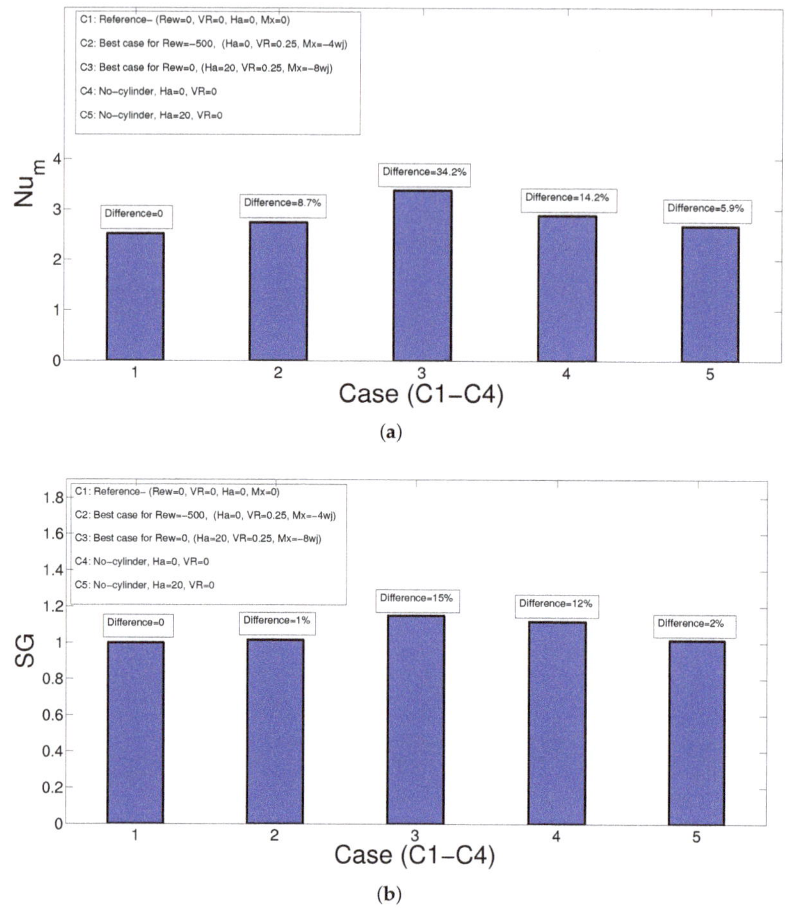

Figure 12. Comparison of best cases in terms of thermal performance (**a**) and EG (**b**) considering both rotating and non-rotating CC case.

4. Conclusions

Convective HT and EG analysis of a SJ-I cooling system for a moving hot wall was conducted under the combined utilization of MGF effects and multiple rotating CCs. Some of the important outcomes can be listed as:

- Rotations of the CCs near the jet inlet have negative impacts of the HT enhancement for both stationary and moving hot wall cases. Reductions in the average Nu of up to 20% and 12.5% are obtained by using rotations at Rew = −500 for stationary (V = 0) and moving wall (VR = 0.25) cases.
- The wall movement contributes positively to the cooling performance while HT enhancements up to 14% are achieved by wall velocity at the highest speed (VR = 0.25).
- Depending upon the activation of cylinder rotations, the impacts of MGF strength on the HT characteristics are different. For non-rotating CCs, cooling performance is reduced by about 28% until Ha = 10, while by using rotations at Rew = −200, it is increased by about 18.6%.
- For the rotating CC case, average Nu variations up to 12% can be achieved by varying the horizontal location of the CCs while away from the inlet, higher cooling performances are obtained.

- When the hot wall starts to move at VR = 0.25, the EG increments up to 6.25% and 5.25% are obtained for non-rotating and rotating CC cases compared with the stationary wall (VR = 0) configuration.
- The MGF acts to reduce the EG by about 23% for non-rotating cylinders while increment of EG by about 66% is obtained for rotating cylinders at Rew = 200. Away from the jet inlet, the EG rises.
- The best configuration for the case of non-rotating CCs is achieved at (Ha, VR, Mx) = (0, 0.25, −4), and HT increment becomes 8.7% compared with the reference case of (Rew = 0, VR = 0, Ha = 0, Mx = 0). For the non-rotating CC case, the optimum set of parameters is achieved at (Ha, VR, Mx) = (20, 0.25, −8) with HT enhancement of 34.2% compared with the reference.
- The maximum EG is obtained for configuration with (Ha, VR, Mx) = (20, 0.25, −8) with non-rotating CCs, while the value is 15% higher than the reference case. When rotations are active at Rew = −500, the case (Ha, VR, Mx) = (0, 0.25, −4) has only 1% higher EG when compared to reference.

Author Contributions: Methodology, F.S. and K.G.; Software, L.K.; Validation, F.S., S.L., H.A. and B.M.A.; Formal analysis, F.S., K.G. and B.M.A.; Investigation, L.K., S.L. and T.L.; Writing—original draft, F.S.; Writing—review & editing, L.K., F.S., K.G. and H.A.; Visualization, T.L. All authors have read and agreed to the published version of the manuscript.

Funding: This research project was funded by the Deanship of Scientific Research, Princess Nourah bint Abdulrahman University, through the Program of Research Project Funding After Publication, grant No (44- PRFA-P-16).

Data Availability Statement: Not applicable.

Conflicts of Interest: The authors declare no conflict of interest.

Nomenclature

B_0	Magnetic field strength
Ha	Hartmann number
H	separating distance
h	heat transfer coefficient
k	thermal conductivity
L	plate length
n	unit normal vector
Nu_s	local Nusselt number
Nu_m	average Nusselt number
p	pressure
Pr	Prandtl number
r	cylinder radius
Re	Reynolds number
Rew	rotational Reynolds number
T	temperature
u, v	x-y velocity components
uc	jet velocity
uw	wall velocity
VR	velocity ratio
wj	slot width
x, y	Cartesian coordinates
Greek Characters	
α	thermal diffusivity
γ	magnetic field inclination
θ	non-dimensional temperature
ν	kinematic viscosity
ρ	density of the fluid
σ	electrical conductivity

ϕ	solid volume fraction
ω	rotational speed
Subscripts	
c	cold
h	hot
m	average
nf	nanofluid
w	wall

References

1. Javidan, M.; Moghadam, A.J. Experimental investigation on thermal management of a photovoltaic module using water-jet impingement cooling. *Energy Convers. Manag.* **2021**, *228*, 113686. [CrossRef]
2. Bahaidarah, H.M. Experimental performance evaluation and modeling of jet impingement cooling for thermal management of photovoltaics. *Sol. Energy* **2016**, *135*, 605–617. [CrossRef]
3. Selimefendigil, F.; Coban, S.O.; Öztop, H.F. Comparative study and hybrid modeling approach with POD for convective drying performance of porous moist object with multi-impinging jet and channel flow configurations. *Int. Commun. Heat Mass Transf.* **2022**, *132*, 105897. [CrossRef]
4. Jambunathan, K.; Lai, E.; Moss, M.; Button, B. A review of heat transfer data for single circular jet impingement. *Int. J. Heat Fluid Flow* **1992**, *13*, 106–115. [CrossRef]
5. Webb, B.; Ma, C. Single-Phase Liquid Jet Impingement Heat Transfer. *Adv. Heat Transf.* **1995**, *26*, 105–217.
6. Tyagi, P.K.; Kumar, R.; Mondal, P.K. A review of the state-of-the-art nanofluid spray and jet impingement cooling. *Phys. Fluids* **2020**, *32*, 121301. [CrossRef]
7. Kakaç, S.; Pramuanjaroenkij, A. Single-phase and two-phase treatments of convective heat transfer enhancement with nanofluids–A state-of-the-art review. *Int. J. Therm. Sci.* **2016**, *100*, 75–97. [CrossRef]
8. Pordanjani, A.H.; Aghakhani, S.; Afrand, M.; Mahmoudi, B.; Mahian, O.; Wongwises, S. An updated review on application of nanofluids in heat exchangers for saving energy. *Energy Convers. Manag.* **2019**, *198*, 111886. [CrossRef]
9. Chamkha, A.J.; Molana, M.; Rahnama, A.; Ghadami, F. On the nanofluids applications in microchannels: A comprehensive review. *Powder Technol.* **2018**, *332*, 287–322. [CrossRef]
10. Sheikholeslami, M.; Rokni, H.B. Simulation of nanofluid heat transfer in presence of magnetic field: A review. *Int. J. Heat Mass Transf.* **2017**, *115*, 1203–1233. [CrossRef]
11. M'hamed, B.; Sidik, N.A.C.; Yazid, M.N.A.W.M.; Mamat, R.; Najafi, G.; Kefayati, G. A review on why researchers apply external magnetic field on nanofluids. *Int. Commun. Heat Mass Transf.* **2016**, *78*, 60–67. [CrossRef]
12. Mohammadpour, J.; Lee, A. Investigation of nanoparticle effects on jet impingement heat transfer: A review. *J. Mol. Liq.* **2020**, *316*, 113819. [CrossRef]
13. Li, P.; Guo, D.; Liu, R. Mechanism analysis of heat transfer and flow structure of periodic pulsating nanofluids slot-jet impingement with different waveforms. *Appl. Therm. Eng.* **2019**, *152*, 937–945. [CrossRef]
14. Manca, O.; Mesolella, P.; Nardini, S.; Ricci, D. Numerical study of a confined slot impinging jet with nanofluids. *Nanoscale Res. Lett.* **2011**, *6*, 188. [CrossRef] [PubMed]
15. Yousefi-Lafouraki, B.; Ramiar, A.; Mohsenian, S. Entropy generation analysis of a confined slot impinging jet in a converging channel for a shear thinning nanofluid. *Appl. Therm. Eng.* **2016**, *105*, 675–685. [CrossRef]
16. Nakharintr, L.; Naphon, P. Magnetic field effect on the enhancement of nanofluids heat transfer of a confined jet impingement in mini-channel heat sink. *Int. J. Heat Mass Transf.* **2017**, *110*, 753–759. [CrossRef]
17. Selimefendigil, F.; Öztop, H.F. Hybrid nano-jet impingement cooling of a curved elastic hot surface under the combined effects of non-uniform magnetic field and upper plate inclination. *J. Magn. Magn. Mater.* **2022**, *561*, 169684. [CrossRef]
18. Hashemi-Tilehnoee, M.; del Barrio, E.P. Magneto laminar mixed convection and entropy generation analyses of an impinging slot jet of Al2O3-water and Novec-649. *Therm. Sci. Eng. Prog.* **2022**, *36*, 101524. [CrossRef]
19. Roslan, R.; Saleh, H.; Hashim, I. Effect of rotating cylinder on heat transfer in a square enclosure filled with nanofluids. *Int. J. Heat Mass Transf.* **2012**, *55*, 7247–7256. [CrossRef]
20. Costa, V.A.F.; Raimundo, A.M. Steady mixed convection in a differentially heated square enclosure with an active rotating circular cylinder. *Int. J. Heat Mass Transf.* **2010**, *53*, 1208–1219. [CrossRef]
21. Selimefendigil, F.; Öztop, H.F. Mixed convection in a PCM filled cavity under the influence of a rotating cylinder. *Sol. Energy* **2020**, *200*, 61–75. [CrossRef]
22. Jiang, L.; Lyu, Y.; Zhu, P.; Gao, W.; Liu, Z. Numerical investigation of conjugate heat transfer on a rotating disk under round liquid jet impingement. *Int. J. Therm. Sci.* **2021**, *170*, 107097. [CrossRef]
23. Lallave, J.C.; Rahman, M.M.; Kumar, A. Numerical analysis of heat transfer on a rotating disk surface under confined liquid jet impingement. *Int. J. Heat Fluid Flow* **2007**, *28*, 720–734. [CrossRef]
24. Iacovides, H.; Kounadis, D.; Launder, B.E.; Li, J.; Xu, Z. Experimental study of the flow and thermal development of a row of cooling jets impinging on a rotating concave surface. *J. Turbomach.* **2005**, *127*, 222–229. [CrossRef]

25. Selimefendigil, F.; Öztop, H.F. Analysis and predictive modeling of nanofluid-jet impingement cooling of an isothermal surface under the influence of a rotating cylinder. *Int. J. Heat Mass Transf.* **2018**, *121*, 233–245. [CrossRef]
26. Bejan, A. Fundamentals of exergy analysis, entropy generation minimization, and the generation of flow architecture. *Int. J. Energy Res.* **2002**, *26*. [CrossRef]
27. Bejan, A. *Entropy Generation Minimization: The Method of Thermodynamic Optimization of Finite-Size Systems and Finite-Time Processes*; CRC Press: Boca Raton, FL, USA, 2013.
28. Narayan, G.P.; Lienhard V, J.H.; Zubair, S.M. Entropy generation minimization of combined heat and mass transfer devices. *Int. J. Therm. Sci.* **2010**, *49*, 2057–2066. [CrossRef]
29. Tayebi, T.; Dogonchi, A.S.; Karimi, N.; Ge-JiLe, H.; Chamkha, A.J.; Elmasry, Y. Thermo-economic and entropy generation analyses of magnetic natural convective flow in a nanofluid-filled annular enclosure fitted with fins. *Sustain. Energy Technol. Assess.* **2021**, *46*, 101274. [CrossRef]
30. Ellahi, R.; Sait, S.M.; Shehzad, N.; Ayaz, Z. A hybrid investigation on numerical and analytical solutions of electro-magnetohydrodynamics flow of nanofluid through porous media with entropy generation. *Int. J. Numer. Methods Heat Fluid Flow* **2020**, *30*, 834–854. [CrossRef]
31. Li, J.; Kleinstreuer, C. Entropy generation analysis for nanofluid flow in microchannels. *J. Heat Transf.* **2010**, *132*, 122401. [CrossRef]
32. Mahian, O.; Kianifar, A.; Kleinstreuer, C.; Mohd A, A.N.; Pop, I.; Sahin, A.Z.; Wongwises, S. A review of entropy generation in nanofluid flow. *Int. J. Heat Mass Transf.* **2013**, *65*, 514–532. [CrossRef]
33. Huminic, G.; Huminic, A. Entropy generation of nanofluid and hybrid nanofluid flow in thermal systems: A review. *J. Mol. Liq.* **2020**, *302*, 112533. [CrossRef]
34. Aghakhani, S.; Pordanjani, A.H.; Afrand, M.; Sharifpur, M.; Meyer, J.P. Natural convective heat transfer and entropy generation of alumina/water nanofluid in a tilted enclosure with an elliptic constant temperature: Applying magnetic field and radiation effects. *Int. J. Mech. Sci.* **2020**, *174*, 105470. [CrossRef]
35. Barnoon, P.; Toghraie, D.; Eslami, F.; Mehmandoust, B. Entropy generation analysis of different nanofluid flows in the space between two concentric horizontal pipes in the presence of magnetic field: Single-phase and two-phase approaches. *Comput. Math. Appl.* **2019**, *77*, 662–692. [CrossRef]
36. Selimefendigil, F.; Öztop, H.F.; Chamkha, A.J. MHD mixed convection and entropy generation of nanofluid filled lid driven cavity under the influence of inclined magnetic fields imposed to its upper and lower diagonal triangular domains. *J. Magn. Magn. Mater.* **2016**, *406*, 266–281. [CrossRef]
37. Reddy, J.N.; Gartling, D.K. *The Finite Element Method in Heat Transfer and Fluid Dynamics*; CRC Press: Boca Raton, FL, USA, 2010.
38. Lewis, R.W.; Nithiarasu, P.; Seetharamu, K.N. *Fundamentals of the Finite Element Method for Heat and Fluid Flow*; John Wiley & Sons: West Sussex, UK, 2004.
39. Das, S.; Biswas, A.; Das, B. Numerical analysis of a solar air heater with jet impingement?comparison of performance between jet designs. *J. Sol. Energy Eng.* **2022**, *144*, 011001. [CrossRef]
40. Froissart, M.; Ziółkowski, P.; Dudda, W.; Badur, J. Heat exchange enhancement of jet impingement cooling with the novel humped-cone heat sink. *Case Stud. Therm. Eng.* **2021**, *28*, 101445. [CrossRef]
41. Selimefendigil, F.; Kolsi, L.; Ayadi, B.; Aich, W.; Alresheedi, F.; Borjini, M.N. Jet impingement cooling using shear thinning nanofluid under the combined effects of inclined separated partition at the inlet and magnetic field. *Eur. Phys. J. Spec. Top.* **2022**, *231*, 2491–2508. [CrossRef]
42. Zhang, T.; Che, D. Double MRT thermal lattice Boltzmann simulation for MHD natural convection of nanofluids in an inclined cavity with four square heat sources. *Int. J. Heat Mass Transf.* **2016**, *94*, 87–100. [CrossRef]
43. Manca, O.; Ricci, D.; Nardini, S.; Di Lorenzo, G. Thermal and fluid dynamic behaviors of confined laminar impinging slot jets with nanofluids. *Int. Commun. Heat Mass Transf.* **2016**, *70*, 15–26. [CrossRef]
44. Chou, Y.; Hung, Y. Impingement cooling of an isothermally heated surface with a confined slot jet. *ASME Trans. J. Heat Transf.* **1994**, *116*, 479–482. [CrossRef]

mathematics

MDPI

Article

Exergy Analysis for Combustible Third-Grade Fluid Flow through a Medium with Variable Electrical Conductivity and Porous Permeability

Samuel O. Adesanya [1,2,3,*], Peace O. Banjo [1] and Ramoshweu S. Lebelo [2,3]

1 Department of Mathematics and Statistics, Faculty of Natural Sciences, Redeemer's University, Ede 232101, Osun State, Nigeria
2 Education Department, Vaal University of Technology, Private Bag X021, Vanderbijlpark 1911, South Africa
3 Hydrodynamics Unit, African Center of Excellence for Water and Environmental Research (ACEWATER), Redeemer's University, Ede 232101, Osun State, Nigeria
* Correspondence: adesanyas@run.edu.ng

Abstract: A mathematical investigation of a thermodynamical system linked with energy management and its impact on the environment, especially climate change, is presented in this study. In this regard, a numerical investigation of the flow and heat transfer of hydromagnetic third-grade liquid through a porous medium. The permeability of the medium and electrical conductivity of the fluid are assumed to be temperature functions. The appropriate mathematical formulations for momentum, energy, and entropy equations are presented in both dimensional and dimensionless forms. We obtained the numerical solutions using the spectral version of the Chebyshev collocation method and compared the result with the shooting Runge–Kutta method. Numerical results for velocity, temperature, entropy, and Bejan profiles are communicated through tables and graphs with adequate physical interpretation. The thermal stability of the thermo-fluid system that guarantees the prevention of spontaneous fluid heating that fuels climate change is also included in the analysis.

Keywords: variable electrical conductivity; third-grade fluid; variable porous permeability; thermal stability; entropy analysis

MSC: 76-10

Citation: Adesanya, S.O.; Banjo, P.O.; Lebelo, R.S. Exergy Analysis for Combustible Third-Grade Fluid Flow through a Medium with Variable Electrical Conductivity and Porous Permeability. *Mathematics* **2023**, *11*, 1882. https://doi.org/10.3390/math11081882

Academic Editor: Efstratios Tzirtzilakis

Received: 22 March 2023
Revised: 4 April 2023
Accepted: 10 April 2023
Published: 15 April 2023

1. Introduction

One typical relationship between some of the Sustainable Development Goals (SDG), including industrialization, a clean environment, and climate change is that of energy usage. In this context, the impact of thermodynamics in thermal engineering and other energy generation settings cannot be overemphasized due to the interconnectivity between heat generation, dissipation, and its net effect on climate change. Over the last few decades, the thermodynamics analysis of third-grade fluid (TGF) flow through a porous medium has been of interest to researchers, scientists, and engineers, due to its numerous and diverse applications in nature. The study finds its application in several branches of agriculture, science, and engineering, to mention just a few. Based on the aforementioned geophysical importance, Adesanya et al. [1] reported thermal analysis for a reactive third-grade liquid through a non-Darcian medium bounded by Riga walls subjected to Newtonian cooling. By applying the rapidly converging homotopy analysis method, Sajid and Hayat [2] presented the solution to a third-grade fluid flow in a porous channel filled with permeable materials by applying a modified Darcy law. Makinde et al. [3] reported a numerical approach to solving unsteady, fully developed flow problems in variable viscous TGF through a porous medium subjected to asymmetrical convective heating in which the fluid undergoes an exothermic chemical reaction. Hayat et al. [4] discussed the steady, fully developed flow of third-grade liquid in a porous space under no-slip and non-moving wall conditions using a homotopy analysis approach. Rundora and Makinde [5] examined the influence of vertical penetration on reactive TGF flow through a Darcian medium under heat-dependent viscosity. Baoku et al. [6] studied numerical solutions to heat and mass transfer in a boundary layer flow of a third-grade fluid flow in an enclosed porous region. Adesanya and Falade [7] analyzed the heat irreversibility inherent in the heat transfer of TGF through a porous medium using the perturbation method. Salawu and Fatunmbi [8] investigated the inherent heat irreversibility in the convective flow of variable, viscous, third-grade combustible liquid experiencing a transverse magnetic field. Magnhsoudi et al. [9] constructed an analytical solution to the heat transfer problem in TGF flowing steadily through a medium with flow barriers by applying the weighted residual least square method. Readers can see other exciting results on TGF through a restricted medium in reference [10–13] and references cited within the work.

Over the last few decades, studies on electrical conducting fluids are becoming more popular due to their wide range of applications in hydroponics, aquaponics, aquaculture, electrolytes, polymers, molten metals, and many more that are too numerous to be listed. Based on a wide application, Rahman et al. [14] used the linear dependence of electrical conductivity on flow velocity to obtain a numerical approximation of a micropolar fluid flow over an infinitely long inclined plane with a variable heat source. Additionally, Makinde and Onyejekwe [15] considered the heat-dependent electrical conductivity of the power law type for the flow and thermal analysis of a time-independent Couette flow. Hossain and Gorla [16] presented another variant of electrical conductivity relation based on free steam and tangential velocity for the developing flow analysis of hydromagnetic liquids. In a related study by Eguia et al. [17], electrical conductivity was assumed to be a linear function of temperature for unsteady dusty flow analysis. Similarly, Sivaraj and Kumar [18] studied the unsteady developing flow of reacting Walter-B fluid along a vertical cone. Eegunjobi and Makinde [19] utilized the power law dependence of electrical conductivity on temperature to study the hydromagnetic slip flow between leaking walls. Salawu et al. [20] presented the heat-dependent electrical conductivity property of an unsteady flow of Eyring–Powell fluid undergoing Arrhenius kinetics in a non-Darcian setting. Obalalu et al. [21] analyzed the convective magnetohydrodynamic flow of Casson nanofluid subjected to an exothermic chemical reaction. Adeosun and Ukaegbu [22] considered the squeezed flow of a reactive fluid experiencing variable electrical conductivity. The literature is inexhaustive when considering the variable electrical conductivity property of hydromagnetic fluid.

Motivated by the studies in [14–22], the first interest is in investigating variable electrical conductivity's influence on the flow of third-grade liquid in the porous medium. Secondly, the studies in [1–9] assumed constant porous permeability. In the real sense, the permeability of any porous medium allowing the passage of viscous fluid depends on temperature, pressure/stress field, and non-homogeneity of the permeable material used. For example, in oil/well engineering, the flows of polymetric fluids in oil recovery/steam injection in petroleum engineering, groundwater, oil in geological flows, some areas involving water seepage in agricultural engineering, and lots more. As a result, the main objective of this paper is to study the steady flow of hydromagnetic third-grade fluid through a porous medium with temperature-dependent porous permeability and electrical conductivity. The problem will be formulated in the following section with some mathematical analysis. Section 3 will be dedicated to the numerical method of solution, and in the Section 4, the results will be presented and discussed while the Section 5 concludes the article.

2. Mathematical Formulation

This work studies the steady, unidirectional, fully developed flow of an electrically conducting, pressure-driven, third-grade fluid through a porous medium with no vertical wall penetration. The third-grade liquid is assumed to undergo a strong exothermic chemical reaction. A magnetic field of intensity B_0 is applied across the horizontal channel.

It is further assumed that both electrical conductivity and porous permeability of the medium are nonlinear functions of temperature as shown in the flow geometry in Figure 1.

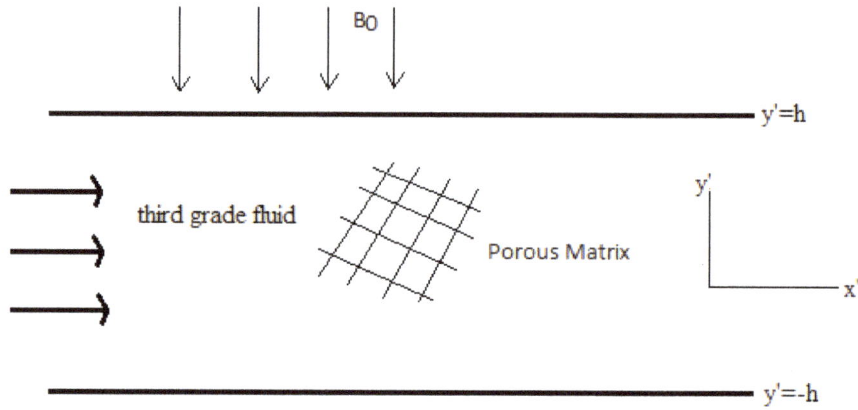

Figure 1. Flow geometry.

Neglecting the unsteadiness or temporal changes and the convective components of acceleration, the appropriate balanced pressure and viscous forces driving the flow can be written as:

$$0 = -\frac{dP'}{dx'} + \frac{d}{dy'}\left(\mu\frac{du'}{dy'} + 2\beta_3\left(\frac{du'}{dy'}\right)^3\right) - \left(\mu + 2\beta_3\left(\frac{du'}{dy'}\right)^2\right)\frac{u'}{K(T)} - \sigma(T)B_0^2 u', \quad (1)$$

and the energy balance equation

$$0 = k\frac{d^2T}{dy'^2} + QC_0A\left(\frac{\hbar T}{\upsilon \iota}\right)^m e^{-\frac{E}{RT}} + \left(\mu + 2\beta_3\left(\frac{du'}{dy'}\right)^2\right)\left(\frac{du'}{dy'}\right)^2 + \left(\left(\mu + 2\beta_3\left(\frac{du'}{dy'}\right)^2\right)\frac{u'}{K(T)} + \sigma(T)B_0^2\right)u'^2, \quad (2)$$

Alongside the non-moving wall and no-slip conditions at the solid boundaries,

$$\begin{array}{ll} u' = 0, \quad T = T_a, & y' = a; \\ u' = 0, \quad T = T_a, & y' = -a. \end{array} \quad (3)$$

While the wall shear stress for the determination of skin friction and heat transfer rate can be written as

$$\tau_{xy} = \mu\frac{du'}{dy'} + 2\beta_3\left(\frac{du'}{dy'}\right)^3\Bigg|_{y=-1}, \quad q_w = -k\frac{dT}{dy}\Bigg|_{y=-1} \quad (4)$$

Entropy changes from the spontaneous process due to heat transfer and viscous interaction can be written as

$$E_G = \frac{k}{T_0^2}\left(\frac{dT'}{dY'}\right)^2 + \frac{1}{T_0}\left\{\left(\mu + 2\beta_3\left(\frac{du'}{dy'}\right)^2\right)\left(\frac{du'}{dy'}\right)^2 + \left(\frac{1}{K(T)}\left(\mu + 2\beta_3\left(\frac{du'}{dy'}\right)^2\right) + \sigma(T)B_0^2\right)u'^2\right\} \quad (5)$$

The first part of (5) is the heat transfer component of heat irreversibility while the other part arises from viscous interaction. The permeability of the porous medium is assumed to vary slightly with temperature, i.e., the thermal effect on permeability is of the form:

$$K(T) = K_0 e^{\overline{\alpha}(T-T_0)} \approx K_0(1 + \overline{\alpha}(T - T_0)) + H.O.T, \quad 0 < \overline{\alpha} << 1 \quad (6)$$

While the dependence of electrical conductivity on temperature is given by [20–22]:

$$\sigma(T) = \sigma_0 \left(\frac{E(T - T_0)}{T_0^2 R} \right)^r, \tag{7}$$

here, $\bar{\alpha}$ represents the coefficient of the temperature difference and r is the exponent of temperature, x', y' represents the Cartesian coordinates of the channel, (P, μ, β_3) are fluid pressure, viscosity, and non-Newtonian material effect, $(u', T, K(T))$ are the dimensional velocity, temperature, and porous permeability. $(\sigma(T), B_0, k)$ are the electrical conductivity, magnetic field intensity, and thermal conductivity, (Q, C_0, A) represents the heat of reaction, initial concentration, and rate constant. (ι, m, v) are Plank's constant, reaction exponent, and frequency of vibration. Using the following dimensionless parameters and variables,

$$\begin{array}{l} y = \frac{y'}{a},\ u = \frac{u'}{UG},\ \theta = \frac{E(T-T_0)}{RT_0^2},\ Ha^2 = \frac{\sigma_0 B_0^2 a^2}{\mu_0},\ \varepsilon = \frac{RT_0}{E}, \\ Da = \frac{K_0}{a^2},\ S^2 = \frac{1}{Da},\ G = -\frac{a^2}{\mu_0 U}\frac{dP}{dx},\ \alpha = \frac{\bar{\alpha} R T_0^2}{E},\ Ns = \frac{a^2 E^2}{kR^2 T_0^2}, \kappa = \frac{\beta_3 U^2 G^2}{\mu_0 a^2}, \\ \lambda = \left(\frac{hT_0}{v\iota}\right)^m \frac{QEAa^2 C_0}{kRT_0^2} e^{-\frac{E}{RT_0}}, \gamma = \left(\frac{v\iota}{hT_0}\right)^m \frac{\mu_0 U^2 G^2}{QAC_0 a^2} e^{\frac{E}{RT_0}} \end{array} \tag{8}$$

we arrive at the following dimensionless nonlinear and coupled boundary-value problem:

$$\left.\begin{array}{l} 0 = 1 + \frac{d}{dy}\left(\frac{du}{dy} + 2\kappa\left(\frac{du}{dy}\right)^3\right) - \left(\left(1 + 2\kappa\left(\frac{du}{dy}\right)^2\right)(1-\alpha\theta)S^2 + H^2\theta^r\right)u;\ \ u(\pm 1) = 0 \\ 0 = \frac{d^2\theta}{dy^2} \\ +\lambda\left((1+\varepsilon\theta)^m e^{\frac{\theta}{1+\varepsilon\theta}} + \gamma\left\{\left(\frac{du}{dy}\right)^2\left(1 + 2\kappa\left(\frac{du}{dy}\right)^2\right) + \left(\left(1 + 2\kappa\left(\frac{du}{dy}\right)^2\right)(1-\alpha\theta)S^2 + H^2\theta^r\right)u^2\right\}\right); \\ \qquad \theta(\pm 1) = 0 \\ N_S = \left(\frac{d\theta}{dy}\right)^2 + \frac{\lambda\gamma}{\varepsilon}\left(\left(\frac{du}{dy}\right)^2\left(1 + 2\kappa\left(\frac{du}{dy}\right)^2\right) + \left(\left(1 + 2\kappa\left(\frac{du}{dy}\right)^2\right)(1-\alpha\theta)S^2 + H^2\theta^r\right)u^2\right). \end{array}\right\} \tag{9}$$

The contribution of each parameter to the entropy profile can be monitored using the ratio:

$$Be = \frac{\left(\frac{d\theta}{dy}\right)^2}{\left(\frac{d\theta}{dy}\right)^2 + \frac{\lambda\gamma}{\varepsilon}\left(\left(\frac{du}{dy}\right)^2\left(1 + 2\kappa\left(\frac{du}{dy}\right)^2\right) + \left(\left(1 + 2\kappa\left(\frac{du}{dy}\right)^2\right)(1-\alpha\theta)S^2 + H^2\theta^r\right)u^2\right)} \tag{10}$$

The dimensionless quantities (u, θ, G) are velocity, temperature, and pressure gradient, (H, ε, Da) are Hartmann number, activation energy parameter, and Darcy number, (S^2, α, Ns) are shape factor, coefficient of electrical conductivity, and dimensionless entropy generation, (κ, λ, Be) are the third-grade parameter, Frank–Kameneskii parameter, and Bejan ratio, respectively.

3. Spectral Collocation Method of Solution

To obtain the solution to the coupled boundary-value problem (9), we apply the idea of the dense set to take a polynomial approximation as suggested in the Weierstrass approximation theorem for the existence of a solution. In this way, we assume that the solution of (9) can be approximated by taking:

$$\left.\begin{array}{l} u(y) \approx u^N(y) = \sum\limits_{j=0}^{N} b_j \Phi_j(y), \\ \theta(y) \approx \theta^N(y) = \sum\limits_{j=0}^{N} c_j \Phi_j(y) \end{array}\right\} \tag{11}$$

where $\Phi_j(y)$ represents spectral Chebyshev polynomials and (b_j, c_j) are the unknown coefficients to be determined. In this way, the residues that denote the difference between the exact solution of (9) and the approximated solutions are given by

$$\left.\begin{array}{l} R_1 = 1 + \left(u_y^N + 2\kappa\left(u_y^N\right)^3\right)_y - \left(\left(1+2\kappa\left(u_y^N\right)^2\right)(1-\alpha\theta^N)S^2 + H^2\theta^{Nr}\right)u^N, \\ R_2 = \theta_{yy}^N + \lambda\left(\left(1+\varepsilon\theta^N\right)^m e^{\frac{\theta^N}{1+\varepsilon\theta^N}} + \gamma\left\{\left(u_y^N\right)^2\left(1+2\kappa\left(u_y^N\right)^2\right) + \left(\left(1+2\kappa\left(u_y^N\right)^2\right)(1-\alpha\theta^N)S^2 + H^2\theta^{Nr}\right)u^{2N}\right\}\right) \end{array}\right\} \tag{12}$$

with

$$u^N(-1) = 0 = u^N(1), \theta_y^N\Big|_{y=-1} = 0 = \theta_y^N\Big|_{y=1}. \tag{13}$$

where y_i are points within $[-1,1] = [y_0, y_N]$. Then the Gauss–Lobato points for the collocation points are

$$y_j = -\cos\left(\frac{j\pi}{N}\right), \quad j = 0,1,2,\ldots,N. \tag{14}$$

Which are evaluated at

$$R_1(y_j) = 0 = R_2(y_j), \quad j = 0,1,2,\ldots,N. \tag{15}$$

The derivatives for dependent variables are obtained as

$$\frac{d^r u}{dy^r} = \sum_{j=0}^{N} b_i \frac{d^r u_j}{dy^r} \text{ and } \frac{d^r \theta}{dy^r} = \sum_{j=0}^{Np} c_j \frac{d^r \theta_j}{dy^r}. \tag{16}$$

The differentiation matrices at each Gauss–Lobato point are

$$\frac{d^r \overline{u}}{d^r y} = D^{(r)}\overline{u} = D^r\overline{u} \;\; r = 1,2,3,\ldots \text{ and } \frac{d^r\overline{\theta}}{d^r y} = D^{(r)}\overline{\theta} = D^r\overline{\theta}, \;\; r = 1,2,3,\ldots \tag{17}$$

So that the vector forms, defined as

$$\left.\begin{array}{r} \overline{u} = (u(y_0), u(y_1), \ldots, u(y_N))^T \\ \overline{\theta} = (\theta(y_0), \theta(y_1), \ldots, \theta(y_N))^T \end{array}\right\}, \tag{18}$$

which are used to convert the coupled, nonlinear, boundary-value problem into a set of algebraic equations. By utilizing the NDSolve algorithm code in Wolfram Mathematica, the spectral collocation result for (9) is confirmed by the shooting Runge–Kutta method as reported in Tables 1 and 2.

Table 1. Numerical validation when $\lambda = 0.5, \epsilon = 0.2, m = 0.5, Bi = 20, \alpha_1 = 0.1 = \alpha_2, \kappa = 0.5, G = \gamma = H = r = S = 1$.

y	$u(y)_{CWRM}$	$u(y)_{RK45}$	$\lvert u(y)_{CWRM} - u(y)_{RK4}\rvert$
0	0.29059333168978096	0.2905933626764475	$3.098666651046855 \times 10^{-8}$
0.1	0.2875531229984736	0.28755315707369733	$3.407522375376004 \times 10^{-8}$
0.2	0.2784580603387541	0.27845809831693963	$3.797818554085452 \times 10^{-8}$
0.3	0.26337977611466257	0.2633798176311869	$4.151652432948793 \times 10^{-8}$
0.4	0.24242311286613746	0.24242315706503897	$4.419890151097228 \times 10^{-8}$
0.5	0.21571064483575425	0.21571069190190728	$4.70661530305172 \times 10^{-8}$
0.6	0.18336924938712212	0.18336929955583225	$5.016871013063806 \times 10^{-8}$
0.7	0.14552094841986923	0.14552100275459876	$5.433472952121043 \times 10^{-8}$
0.8	0.10227800886883381	0.10227806869495099	$5.982611717136876 \times 10^{-8}$
0.9	0.05374120503311206	0.053741270769143465	$6.573603140297424 \times 10^{-8}$
1.0	$-2.306642251952988 \times 10^{-20}$	$7.112594643000738 \times 10^{-8}$	$7.112594643003045 \times 10^{-8}$

Table 2. Numerical validation when $\lambda = 0.5, \epsilon = 0.2, m = 0.5, Bi = 20, \alpha = 0.1, \kappa = 0.5, G = \gamma = H = r = S = 1$.

y	$\theta(y)_{CWRM}$	$\theta(y)_{RK45}$	$\lvert\theta(y)_{CWRM} - \theta(y)_{RK4}\rvert$
0	0.38590264401948615	0.38590263681021386	$7.209272290253921 \times 10^{-9}$
0.1	0.38190747745163284	0.3819074704013799	$7.050252937013113 \times 10^{-9}$
0.2	0.36994346490495467	0.36994345794711025	$6.957844411736858 \times 10^{-9}$
0.3	0.3500734997990253	0.35007349292163265	$6.877392655368908 \times 10^{-9}$
0.4	0.322397158501689	0.3223971517395266	$6.762162385598458 \times 10^{-9}$
0.5	0.28704280983610114	0.28704280313984776	$6.696253385118922 \times 10^{-9}$
0.6	0.24415664136968146	0.24415663474092741	$6.628754045667762 \times 10^{-9}$
0.7	0.1938888308327938	0.19388882426219917	$6.570594623944714 \times 10^{-9}$
0.8	0.13637721443702794	0.13637720796479974	$6.472228197829111 \times 10^{-9}$
0.9	0.07172886591491089	0.07172885987750592	$6.037404964853721 \times 10^{-9}$
1.0	$-2.853996136427998 \times 10^{-17}$	$-5.010838847582127 \times 10^{-9}$	$5.010838819042166 \times 10^{-9}$

4. Results and Discussion

Tables 1 and 2 reveal the results of comparing the two numerical methods used to solve (9) with parameter values used for the computation. The two results point to a unique numerical approximation. Table 3 shows the rapid convergence of the weighted residual method based on spectral collocation. Table 4 presents the effects of various parameters on thermal flow stability. As seen from the table, porous permeability extends the critical value of the Frank–Kameneskii parameter, thus stabilizing the flow. Similarly, the third-grade parameter also delays the early occurrence of instability in the flow field. However, increasing magnetic field intensity and shape factor parameter values encourage thermal instability in the flow field.

Table 3. Convergence of critical values $\alpha = \kappa = H = S = 0.1, \gamma = 1 = G = r$.

N	$Nu(\varepsilon=0.1, m=0)$	$\lambda_c(\varepsilon=0.1, m=0)$	$Nu(\varepsilon=0.2, m=0.5)$	$\lambda_c(\varepsilon=0.2, m=0.5)$
5	2.846627970782652	0.9374126233474899	3.3625398187527153	0.9572961968391648
10	2.8501958216533407	0.9361333473082082	3.3789867582903286	0.9568015766586552
15	2.8501681280213513	0.9361332230608462	3.378984457190358	0.9568015916975717
20	2.8501680805234675	0.9361332229338021	3.3789844559157363	0.9568015917674281
25	2.8501680600340458	0.9361332229334971	3.3789844559157363	0.9568015917671002
30	2.850168051013584	0.9361332229334965	3.3789844559157363	0.9568015917670978

Table 4. Numerical result for stability analysis $r = 1, \in = 0.2$.

α	κ	γ	H	S	λ_c
0.1	0.1	1	0.1	0.1	0.9568015917674281
0.3	0.1	1	0.1	0.1	0.9568150977269292
0.5	0.1	1	0.1	0.1	0.9568287821788171
0.1	0.3	1	0.1	0.1	0.9602351070037033
0.1	0.5	1	0.1	0.1	0.9624900150111185
0.1	0.1	2	0.1	0.1	0.9182714895582226
0.1	0.1	1	0.3	0.1	0.9563995116349202
0.1	0.1	1	0.5	0.1	0.9562084981492455
0.1	0.1	1	0.1	0.5	0.9563131736440517
0.1	0.1	1	0.1	1	0.9571020598593034

Figure 2 reveals the Frank–Kameneskii parameter's effect on the third-grade fluid's temperature-dependent electrical conductivity. From the plot, an increase in the heat of the reaction from the initial liquid concentration produces a slight drop in the maximum flow velocity. This decline in flow peak is directly connected with the electrical resistance of the fluid to allow for passage of the electric current due to reduced ion formation in the liquid. In Figure 3, the rise in the Frank–Kameneskii parameter shows that the temperature distribution within the flow domain increases. This positive rise in temperature is due to a rise in the heat of the reaction, indicating that heat significantly flows into the flow domain from the surroundings. The significant increase in fluid temperature distribution (as shown in Figure 3) and the almost negligible decrease in the velocity maximum, as shown in Figure 2, reveal that the system's entropy mainly depends on heat transfer rather than viscous interactions, as shown in Figure 4. Moreover, Figure 5 represents the effect of the Frank–Kameneskii parameter on the Bejan profile. From the plot, it is evident that the viscosity of the fluid becomes infinite at the center, thus $BE(\lambda, y) = 0$ at the core center. Beyond this point, towards the channel walls, the magnitude of attains $BE(\lambda, y) = 0.25$. This means that, at the walls, heat transfer irreversibility contributes to heat irreversibility.

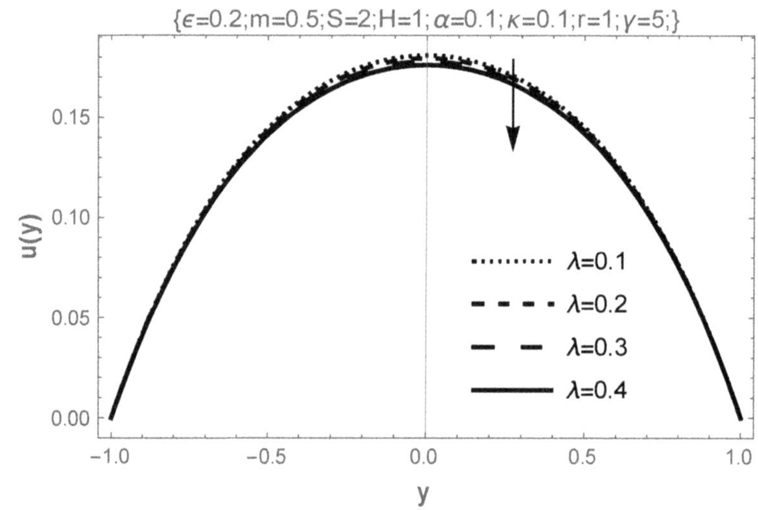

Figure 2. Effect of the Frank–Kameneskii parameter on flow velocity.

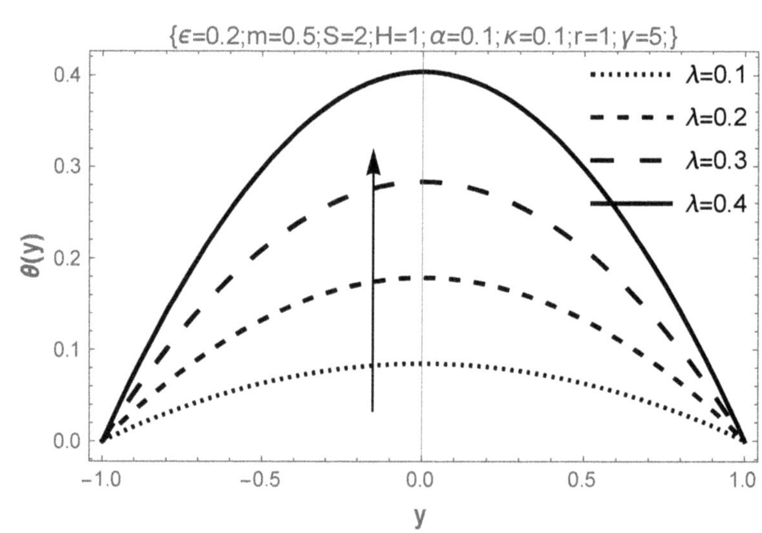

Figure 3. Effect of the Frank–Kameneskii parameter on fluid temperature.

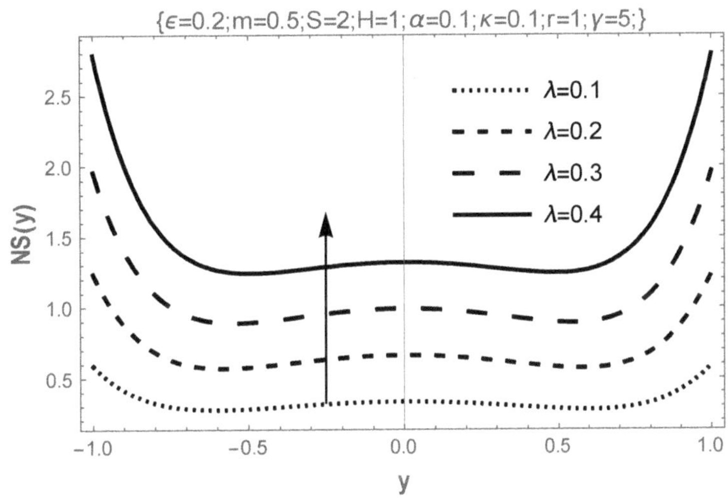

Figure 4. Effect of the Frank–Kameneskii parameter on entropy profile.

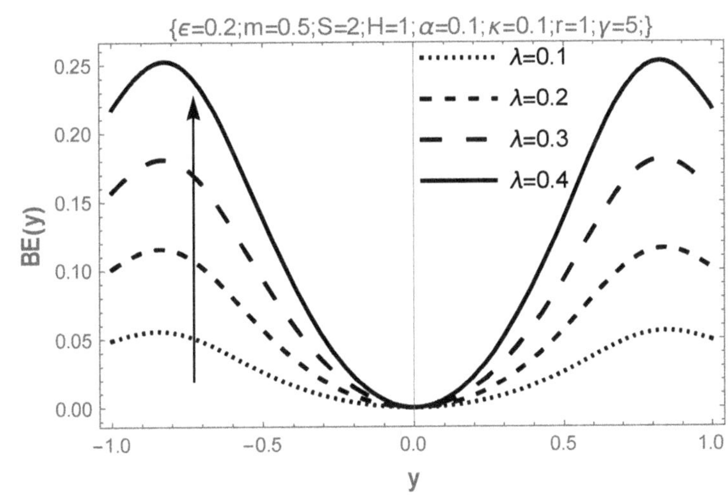

Figure 5. Effect of the Frank–Kameneskii parameter on Bejan profile.

Figure 6 represents the effect of the porous permeability variation parameter on flow velocity. The plot reveals that the flow velocity maximum rises with increasing values of the passable permeability parameter. This is because as temperature increases, there is a reduction in the viscosity of the third-grade fluid. This encourages flow due to the increased permeability of the porous matrix. The increase in porous permeability with the temperature of the fluid is presented in Figure 7. The result shows that the fluid's porous permeability improves temperature distribution within the flow channel. This is connected with the reduced activation energy of the combustible liquid. Figure 8 shows the effect of variations in porous permeability on the entropy generation rate. The fact that flow velocity and temperature distribution increases with this parameter indicates that frictional interaction within the fluid layer is negligible. Therefore, the entropy profile is on the rise across the channel. Finally, fluid viscosity-related irreversibility dominates significantly over heat-transfer irreversibility at the core center of the flow channel. In contrast, heat transfer irreversibility is more prominent at the walls, as seen in Figure 9.

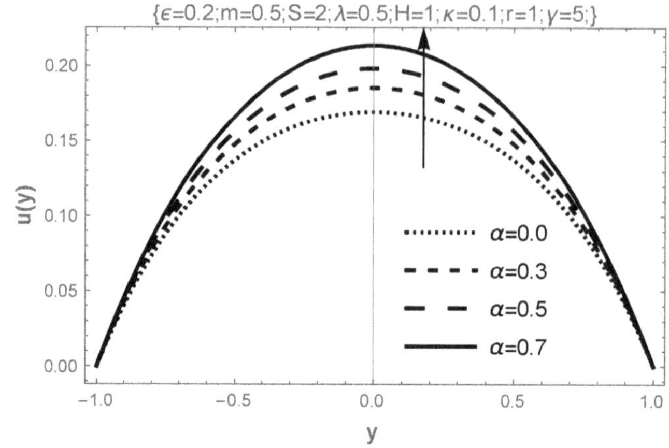

Figure 6. Effect of the porous permeability variation parameter on flow velocity.

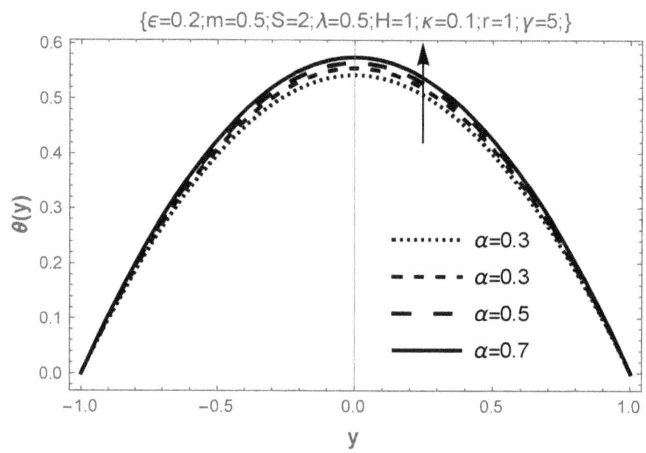

Figure 7. Effect of the porous permeability variation parameter on fluid temperature.

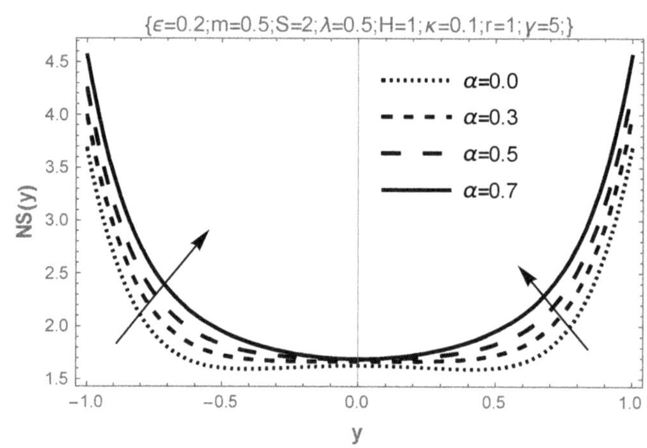

Figure 8. Effect of the porous permeability variation parameter on entropy profile.

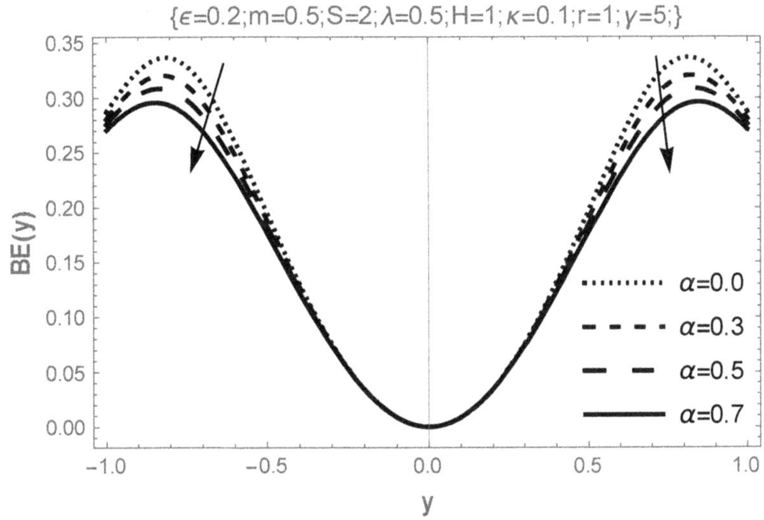

Figure 9. Effect of the porous permeability variation parameter on Bejan profile.

Figure 10 shows the effect of magnetization on the flow of third-grade fluid. The velocity peak declines with increasing magnetic field strength, as seen in the plot. This decline is physically correct since the spinning of fluid particles encourages fluid thickening; therefore, flow velocity declines with the increasing intensity of the magnetic field. Similarly, the kinetic energy of the fluid particles is expected to decrease due to fluid thickening. Therefore, the liquid temperature distribution declines, as seen in Figure 11. The combined effect of reducing flow velocity and temperature shows that heat transfer and fluid friction irreversibility will uniformly decrease across the flow channel, as observed in Figure 12. The Bejan ratio signifies the dominating viscous effect over heat transfer irreversibility at the core region of the flow channel, while irreversibility from heat transfer is more at the walls as seen in Figure 13. The bifurcation study in Figure 14a,b shows the variation of the Nusselt number with the Frank–Kameneskii parameter. The reaction exponent, m, has a stabilizing effect on third-grade fluid's thermal stability.

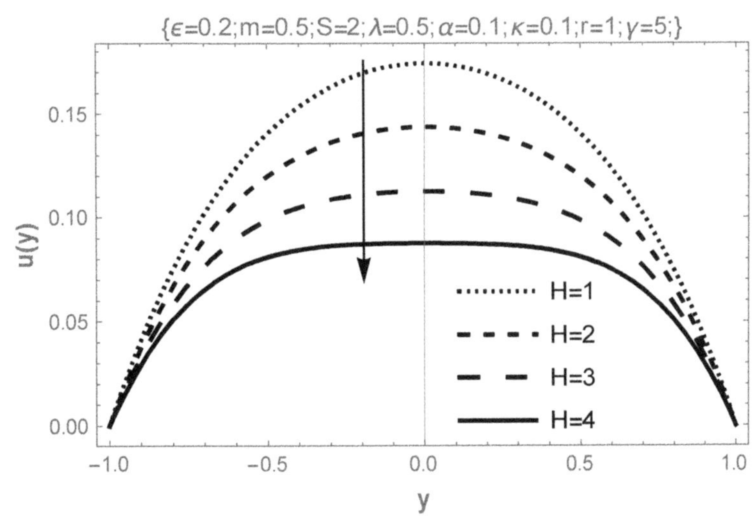

Figure 10. Effect of Hartmann number on flow velocity.

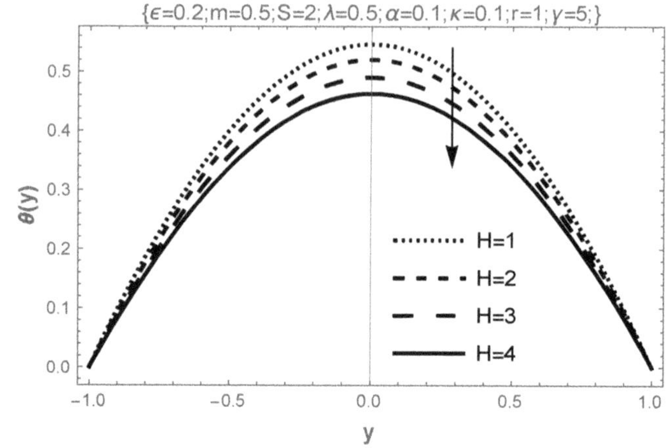

Figure 11. Effect of Hartmann number on fluid temperature.

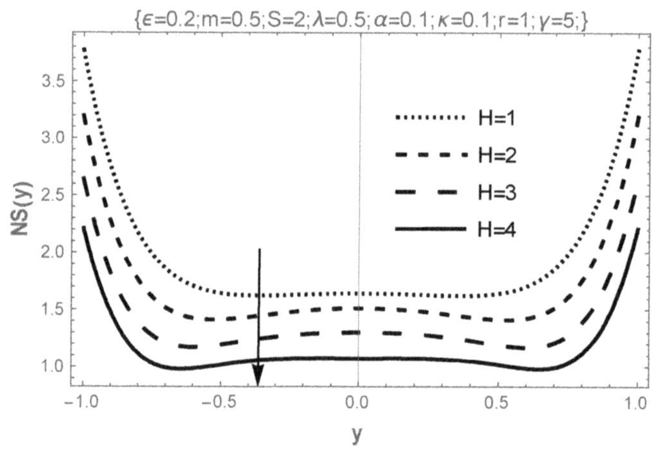

Figure 12. Effect of Hartmann number on entropy profile.

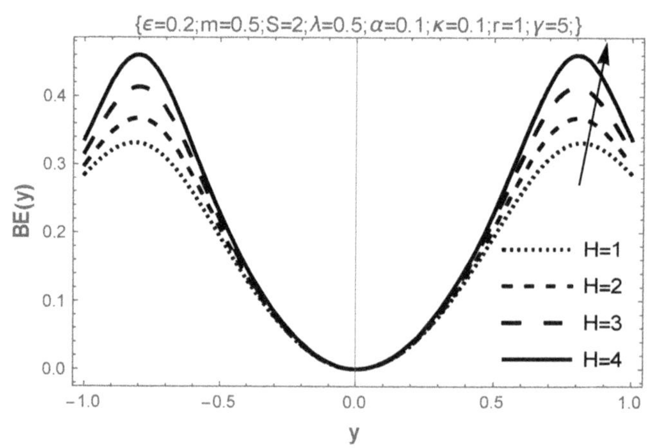

Figure 13. Effect of Hartmann number on Bejan profile.

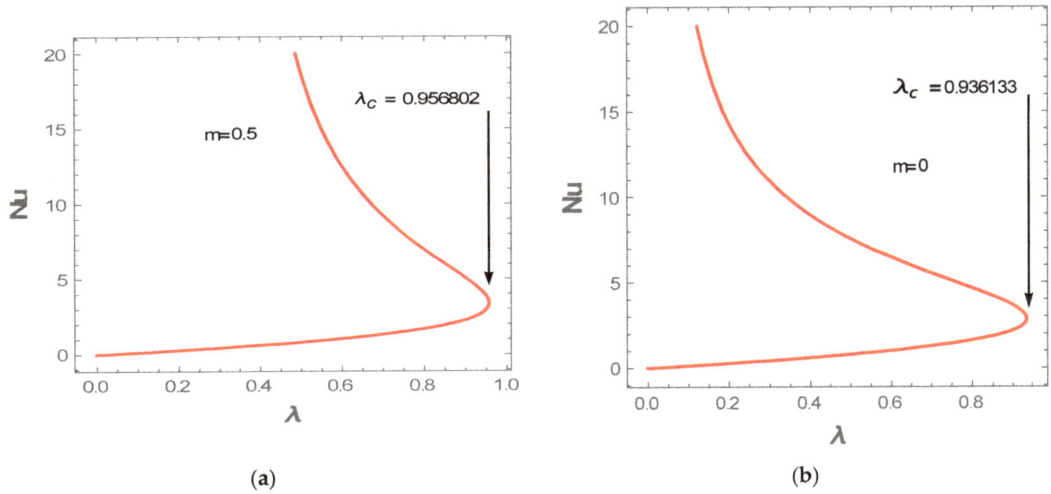

Figure 14. (**a**)-Bifurcation for bimolecular reaction. (**b**)-Bifurcation for Arrhenius kinetics.

5. Conclusions

In this work, numerical simulations have been conducted to study the heat irreversibility inherent in the steady flow of hydromagnetic third-grade fluid through a porous medium with temperature-dependent porous permeability and electrical conductivity. The spectral collocation method solved the dimensionless nonlinear equations and validated the results using the shooting Runge–Kutta method. The agreement between the two numerical results suggests the accuracy of the two numerical methods in handling the coupled nonlinear boundary-value problem. The significant contributions to knowledge from this study are:

i. The effect of the increasing values of temperature-dependent porous permeability in the present study reveal that it stabilizes the flow and elevates both velocity and temperature while encouraging entropy generation;
ii. The influence of rising temperature-dependent electrical conductivity parameters destabilizes the flow, lowering both flow and temperature peaks while discouraging entropy generation in the flow field.

Author Contributions: Conceptualization, S.O.A.; Methodology, S.O.A.; Validation, P.O.B.; Formal analysis, P.O.B.; Investigation, R.S.L.; Data curation, P.O.B.; Writing—original draft, S.O.A.; Supervision, R.S.L.; Funding acquisition, R.S.L. All authors have read and agreed to the published version of the manuscript.

Funding: This research received no external funding.

Conflicts of Interest: The authors declare no conflict of interest.

References

1. Adesanya, S.O.; Rundora, L.; Thosago, K.F. Numerical evaluation of heat irreversiblity in porous medium combustion of third-grade fluid subjected to Newtonian cooling. *Numer. Heat Transf. Part A Appl.* **2023**. [CrossRef]
2. Sajid, M.; Hayat, T. Series solution for steady flow of a third grade fluid through porous space. *Transp. Porous Media* **2008**, *71*, 173–183. [CrossRef]
3. Makinde, O.; Chinyoka, T.; Rundora, L. Unsteady flow of a reactive variable viscosity non-Newtonian fluid through a porous saturated medium with asymmetric convective boundary conditions. *Comput. Math. Appl.* **2011**, *62*, 3343–3352. [CrossRef]
4. Hayat, T.; Naz, R.; Abbasbandy, S. Poiselle flow of a Third-grade fluid in a porous medium. *Trans. Porous. Media* **2011**, *87*, 355–366. [CrossRef]
5. Rundora, L.; Makinde, O. Effects of suction/injection on unsteady reactive variable viscosity non-Newtonian fluid flow in a channel filled with porous medium and convective boundary conditions. *J. Pet. Sci. Eng.* **2013**, *108*, 328–335. [CrossRef]

6. Baoku, I.; Olajuwon, B.; Mustapha, A. Heat and mass transfer on a MHD third grade fluid with partial slip flow past an infinite vertical insulated porous plate in a porous medium. *Int. J. Heat Fluid Flow* **2013**, *40*, 81–88. [CrossRef]
7. Adesanya, S.O.; Falade, J.A. Thermodynamics analysis of hydromagmetic third grade fluid flow through a channel filled with porous medium. *Alex. Eng. J.* **2015**, *54*, 615–622. [CrossRef]
8. Salawu, S.O.; Fatunmbi, E.O. Inherent Irreversibility of hydromagnetic third grade reactive poiseille flow of variable viscosity in porous media with convective cooling. *J. Serb. Soc. Comput. Mech.* **2017**, *11*, 46–58. [CrossRef]
9. Maghsoudi, P.; Shabriari, G.; Mirzaei, M.; Mirzaei, M. Natural convection of third grade non-Newtonian fluid flow in a porous medium with heat source: Analytical solution. *Eur. Phys. J. Plus* **2018**, *133*, 502. [CrossRef]
10. Hayat, T.; Shahzad, F.; Ayub, M. Analytical solution for the steady flow of the third grade fluid in a porous half space. *Appl. Math. Model.* **2007**, *31*, 2424–2432. [CrossRef]
11. Chinyoka, T.; Makinde, O. Analysis of non-Newtonian flow with reacting species in a channel filled with a saturated porous medium. *J. Pet. Sci. Eng.* **2014**, *121*, 1–8. [CrossRef]
12. Akinshilo, A.T. Steady flow and heat transfer analysis of third grade fluid with porous medium and heat generation. *Eng. Sci. Technol. Int. J.* **2017**, *20*, 1602–1609. [CrossRef]
13. Rahman, S.; Hayat, T.; Muneer, M.; Ahmad, B. Global existence of solutions for MHD third grade flow equations saturating porous medium. *Comput. Math. Appl.* **2018**, *76*, 2360–2374. [CrossRef]
14. Rahman, M.M.; Uddin, M.J.; Aziz, A. Effects of variable electric conductivity and non-uniform heat source (or sink) on convective micropolar fluid flow along an inclined flat plate with surface heat flux. *Int. J. Therm. Sci.* **2009**, *48*, 2331–2340. [CrossRef]
15. Makinde, O.; Onyejekwe, O. A numerical study of MHD generalized Couette flow and heat transfer with variable viscosity and electrical conductivity. *J. Magn. Magn. Mater.* **2011**, *323*, 2757–2763. [CrossRef]
16. Hossain, A.; Gorla, R.S.R. Joule heating effect on magnetohydrodynamic mixed convection boundary layer flow with variable electrical conductivity. *Int. J. Numer. Methods Heat Fluid Flow* **2013**, *23*, 275–288. [CrossRef]
17. Eguía, P.; Zueco, J.; Granada, E.; Patiño, D. NSM solution for unsteady MHD Couette flow of a dusty conducting fluid with variable viscosity and electric conductivity. *Appl. Math. Model.* **2011**, *35*, 303–316. [CrossRef]
18. Sivaraj, R.; Kumar, B.R. Viscoelastic fluid flow over a moving vertical cone and flat plate with variable electric conductivity. *Int. J. Heat Mass Transf.* **2013**, *61*, 119–128. [CrossRef]
19. Eegunjobi, A.S.; Makinde, O.D. Second law analysis for MHD permeable channel flow with variable electrical conductivity and asymmetric Navier slips. *Open Phys.* **2015**, *13*, 100–110. [CrossRef]
20. Salawu, S.O.; Kareem, R.A.; Shonola, S.A. Radiative thermal criticality and entropy generation of hydromagnetic reactive Powell-Erying fluid in saturated porous media with variable conductivity. *Energy Rep.* **2019**, *5*, 480–488. [CrossRef]
21. Obalalu, A.M.; Ajala, O.; Adeosun, A.T.; Akindele, A.O.; Oladapo, O.A.; Olajide, O.A.; Peter, A. *Partial Differential Equations in Applied Mathematics*; John Wiley & Sons: Hoboken, NJ, USA, 2021; Volume 4, p. 100184.
22. Adeosun, A.T.; Ukaegbu, J.C. Effect of the variable electrical conductivity on the thermal stability of the MHD reactive squeezed fluid flow through a channel by a spectral collocation approach. *Partial. Differ. Equ. Appl. Math.* **2022**, *5*, 100256. [CrossRef]

mathematics

Article

Developments of Electro-Osmotic Two-Phase Flows of Fourth-Grade Fluid through Convergent and Divergent Channels

Nahid Fatima [1], Mubbashar Nazeer [2], Maha M. A. Lashin [3], M. M. Ghafar [4], M. R. Gorji [5,*] and M. K. Hameed [4]

[1] Department of Mathematics and Sciences, Prince Sultan University, Riyadh 11586, Saudi Arabia; nfatima@psu.edu.sa
[2] Department of Mathematics, Institute of Arts and Sciences, Government College University Faisalabad Chiniot Campus, Chiniot 35400, Pakistan; mubbasharnazeer@gcuf.edu.pk
[3] Electrical Engineering Department, College of Engineering, Princess Nourah bint Abdulrahman University, P.O. Box 84428, Riyadh 11671, Saudi Arabia; mmlashin@pnu.edu.sa
[4] Department of Mathematica, Riphah International University Faisalabad Campus, Faisalabad 38000, Pakistan
[5] Faculty of Medicine and Health Sciences, Ghent University, 9000 Ghent, Belgium
* Correspondence: mohammad.rahimigorji@ugent.be

Abstract: This paper discusses the development of two different bi-phase flows. Fourth-grade fluid exhibiting the non-Newtonian fluid nature is taken as the base liquid. Two-phase suspension is obtained by using the spherically homogeneous metallic particle. Owing to the intense application of mechanical and chemical multiphase flows through curved and bent configurations effectively transforms the flow dynamics of the fluid. Differential equations for electro-osmotically driven fluid are modeled and solved with the help of the regular perturbation method. The obtained theoretical solution is further compared with the ones obtained by using two different numerical techniques and found to be in full agreement.

Keywords: fourth-grade fluid; homogeneous; configurations; perturbation method suspension

MSC: 35Q35

Citation: Fatima, N.; Nazeer, M.; Lashin, M.M.A.; Ghafar, M.M.; Gorji, M.R.; Hameed, M.K. Developments of Electro-Osmotic Two-Phase Flows of Fourth-Grade Fluid through Convergent and Divergent Channels. *Mathematics* **2023**, *11*, 1832. https://doi.org/10.3390/math11081832

Academic Editors: Ramoshweu Solomon Lebelo, Antonio Lamura and Efstratios Tzirtzilakis

Received: 3 January 2023
Revised: 21 March 2023
Accepted: 10 April 2023
Published: 12 April 2023

1. Introduction

In various applications, the flow of non-Newtonian fluids (such as blood, greases, drilling muds, and suspension, etc.) cannot be expressed by the classical Navier–Stokes theory, and these fluids are categorized as tangent hyperbolic fluids, power-law fluids, generalized Newtonian fluids, Ellis fluids, Williamson fluids, Burgers fluids, Johnson–Segalman fluids, Sisko fluid model, Eyring–Powell fluid, third grade fluid, etc. Due to complex rheological properties and behavior, the fourth-grade fluid [1] is a special type of non-Newtonian fluid that describe the shear thinning and shear thickening phenomena which cannot be expressed by the classical Navier–Stokes equations. The applications of fourth-grade fluids in industry, petroleum and food manufacturing, etc., have significant involvement of diffusion reaction [2] and thermal transports in parallel flows. The constitutive relation [3] of fourth-grade fluid is more complex as compared to second- and third-grade fluids due to more material parameters. So, the study of such highly viscous fluids is hard to model and predict the flow properties, due to scores of parameters. Salawu et al. [4] reported important results on fourth-grade fluid. The investigation is carried out for a parallel flow that obeys the fundamentals of Couette flow mechanism. The numerical results are reached via finite semi-discretization difference method.

Fourth-grade fluid is treated as biological flow in [5] in the curved artery channel by Khan et al. with the help of numerical technique. An approximate study of circular flows with temperature-dependent viscosity of fourth-grade fluid through is the focal point of different authors in [6,7]. Aziz and Mahomed [8] present a theoretical analysis of fourth-grade fluid over a porous plate.

The flow of bulk fluids through the membrane, porous channel, capillary tube, microchannel, or any other fluid channel under the action of the electric field applied at the end of the conduit is termed electro-osmosis flow. The electroosmotic flow getting the attention of researchers and authors due to its wider applications in medical science, natural chemistry, industrial processes [9], etc. The electro-osmotic flow in non-Newtonian fluids, namely, colloidal suspension, blood, polymeric and protein arrangements, have significant usages. Currently, various studies on the electro-osmotic flow of non-Newtonian fluids have been reported by researchers by considering different constitutive models such as Eyring–Powell fluid [10], Williamson fluid [11], Casson fluid [12], Sutterby fluid [13], generalized Newtonian fluid [14], fractional Maxwell fluid [15], Walters'-B fluid [16], Phan Thien Tanner fluid [17], Power-law fluid [18], Oldroyd-B fluid [19] and third-grade fluid [20], etc.

In addition to the above literature, close analysis of some recent studies on the multiphase flow of fourth-grade fluid under the action of the electric field in two complex configurations, namely, convergent and divergent channels, is worthful investigation. The analysis of this study is a significant contribution to understanding the behavior of the multiphase flow of fourth-grade fluid in terms of physical and mathematical point of view. The modeled highly nonlinear differential equations are dealt with "*Perturbation technique*" to achieve an approximate solution.

2. Development of a Mathematical Model of Multiphase Flow of Non-Newtonian Fluid with Electro-Osmotic Phenomena

Consider a two-phase flow of fourth-grade fluid through channels as shown in Figures 1 and 2, respectively. The configuration of convergent [21] and divergent [22] channels can be defined as:

Geometry 1:

$$H(x) = \begin{cases} a - b\sqrt{1-\cos\left(\frac{\pi x}{\lambda}\right)}; & \text{When } \frac{11\lambda}{7} < x < \frac{33\lambda}{7}, \\ 0.5a; & \text{Othwewise} \end{cases} \tag{1}$$

Geometry 2:

$$H(x) = \begin{cases} a - b\,\sin^2\left(\frac{\pi x}{\lambda}\right) & \text{When } \frac{11\lambda}{7} < x < \frac{33\lambda}{7}, \\ 0.5a; & \text{Othwewise.} \end{cases} \tag{2}$$

If $\mathbf{V}_{vf} = \left[u_{vf}(x,y), 0, 0\right]$ and $\mathbf{V}_{vp} = \left[u_{vp}(x,y), 0, 0\right]$ denote the velocity of fluid and particle phase, respectively. The governing equations for this dissemination of fluid and particle phases are:

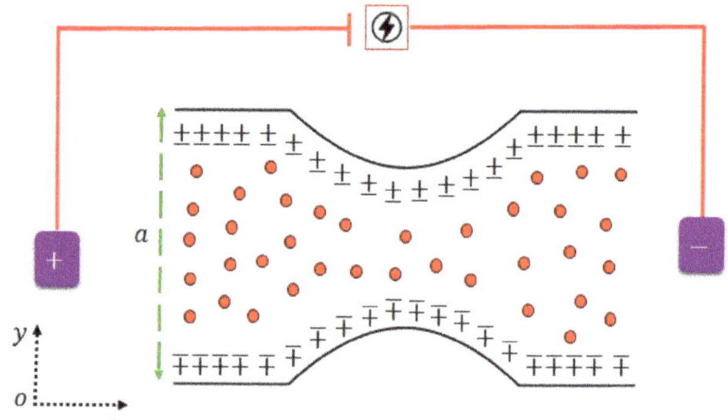

Figure 1. Convergent geometry.

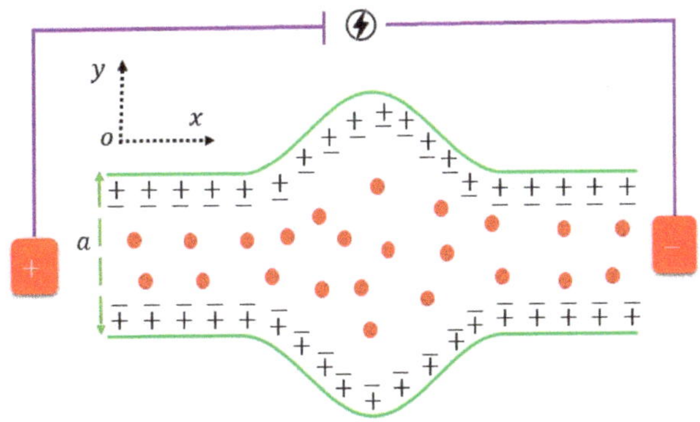

Figure 2. Divergent geometry.

2.1. Flow Equations for Fluid Phase

The equation of continuity which governs the conservation of mass of the flow is

$$\nabla.\mathbf{V}_{vf} = 0, \tag{3}$$

similarly, the conservation of momentum [23,24] for the fluid phase of the considered problem is given as

$$\rho_f(1-C)\frac{D\,\mathbf{V}_{vf}}{D\,t} = -(1-C)\nabla.p + (1-C)\nabla.\mathbf{T} - SC\left(\mathbf{V}_{vp} - \mathbf{V}_{vf}\right) + \mathbf{J}\times\mathbf{B} + g\rho_f. \tag{4}$$

The mathematical expression of "**T**" is defined as [2]

$$\mathbf{T} = S_1 + S_2 + S_3 + S_4, \tag{5}$$

$$S_1 = \mu\mathbf{A}_1, \tag{6}$$

$$S_2 = \alpha_1\mathbf{A}_2 + \alpha_1\mathbf{A}_1^2, \tag{7}$$

$$S_3 = \beta_1\mathbf{A}_3 + \beta_2(\mathbf{A}_2\mathbf{A}_1 + \mathbf{A}_1\mathbf{A}_2) + \beta_3(tr\mathbf{A}_1^2)\mathbf{A}_1, \tag{8}$$

$$\begin{gathered} S_4 = \gamma_1\mathbf{A}_4 + \gamma_2(\mathbf{A}_3\mathbf{A}_1 + \mathbf{A}_1\mathbf{A}_3) + \gamma_3\mathbf{A}_2^2 + \gamma_4(\mathbf{A}_2\mathbf{A}_1^2 + \mathbf{A}_1^2\mathbf{A}_2) \\ +\gamma_5(tr\mathbf{A}_2)\mathbf{A}_2 + \gamma_6(tr\mathbf{A}_2)\mathbf{A}_1^2 + \gamma_7(tr\mathbf{A}_3) + \gamma_8(tr(\mathbf{A}_2\mathbf{A}_1))\mathbf{A}_1, \end{gathered} \tag{9}$$

$$\mathbf{A}_1 = \mathbf{L} + \mathbf{L}^{\mathbf{T}}, \tag{10}$$

$$\mathbf{A}_n = \frac{d\mathbf{A}_{n-1}}{dt} + \mathbf{A}_{n-1}\mathbf{L} + \mathbf{L}^{\mathbf{T}}\mathbf{A}_{n-1},\ n\geq 2, \tag{11}$$

$$\mathbf{L} = \nabla\mathbf{V}_{vf}. \tag{12}$$

In the above one can identify

$$\mathbf{J} = \sigma(\mathbf{E} + \mathbf{V}_{vf}\times\mathbf{B}), \tag{13}$$

The equation of continuity and momentum equations are defined in the following manner

$$\nabla.\mathbf{V}_{vp} = 0, \tag{14}$$

$$\rho_p C \frac{D\,\mathbf{V}_{vp}}{D\,t} = -C\nabla.p + -SC\left(V_{vp} - V_{vf}\right). \tag{15}$$

It is presumed that the velocity of the bi-phase fluid is zero and the particle concentration remains the same during the study, so the Equations (3), (4), (13) and (15) in the component's forms can be written as

$$\frac{\partial u_{vf}}{\partial y} = 0. \tag{16}$$

The momentum of the fluid phase can be obtained as

$$\begin{gathered}\rho_f(1-C)\left(\frac{\partial u_{vf}}{\partial t} + u_{vf}\frac{\partial u_{vf}}{\partial x} + v_{vf}\frac{\partial u_{vf}}{\partial y}\right) = -(1-C)\frac{\partial p}{\partial x} + \\ (1-C)\left\{\mu\frac{\partial^2 u_{vf}}{\partial y^2} + 6\beta\left(\frac{\partial u_{vf}}{\partial y}\right)^2\left(\frac{\partial^2 u_{vf}}{\partial y^2}\right)\right\} - SC\left(u_{vp} - u_{vf}\right) + \left(\frac{\partial^2\phi}{\partial x^2} + \frac{\partial^2\phi}{\partial y^2}\right)\vec{E}_x, \text{ where } \beta = \beta_2 + \beta_3.\end{gathered} \tag{17}$$

The overhead expression has a lot of significance because if β_3 is zero, the Equation (17) turns into a momentum equation for third-grade fluid and if both are equal to zero, then the resulting equation is also a momentum equation of second-grade Newtonian fluids and if both are not turned into zero the result will be momentum equation of four grade.

$$\begin{gathered}\rho_f(1-C)\left(\frac{\partial v_{vf}}{\partial t} + u_{vf}\frac{\partial v_{vf}}{\partial x} + v_{vf}\frac{\partial v_{vf}}{\partial y}\right) = -(1-C)\frac{\partial p}{\partial y} + (1-C) \\ \left\{\alpha\left(\frac{\partial u_{vf}}{\partial y}\right)\left(\frac{\partial^2 u_{vf}}{\partial y^2}\right) + \gamma\left(\frac{\partial u_{vf}}{\partial y}\right)^3\left(\frac{\partial^2 u_{vf}}{\partial y^2}\right)\right\} - SC\left(v_{vp} - v_{vf}\right).\end{gathered} \tag{18}$$

where $\alpha = 4\alpha_1 + 2\alpha_2$ and $\gamma = 16(\gamma_3 + \gamma_4 + \gamma_5 + 0.5\gamma_6)$.

2.2. Governing Equations (Particle Phase)

The Equations (14) and (15) can be expressed in the following form as

$$\frac{\partial u_{vp}}{\partial y} = 0. \tag{19}$$

$$\rho_f C\left(\frac{\partial u_{vp}}{\partial t} + u_{vp}\frac{\partial u_{vp}}{\partial x} + v_{vp}\frac{\partial u_{vp}}{\partial y}\right) = -C\frac{\partial p}{\partial x} + SC\left(u_{vp} - u_{vf}\right). \tag{20}$$

$$\rho_f C\left(\frac{\partial v_{vp}}{\partial t} + u_{vp}\frac{\partial v_{vp}}{\partial x} + v_{vp}\frac{\partial v_{vp}}{\partial y}\right) = -C\frac{\partial p}{\partial y} + SC\left(v_{vp} - v_{vf}\right). \tag{21}$$

For steady flow Equations (17), (18), (20) and (21) gained the shape

$$\left\{\mu\frac{\partial^2 u_{vf}}{\partial y^2} + 6\beta\left(\frac{\partial u_{vf}}{\partial y}\right)^2\left(\frac{\partial^2 u_{vf}}{\partial y^2}\right)\right\} - \frac{1}{(1-C)}\frac{\partial p}{\partial x} + \frac{1}{(1-C)}\left(\frac{\partial^2\phi}{\partial x^2} + \frac{\partial^2\phi}{\partial y^2}\right)\vec{E}_x = 0, \tag{22}$$

$$\frac{\partial p}{\partial y} - \left\{\alpha\left(\frac{\partial u_{vf}}{\partial y}\right)\left(\frac{\partial^2 u_{vf}}{\partial y^2}\right) + \gamma\left(\frac{\partial u_{vf}}{\partial y}\right)^3\left(\frac{\partial^2 u_{vf}}{\partial y^2}\right)\right\} = 0, \tag{23}$$

$$C\frac{\partial p}{\partial x} = SC\left(u_{vp} - u_{vf}\right). \tag{24}$$

Equation (23), is solved for modified pressure, which gives

$$\frac{\partial p}{\partial y} = 0. \tag{25}$$

The boundary conditions are given as

$$u_{vf}(y) = u_{vf_{at\,wall}};\ When\ y = -H(x), \tag{26}$$

$$u_{vf}(y) = u_{vf_{at\,wall}};\ When\ y = H(x). \tag{27}$$

3. Dimensionalization of the Problem

To predict the contribution of the most significant variables and parameters, it is mandatory to reduce or accumulate certain quantities which are of the least importance. Therefore, the following dimensionless transformation is of effective use.

$$\left.\begin{array}{l} \overline{x} = \frac{x}{L},\ \overline{y} = \frac{y}{L},\ \overline{\mathrm{u}}_{vf} = \frac{\mathrm{u}_{vf}}{u^*},\ \overline{\mathrm{u}}_{vp} = \frac{\mathrm{u}_{vp}}{u^*},\ \rho_{rel} = \frac{\rho_f}{\rho_p}, \\ \overline{p} = \frac{pL}{\mu_s u^*},\ \varpi = \frac{\beta u^{*2}}{\mu L^2},\ M = B_o L\sqrt{\frac{\sigma}{\mu}}. \end{array}\right\}. \tag{28}$$

The dimensionless form of Equations (20)–(27) is achieved by using the expression defined in Equation (28) in the following form as (bars are omitted)

$$\frac{\partial^2 \mathrm{u}_{vf}}{\partial y^2} + 6\varpi\left(\frac{\partial \mathrm{u}_{vf}}{\partial y}\right)^2\left(\frac{\partial^2 \mathrm{u}_{vf}}{\partial y^2}\right) - \frac{M^2}{(1-C)}\mathrm{u}_{vf} - \frac{1}{(1-C)}\frac{\partial p}{\partial x} + \left(\frac{m^2 U_{HS}}{(1-C)}\right)\frac{\cosh(mx)}{\cosh(mh)} = 0, \tag{29}$$

$$\mathrm{u}_{vp} = \mathrm{u}_{vf} - \mathrm{m}_2\frac{\partial p}{\partial x}, \tag{30}$$

$$u_{vf}(y) = 0;\ When\ y = -h(x), \tag{31}$$

$$u_{vf}(y) = 0;\ When\ y = h(x). \tag{32}$$

Similarly, the dimensionless form of the relations described in Equations (1) and (2) are narrated as

$$h(x) = \begin{cases} a - \beta\sqrt{1-\cos(\pi x)}; & \text{When } 0.5 < x < 4.5, \\ 0.5; & \text{Othwewise.} \end{cases} \tag{33}$$

$$h(x) = \begin{cases} 1 - \beta\sin^2(\pi x) & \text{When } 0.5 < x < 4.5, \\ 0.5; & \text{Othwewise.} \end{cases} \tag{34}$$

We assume that

$$\frac{dp}{dx} = P. \tag{35}$$

Then, Equations (29) and (30) become,

$$\frac{\partial^2 \mathrm{u}_{vf}}{\partial y^2} + 6\varpi\left(\frac{\partial \mathrm{u}_{vf}}{\partial y}\right)^2\left(\frac{\partial^2 \mathrm{u}_{vf}}{\partial y^2}\right) - \frac{M^2}{(1-C)}\mathrm{u}_{vf} - \frac{1}{(1-C)}P + \left(\frac{m^2 U_{HS}}{(1-C)}\right)\frac{\cosh(mx)}{\cosh(mh)} = 0, \tag{36}$$

$$\mathrm{u}_{vp} = \mathrm{u}_{vf} - \mathrm{m}_2 P. \tag{37}$$

4. Perturbation Solution

To find the approximate analytical solution to Equation (36) can easily be achieved due to the nonlinear term. Therefore, the most effective and reliable solution with the least margin of error can be obtained if the perturbation technique is applied. For this purpose, we assume that:

$$u_{vf} = u_{vf_0} + \varepsilon u_{vf_1} + \varepsilon^2 u_{vf_2} + o(\varepsilon^3), \tag{38}$$

and more suppose that,

$$\varpi = \lambda\varepsilon. \tag{39}$$

The above equation ϵ is known as the perturbation parameter. In view of Equations (38) and (39), Equation (36) becomes

$$\begin{aligned}\frac{\partial^2\left(u_{vf_0}+\varepsilon u_{vf_1}+\varepsilon^2 u_{vf_2}\right)}{\partial y^2} + 6\lambda\varepsilon\left(\frac{\partial\left(u_{vf_0}+\varepsilon u_{vf_1}+\varepsilon^2 u_{vf_2}\right)}{\partial y}\right)^2\left(\frac{\partial^2\left(u_{vf_0}+\varepsilon u_{vf_1}+\varepsilon^2 u_{vf_2}\right)}{\partial y^2}\right) \\ -\frac{M^2\left(u_{vf_0}+\varepsilon u_{vf_1}+\varepsilon^2 u_{vf_2}\right)}{(1-C)} - \frac{1}{(1-C)}P + \left(\frac{m^2 U_{HS}}{(1-C)}\right)\frac{\cosh(mx)}{\cosh(mh)} = 0.\end{aligned} \tag{40}$$

Equating and determining the equation of each order of ε^0, ε^1 and ε^2:

$$\varepsilon^0: \quad \frac{\partial^2 u_{vf_0}}{\partial(y)^2} - \frac{M^2}{(1-C)}u_{vf_0} - \frac{1}{(1-C)}P + \left(\frac{m^2 U_{HS}}{(1-C)}\right)\frac{\cosh(mx)}{\cosh(mh)} = 0, \tag{41}$$

$$u_{vf_0}(\pm h(x)) = 0. \tag{42}$$

Similarly,

$$\varepsilon^1: \quad \frac{\partial^2 u_{vf_1}}{\partial y^2} + 6\lambda\left(\frac{\partial u_{vf_0}}{\partial y}\right)^2\left(\frac{\partial^2 u_{vf_0}}{\partial y^2}\right) - \frac{M^2 u_{vf_1}}{(1-C)} = 0, \tag{43}$$

$$u_{vf_1}(\pm h(x)) = 0, \tag{44}$$

$$\varepsilon^2: \quad \frac{\partial^2 u_{vf_1}}{\partial y^2} + 6\lambda\left(\frac{\partial u_{vf_0}}{\partial y}\right)\left[\left(\frac{\partial u_{vf_0}}{\partial y}\right)\left(\frac{\partial^2 u_{vf_1}}{\partial y^2}\right) + 2\left(\frac{\partial^2 u_{vf_0}}{\partial y^2}\right)\left(\frac{\partial u_{vf_1}}{\partial y}\right)\right] - \frac{M^2 u_{vf_2}}{(1-C)} = 0, \tag{45}$$

$$u_{vf_2}(\pm h(x)) = 0. \tag{46}$$

The solution to Equation (41) is given below

$$u_{vf_0} = (a_4\cosh[my] + P(a_5 - a_6\cosh[ya_1]) + a_7\cosh[ya_1]). \tag{47}$$

The solution to Equation (43) is given below

$$\left.\begin{aligned} u_{vf_1} = &\left(\begin{aligned}a_{34} + a_{35}P + \\ a_{36}P^2 + a_{37}P^3\end{aligned}\right)\left(\begin{aligned}\cosh[a_1 y] - \\ \sinh[a_1 y]\end{aligned}\right) + \left(\begin{aligned}a_{38} + a_{39}P + \\ a_{40}P^2 + a_{41}P^3\end{aligned}\right)\left(\begin{aligned}\cosh[a_1 y] + \\ \sinh[a_1 y]\end{aligned}\right) + \\ &\left(\begin{aligned}a_8 y\sinh[ya_1] + a_9\cosh[ya_1] + a_{10}\cosh[3ya_1] + a_{11}\sinh[2ya_1] \\ \sinh[my] + a_{12}\cosh[my](9 - 5\cosh[2ya_1]) + a_{13}\cosh[my] \\ \sinh[ya_1]^2 + a_{14}\sinh[2my]\sinh[ya_1] + a_{15}\cosh[my]^2\cosh[ya_1]\end{aligned}\right) \\ &+P\left(\begin{aligned}a_{16}y\sinh[ya_1] + a_{17}\cosh[ya_1] + a_{18}\cosh[3ya_1] + a_{19}\sinh[my] \\ \sinh[2ya_1] + a_{20}\cosh[my](9 - 5\cosh[2ya_1]) + a_{21}\cosh[my] \\ \sinh[ya_1]^2 + a_{22}\sinh[2my]\sinh[ya_1] + a_{23}\cosh[my]^2\cosh[ya_1]\end{aligned}\right) \\ &+P^2\left(\begin{aligned}a_{24}y\sinh[ya_1] + a_{25}\cosh[3ya_1] + a_{26}\cosh[ya_1] + a_{27}\sinh[my] \\ \sinh[2ya_1] + a_{28}\cosh[my]\cosh[2ya_1] + a_{29}\cosh[my] + a_{30}\cosh \\ [my]\sinh[ya_1]^2\end{aligned}\right) \\ &+P^3(a_{31}y\sinh[ya_1] + a_{32}\cosh[ya_1] + a_{33}\cosh[3ya_1]).\end{aligned}\right\}. \tag{48}$$

The solution of Equation (45) is not presented here due to lengthy expressions that appeared after solving it. The final expression of the velocity can be obtained from Equation (48), i.e.,

$$\left.\begin{array}{l} A_{11} = (a_4\cosh[my] + P(a_5 - a_6\cosh[ya_1]) + a_7\cosh[ya_1]) + \\ \omega\left(\begin{array}{l} (a_{34} + a_{35}P + a_{36}P^2 + a_{37}P^3)(\cosh[a_1y] - \sinh[a_1y]) \\ +(a_{38} + a_{39}P + a_{40}P^2 + a_{41}P^3)(\cosh[a_1y] + \sinh[a_1y]) \\ +(a_8 y\sinh[ya_1] + a_9\cosh[ya_1] + a_{10}\cosh[3ya_1] + a_{11} \\ \sinh[2ya_1]\sinh[my] + a_{12}\cosh[my](9 - 5\cosh[2ya_1]) \\ +a_{13}\cosh[my]\sinh[ya_1]^2 + a_{14}\sinh[2my]\sinh[ya_1] + \\ a_{15}\cosh[my]^2\cosh[ya_1] \end{array}\right) \end{array}\right\} \tag{49}$$

$$\left.\begin{array}{l} \mathrm{A}_{12} = P\left(\begin{array}{l} a_{16}y\sinh[ya_1] + a_{17}\cosh[ya_1] + a_{18}\cosh[3ya_1] + a_{19} \\ \sinh[my]\sinh[2ya_1] + a_{20}\cosh[my](9 - 5\cosh[2ya_1]) \\ +a_{21}\cosh[my]\sinh[ya_1]^2 + a_{22}\sinh[2my]\sinh[ya_1] + \\ a_{23}\cosh[my]^2\cosh[ya_1] \end{array}\right) + \\ P^2\left(\begin{array}{l} a_{24}y\sinh[ya_1] + a_{25}\cosh[3ya_1] + a_{26}\cosh[ya_1] + a_{27} \\ \sinh[my]\sinh[2ya_1] + a_{28}\cosh[my]\cosh[2ya_1] + a_{29} \\ \cosh[my] + a_{30}\cosh[my]\sinh[ya_1]^2 \end{array}\right) + \\ P^3(a_{31}y\sinh[ya_1] + a_{32}\cosh[ya_1] + a_{33}\cosh[3ya_1]) \end{array}\right\}. \tag{50}$$

$$u_{vf} = \mathrm{A} + \mathrm{B} + \ldots \tag{51}$$

Similarly, we can get the expression for the velocity of the particulate phase u_{vp}.

$$\left.\begin{array}{l} u_p = (a_4\cosh[my] + P(a_5 - a_6\cosh[ya_1]) + a_7\cosh[ya_1]) + \\ \omega\left(\begin{array}{l} (a_{34} + a_{35}P + a_{36}P^2 + a_{37}P^3)(\cosh[a_1y] - \sinh[a_1y]) \\ +(a_{38} + a_{39}P + a_{40}P^2 + a_{41}P^3)(\cosh[a_1y] + \sinh[a_1y]) \\ +(a_8 y\sinh[ya_1] + a_9\cosh[ya_1] + a_{10}\cosh[3ya_1] + a_{11} \\ \sinh[2ya_1]\sinh[my] + a_{12}\cosh[my](9 - 5\cosh[2ya_1]) \\ +a_{13}\cosh[my]\sinh[ya_1]^2 + a_{14}\sinh[2my]\sinh[ya_1] + \\ a_{15}\cosh[my]^2\cosh[ya_1] \end{array}\right) + \\ P\left(\begin{array}{l} a_{16}y\sinh[ya_1] + a_{17}\cosh[ya_1] + a_{18}\cosh[3ya_1] + a_{19} \\ \sinh[my]\sinh[2ya_1] + a_{20}\cosh[my](9 - 5\cosh[2ya_1]) \\ +a_{21}\cosh[my]\sinh[ya_1]^2 + a_{22}\sinh[2my]\sinh[ya_1] + \\ a_{23}\cosh[my]^2\cosh[ya_1] \end{array}\right) + \\ P^2\left(\begin{array}{l} a_{24}y\sinh[ya_1] + a_{25}\cosh[3ya_1] + a_{26}\cosh[ya_1] + a_{27} \\ \sinh[my]\sinh[2ya_1] + a_{28}\cosh[my]\cosh[2ya_1] + a_{29} \\ \cosh[my] + a_{30}\cosh[my]\sinh[ya_1]^2 \end{array}\right) + \\ P^3(a_{31}y\sinh[ya_1] + a_{32}\cosh[ya_1] + a_{33}\cosh[3ya_1]) - \left(\frac{\mu_s}{a\delta\lambda S}\right)P + \ldots \end{array}\right\} \tag{52}$$

The volumetric flow rates (fluid and particle phases) can be determined from the following expressions:

$$Q_f = \int_0^{\overline{h}} u_f dy, \tag{53}$$

$$Q_p = \int_0^{\overline{h}} u_p dy. \tag{54}$$

The mathematical expression for the total volumetric flow rate is defined as

$$Q = Q_f + Q_p. \tag{55}$$

The expression for pressure P can be obtained by solving the above Equation (55).

5. Comparative Analysis

The comparison between numerical and perturbation solutions is displayed in Table 1. The perturbation solution is obtained in second order while the numerical solution is obtained through the spectral collocation method. In this method, we discretize the derivatives by using the Jacobi orthogonal polynomials or Chebyshev. The nonlinearity is handled through the Newton–Raphson method and finite difference approximation of the Jacobean (discrete Jacobean). Both solutions are obtained in convergent geometry. For this comparison, we obtained the numerical values of fluid velocity and particle velocity against the variation of the Hartmann number. From Table 1, it can be observed that both solutions are well-matched with each other. To validate the numerical results, we used another scheme, namely, the shooting method, and noted that both numerical results are accurate, as listed in Table 2.

Table 1. Absolute error between perturbation and numerical solutions.

	Perturbation Solution		Numerical Solution		Absolute Error	
M	u_{vf}	u_{vp}	u_{vf}	u_{vp}	u_{vf}	u_{vp}
1.0	1.45064	1.45089	1.45198	1.45001	0.134%	0.088%
2.0	1.35730	1.35744	1.35598	1.35671	0.132%	0.073%
3.0	1.21666	1.21680	1.21549	1.21612	0.117%	0.068%
4.0	1.04626	1.04638	1.04519	1.04590	0.107%	0.048%
5.0	0.86436	0.86446	0.86332	0.86399	0.104%	0.047%

Table 2. Absolute error between shooting method and pseudo-spectral collocation method.

	Pseudo-Spectral Collocation Method		Shooting Method	
C	u_{vf}	u_{vp}	u_{vf}	u_{vp}
0.1	1.41085	1.41110	1.41090	1.41001
0.2	1.33491	1.33506	1.33231	1.33325
0.3	1.23917	1.23932	1.23523	1.23567

6. Results and Discussion

A comprehensive parametric study is carried out in this section. The momentum of the particulate flow is predicted via the change in the numerical values of fourth-grade parameter "ϖ", electro-osmotic "m", particle concentration "C" and volumetric flow rate "Q". Because of the diverse shapes and layout, the velocity acts entirely differently. In Figures 3 and 4, the graphs of the most significant parameter ϖ, the fourth-grade parameter are drawn against the different values ϖ in both channels. It is of great interest that the velocity of both phases inclines with the respect to the variation in the dimensionless quantity. However, both geometries affect the flow quite differently. This opposite impression of the geometry of the multiphase flow can easily be apprehended due to Bernoulli's principle of fluid dynamics.

The variation of the electro-osmotic parameter "m" on fluid and particle phases is shown in Figures 5 and 6. It can be viewed from the plotted graphs that the electro-osmotic parameter inversely impacts the velocity profile of the fluid and particle phases, respectively. This inverse relationship introduces a force of hindrance across the flow. Therefore, the momentum of the fluid and particles diminishes gradually. Variation in the concentration of metallic particles is depicted in Figures 7 and 8. Unlike previous graphs, the impact of C is quite different, all depending on the configuration of the geometry through which the bi-phase suspension is transported. The momentum of both phases declines by opting for the convergent channel. On the other hand, there is a tremendous enhancement in the velocity of fluid and particle phases when the channel is considered to be the divergent one.

The volumetric flow rate is also a pivotal parameter of this analytic study. In the most recent decade, when every appliance has reduced in size, the need of the hour is to conduct such research where micro-size geometries and channels are considered; so, in this regard, the volumetric flow rate is especially important to measure. The volumetric flow rate is also known as the rate of fluid flow or volume velocity. This is the volume of the fluid which is passing through the considered geometry per unit of time. Its units in system international (SI) are cubic meter/second; however, cubic centimeters per minute is also in practice. The volumetric flow rate is mathematically defined as

$$Q = \dot{V} = \operatorname*{Lim}_{\Delta t \to \to 0} \frac{\Delta V}{\Delta t} = \frac{dV}{dt} \tag{56}$$

and this is the scalar quantity. In Figures 9 and 10, the graphs of the volumetric flow rate are plotted against the different values of the parameters Q for both the phases in convergent and divergent channels. As the values of Q enhanced, the velocity profile of fluidic and particulate phases increased in convergent and divergent channels. The same behavior of the graph has been seen in both phases and the simultaneous effect is observed for diversely shaped convergent and divergent channels. This is because when the volumetric flow rate is increased, the velocity in the geometries experienced pressure and the velocity is enhanced due to extra pressure of the flowing fluid, the fluid entering the channel and gaining high velocity.

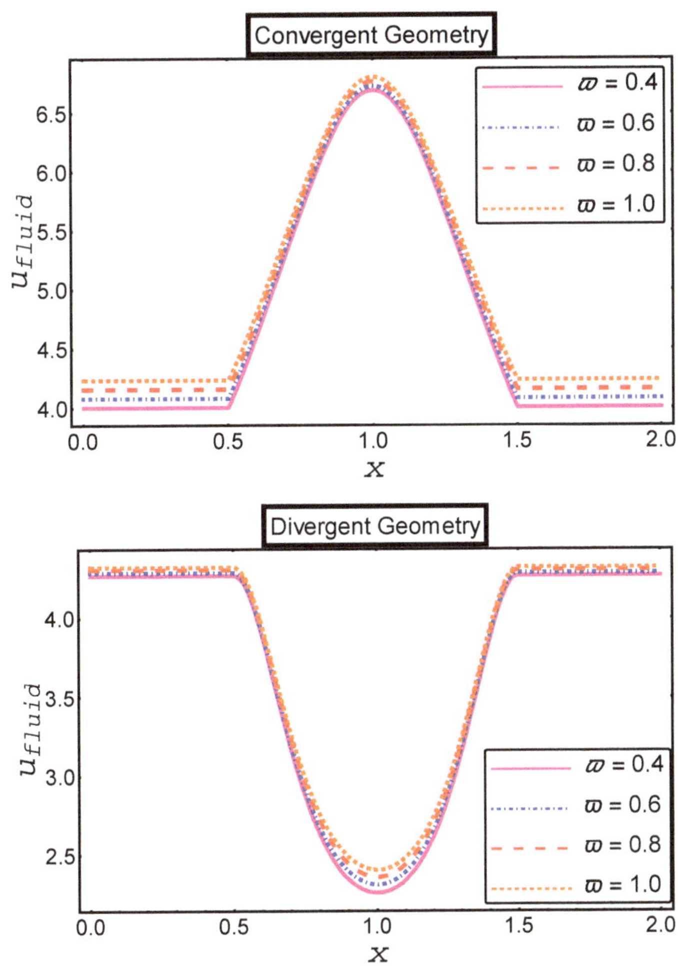

Figure 3. Impact of fourth-grade parameter on fluid velocity.

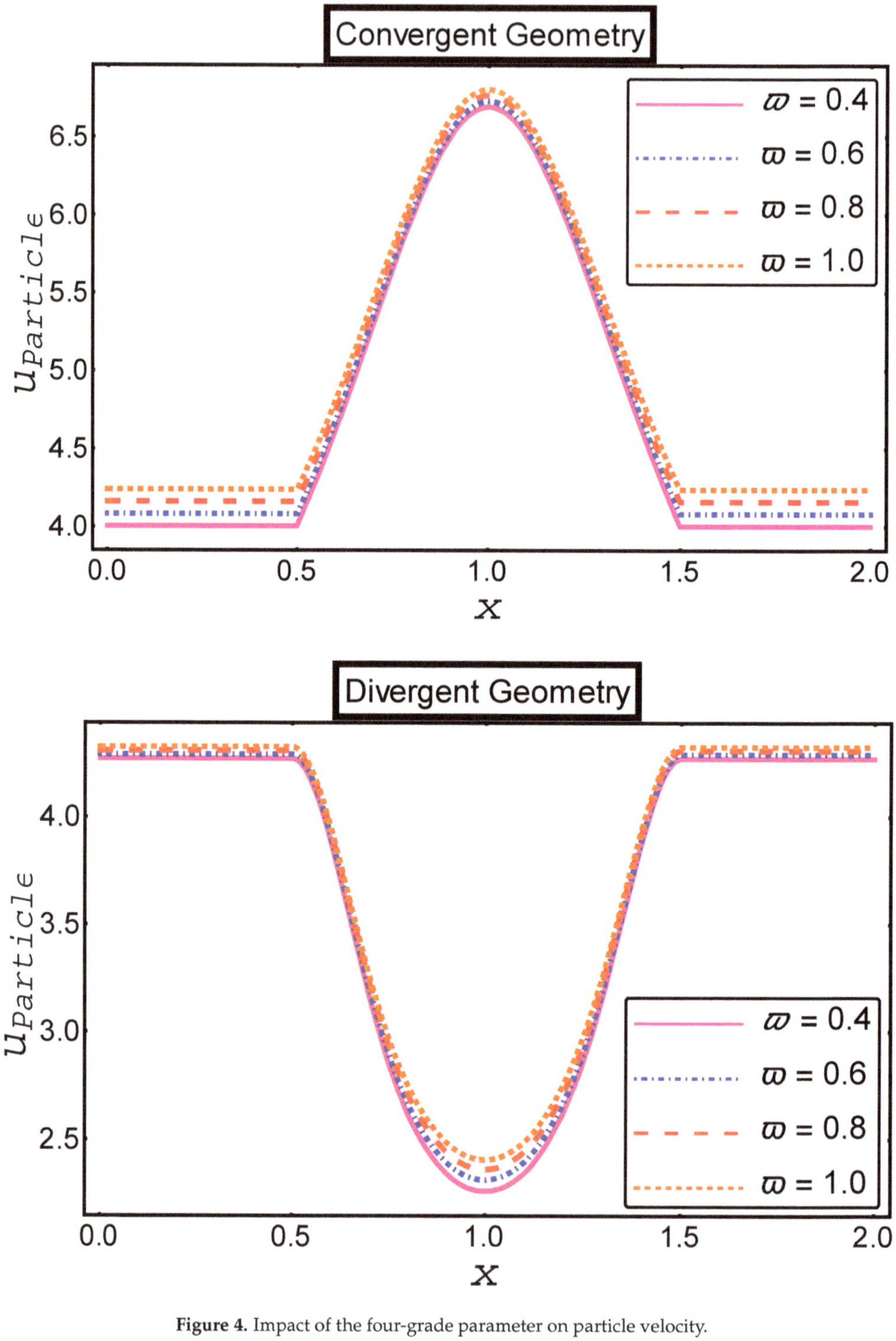

Figure 4. Impact of the four-grade parameter on particle velocity.

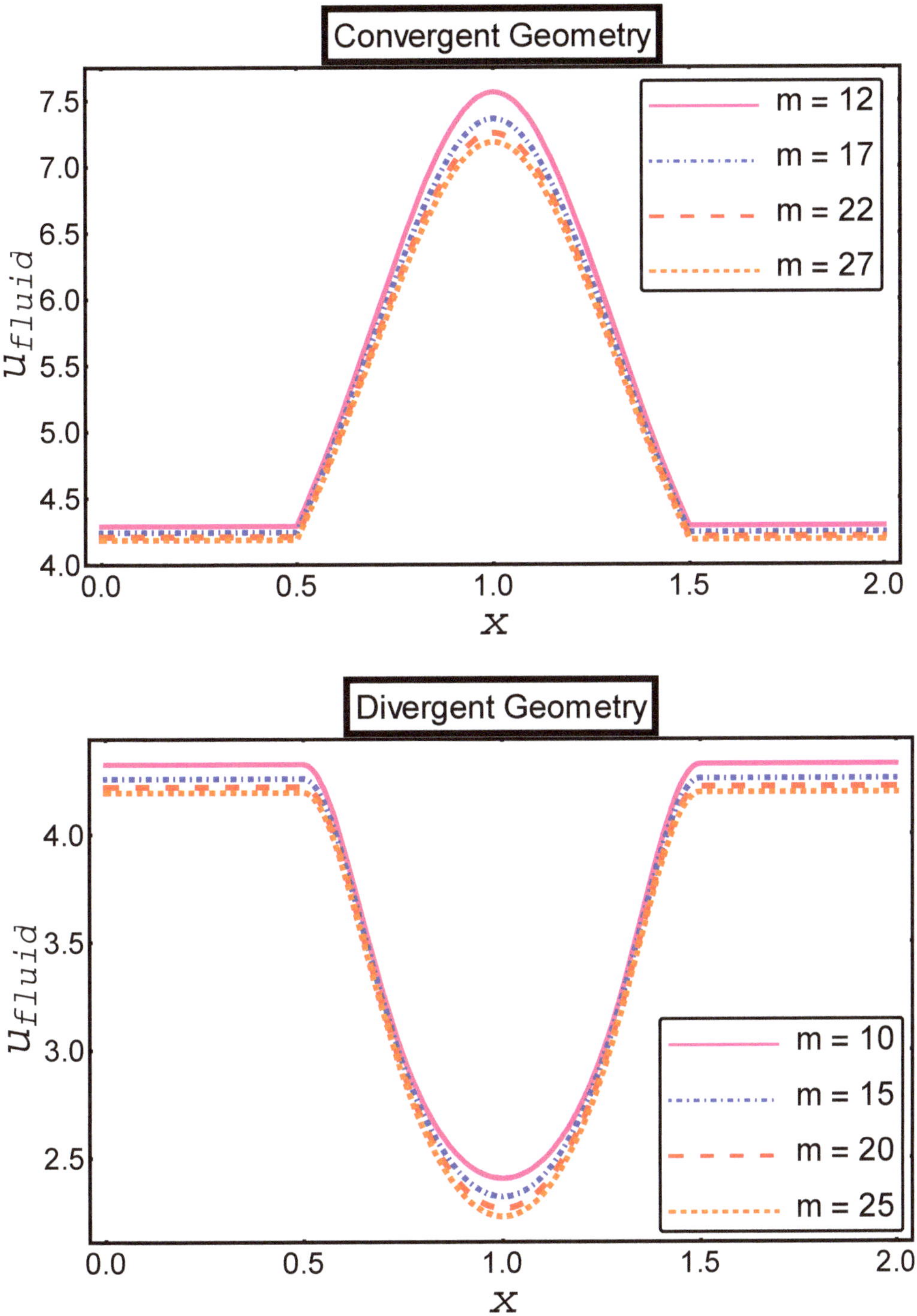

Figure 5. Impact of electro-osmotic parameters on fluid velocity.

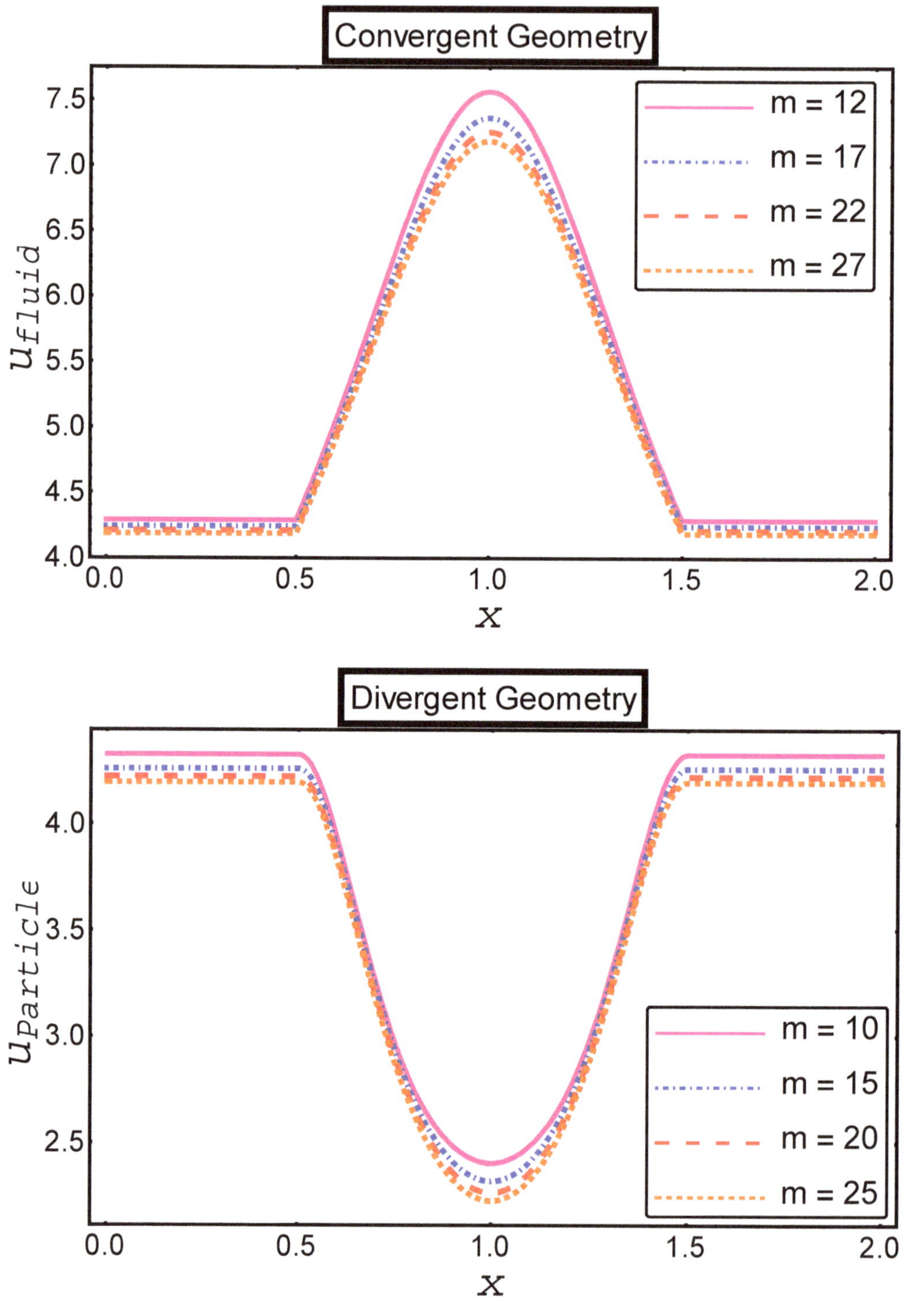

Figure 6. Impact of electro-osmotic parameters on particle velocity.

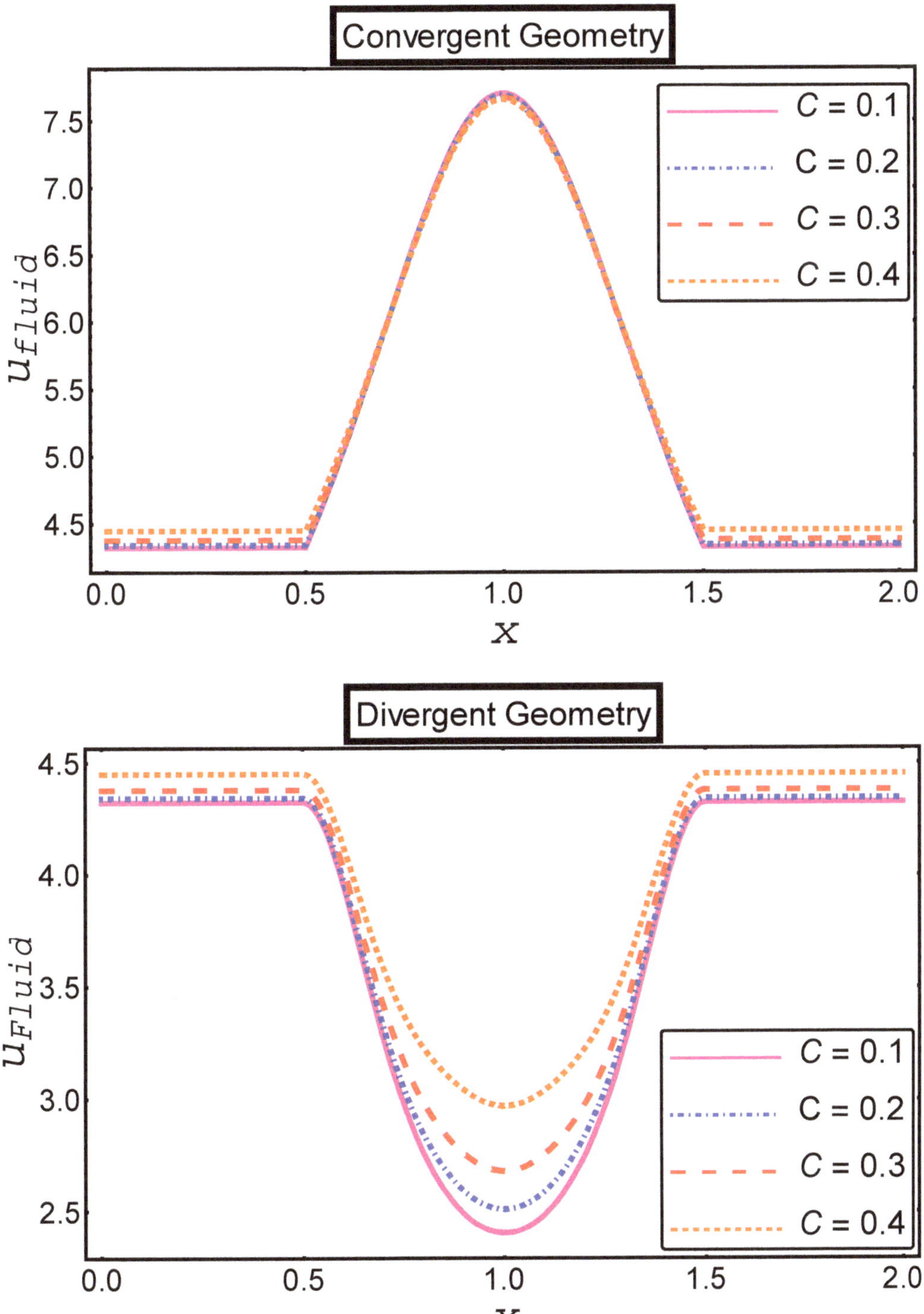

Figure 7. Impact of particle concentration on fluid velocity.

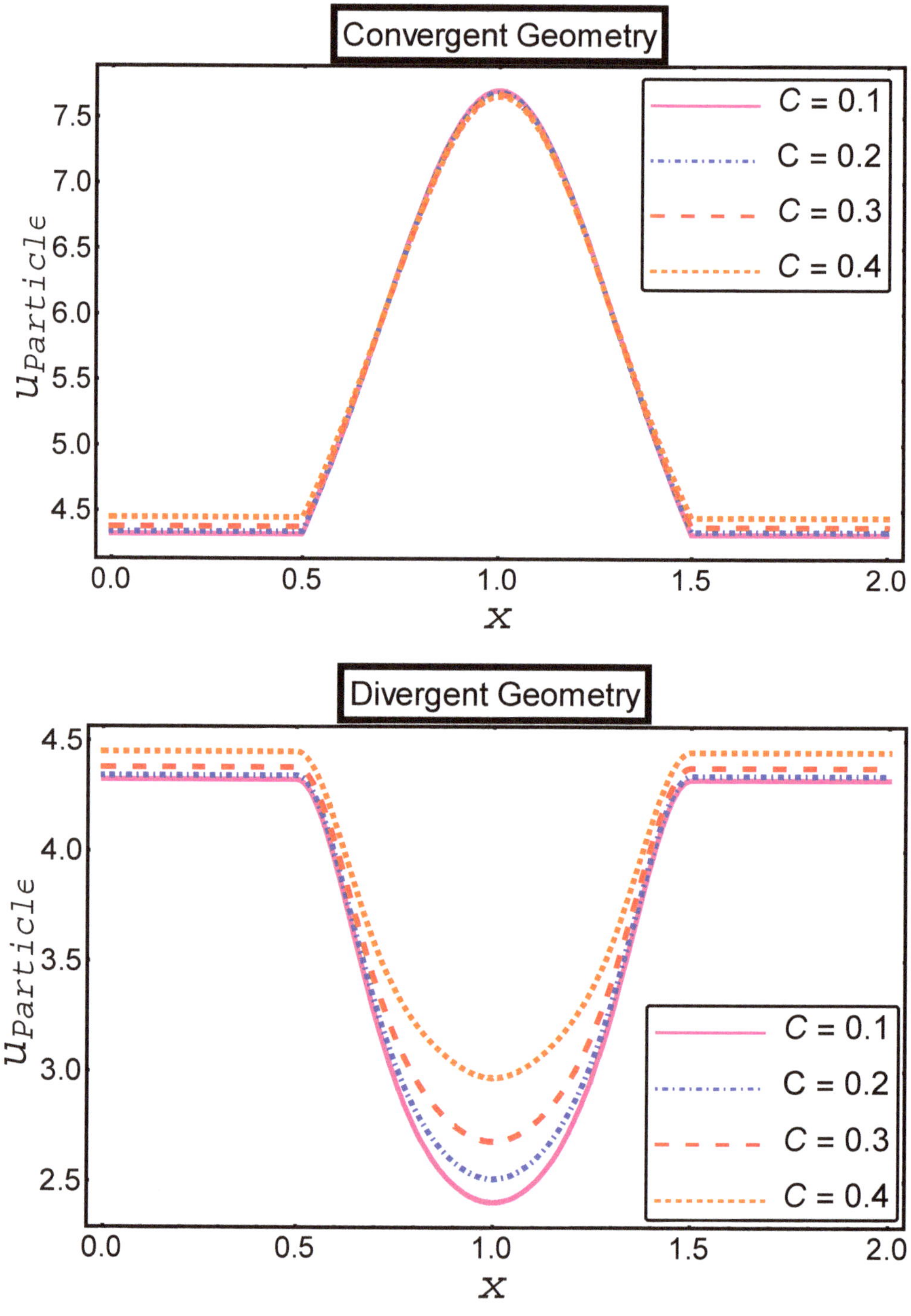

Figure 8. Impact of particle concentration on particle velocity.

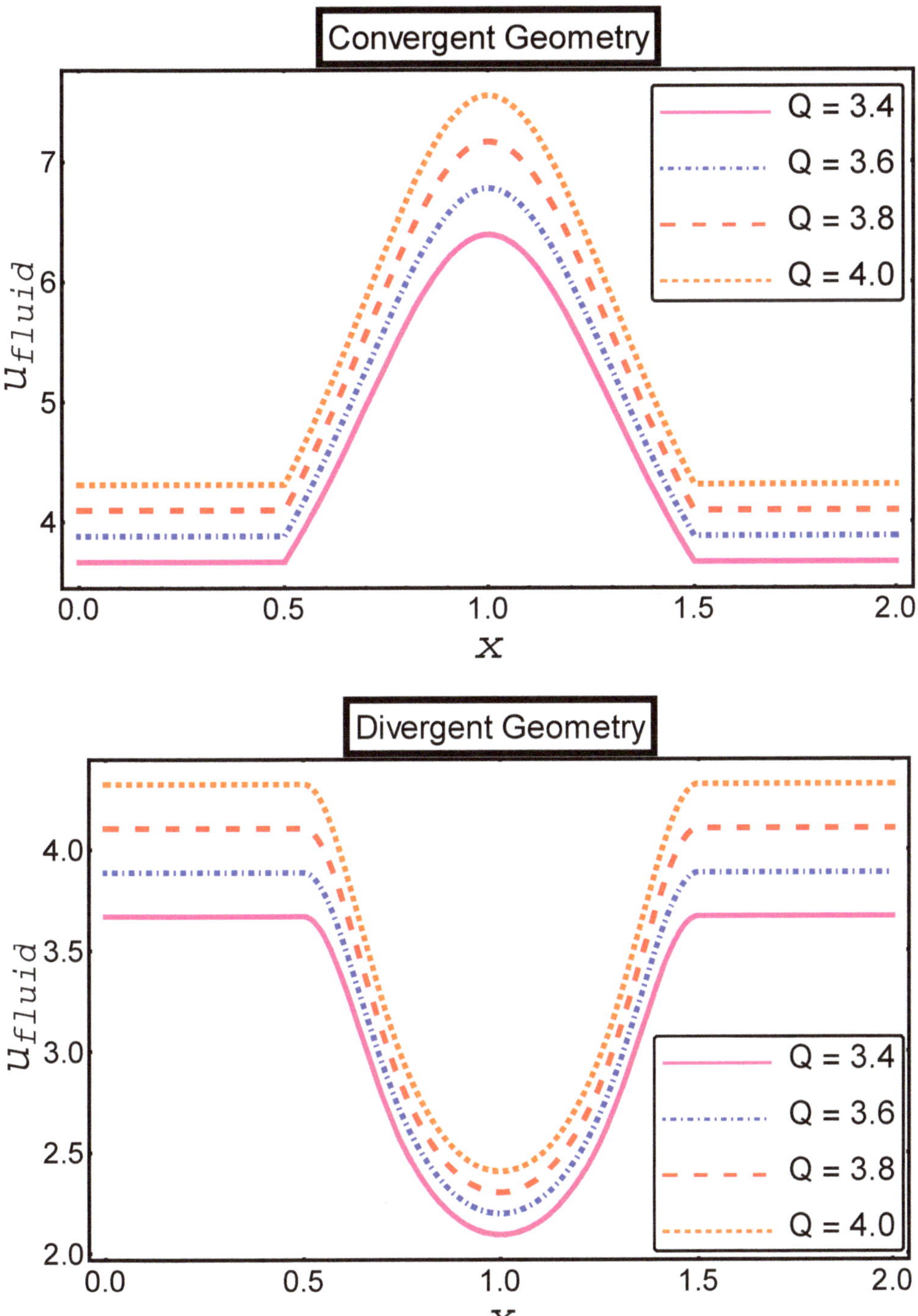

Figure 9. Impact of volumetric flow rate on fluid velocity.

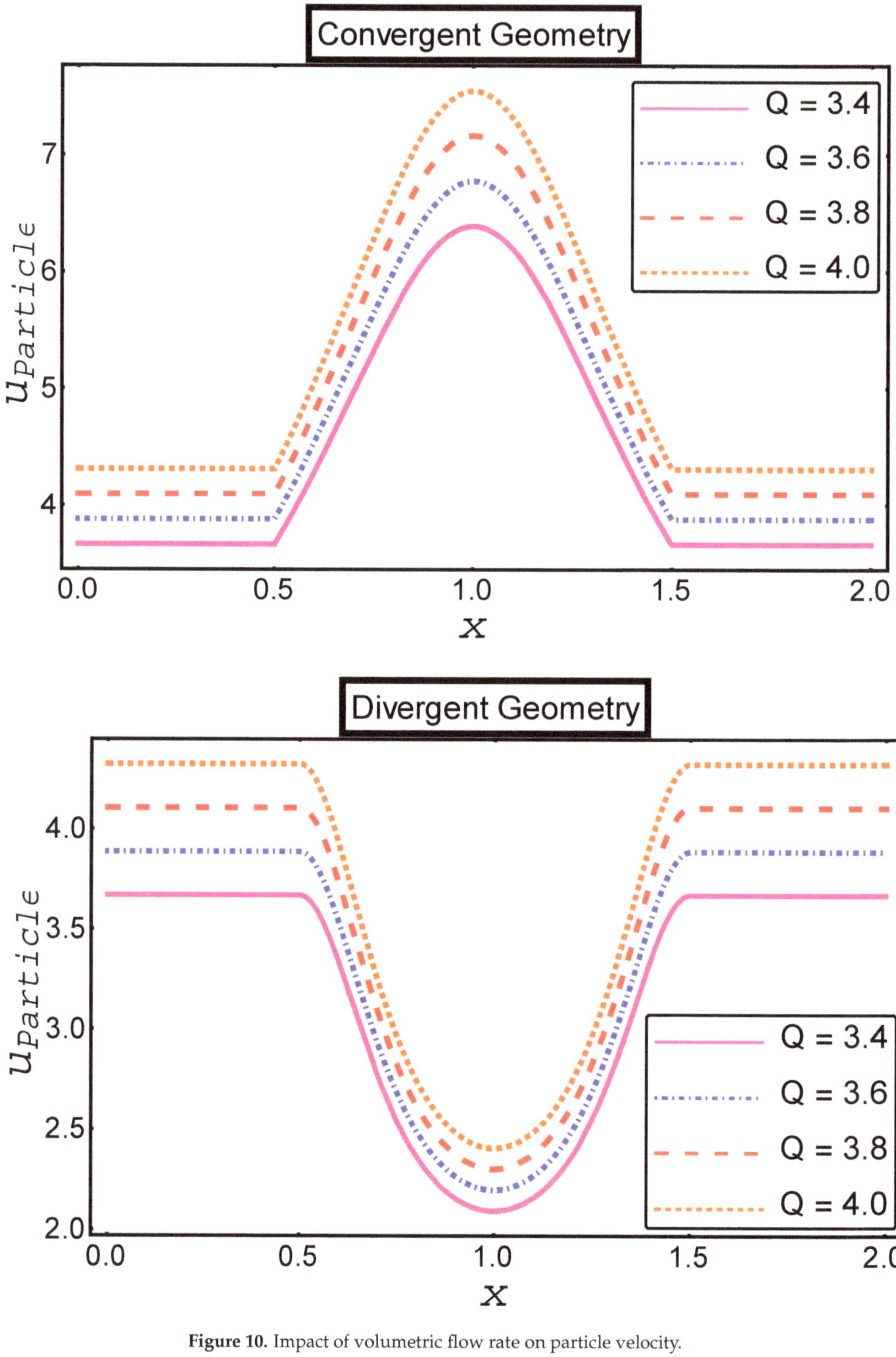

Figure 10. Impact of volumetric flow rate on particle velocity.

7. Concluding Remarks

A closed-form pronouncement for the velocity dispersal of utterly evolve flow of hafnium particles and fourth-grade base fluid adjournment via two different geometries diverse in shape are dispensed. The impact of germane parameters such as fourth-grade parameters, electro-osmotic parameter, the concentration of nanoparticles and volumetric flow rate in a couple of channels such as convergent and divergent flow has been exhibited and inspected graphically. The most noteworthy remarks itemized are:

- ❖ An increase in the behavior of both particle and fluid phase velocities is viewed in convergent and divergent geometries when enhancement is made in the fourth-grade parameter;
- ❖ A remarkable decrease in the velocity profiles of fluid and particle phases in both channels is noted when the value of the electro-osmotic parameter is enhanced;
- ❖ The credible incline is measured in the velocity profile of both phases in the divergent channel when the value of particle concentration is increased, and a very dubious decline has been seen in the velocities of both phases in the convergent channel;
- ❖ When the volumetric flow rate upraised in both channels the velocity profile of fluid and particle phases improved as the volumetric flow rate more in velocities.

Author Contributions: Conceptualization, M.N., M.M.G., M.R.G. and M.K.H.; Methodology, N.F. and M.M.A.L.; Software, M.N. and M.K.H.; Validation, N.F., M.N., M.M.G., M.R.G. and M.K.H.; Formal analysis, M.M.A.L.; Investigation, N.F. and M.R.G.; Data curation, M.M.A.L.; Writing—original draft, M.N., M.M.G. and M.K.H. All authors have read and agreed to the published version of the manuscript.

Funding: This research was funded by [Princess Nourah bint Abdulrahman University, Riyadh, Saudi Arabia] grant number [PNURSP2023R152].

Data Availability Statement: The datasets used and/or analyzed during the current study are avail-able from the corresponding author upon reasonable request.

Acknowledgments: The authors would like to acknowledge Princess Nourah bint Abdulrahman University Researchers Supporting Project number (PNURSP2023R152), Princess Nourah bint Abdulrahman University, Riyadh, Saudi Arabia for their support.

Conflicts of Interest: The authors declare no conflict of interest.

References

1. Islam, S.; Bano, Z.; Siddique, I.; Siddiqui, A. The optimal solution for the flow of a fourth-grade fluid with partial slip. *Comput. Math. Appl.* **2011**, *61*, 1507–1516. [CrossRef]
2. Hayat, T.; Noreen, S. Peristaltic transport of fourth grade fluid with heat transfer and induced magnetic field. *Comptes Rendus Mécanique* **2010**, *338*, 518–528. [CrossRef]
3. Sajid, M.; Hayat, T.; Asghar, S. On the analytic solution of the steady flow of a fourth grade fluid. *Phys. Lett. A* **2006**, *355*, 18–26. [CrossRef]
4. Salawu, S.; Fatunmbi, E.; Ayanshola, A. On the diffusion reaction of fourth-grade hydromagnetic fluid flow and thermal criticality in a plane Couette medium. *Results Eng.* **2020**, *8*, 100169. [CrossRef]
5. Khan, A.A.; Masood, F.; Ellahi, R.; Bhatti, M. Mass transport on chemicalized fourth-grade fluid propagating peristaltically through a curved channel with magnetic effects. *J. Mol. Liq.* **2018**, *258*, 186–195. [CrossRef]
6. Sobamowo, M.; Akinshilo, A. Analysis of flow, heat transfer and entropy generation in a pipe conveying fourth grade fluid with temperature dependent viscosities and internal heat generation. *J. Mol. Liq.* **2017**, *241*, 188–198. [CrossRef]
7. Nadeem, S.; Ali, M. Analytical solutions for pipe flow of a fourth grade fluid with Reynold and Vogel's models of viscosities. *Commun. Nonlinear Sci. Numer. Simul.* **2009**, *14*, 2073–2090. [CrossRef]
8. Aziz, T.; Mahomed, F. Reductions and solutions for the unsteady flow of a fourth grade fluid on a porous plate. *Appl. Math. Comput.* **2013**, *219*, 9187–9195. [CrossRef]
9. Yan, S.-R.; Toghraie, D.; Abdulkareem, L.A.; Alizadeh, A.; Barnoon, P.; Afrand, M. The rheological behavior of MWCNTs–ZnO/Water–Ethylene glycol hybrid non-Newtonian nanofluid by using of an experimental investigation. *J. Mater. Res. Technol.* **2020**, *9*, 8401–8406. [CrossRef]
10. Palencia, J.L.D.; Rahman, S.U.; Redondo, A.N. Regularity and reduction to a Hamilton-Jacobi equation for a MHD Eyring-Powell fluid. *Alex. Eng. J.* **2022**, *61*, 12283–12291. [CrossRef]

11. Subbarayudu, K.; Suneetha, S.; Ankireddy, B. The assessment of time dependent flow of Williamson fluid with radiative blood flow against a wedge. *Propuls. Power Res.* **2020**, *9*, 87–99. [CrossRef]
12. Majeed, A.H.; Mahmood, R.; Shahzad, H.; Pasha, A.A.; Raizah, Z.; Hosham, H.A.; Reddy, D.S.K.; Hafeez, M.B. Heat and mass transfer characteristics in MHD Casson fluid flow over a cylinder in a wavy channel: Higher-order FEM computations. *Case Stud. Therm. Eng.* **2023**, *42*, 102730. [CrossRef]
13. Asfour, H.A.H.; Ibrahim, M.G. Numerical simulations and shear stress behavioral for electro-osmotic blood flow of magneto Sutterby nanofluid with modified Darcy's law. *Therm. Sci. Eng. Prog.* **2023**, *37*, 101599. [CrossRef]
14. Asghar, Z.; Waqas, M.; Gondal, M.A.; Khan, W.A. Electro-osmotically driven generalized Newtonian blood flow in a divergent micro-channel. *Alex. Eng. J.* **2022**, *61*, 4519–4528. [CrossRef]
15. Wang, X.; Xu, H.; Qi, H. Numerical analysis for rotating electro-osmotic flow of fractional Maxwell fluids. *Appl. Math. Lett.* **2020**, *103*, 106179. [CrossRef]
16. Ali, F.; Iftikhar, M.; Khan, I.; Sheikh, N.A.; Aamina; Nisar, K.S. Time fractional analysis of electro-osmotic flow of Walters's-B fluid with time-dependent temperature and concentration. *Alex. Eng. J.* **2019**, *59*, 25–38. [CrossRef]
17. Trivedi, M.; Maurya, S.; Nirmalkar, N. Numerical simulations for electro-osmotic flow of PTT fluids in diverging microchannel. *Mater. Today Proc.* **2022**, *57*, 1765–1769. [CrossRef]
18. Miao, H.; Dokhani, V.; Ma, Y.; Zhang, D. Numerical modeling of laminar and turbulent annular flows of power-law fluids in partially blocked geometries. *Results Eng.* **2023**, *17*, 100930. [CrossRef]
19. Pan, T.-W.; Chiu, S.-H. A DLM/FD method for simulating balls settling in Oldroyd-B viscoelastic fluids. *J. Comput. Phys.* **2023**, *484*, 112071. [CrossRef]
20. Tahraoui, Y.; Cipriano, F. Optimal control of two dimensional third grade fluids. *J. Math. Anal. Appl.* **2023**, *523*, 127032. [CrossRef]
21. Mekheimer, K.S.; El Shehawey, E.F.; Elaw, A.M. Peristaltic Motion of a Particle-Fluid Suspension in a Planar Channel. *Int. J. Theor. Phys.* **1998**, *37*, 2895–2920. [CrossRef]
22. Hussain, F.; Ellahi, R.; Zeeshan, A. Mathematical Models of Electro-Magnetohydrodynamic Multiphase Flows Synthesis with Nano-Sized Hafnium Particles. *Appl. Sci.* **2018**, *8*, 275. [CrossRef]
23. Ellahi, R.; Zeeshan, A.; Hussain, F.; Abbas, T. Thermally Charged MHD Bi-Phase Flow Coatings with Non-Newtonian Nanofluid and Hafnium Particles along Slippery Walls. *Coatings* **2019**, *9*, 300. [CrossRef]
24. Ellahi, R.; Zeeshan, A.; Hussain, F.; Abbas, T. Two-Phase Couette Flow of Couple Stress Fluid with Temperature Dependent Viscosity Thermally Affected by Magnetized Moving Surface. *Symmetry* **2019**, *11*, 647. [CrossRef]

mathematics

Article

Heat and Mass Transfer Analysis for the Viscous Fluid Flow: Dual Approximate Solutions

Remus-Daniel Ene [1,†], Nicolina Pop [2,*,†] and Rodica Badarau [3,†]

1. Department of Mathematics, Politehnica University of Timisoara, 2 Victoria Square, 300006 Timisoara, Romania
2. Department of Physical Foundations of Engineering, Politehnica University of Timisoara, 2 Vasile Parvan Blvd, 300223 Timisoara, Romania
3. Department of Mechanical Machines, Equipment and Transportation, Politehnica University of Timisoara, 1 Mihai Viteazul Blvd., 300222 Timisoara, Romania

* Correspondence: nicolina.pop@upt.ro
† These authors contributed equally to this work.

Abstract: The aim of this paper is to investigate effective and accurate dual analytic approximate solutions, while taking into account thermal effects. The heat and mass transfer problem in a viscous fluid flow are analytically explored by using the modified Optimal Homotopy Asymptotic Method (OHAM). By using similarity transformations, the motion equations are reduced to a set of nonlinear ordinary differential equations. Based on the numerical results, it was revealed that there are dual analytic approximate solutions within the mass transfer problem. The variation of the physical parameters (the Prandtl number and the temperature distribution parameter) over the temperature profile is analytically explored and graphically depicted for the first approximate and the corresponding dual solution, respectively. The advantage of the proposed method arises from using only one iteration for obtaining the dual analytical solutions. The presented results are effective, accurate and in good agreement with the corresponding numerical results with relevance for further engineering applications of heat and mass transfer problems.

Keywords: Optimal Homotopy Asymptotic Method; boundary layer flow; viscous fluid flow; heat transfer; exponential stretching sheet

MSC: 65L60; 76A10; 76D10; 76D05; 76M55

Citation: Ene, R.-D.; Pop, N.; Badarau, R. Heat and Mass Transfer Analysis for the Viscous Fluid Flow: Dual Approximate Solutions. *Mathematics* **2023**, *11*, 1648. https://doi.org/10.3390/math11071648

Academic Editor: Ramoshweu Solomon Lebelo

Received: 15 February 2023
Revised: 11 March 2023
Accepted: 20 March 2023
Published: 29 March 2023

1. Introduction

Boundary layer behaviour over a moving continuous solid surface can be observed in many important technological processes and involves thermal effects, which show the characteristics of non-Newtonian fluids.

An important effect is viscous dissipation when the velocity gradient is high. The analysis of the temperature field as modified by the generation or absorption of heat in moving fluids is relevant for some physical problems, as presented by Sparrow and Cess [1], Topper [2], and Khashi et al. [3]. Further, the contributions of the suction parameter, Prandtl number, the heat source/sink parameter and the Eckert number to the heat transfer characteristics are found to be quite significant in [4].

In recent years, many the analytical methods have attempted to provide the solutions of different nonlinear models involving thermal effects.

Xu [5] analytically solved the mixed convection flow of a hybrid nanofluid in an inclined channel with top wall-slip due to wall stripe and constant heat flux conditions. Hayat et al. [6] analytically examined the melting phenomenon in the two-dimensional (2D) flow of fourth-grade material over a stretching surface, while taking into account the existence of the Cattaneo–Christov (C-C) heat flux. The heat and mass transfer characteristics for a self-similarity boundary layer of an exponentially stretching surface were investigated by [7] using the Homotopy Analysis Method (HAM). This method is performed by several researchers, such as Khan et al. [8], Khan et al. [9], Khan et al. [10], Khan et al. [11], Zuhra et al. [12], Bilal et al. [13], and Shehzad et al. [14], who examine the thermal effect. Alizadeh et al. [15] solved the transient flow and heat transfer of a non-newtonian fluid (Casson fluid) between parallel disks in the presence of an external magnetic field semi-analytically using Least Square Method. Huaxing et al. [16] combined the effects of molecular and thermal diffusion processes by means of a generalized integral transform technique (GITT).

Some methods provide numerical solutions, such as those of Nadeem et al. [17], Abbasi et al. [18], Xie et al. [19], Abdelaziz et al. [20], Muhammad et al. [21], Mabood et al. [22], and Eid et al. [23], who numerically analyzed the flow and heat transfer resulting from an exponentially decreased sheet of hybrid nanoparticles, using the Runge–Kutta–Fehlberg method (RKF45) with the shooting technique. Boumaiza et al. [24] numerically investigated the effects of variable thermal conductivity in mixed convection in the presence of an external magnetic field using the Runge–Kutta–Fehlberg method (RKF) based on the shooting technique, and analytically by using the differential transform method (DTM). Gireesha et al. [25] numerically explored the thermal performance of a fully wet stretching/shrinking longitudinal fin with an exponential profile. Waini et al. [26] numerically solved the magnetohydrodynamic (MHD) mixed convection flow by considering thermal radiation. Tang et al. [27] applied some parallel finite element (FE) iterative methods for stationary incompressible magnetohydrodynamics (MHD).

For the analysis of many physical phenomena, numerical schemes or analytical/geometrical methods are applied in [28–35].

The Optimal Homotopy Asymptotic Method (OHAM) developed by Marinca et al. [36–41], and successfully applied to solve nonlinear equations arising in heat transfer [42–49], is used in the present paper to obtain effective and accurate dual analytic approximate solutions while taking into account the thermal effects.

The advantages of this procedure in comparison with HAM include the independence of small or large parameters, and the ease of optimally controlling the convergence of the approximate solutions.

Based on the mathematical model development in [7], in the present work, the OHAM technique is used to obtain effective and accurate dual analytic approximate solutions, while taking into account the thermal effects. Therefore, the novelty of our work is represented by the dual solutions of the mathematical model with the OHAM technique using only one iteration in comparison with [7], where only one solution is presented with the HAM method. Furthermore, ref. [7] did not elaborate on the possibility of dual solutions.

The paper is organized as follows: The Introduction is followed by a brief description of the two-dimensional flow of an incompressible viscous fluid passing a continuous stretching surface, taking into account the thermal effect. The steps of the OHAM technique are presented in Section 3. Section 4 presents the heat and mass transfer problem by the modified OHAM. Our results and some interesting behaviours of the effects of nonlinear stretching on flow and heat transfer characteristics are discussed in Section 5. The paper ends with conclusions.

2. Equations of Motion

In this section, the two-dimensional flow of an incompressible viscous fluid passes a continuous stretching surface in the half-plane, $y > 0$, taking into account the thermal effect. Additionally, the occurrence of the flow without suction/blowing and without partial slip is explored.

The schematic of the physical model is presented in Figure 1.

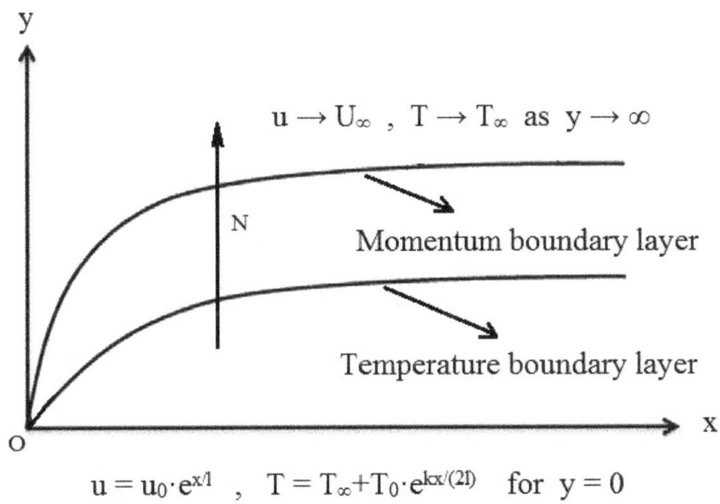

Figure 1. Schematic diagram of the physical model.

For the constant pressure at the boundary layer, the continuity, momentum and temperature equations governing the fluid flow are given by [7]:

$$\begin{aligned} &\frac{\partial u}{\partial x} + \frac{\partial v}{\partial y} = 0, \\ &\rho \cdot \left(u\frac{\partial u}{\partial x} + v\frac{\partial u}{\partial y} \right) = \mu \frac{\partial^2 u}{\partial y^2}, \end{aligned} \tag{1}$$

$$u\frac{\partial T}{\partial x} + v\frac{\partial T}{\partial y} = \alpha \frac{\partial^2 T}{\partial y^2} . \tag{2}$$

The physical initial/boundary conditions can be written in the following form [7]:

$$\begin{aligned} &y = 0 \, : \, u = u_0 \cdot e^{\frac{x}{l}} \, , \; v = v_w \, , \; T = T_\infty + T_0 \cdot e^{\frac{k \cdot x}{2l}} , \\ &y \to \infty \, : \, u \to U_\infty \, , \quad T \to T_\infty \; . \end{aligned} \tag{3}$$

By means of the similarity transformations,

$$\begin{aligned} &\eta = \sqrt{\frac{u_0 \cdot \rho \cdot l}{2\mu}} \cdot \frac{y}{l} \cdot e^{(k \cdot x)/(2l)} , \quad u = u_0 e^{x/l} f'(\eta), \\ &v = -\frac{\mu}{\rho l} \cdot \sqrt{\frac{u_0 \cdot \rho \cdot l}{2\mu}} \cdot e^{x/(2l)} [f(\eta) + \eta \cdot f'(\eta)], \\ &T = T_\infty + T_0 \cdot e^{(k \cdot x)/(2l)} \theta(\eta) \end{aligned} \tag{4}$$

and by inserting Equation (4) into Equations (1)–(3), we obtain:

$$f''' + ff'' - 2(f')^2 = 0, \tag{5}$$

$$\theta'' + pr(f\theta' - k \cdot f'\theta) = 0 \; . \tag{6}$$

with the initial/boundary conditions:

$$f(0) = 0\,, \quad f'(0) = 1\,, \quad \theta(0) = 1\,,$$
$$f' \to 0\,, \quad \theta \to 0 \quad \text{for}\ \eta \to \infty, \tag{7}$$

where the prime denotes differentiation with respect to η.

3. The Modified Optimal Homotopy Asymptotic Method (OHAM)

The steps of the modified OHAM technique [36] are presented in detail below:

(i) The nonlinear differential equation has the following general form:

$$\mathcal{L}_\varphi\Big(\varphi(\eta)\Big) + \mathcal{N}_\varphi\Big(\varphi(\eta)\Big) = 0, \tag{8}$$

under the boundary/initial conditions

$$\mathcal{B}_\varphi\left(\varphi(\eta), \frac{d\varphi(\eta)}{d\eta}\right) = 0, \tag{9}$$

where $\mathcal{L}_\varphi$ is an arbitrary linear operator, $\mathcal{N}_\varphi$ is the corresponding nonlinear operator and $\mathcal{B}_\varphi$ is an operator describing the boundary conditions.

(ii) The homotopic relation is given by:

$$\mathcal{H}_\varphi\Big[\mathcal{L}_\varphi\Big(\varphi(\eta,p)\Big),\ H(\eta,C_i),\ \mathcal{N}_\varphi\Big(\varphi(\eta,p)\Big)\Big] =$$
$$= \mathcal{L}_\varphi\Big(\varphi_0(\eta)\Big) + G_0(\eta) + p\Big[\mathcal{L}_\varphi\Big(\varphi_1(\eta,C_i)\Big) - H(\eta,C_i)\mathcal{N}_\varphi\Big(\varphi_0(\eta)\Big)\Big], \tag{10}$$

where $G_0(\eta)$ is a given continuous function, $p \in [0,\ 1]$ is the embedding parameter and $H(\eta, C_i) \neq 0$ is an auxiliary convergence–control function depending on the variable η and of the convergence–control parameters $C_1, C_2, \ldots, C_s$, and choosing the unknown function $\varphi(\eta)$ in the following form:

$$\varphi(\eta,p) = \varphi_0(\eta) + p\varphi_1(\eta, C_i), \tag{11}$$

and by equating the coefficients of p^0 and p^1, respectively, we obtain:

- the zeroth-order deformation problem

$$\mathcal{L}_\varphi\Big(\varphi_0(\eta)\Big) + G_0(\eta) = 0, \quad \mathcal{B}_\varphi\left(\varphi_0(\eta), \frac{d\varphi_0(\eta)}{d\eta}\right) = 0, \tag{12}$$

- the first-order deformation problem

$$\mathcal{L}_\varphi\Big(\varphi_1(\eta,C_i)\Big) = H(\eta,C_i)\mathcal{N}_\varphi\Big(\varphi_0(\eta)\Big),$$
$$\mathcal{B}_\varphi\left(\varphi_1(\eta,C_i), \tfrac{d\varphi_1(\eta,C_i)}{d\eta}\right) = 0, \ i = 1,2,\ldots,s. \tag{13}$$

(iii) $\varphi_0(\eta)$ could be obtained by solving the linear Equation (12).

(iv) In Equation (13), the expression $\mathcal{N}_\varphi$ has the following general form:

$$\mathcal{N}_\varphi(\varphi_0(\eta)) = \textstyle\sum_{i=1}^{n} h_i(\eta) g_i(\eta), \tag{14}$$

where n is a positive integer, and $h_i(\eta)$ and $g_i(\eta)$ are known elementary functions that depend on $\varphi_0(\eta)$ and on $\mathcal{N}_\varphi$.

The Equation (13) is a non-homogenous differential equation.

By means of the general theory of the differential equations, the computation of the function $\varphi_1(\eta, C_i)$ has the following form:

$$\varphi_1(\eta, C_i) = \sum_{i=1}^{m} H_i(\eta, h_j(\eta), C_j) g_i(\eta), \quad j = 1, \ldots, s, \tag{15}$$

or

$$\varphi_1(\eta, C_i) = \sum_{i=1}^{m} H_i(\eta, g_j(\eta), C_j) h_i(\eta), \quad j = 1, \ldots, s,$$

$$\mathcal{B}_\varphi\left(\varphi_1(\eta, C_i), \frac{d\varphi_1(\eta, C_i)}{d\eta}\right) = 0, \tag{16}$$

where $m \in \mathbf{N}^*$ is an arbitrary number.

The above expressions of $H_i(\eta, h_j(\eta), C_j)$ contain linear combinations of the elementary functions $h_j, j = 1, \ldots, s$ and the parameters $C_j, j = 1, \ldots, s$.

(v) By means of Equation (11) for $p = 1$, the first-order analytical approximate solution of Equations (8) and (9), namely the OHAM-solution, is:

$$\overline{\varphi}(\eta, C_i) = \varphi(\eta, 1) = \varphi_0(\eta) + \varphi_1(\eta, C_i). \tag{17}$$

The parameters $C_1, C_2, \ldots, C_s$ can be optimally identified by means of various methods, such as the Galerkin method, the collocation method, the Kantorowich method, the least square method or the weighted residual method.

Thus, the first-order approximate solution (17) is well-determined.

4. Heat and Mass Transfer Problem

Based on a previous paper [50], the dual approximate solutions $\bar{f}(\eta)$ for Equation (5) are established.

The skin-friction coefficient is $\bar{f}''(0) = -1.2818085481$ for the first solution and $\bar{f}''(0) = -1.2916563038$ for the corresponding dual solution, respectively.

Using the same modified OHAM procedure, the approximate solutions, denoted by $\bar{\theta}$ of Equations (6) and (7) (for the unknown function θ), were obtained.

The expression of the linear operator $L_\theta(\eta)$ could be:

$$L_\theta(\eta) = \theta'' + K_1\theta', \tag{18}$$

where $K_1 > 0$ is an unknown parameter at this moment.

From Equation (6), the nonlinear operator N_θ corresponding to the unknown function θ becomes:

$$N_\theta(\eta) = -K_1\theta' + pr(f\theta' - k \cdot f'\theta). \tag{19}$$

There are a number of possibilities to choose from for the known function $G_0(\eta)$, including the following:

$$G_0(\eta) = (a_0 + a_1\eta + a_2\eta^2) \cdot e^{-K_2\eta}, \tag{20}$$

or

$$G_0(\eta) = (a_0 + a_1\eta + a_2\eta^2 + a_3\eta^3) \cdot e^{-K_2\eta},$$

or

$$G_0(\eta) = (a_0 + a_1\eta) \cdot e^{-K_2\eta} + (b_0 + b_1\eta + b_2\eta^2) \cdot e^{-2 \cdot K_2\eta},$$

or

$$G_0(\eta) = (a_0 + a_1\eta + a_2\eta^2) \cdot e^{-K_2\eta} + (b_0 + b_1\eta + b_2\eta^2) \cdot e^{-K_3\eta},$$

and so on.

4.1. The Zeroth-Order Deformation Problem

Choosing for $G(\eta)$ the expression given by Equation (20), Equation (12) becomes:

$$\theta_0'' + K_1\theta_0' + (a_0 + a_1\eta + a_2\eta^2)\cdot e^{-K_2\eta} = 0,\quad \theta_0(0) = 1,\quad \theta_0(\infty) = 0 \tag{21}$$

with the solution

$$\theta_0(\eta) = (1 - b_0)e^{-K_1\eta} + (b_0 + b_1\eta + b_2\eta^2)\cdot e^{-K_2\eta}, \tag{22}$$

where b_0, b_1, b_2 depend on a_0, a_1, a_2, K_1, K_2 and will be optimally identified.

4.2. The First-Order Deformation Problem

Taking into account the function $\theta_0(\eta)$ (22), the nonlinear operator $N_{\theta_0}(\eta)$ from Equation (19) is:

$$\begin{aligned} N_{\theta_0}(\eta) &= m_0e^{-K_1\eta} + m_1e^{-(K+K_1)\eta} + \\ &+(n_0 + n_1\eta + n_2\eta^2)\cdot e^{-K_2\eta} + (p_0 + p_1\eta + p_2\eta^2)\cdot e^{-(K+K_2)\eta}, \end{aligned} \tag{23}$$

where the unknown convergence-control parameters m_0, m_1, n_0, n_1, n_2, p_0, p_1, p_2, K_1, K_2 will be optimally identified and they depend on b_0, b_1, b_2, K ($K = -1.1041868797$, for the first solution and $K = -3.2611576654$, for the corresponding dual solution, respectively [50]) and the physical parameters pr, k, respectively.

The comparison between the Equations (14) and (23) yields:

$$\begin{aligned} h_1^*(\eta) &= m_0, & g_1^*(\eta) &= e^{-K_1\eta}, \\ h_2^*(\eta) &= m_1, & g_2^*(\eta) &= e^{-(K+K_1)\eta}, \\ h_3^*(\eta) &= n_0 + n_1\eta + n_2\eta^2, & g_3^*(\eta) &= e^{-K_2\eta}, \\ h_4^*(\eta) &= p_0 + p_1\eta + p_2\eta^2, & g_4^*(\eta) &= e^{-(K+K_2)\eta}. \end{aligned} \tag{24}$$

For the first-order deformation problem given by Equation (13), the first approximation $\theta_1(\eta, D_i)$, from Equation (15), becomes:

$$\begin{aligned} \theta_1(\eta, D_i) &= H_1^*(\eta, D_i)e^{-K_1\eta} + H_2^*(\eta, D_i)e^{-(K+K_1)\eta} + \\ &+H_3^*(\eta, D_i)e^{-K_2\eta} + H_4^*(\eta, D_i)e^{-(K+K_2)\eta}, \end{aligned} \tag{25}$$

where D_i are the unknown real numbers and the unknown auxiliary functions $H_1^*(\eta, D_i)$, ..., $H_4^*(\eta, D_i)$ could be written in the form:

$$\begin{aligned} &H_1^*(\eta, D_i) = D_0\eta,\ \ H_2^*(\eta, D_i) = D_7,\ \ H_3^*(\eta, D_i) = D_1 + D_2\eta + D_3\eta^2, \\ &H_4^*(\eta, D_i) = D_4 + D_5\eta + D_6\eta^2, \end{aligned} \tag{26}$$

where $D_7 = -D_1 - D_4$.

Substituting Equation (26) into Equation (25) one can obtain:

$$\begin{aligned} \theta_1(\eta, D_i) &= D_0\eta e^{-K_1\eta} + D_7e^{-(K+K_1)\eta} + \\ &+(D_1 + D_2\eta + D_3\eta^2)e^{-K_2\eta} + (D_4 + D_5\eta + D_6\eta^2)e^{-(K+K_2)\eta}. \end{aligned} \tag{27}$$

4.3. The First-Order Analytical Approximate Solution $\bar{\theta}$

From Equations (22) and (27) the first-order approximate solution given by Equation (17) is obtained:

$$\bar{\theta}(\eta, D_i) = \theta_0(\eta) + \theta_1(\eta, D_i) = (1 - b_0 + D_0\eta)e^{-K_1\eta} + D_7 e^{-(K+K_1)\eta} + \\ + \left[b_0 + D_1 + (b_1 + D_2)\eta + (b_2 + D_3)\eta^2\right]e^{-K_2\eta} + (D_4 + D_5\eta + D_6\eta^2)e^{-(K+K_2)\eta}. \quad (28)$$

5. Results and Discussion

The accuracy of the obtained results is shown by comparison of the above obtained approximate solutions with the corresponding numerical integration results, computed by means of the shooting method combined with the fourth-order Runge-Kutta method using Wolfram Mathematica 9.0 software. The goal of this section is to compute the convergence-control parameters K_1, K_2, b_0, b_1, b_2 and D_i, which appear in Equation (28), by the least square method for different values of the known parameters k and pr.

For fixed value of the parameter k and different values of the Prandtl number pr, four approximate solutions $\bar{\theta}(\eta)$ for temperature obtained from Equation (28), are presented below:

(a_1) the parameter $k = 0.25$, the Prandtl number $pr = 0.5$.

The first-order approximate solution is:

$$\bar{\theta}(\eta) = 6.4270002120 \cdot e^{-0.5270407713 \cdot \eta} + (1 + 1.0977591304 \cdot \eta) \cdot e^{-1.6312276511 \cdot \eta} + \\ +(-0.4903842473 - 0.8677675300 \cdot \eta - 0.0655183831 \cdot \eta^2) \cdot e^{-1.7700844051 \cdot \eta} + \\ +(-5.6792005322 - 0.5323882885 \cdot \eta - 0.0584055951 \cdot \eta^2) \cdot e^{-0.6658975254 \cdot \eta} + \\ +(-0.2574154323 - 0.1860126005 \cdot \eta - 0.0354344370 \cdot \eta^2) \cdot e^{-1.7700844051 \cdot \eta} \quad (29)$$

and the corresponding dual approximate solution becomes:

$$\bar{\theta}(\eta) = 0.4203972946 \cdot e^{-0.1531387065 \cdot \eta} + (1 + 0.0224661936 \cdot \eta) \cdot e^{-1.2573255862 \cdot \eta} + \\ +(-0.1188278766 - 0.1003865016 \cdot \eta - 0.0318369145 \cdot \eta^2) \cdot e^{-3.1410563458 \cdot \eta} + \\ +(-0.8375405893 - 0.3581935944 \cdot \eta - 0.0048598683 \cdot \eta^2) \cdot e^{-1.5705281729 \cdot \eta} + \\ +(0.5359711713 - 0.0282283101 \cdot \eta + 0.0012582354 \cdot \eta^2) \cdot e^{-0.4663412931 \cdot \eta}. \quad (30)$$

Other cases (a_2–a_6) for different values of the physical parameters k and pr are treated in Appendix A.

Tables 1 and 2 provides a comparison between the OHAM approximate solutions $\bar{\theta}_{OHAM}$ (temperature) given by Equations (29), (A1) and (A3) for the first solution, and the corresponding dual approximate solutions $\bar{\theta}_{OHAM}$ (temperature) given by Equations (30), (A2) and (A4), and numerical results for $k = 0.15$ for different values of the Prandtl number pr.

In Tables 3 and 4, respectively, the effect of the mass transfer coefficient $\theta'(0)$ obtained from Equations (29), (A1), (A3) and (A5) for both approximate solutions $\bar{\theta}(\eta)$ and corresponding numerical values are presented.

In the case of the approximate solution $\bar{\theta}(\eta)$ given by Equation (28), the residual from Equation (6) becomes:

$$R_{\bar{\theta}}(\eta) = \bar{\theta}''(\eta) + pr \cdot \left(\bar{f}(\eta)\bar{\theta}'(\eta) - k \cdot \bar{f}'(\eta)\bar{\theta}(\eta)\right). \quad (31)$$

The numerical values of the integral of the square residual given by Equation (31) are shown in Table 5.

Table 1. Comparison between the first-order approximate solutions $\bar{\theta}$ given by Equations (29), (A1) and (A3) and the corresponding numerical results for $k = 0.25$ and different values of the Prandtl parameter pr (absolute errors: $\epsilon_\theta = |\theta_{numerical} - \bar{\theta}_{OHAM}|$).

	$pr = 0.5$	$pr = 1$	$pr = 2.5$
η	$\theta_{numerical}$	$\theta_{numerical}$	$\theta_{numerical}$
0	1	1	1.
7/10	0.7489771062	0.6049297342	0.3645655020
7/5	0.5542053824	0.3451592660	0.1026555405
14/5	0.2979176957	0.1031937444	0.0055540922
7/2	0.2175152070	0.0553624933	0.0011533541
21/5	0.1586233161	0.0295455695	0.0002005939
28/5	0.0842151436	0.0083554452	−0.000042451
7	0.0446648315	0.0023547811	−0.000052955
	$\bar{\theta}_{OHAM}$	$\bar{\theta}_{OHAM}$	$\bar{\theta}_{OHAM}$
η	given by Equation (29)	given by Equation (A1)	given by Equation (A3)
0	1	1	1
7/10	0.7489754987	0.6049296367	0.3645643475
7/5	0.5542074180	0.3451592418	0.1026577273
14/5	0.2979183855	0.1031938730	0.0055556103
7/2	0.2175209607	0.0553624331	0.0011544802
21/5	0.1586257234	0.0295454423	0.0001987361
28/5	0.0842093003	0.0083555208	−0.0000415991
7	0.0446671512	0.0023548325	−0.0000527572
	ϵ_θ	ϵ_θ	ϵ_θ
η	for Equation (29)	for Equation (A1)	for Equation (A3)
0	0	0	0
7/10	$1.607455882179920 \times 10^{-6}$	$9.742094708720117 \times 10^{-8}$	$1.154538486203282 \times 10^{-6}$
7/5	$2.035591219806676 \times 10^{-6}$	$2.419851896640068 \times 10^{-8}$	$2.186829855102545 \times 10^{-6}$
14/5	$6.897985110887461 \times 10^{-7}$	$1.286033818881371 \times 10^{-7}$	$1.518143775084551 \times 10^{-6}$
7/2	$5.753703639754804 \times 10^{-6}$	$6.013523504849738 \times 10^{-8}$	$1.126029875720986 \times 10^{-6}$
21/5	$2.407287287453652 \times 10^{-6}$	$1.271883639693272 \times 10^{-7}$	$1.857745851002750 \times 10^{-6}$
28/5	$5.843302489164093 \times 10^{-6}$	$7.561148052809274 \times 10^{-8}$	$8.526737934899268 \times 10^{-7}$
7	$2.319736837258501 \times 10^{-6}$	$5.142016535871277 \times 10^{-8}$	$1.977883636694968 \times 10^{-7}$

Table 2. Comparison between the corresponding dual approximate solutions $\bar{\theta}$ given by Equations (30), (A2) and (A4) and the corresponding numerical results for $k = 0.25$ and different values of the Prandtl parameter pr (absolute errors: $\epsilon_\theta = |\theta_{numerical} - \bar{\theta}_{OHAM}|$).

	$pr = 0.5$	$pr = 1$	$pr = 2.5$
η	$\theta_{numerical}$	$\theta_{numerical}$	$\theta_{numerical}$
0	1	1	1
7/10	0.7858094356	0.6190556769	0.3688458421
7/5	0.6228249648	0.3691930827	0.1074930546
14/5	0.4085135781	0.1323390363	0.0072619140
7/2	0.3388566718	0.0822984147	0.0019692645
21/5	0.2852286001	0.0530118448	0.0006029330
28/5	0.2095004124	0.0244710236	0.0001201557
7	0.1593196243	0.0127723199	0.0000713911

Table 2. *Cont.*

η	$\bar{\theta}_{OHAM}$ given by Equation (30)	$\bar{\theta}_{OHAM}$ given by Equation (A2)	$\bar{\theta}_{OHAM}$ given by Equation (A4)
0	1	1	1
7/10	0.7858136455	0.6190546849	0.3688476879
7/5	0.6228223796	0.3691937408	0.1074929959
14/5	0.4085152511	0.1323390010	0.0072605998
7/2	0.3388541904	0.0822992211	0.0019705677
21/5	0.2852266049	0.0530118072	0.0006039728
28/5	0.2095026184	0.0244705609	0.0001187666
7	0.1593190414	0.0127728948	0.0000719311

η	ϵ_θ for Equation (30)	ϵ_θ for Equation (A2)	ϵ_θ for Equation (A4)
0	0	0	0
7/10	$4.209839553404038 \times 10^{-6}$	$9.919631346333446 \times 10^{-7}$	$1.845765404961952 \times 10^{-6}$
7/5	$2.585204789018469 \times 10^{-6}$	$6.580974769021530 \times 10^{-7}$	$5.869076206976853 \times 10^{-8}$
14/5	$1.673009802138914 \times 10^{-6}$	$3.525957825711856 \times 10^{-8}$	$1.314288065634369 \times 10^{-6}$
7/2	$2.481370770357482 \times 10^{-6}$	$8.064145169128789 \times 10^{-7}$	$1.303131223448148 \times 10^{-6}$
21/5	$1.995245388297650 \times 10^{-6}$	$3.768198066078643 \times 10^{-8}$	$1.039826044727616 \times 10^{-6}$
28/5	$2.205963331475269 \times 10^{-6}$	$4.627662200211435 \times 10^{-7}$	$1.389042267445562 \times 10^{-6}$
7	$5.828584578315699 \times 10^{-7}$	$5.749493005424017 \times 10^{-7}$	$5.400103917853105 \times 10^{-7}$

Table 3. Comparison between the heat transfer coefficient $\bar{\theta}'(0)$ obtained by means of the OHAM for different values of the Prandtl number pr and the parameter k, respectively, in the case of the first-order approximate solution.

		Numerical Solution	**OHAM Solution**	**Absolute Errors**
pr	k	$\theta'_{numerical}(0)$	$\bar{\theta}'_{OHAM}(0)$	$\epsilon_{\theta'(0)} = \lvert\theta'_{numerical}(0) - \bar{\theta}'_{OHAM}(0)\rvert$
0.5	0.15	−0.3727417350	−0.3727417250	$1.000000127149292 \times 10^{-8}$
0.5	0.25	−0.4014940569	−0.4014939569	$9.999999639465074 \times 10^{-8}$
0.5	0.5	−0.4686586964	−0.4686585964	$9.999899136525770 \times 10^{-8}$
1	0.15	−0.6171741875	−0.6171740875	$9.999999328602627 \times 10^{-8}$
1	0.25	−0.6608537627	−0.6608537527	$1.000000671158574 \times 10^{-8}$
1	0.5	−0.7647932545	−0.7647931545	$9.999999373011548 \times 10^{-8}$
2.5	0.15	−1.1185512466	−1.1185511466	$9.999999783794067 \times 10^{-8}$
2.5	0.25	−1.1923711840	−1.1923710840	$9.999999694976225 \times 10^{-8}$
2.5	0.5	−1.3666535048	−1.3666534948	$1.000000171558213 \times 10^{-8}$

Table 4. Comparison between the heat transfer coefficient $\bar{\theta}'(0)$ obtained by means of the OHAM for different values of the Prandtl number pr and the parameter k, respectively, in the case of the corresponding dual approximate solution.

		Numerical Solution	**OHAM Solution**	**Absolute Errors**
pr	k	$\theta'_{numerical}(0)$	$\bar{\theta}'_{OHAM}(0)$	$\epsilon_{\theta'(0)}$
0.5	0.15	−0.3238611974	−0.3238611874	$1.000000138251522 \times 10^{-8}$
0.5	0.25	−0.3473663384	−0.3473662384	$9.999999683873995 \times 10^{-8}$
0.5	0.5	−0.4014554630	−0.4014554530	$1.000000332540551 \times 10^{-8}$

Table 4. *Cont.*

		Numerical Solution	OHAM Solution	Absolute Errors
pr	k	$\theta'_{numerical}(0)$	$\bar{\theta}'_{OHAM}(0)$	$\epsilon_{\theta'(0)}$
1	0.15	−0.5929179987	−0.5929179887	$9.999954753148188 \times 10^{-9}$
1	0.25	−0.6393617637	−0.6393617537	$9.999972516716582 \times 10^{-9}$
1	0.5	−0.7402284508	−0.7402283508	$9.999999661669534 \times 10^{-8}$
2.5	0.15	−1.1110208487	−1.1110208387	$1.000000215967134 \times 10^{-8}$
2.5	0.25	−1.1848129415	−1.1848129315	$1.000000104944831 \times 10^{-8}$
2.5	0.5	−1.3591246415	−1.3591246315	$1.000000637851883 \times 10^{-8}$

Table 5. Integral of the square residual given by Equation (31) respectively, for different values of the parameters k and pr.

		The First Solution	The Corresponding Dual Solution
k	pr	$\int_0^\infty R_{\bar{\theta}}^2(\eta)\, d\eta$	$\int_0^\infty R_{\bar{\theta}}^2(\eta)\, d\eta$
0.15	0.5	$6.575432601542083 \times 10^{-9}$	$2.908978433213571 \times 10^{-10}$
0.25	0.5	$2.692683749426807 \times 10^{-8}$	$6.130825386312505 \times 10^{-8}$
0.5	0.5	$2.877470397657074 \times 10^{-10}$	$2.329753093802392 \times 10^{-7}$
0.15	1	$4.686687280794850 \times 10^{-9}$	$4.935777384019864 \times 10^{-8}$
0.25	1	$2.769513856968707 \times 10^{-10}$	$3.168740967468855 \times 10^{-9}$
0.5	1	$1.390428703762422 \times 10^{-9}$	$9.337205151229265 \times 10^{-6}$
0.15	2.5	$2.424695938193004 \times 10^{-6}$	$1.109703562037575 \times 10^{-6}$
0.25	2.5	$2.547691207587611 \times 10^{-7}$	$8.786794816590718 \times 10^{-8}$
0.5	2.5	$3.953512700816478 \times 10^{-10}$	$5.589488626508835 \times 10^{-6}$

5.1. *Influence of the Prandtl Number pr*

From Figures 2–5 we can notice that the variation of the temperature $\bar{\theta}(\eta)$ decreases with the increasing of the Prandtl number pr, for some fixed values of the parameter k.

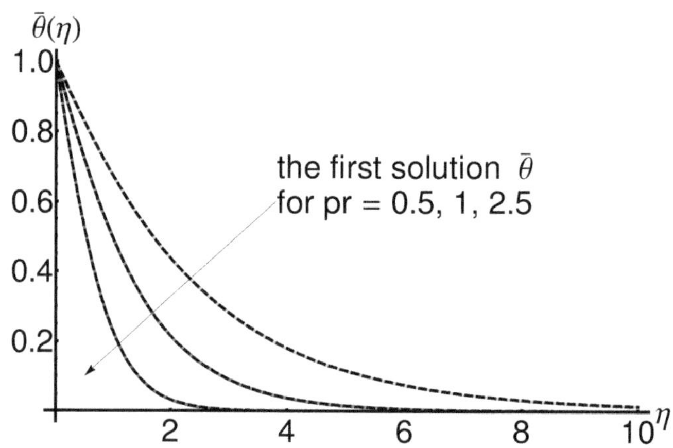

Figure 2. Variation of the temperature $\bar{\theta}(\eta)$ given by Equations (29), (A1) and (A3) with the Prandtl number $pr = 0.5,\ 1,\ 2.5$ for $k = 0.15$: OHAM solution (with lines) and numerical solution (dashing lines), respectively.

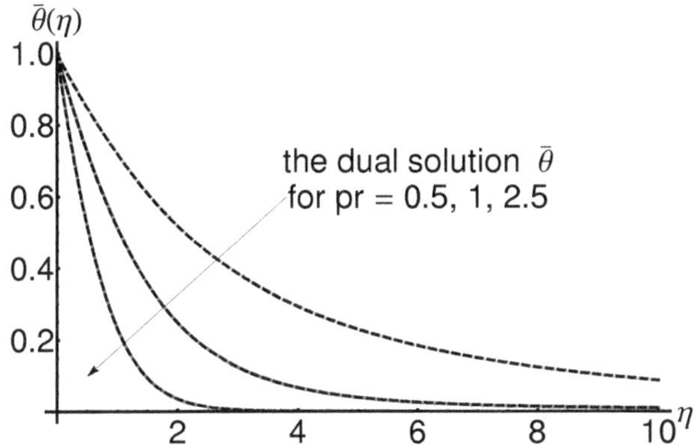

Figure 3. Variation of the temperature $\bar{\theta}(\eta)$ given by Equations (30), (A2) and (A4) with the Prandtl number $pr = 0.5,\ 1,\ 2.5$ for $k = 0.15$: OHAM solution (with lines) and numerical solution (dashing lines), respectively.

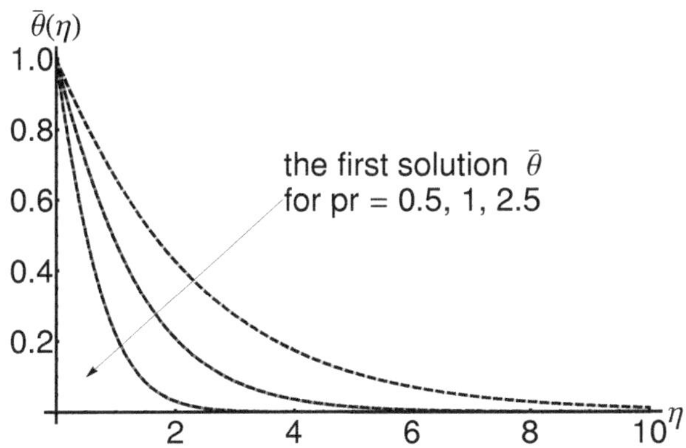

Figure 4. Variation of the temperature $\bar{\theta}(\eta)$ given by Equations (29), (A1) and (A3) with the Prandtl number $pr = 0.5,\ 1,\ 2.5$ for $k = 0.25$: OHAM solution (with lines) and numerical solution (dashing lines), respectively.

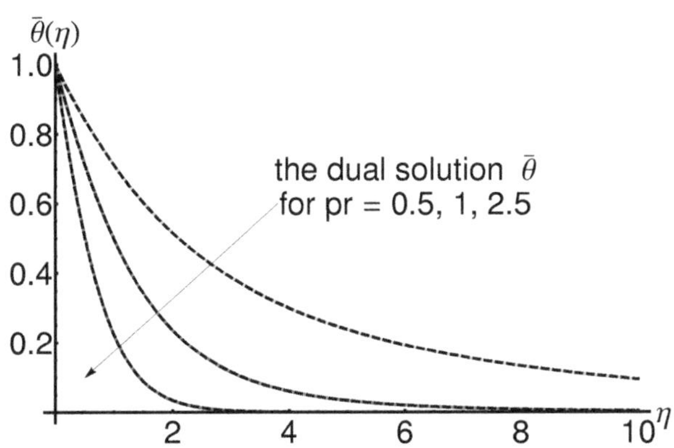

Figure 5. Variation of the temperature $\bar{\theta}(\eta)$ given by Equations (30), (A2) and (A4) with the Prandtl number $pr = 0.5,\ 1,\ 2.5$ for $k = 0.25$: OHAM solution (with lines) and numerical solution (dashing lines), respectively.

5.2. Influence of the Temperature Distribution Parameter k

Additionally, Figures 6 and 7 show that the variation of the temperature $\bar{\theta}(\eta)$ decreases with the increase in the parameter k for some fixed values of the Prandtl number pr.

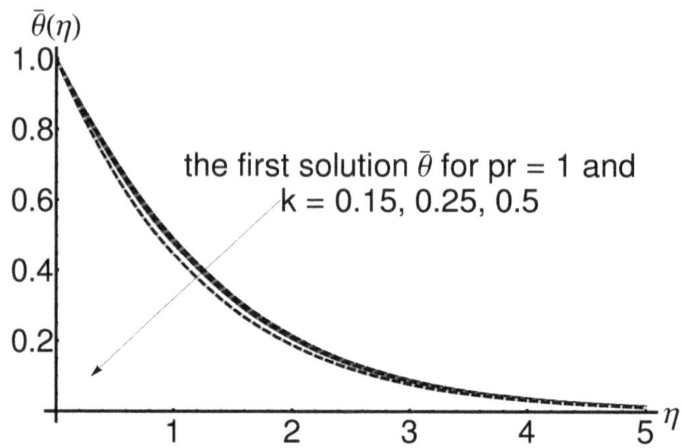

Figure 6. Variation of the temperature $\bar{\theta}(\eta)$ given by Equations (A1), (A7) and (A9) with the parameter $k = 0.15,\ 0.25,\ 0.5$ for $pr = 1$: OHAM solution (with lines) and numerical solution (dashed lines), respectively.

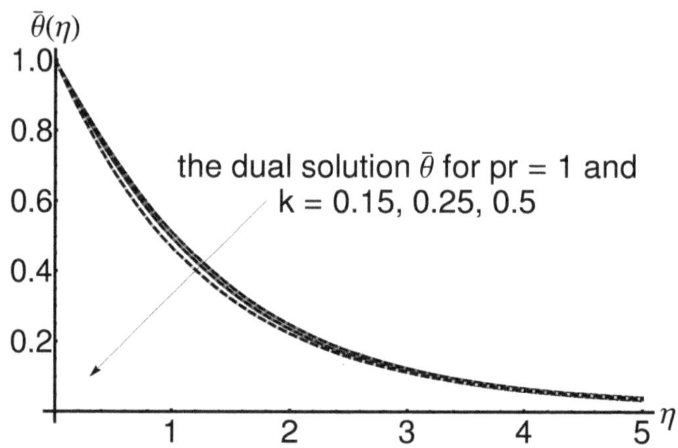

Figure 7. Variation of the temperature $\bar{\theta}(\eta)$ given by Equations (A2), (A8) and (A10) with the parameter $k = 0.15,\ 0.25,\ 0.5$ for $pr = 1$: OHAM solution (with lines) and numerical solution (dashed lines), respectively.

From all the Tables 1–5 and Figures 2–7 we can summarize that the OHAM solutions are effective and very accurate.

The advantages of the modified OHAM technique by comparison of the OHAM-solutions with the corresponding iterative solutions obtained by means of the iterative method developed in [51] are presented below.

The Equations (5) and (6) convert in the following system:

$$\begin{cases} f_1'(\eta) = f_2(\eta) \\ f_2'(\eta) = f_3(\eta) \\ f_3'(\eta) = 2f_2^2(\eta) - f_1(\eta)f_3(\eta) \\ \theta_1'(\eta) = \theta_2(\eta) \\ \theta_2'(\eta) = pr \cdot \Big(k \cdot f_2(\eta)\theta_1(\eta) - f_1(\eta)\theta_2(\eta)\Big) \end{cases} , \quad (32)$$

where $f_1(\eta) = f(\eta), f_2(\eta) = f'(\eta), f_3(\eta) = f''(\eta), \theta_1(\eta) = \theta(\eta), \theta_2(\eta) = \theta'(\eta)$.

By integration of the system (32) over the interval $[0,\ \eta]$, the following expressions are obtained:

$$\begin{aligned} f_1(\eta) &= f_1(0) + \int_0^\eta f_2(s)\, ds \\ f_2(\eta) &= f_2(0) + \int_0^\eta f_3(s)\, ds \\ f_3(\eta) &= f_3(0) + \int_0^\eta \Big(2f_2^2(s) - f_1(s)f_3(s)\Big)\, ds \\ \theta_1(\eta) &= \theta_1(0) + \int_0^\eta \theta_2(s)\, ds \\ \theta_2(\eta) &= \theta_2(0) + \int_0^\eta pr \cdot \Big(k \cdot f_2(s)\theta_1(s) - f_1(s)\theta_2(s)\Big)\, ds \end{aligned} . \quad (33)$$

The iterative algorithm is written as:

$$\begin{aligned}
&f_{1,0}(\eta) = f_1(0)\,, \quad f_{1,1}(\eta) = N_1(f_{1,0}, f_{2,0}, f_{3,0}, \theta_{1,0}, \theta_{2,0}) = \int_0^\eta f_{2,0}(s)\, ds\,, \\
&f_{2,0}(\eta) = f_2(0)\,, \quad f_{2,1}(\eta) = N_2(f_{1,0}, f_{2,0}, f_{3,0}, \theta_{1,0}, \theta_{2,0}) = \int_0^\eta f_{3,0}(s)\, ds\,, \\
&f_{3,0}(\eta) = f_3(0)\,, \quad f_{3,1}(\eta) = N_3(f_{1,0}, f_{2,0}, f_{3,0}, \theta_{1,0}, \theta_{2,0}) = \int_0^\eta \Big(2f_{2,0}^2(s) - f_{1,0}(s)f_{3,0}(s)\Big)\, ds\,, \\
&\theta_{1,0}(\eta) = \theta_1(0)\,, \quad \theta_{1,1}(\eta) = N_4(f_{1,0}, f_{2,0}, f_{3,0}, \theta_{1,0}, \theta_{2,0}) = \int_0^\eta \theta_{2,0}(s)\, ds\,, \\
&\theta_{2,0}(\eta) = \theta_2(0)\,, \quad \theta_{2,1}(\eta) = N_5(f_{1,0}, f_{2,0}, f_{3,0}, \theta_{1,0}, \theta_{2,0}) = \int_0^\eta pr \cdot \Big(k \cdot f_{2,0}(s)\theta_{1,0}(s) - f_{1,0}(s)\theta_{2,0}(s)\Big)\, ds\,, \\
&\ldots \\
&f_{1,m}(\eta) = N_1\left(\sum_{i=0}^{m-1} f_{1,i}, \sum_{i=0}^{m-1} f_{2,i}, \sum_{i=0}^{m-1} f_{3,i}, \sum_{i=0}^{m-1} \theta_{1,i}, \sum_{i=0}^{m-1} \theta_{2,i}\right) - N_1\left(\sum_{i=0}^{m-2} f_{1,i}, \sum_{i=0}^{m-2} f_{2,i}, \sum_{i=0}^{m-2} f_{3,i}, \sum_{i=0}^{m-2} \theta_{1,i}, \sum_{i=0}^{m-2} \theta_{2,i}\right), \\
&f_{2,m}(\eta) = N_2\left(\sum_{i=0}^{m-1} f_{1,i}, \sum_{i=0}^{m-1} f_{2,i}, \sum_{i=0}^{m-1} f_{3,i}, \sum_{i=0}^{m-1} \theta_{1,i}, \sum_{i=0}^{m-1} \theta_{2,i}\right) - N_2\left(\sum_{i=0}^{m-2} f_{1,i}, \sum_{i=0}^{m-2} f_{2,i}, \sum_{i=0}^{m-2} f_{3,i}, \sum_{i=0}^{m-2} \theta_{1,i}, \sum_{i=0}^{m-2} \theta_{2,i}\right), \\
&f_{3,m}(\eta) = N_3\left(\sum_{i=0}^{m-1} f_{1,i}, \sum_{i=0}^{m-1} f_{2,i}, \sum_{i=0}^{m-1} f_{3,i}, \sum_{i=0}^{m-1} \theta_{1,i}, \sum_{i=0}^{m-1} \theta_{2,i}\right) - N_3\left(\sum_{i=0}^{m-2} f_{1,i}, \sum_{i=0}^{m-2} f_{2,i}, \sum_{i=0}^{m-2} f_{3,i}, \sum_{i=0}^{m-2} \theta_{1,i}, \sum_{i=0}^{m-2} \theta_{2,i}\right), \\
&\theta_{1,m}(\eta) = N_4\left(\sum_{i=0}^{m-1} f_{1,i}, \sum_{i=0}^{m-1} f_{2,i}, \sum_{i=0}^{m-1} f_{3,i}, \sum_{i=0}^{m-1} \theta_{1,i}, \sum_{i=0}^{m-1} \theta_{2,i}\right) - N_4\left(\sum_{i=0}^{m-2} f_{1,i}, \sum_{i=0}^{m-2} f_{2,i}, \sum_{i=0}^{m-2} f_{3,i}, \sum_{i=0}^{m-2} \theta_{1,i}, \sum_{i=0}^{m-2} \theta_{2,i}\right), \\
&\theta_{2,m}(\eta) = N_5\left(\sum_{i=0}^{m-1} f_{1,i}, \sum_{i=0}^{m-1} f_{2,i}, \sum_{i=0}^{m-1} f_{3,i}, \sum_{i=0}^{m-1} \theta_{1,i}, \sum_{i=0}^{m-1} \theta_{2,i}\right) - N_5\left(\sum_{i=0}^{m-2} f_{1,i}, \sum_{i=0}^{m-2} f_{2,i}, \sum_{i=0}^{m-2} f_{3,i}, \sum_{i=0}^{m-2} \theta_{1,i}, \sum_{i=0}^{m-2} \theta_{2,i}\right), \\
&m \geq 2\,.
\end{aligned} \quad (34)$$

By carrying out the iterative method, the solutions of the Equations (5) and (6) have the form:

$$f_{1_{iter}}(\eta) = \sum_{m=0}^{\infty} f_{1,m}(t)\,,\; f_{2_{iter}}(\eta) = \sum_{m=0}^{\infty} f_{2,m}(\eta)\,,\; f_{3_{iter}}(\eta) = \sum_{m=0}^{\infty} f_{3,m}(\eta)\,,\; \theta_{1_{iter}}(\eta) = \sum_{m=0}^{\infty} \theta_{3,m}(\eta)\,, \theta_{2_{iter}}(\eta) = \sum_{m=0}^{\infty} \theta_{2,m}(\eta)\,.$$

Using five iterations, with the initial conditions $f_1(0) = 0$, $f_2(0) = 1$, $f_3(0) = -1.2818085481$, $\theta_1(0) = 1$, $\theta_2(0) = -0.6608537627$ (presented in the Table 3) and the physical constants $k = 0.25$, $pr = 1$, taking into account of the algorithm (33), the iterative solutions become:

$$\begin{aligned} f_{1_{iter}}(\eta) &= \sum_{m=0}^{5} f_{1,m}(\eta) = \eta - 0.6409042740\eta^2 + 0.3333333333\eta^3 - 0.1602260685\eta^4 + \\ &+ 0.0744091621\eta^5 - 0.0160226068\eta^6 + 0.0016144588\eta^7 + 0.0008177037\eta^8\,, \\ \theta_{1_{iter}}(\eta) &= \sum_{m=0}^{5} \theta_{1,m}(\eta) = 1 - 0.6608537627\eta - 0.3304268813\eta^2 - 0.1101422937\eta^3 - \\ &- 0.0275355734\eta^4 - 0.0055071146\eta^5\,. \end{aligned} \tag{35}$$

A comparison between the OHAM solutions $\bar{f}_{OHAM}$, $\bar{\theta}_{OHAM}$ and the corresponding iterative solutions $f_{1_{iter}}$, $\theta_{1_{iter}}$ given in Equation (35) is highlighted graphically in Figures 8 and 9 and tabularly in Table 6, respectively.

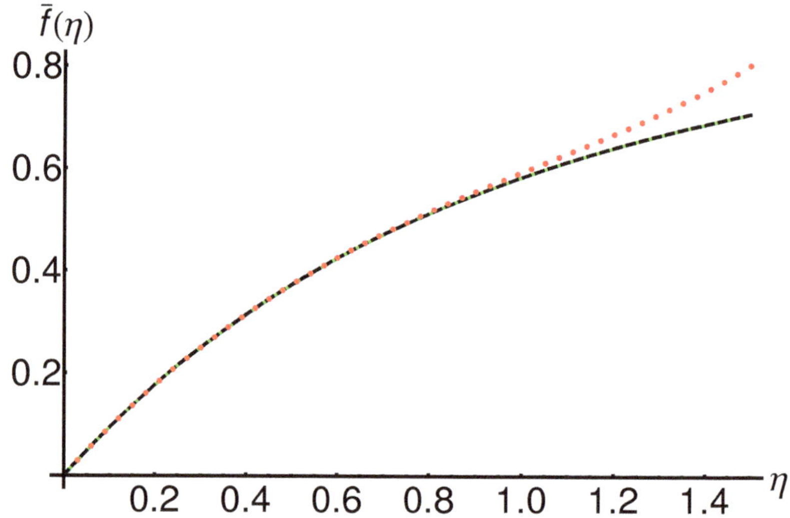

Figure 8. Comparison between the approximate analytical solution $\bar{f}(\eta)$, of the Equation (5) given by Equation [50], the iterative solution $f_{1_{iter}}(\eta)$ given by Equation (35) and the corresponding numerical solution: numerical solution (with lines), OHAM solution (dashed lines), and iterative solution (dotted curve), respectively.

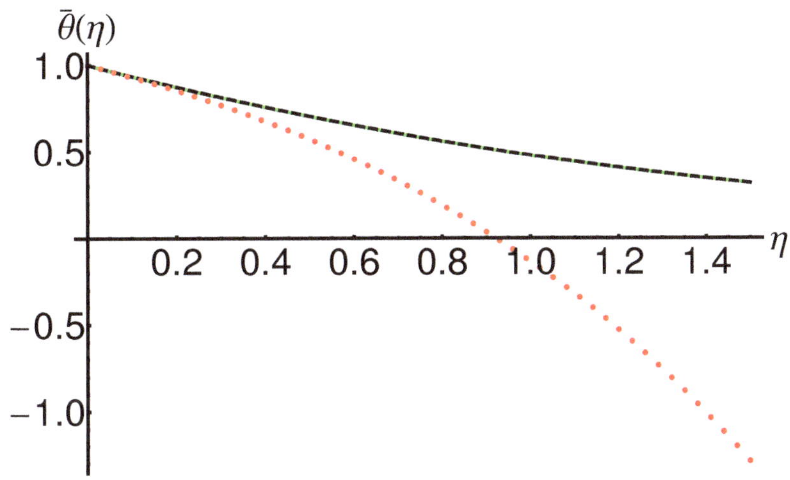

Figure 9. Comparison between the approximate analytical solution $\bar{\theta}(\eta)$, of the Equation (6) given by Equation (A1), the corresponding numerical solution and the iterative solution $\theta_{1_{iter}}(\eta)$ given by Equation (35): numerical solution (with lines), OHAM solution (dashed lines), and iterative solution (dotted curve), respectively.

The precision and efficiency of the OHAM method (using just one iteration) against to the iterative method described in [51] (using five iterations) arising from the presented comparison.

Table 6. Comparison between the approximate analytical solution $\bar{f}(\eta)$ given by Equation [50], the iterative solution $f_{1_{iter}}(\eta)$ given by Equation (35) and the corresponding numerical solution.

η	$f_{numerical}$	$\bar{f}_{OHAM}$ [50]	$f_{1_{iter}}$
0	0	0	0
1/10	0.0939089690	0.0939087919	0.0939089962
1/5	0.1767959477	0.1767950192	0.1767969422
3/10	0.2501798542	0.2501779276	0.2501903246
2/5	0.3153313350	0.3153286203	0.3153863643
1/2	0.3733200865	0.3733170634	0.3735172090
3/5	0.4250519374	0.4250491302	0.4256065973
7/10	0.4712981406	0.4712959627	0.4726209608
4/5	0.5127187008	0.5127173844	0.5155172621
9/10	0.5498810935	0.5498806830	0.5552901957
1	0.5832753856	0.5832757722	0.5930217087

Case Study

In the following we apply our analytical results in the case of the hydraulic oil with a large application at the hydraulic drive systems as turbines, pumps, naval propellers.

We consider the fluid flow scenario from a hydraulic installation with the following values of the characteristic quantities: the reference velocity $u_0 = 0.05$ [m/s], the reference temperature $T_0 = 40$, the kinematical viscosity $\nu = 46 \cdot 10^{-6}$ [m^2/s] and the environmental temperature $T_\infty = 22$, respectively.

The analytical obtained results in our paper using the specific physical sizes, are presented in Figures 10 and 11 for the first solution and in Figures 12 and 13 for the corresponding dual solution, respectively.

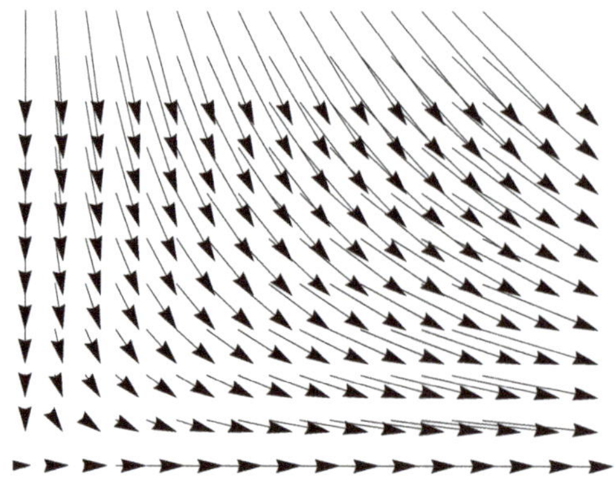

Figure 10. The vector field (u, v) from Equation (4) for hydraulic oil at a temperature of 40 °C, in the case of the first-solution given by Equation [50].

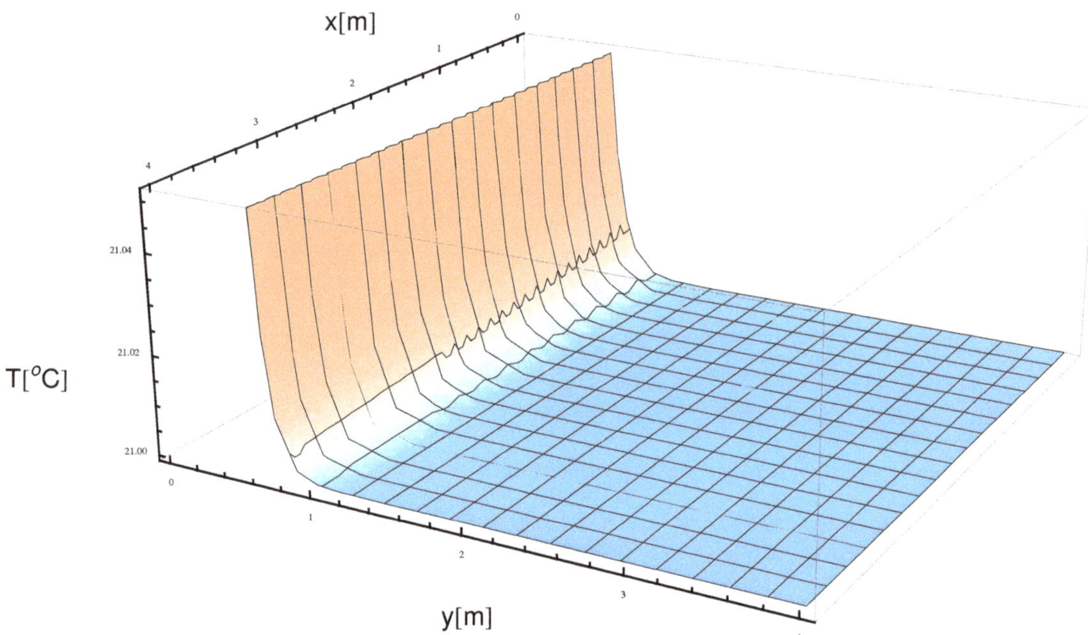

Figure 11. The 3D-profile of the temperature T from Equation (4) for $k = 0.25$, $pr = 1$ for hydraulic oil at a temperature of 40 °C, in the case of the first-solution given by Equation (A1).

The obtained results are in agreement with the Fluid Mechanics [52,53].

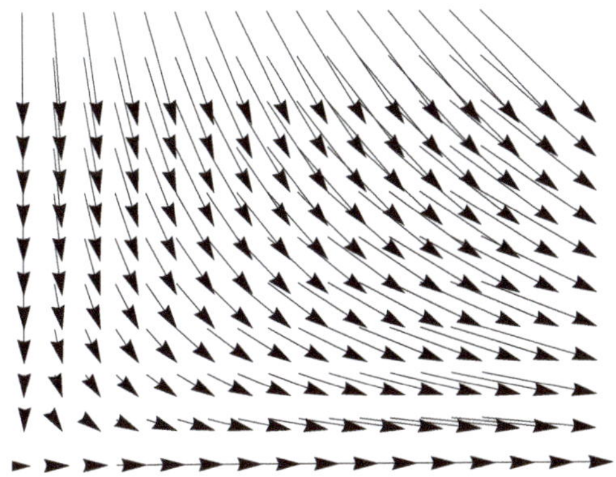

Figure 12. The vector field (u, v) from Equation (4) for hydraulic oil at a temperature of 40 °C, in the case of the corresponding dual solution given by Equation [50].

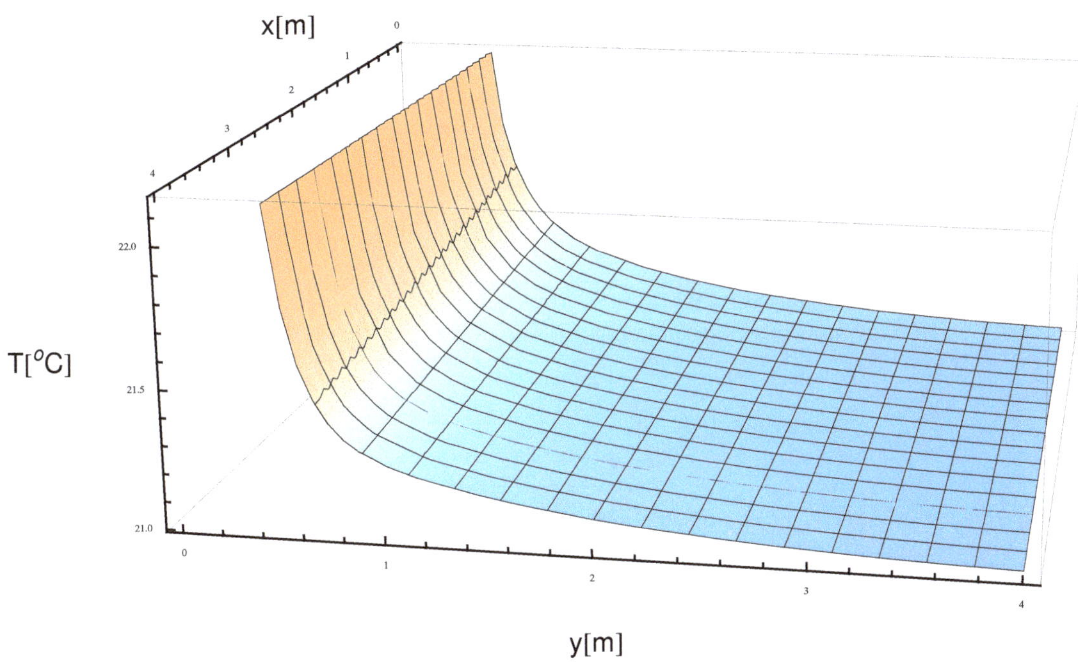

Figure 13. The 3D-profile of the temperature T from Equation (4) for $k = 0.25$, $pr = 1$ for hydraulic oil at a temperature of 40 °C, in the case of the corresponding dual solution given by Equation (A2).

6. Conclusions

The steady boundary layer flow and heat transfer over a stretching sheet were analyzed by using a nonlinear differential equation. The variation of the temperature $\theta(\eta)$ decreases with the increase in the Prandtl number pr for some fixed values of the parameter k. As a result, we can observe a decrease in the fluid temperature. This shows that more heat is released from the sheet and the Prandtl number decreases in the boundary layer thickness. Therefore, the heat transfer rate increases.

The processes with strongly nonlinear behaviors appear in different technological applications. Thus, an approximate analytical solution is a more realistic option.

The OHAM treatment related to the heat and mass transfer problem without partial slip in the flow of a viscous fluid over an exponentially stretching sheet without suction/blowing is considered and provides an accurate solution for the nonlinear differential equation with initial and boundary conditions.

In this paper, the thermal effects of the Prandtl number and the temperature distribution parameter are analytically analyzed. The variations of the dimensionless surface temperature and heat transfer characteristics with the governing parameters are graphed and tabulated. In particular, the analytically obtained results are applied from the hydraulic system.

The advantage of the method applied in this work is the efficiency by only one iteration. Other advantages, including accuracy, flexibility, validity and convergence, of the approximate solutions are highlighted by comparing the OHAM solutions with the corresponding iterative solutions.

Some characteristics of the heat and mass transfer, such as the vector field (u, v) and the temperature profile T are graphically depicted in a case study of the hydraulic oil using the obtained approximate solutions via OHAM.

This study is useful for many engineering applications of heat and mass transfer problems such as strand casting processes, polymeric liquids, the extraction of metals and polymers, glass-fiber production, and physiological fluid dynamics.

Author Contributions: Conceptualization, N.P.; data curation, R.-D.E. and N.P.; formal analysis, N.P.; investigation, R.-D.E. and R.B.; methodology, R.-D.E. and R.B.; software, R.-D.E.; supervision, N.P.; validation, R.-D.E. and N.P.; visualization, R.-D.E. and N.P.; writing—original draft, R.-D.E., R.B. and N.P. All authors have read and agreed to the published version of the manuscript.

Funding: This research received no external funding.

Institutional Review Board Statement: Not applicable.

Informed Consent Statement: Not applicable.

Data Availability Statement: Not applicable.

Conflicts of Interest: The authors declare no conflict of interest.

Nomenclature/Notation

Symbols	Names
$u,\ v$	Velocity components (m/s)
$x,\ y$	Cartesian coordinates (m)
ν	Kinematical viscosity (m^2/s)
μ	Viscosity
ρ	Fluid density
U_∞	Velocity of uniform flow
$u_0,\ T_0$	Reference velocity and reference temperature
N	Velocity slip factor
l	Characteristic length
pr	Prandtl number
k	Parameter of temperature distribution
T_∞	Environment temperature (K)
η	Independent dimensionless variable
$f(\eta)$	Stream function
$\theta(\eta)$	Temperature
OHAM solution	approximate analytical solution by means of the modified Optimal Homotopy Asymptotic Method

Appendix A

In this section there are presented in details the first-order approximate solution given by Equation (28) and the corresponding dual solution for different values of the physical parameters.

(a$_2$) the parameter $k = 0.25$, the Prandtl number $pr = 1$.

$$\begin{aligned}
\bar{\theta}(\eta) &= 1.1600261597 \cdot e^{-0.9234397221 \cdot \eta} + (1 - 0.3362966408 \cdot \eta)e^{-2.0276266018 \cdot \eta} + \\
&+ (-1.3520589334 + 0.2475145276 \cdot \eta - 0.0094831510 \cdot \eta^2)e^{-2.0138501698 \cdot \eta} + \\
&+ (0.1720819672 + 0.0218466525 \cdot \eta - 0.0001162030 \cdot \eta^2)e^{-0.9096632901 \cdot \eta} + \\
&+ (0.0199508065 + 0.0189709502 \cdot \eta + 0.0059452174 \cdot \eta^2)e^{-4.0277003397 \cdot \eta}
\end{aligned} \tag{A1}$$

and the corresponding dual approximate solution becomes:

$$\begin{aligned}
\bar{\theta}(\eta) &= 0.2369245422 \cdot e^{-0.5134938585 \cdot \eta} + (1 + 1.0513455728 \cdot \eta)e^{-1.6176807383 \cdot \eta} + \\
&+ (0.0238401179 - 0.0020255651 \cdot \eta + 0.0000568573 \cdot \eta^2)e^{-0.1207882199 \cdot \eta} + \\
&+ (0.8098858748 + 0.2652845586 + 0.0285288295 \cdot \eta^2)e^{-2.4499501993 \cdot \eta} + \\
&+ (-1.0706505351 + 0.4609131356 \cdot \eta + 0.0444370182 \cdot \eta^2)e^{-1.2249750996 \cdot \eta}
\end{aligned} \tag{A2}$$

(a$_3$) the parameter $k = 0.25$, the Prandtl number $pr = 2.5$.

$$\begin{aligned}
\bar{\theta}(\eta) &= -0.0000694691 \cdot e^{-0.0373028427 \cdot \eta} + (1 - 0.6822337697 \cdot \eta)e^{-1.1414897224 \cdot \eta} + \\
&+ (-0.4228006041 - 0.1276408784 \cdot \eta - 0.7048478809 \cdot \eta^2)e^{-2.0265756010 \cdot \eta} + \\
&+ (-0.2530559522 - 0.5915096926 \cdot \eta - 0.3170935172 \cdot \eta^2)e^{-4.0531512020 \cdot \eta} + \\
&+ (0.6759260256 + 0.0914555047 \cdot \eta - 0.0099048402 \cdot \eta^2)e^{-0.9223887212 \cdot \eta}
\end{aligned} \tag{A3}$$

and the corresponding dual approximate solution becomes:

$$\begin{aligned}
\bar{\theta}(\eta) &= 0.0001260254 \cdot e^{-0.0745075617 \cdot \eta} + (1 - 0.0964643097 \cdot \eta)e^{-1.1786944415 \cdot \eta} + \\
&+ (-0.3203284037 + 1.8308083903 \cdot \eta - 0.0762347077 \cdot \eta^2)e^{-2.6623999026 \cdot \eta} + \\
&+ (-0.3417840921 - 0.4098840584 \cdot \eta - 0.0321843260 \cdot \eta^2)e^{-1.5582130229 \cdot \eta} + \\
&+ (0.6619864704 + 0.8089615723 \cdot \eta + 0.3222124279 \cdot \eta^2)e^{-5.3247998053 \cdot \eta}
\end{aligned} \tag{A4}$$

(a$_4$) the parameter $k = 0.15$, the Prandtl number $pr = 0.5$.

$$\begin{aligned}
\bar{\theta}(\eta) &= 1.0983077175 \cdot e^{-0.4534958469 \cdot \eta} + (1 + 0.4654950159 \cdot \eta)e^{-1.5576827266 \cdot \eta} + \\
&+ (-1.5363389724 + 0.3221215358 \cdot \eta - 0.0234501255 \cdot \eta^2)e^{-1.1308627520 \cdot \eta} + \\
&+ (0.0026808712 + 0.0000481427 \cdot \eta + 1.652544 \cdot 10^{-6} \cdot \eta^2)e^{-0.0266758723 \cdot \eta} + \\
&+ (0.4353503835 + 0.1426803574 \cdot \eta + 0.0157885694 \cdot \eta^2)e^{-2.2617255040 \cdot \eta}
\end{aligned} \tag{A5}$$

and the corresponding dual approximate solution becomes:

$$\begin{aligned}
\bar{\theta}(\eta) &= 0.5391981597 \cdot e^{-0.4822945356 \cdot \eta} + (1 + 0.5250981944 \cdot \eta)e^{-1.5864814153 \cdot \eta} + \\
&+ (0.3818837053 - 0.0092643162 \cdot \eta + 0.0000912270 \cdot \eta^2)e^{-0.1278543543 \cdot \eta} + \\
&+ (0.3480470015 + 0.1414195195 \cdot \eta + 0.0192266212 \cdot \eta^2)e^{-2.4640824681 \cdot \eta} + \\
&+ (-1.2691288666 + 0.2082420703 \cdot \eta + 0.0374340788 \cdot \eta^2)e^{-1.2320412340 \cdot \eta}
\end{aligned} \tag{A6}$$

The influence of the temperature distribution parameter k on the heat transfer is presented below. In this way, we provide the approximate analytical solutions for the case of $pr = 1.5$ and different values for k.

(a$_5$) In this case, we consider $k = 0.5$ and $pr = 0.5$.

$$\begin{aligned}
\bar{\theta}(\eta) &= 0.0007206409 \cdot e^{-0.0000131202 \cdot \eta} + (1 - 0.1316345278 \cdot \eta)e^{-1.1042000000 \eta} + \\
&+ (-1.0617752958 - 0.2671398226 \cdot \eta - 0.0590973088 \cdot \eta^2)e^{-1.5423591909 \cdot \eta} + \\
&+ (0.9992265151 - 0.0158696110 \cdot \eta + 0.0001546813 \cdot \eta^2)e^{-0.4381723112 \cdot \eta} + \\
&+ (0.0618281397 + 0.0411022791 \cdot \eta + 0.0096407838 \cdot \eta^2)e^{-3.0847183818 \cdot \eta}
\end{aligned} \tag{A7}$$

and the corresponding dual approximate solution is:

$$\begin{aligned}
\bar{\theta}(\eta) &= 0.5513428358 \cdot e^{-0.1531064052 \cdot \eta} + (1 + 2.1654964846 \cdot \eta)e^{-1.2572932849 \cdot \eta} + \\
&+ (-0.5837332313 + 0.1717462354 \cdot \eta - 0.3926304130 \cdot \eta^2)e^{-2.2051877057 \cdot \eta} + \\
&+ (0.4350443625 + 0.4520872315 \cdot \eta + 0.1673037370 \cdot \eta^2)e^{-4.4103754114 \cdot \eta} + \\
&+ (-0.4026539670 - 1.6609327359 \cdot \eta + 0.2537560261 \cdot \eta^2)e^{-1.1010008259 \cdot \eta}
\end{aligned} \tag{A8}$$

(a$_6$) In the second case, if $k = 0.15$ and $pr = 1$, then:

$$\begin{aligned}\bar{\theta}(\eta) &= 2.0683347602 \cdot e^{-0.9155012717\cdot\eta} + (1 + 0.6833539452 \cdot \eta)e^{-2.0196881514\cdot\eta} + \\ &+(-0.8031347222 - 0.7492677018 \cdot \eta - 0.1062182432 \cdot \eta^2)e^{-2.2813648143\cdot\eta} + \\ &+(-1.2654357598 + 0.0862848192 \cdot \eta - 0.0427535328 \cdot \eta^2)e^{-1.1406824071\cdot\eta} + \\ &+(0.0002357217 + 0.0000111023 \cdot \eta + 2.111508 \cdot 10^{-7} \cdot \eta^2)e^{-0.0364955274\cdot\eta}\end{aligned} \tag{A9}$$

and the corresponding dual approximate solution is:

$$\begin{aligned}\bar{\theta}(\eta) &= 0.1878857950 \cdot e^{-0.4637756403\cdot\eta} + (1 + 1.2766230888 \cdot \eta)e^{-1.5679625200\cdot\eta} + \\ &+(0.0201352353 - 0.0009567685 \cdot \eta + 0.0000525361 \cdot \eta^2)e^{-0.0600476542\cdot\eta} + \\ &+(1.1008756684 + 0.3253629902 \cdot \eta + 0.0338366627 \cdot \eta^2)e^{-2.3284690680\cdot\eta} + \\ &+(-1.3088966989 + 0.5018533528 \cdot \eta + 0.0380250588 \cdot \eta^2)e^{-1.1642345340\cdot\eta}\end{aligned} \tag{A10}$$

(a$_7$) In the third case, if $k = 0.15$ and $pr = 2.25$:

$$\begin{aligned}\bar{\theta}(\eta) &= 0.0752690301 \cdot e^{-1.3036641988\cdot\eta} + (1 - 0.7216448042 \cdot \eta)e^{-2.4078510785\cdot\eta} + \\ &+(-0.3478069740 - 0.4476541650 \cdot \eta - 0.2238753041 \cdot \eta^2)e^{-6.1184936238\cdot\eta} + \\ &+(0.6232764356 + 0.4531574839 \cdot \eta - 0.1414849066 \cdot \eta^2)e^{-1.9550599322\cdot\eta} + \\ &+(-0.3507384917 + 0.1210593773 \cdot \eta + 1.3248866865 \cdot \eta^2)e^{-3.0592468119\cdot\eta}\end{aligned} \tag{A11}$$

and the corresponding dual approximate solution is:

$$\begin{aligned}\bar{\theta}(\eta) &= -0.0094833474 \cdot e^{-1.8199222326\cdot\eta} + (1 + 4.8903828049 \cdot \eta) \cdot e^{-2.9241091123\cdot\eta} + \\ &+(-0.2794803592 - 3.8038471274 \cdot \eta - 1.1301670707 \cdot \eta^2)e^{-3.4371389008\cdot\eta} + \\ &+(0.2022965663 + 0.1763166133 \cdot \eta - 0.0935804838 \cdot \eta^2)e^{-1.7185694504\cdot\eta} + \\ &+(0.0866671403 - 0.0267283075 \cdot \eta + 0.0022982197 \cdot \eta^2)e^{-0.6143825706\cdot\eta}\end{aligned} \tag{A12}$$

In this way, we can construct other accurate approximate solutions.

References

1. Sparrow, E.M.; Cess, R.D. Temperature dependent heat sources or sinks in a stagnation point flow. *Appl. Sci. Res.* **1961**, *A10*, 185. [CrossRef]
2. Topper, L. Heat transfer in cylinders with heat generation. *Am. Inst. Chem. Fagug. J.* **1955**, *463*, 463–466. [CrossRef]
3. Khashi, N.S.; Waini, I.; Kasim, A.R.M.; Zainal, N.A.; Ishak, A.; Pop, I. Magnetohydrodynamic and viscous dissipation effects on radiative heat transfer of non-Newtonian fluid flow past a nonlinearly shrinking sheet: Reiner–Philippoff model. *Alex. Eng. J.* **2022**, *61*, 7605–7617. [CrossRef]
4. Vajravelu, K.; Hadjinicolaou, A. Heat transfer in a viscous fluid over a stretching sheet with viscous dissipation and internal heat generation. *Int. Comm. Heat Mass Transf.* **1993**, *20*, 417–430. [CrossRef]
5. Xu, H. Mixed convective flow of a hybrid nanofluid between two parallel inclined plates under wall-slip condition. *Appl. Math. Mech.-Engl. Ed.* **2022**, *43*, 113–126. [CrossRef]
6. Hayat, T.; Muhammad, K.; Alsaedi, A. Melting effect and Cattaneo-Christov heat flux in fourth-grade material flow through a Darcy-Forchheimer porous medium. *Appl. Math. Mech.-Engl. Ed.* **2021**, *42*, 1787–1798. [CrossRef]
7. Bararnia, H.; Gorji, M.; Domairry, G. An Analytical Study of Boundary Layer Flows on a Continuous Stretching Surface. *Acta Appl. Math.* **2009**, *106*, 125–133. [CrossRef]
8. Khan, N.S.; Islam, S.; Gul, T.; Khan, I.; Khan, W.; Ali, L. Thin film flow of a second grade fluid in a porous medium past a stretching sheet with heat transfer. *Alex. Eng. J.* **2018**, *57*, 1019–1031. [CrossRef]
9. Khan, N.S.; Gul, T.; Islam, S.; Khan, I.; Alqahtani, A.M.; Alshomrani, A.S. Magnetohydrodynamic Nanoliquid Thin Film Sprayed on a Stretching Cylinder with Heat Transfer. *Appl. Sci.* **2017**, *7*, 271. [CrossRef]
10. Khan, N.S.; Gul, T.; Kumam, P.; Shah, Z.; Islam, S.; Khan, W.; Zuhra, S.; Arif, S. Influence of Inclined Magnetic Field on Carreau Nanoliquid Thin Film Flow and Heat Transfer with Graphene Nanoparticles. *Energies* **2019**, *12*, 1459. [CrossRef]
11. Khan, N.S.; Zuhra, S. Boundary layer flow and heat transfer in a thin-film second-grade nanoliquid embedded with graphene nanoparticles. *Adv. Mech. Eng.* **2019**, *11*, 1–11. [CrossRef]
12. Zuhra, S.; Khan, N.S.; Khan, M.A.; Islam, S.; Khan, W.; Bonyah, E. Flow and heat transfer in water based liquid film fluids dispensed with graphene nanoparticles. *Results Phys.* **2018**, *8*, 1143–1157. [CrossRef]
13. Bilal, A.M.; Alsaedi, A.; Hayat, T.; Shehzad, S.A. Convective Heat and Mass Transfer in Three-Dimensional Mixed Convection Flow of Viscoelastic Fluid in Presence of Chemical Reaction and Heat Source/Sink. *Comp. Math. Math. Phys.* **2017**, *57*, 1066–1079. [CrossRef]

14. Shehzad, S.A.; Hayat, T.; Alsaedi, A. Flow of a thixotropic fluid over an exponentially stretching sheet with heat transfer. *J. Appl. Mech. Tech. Phy.* **2016**, *57*, 672–680. [CrossRef]
15. Alizadeh, Y.; Mosaddeghi, M.R.; Khazayinejad, M. Semi-analytical assessment of heat transfer rate for MHD transient flow in a semi-porous channel considering heat source and slip effect. *Waves Random Complex Media* **2021**, 1–23. [CrossRef]
16. Yan, H.; Sedighi, M.; Xie, H. Thermally induced diffusion of chemicals under steady-state heat transfer in saturated porous media. *Int. J. Heat Mass Tran.* **2020**, *153*, 119664. [CrossRef]
17. Nadeem, J.; Marwat, D.N.K.; Khan, T.S. Heat transfer in viscous flow over a heated cylinder of nonuniform radius. *Ain Shams Eng. J.* **2021**, *12*, 4189–4199.
18. Abbasi, A.; Khan, S.U.; Farooq, W.; Mughal, F.M.; Khan, M.I.; Prasannakumara, B.C.; El-Wakad, M.T.; Guedri, K.; Galal, A.M. Peristaltic flow of chemically reactive Ellis fluid through an asymmetric channel: Heat and mass transfer analysis. *Ain Shams Eng. J.* **2023**, *14*, 101832. [CrossRef]
19. Xie, W.A.; Xi, G.N. Flow instability and heat transfer enhancement of unsteady convection in a step channel. *Alex. Eng. J.* **2022**, *61*, 7377–7391. [CrossRef]
20. Abdelaziz, A.H.; El-Maghlany, W.M.; El-Din, A.A.; Alnakeeb, M.A. Mixed convection heat transfer utilizing Nanofluids, ionic Nanofluids, and hybrid nanofluids in a horizontal tube. *Alex. Eng. J.* **2022**, *61*, 9495–9508. [CrossRef]
21. Muhammad, K.; Hayat, T.; Alsaedi, A. OHAM analysis of fourth-grade nanomaterial in the presence of stagnation point and convective heat-mass conditions. *Waves Random Complex Media* **2021**, 1–17. [CrossRef]
22. Mabood, F.; Lorenzini, G.; Pochai, N.; Ibrahim, S.M. Effects of prescribed heat flux and transpiration on MHD axisymmetric flow impinging on stretching cylinder. *Contin. Mech. Therm.* **2016**, *28*, 1925–1932. [CrossRef]
23. Eid, M.R.; Nafe, M.A. Thermal conductivity variation and heat generation effects on magneto-hybrid nanofluid flow in a porous medium with slip condition. *Waves Random Complex Media* **2020**. [CrossRef]
24. Boumaiza, N.; Kezzar, M.; Eid, M.R.; Tabet, I. On numerical and analytical solutions for mixed convection Falkner-Skan flow of nanofluids with variable thermal conductivity. *Waves Random Complex Media* **2019**. [CrossRef]
25. Gireesha, B.J.; Keerthi, M.L.; Sowmya, G. Effects of stretching/shrinking on the thermal performance of a fully wetted convective-radiative longitudinal fin of exponential profile. *Appl. Math. Mech.-Engl. Ed.* **2022**, *43*, 389–402. [CrossRef]
26. Waini, I.; Ishak, A.; Pop, I. Magnetohydrodynamic flow past a shrinking vertical sheet in a dusty hybrid nanofluid with thermal radiation. *Appl. Math. Mech.-Engl. Ed.* **2022**, *43*, 127–140. [CrossRef]
27. Tang, Q.; Huang, Y. Parallel finite element computation of incompressible magnetohydrodynamics based on three iterations. *Appl. Math. Mech.-Engl. Ed.* **2022**, *43*, 141–154. [CrossRef]
28. Cveticanin, L. Exact Closed-Form Solution for the Oscillator with a New Type of Mixed Nonlinear Restitution Force. *Mathematics* **2023**, *11*, 596. [CrossRef]
29. Raduca, M.; Hatiegan, C.; Pop, N.; Raduca, E.; Gillich, G.-R. Finite element analysis of heat transfer in transformers from high voltage stations. *J. Therm. Anal. Calorim.* **2014**, *18*, 1355–1360. [CrossRef]
30. Martin, M.J.; Boyd, I.D. Momentum and heat transfer in a laminar boundary layer with slip flow. *J. Thermo Heat Trans.* **2006**, *20*, 710–719. [CrossRef]
31. Anderson, H.I. Slip flow past a stretching surface. *Acta Mech.* **2002**, *158*, 121–125. [CrossRef]
32. Khan, N.S.; Zuhra, S.; Shah, Z.; Bonyah, E.; Khan, W.; Islam, S.; Khan, A. Hall current and thermophoresis effects on magnetohydrodynamic mixed convective heat and mass transfer thin film flow. *J. Phys. Commun.* **2019**, *3*, 035009. [CrossRef]
33. Nayak, M.K.; Mabood, F.; Dogonchi, A.S.; Ramadan, K.M.; Tlili, I.; Khan, W.A. Entropy optimized assisting and opposing non-linear radiative flow of hybrid nanofluid. *Waves Random Complex Media* **2022**, 1–22. [CrossRef]
34. Cortell, R. Viscous flow and heat transfer over a nonlinearly stretching sheet. *Appl. Math. Comput.* **2007**, *184*, 864–873. [CrossRef]
35. Alam, A.; Marwat, D.N.K. Heat and mass transfer on a stretching/shrinking and porous sheet of variable thickness with suction and injection. *Proc. Inst. Mechanical Eng. Part C J. Mech. Eng.* **2021**, *235*, 5297–5308. [CrossRef]
36. Marinca, V.; Herisanu, N. *The Optimal Homotopy Asymptotic Method-Engineering Applications*; Springer: Berlin/Heidelberg, Germany, 2015.
37. Marinca, V.; Ene, R.D.; Marinca, B.; Negrea, R. Different approximations to the solution of upper-convected Maxwell fluid over a porous stretching plate. *Abstr. Appl. Anal.* **2014**, *2014*, 139314. [CrossRef]
38. Ene, R.D.; Marinca, V. Approximate solutions for steady boundary layer MHD viscous flow and radiative heat transfer over an exponentially porous stretching sheet. *Appl. Math. Comput.* **2015**, *269*, 389–401. [CrossRef]
39. Ene, R.D.; Szabo, M.A.; Danoiu, S. Viscous flow and heat transfer over a permeable shrinking sheet with partial slip. *Mater Plast.* **2015**, *52*, 408–412.
40. Marinca, V.; Ene, R.D. Dual approximate solutions of the unsteady viscous flow over a shrinking cylinder with Optimal Homotopy Asymptotic Method. *Adv. Math. Phys.* **2014**, 417643. [CrossRef]
41. Ene, R.D.; Pop, N. Dual approximate solutions for the chemically reactive solute transfer in a viscous fluid flow. *Waves Random Complex Media* **2021**, 1–23. [CrossRef]
42. Ullah, H.; Nawaz, R.; Islam, S.; Idrees, M.; Fiza, M. The optimal homotopy asymptotic method with application to modified Kawahara equation. *J. Assoc. Arab. Univ. Basic Appl. Sci.* **2015**, *18*, 82–88. [CrossRef]
43. Almousa, M.; Ismail, A. Optimal Homotopy Asymptotic Method for Solving the Linear Fredholm Integral Equations of the First Kind. *Abstr. Appl. Anal.* **2013**, *2013*, 278097. [CrossRef]

44. Golbabai, A.; Fardi, M.; Sayevandc, K. Application of the optimal homotopy asymptotic method for solving a strongly nonlinear oscillatory system. *Math. Comput. Model.* **2013**, *58*, 1837–1843. [CrossRef]
45. Alomari, A.K.; Anakira, N.R.; Hashim, I. Multiple Solutions of Problems in Fluid Mechanics by Predictor Optimal Homotopy Asymptotic Method. *Adv. Mech. Eng.* **2014**, *6*, 372537. [CrossRef]
46. Marinca, V.; Herisanu, N. Application of Optimal Homotopy Asymptotic Method for solving nonlinear equations arising in heat transfer. *Int. Commun. Heat Mass.* **2008**, *35*, 710–715 [CrossRef]
47. Ullah, H.; Islam, S.; Idrees, M.; Arif, M. Solution of Boundary Layer Problems with Heat Transfer by Optimal Homotopy Asymptotic Method. *Abstr. Appl. Anal.* **2013**, *2013*, 324869. [CrossRef]
48. Waqar, K. Optimal homotopy asymptotic method for heat transfer in hollow sphere with robin boundary conditions. *Heat Transf. Asian Res.* **2014**, *43*, 124–133.
49. Ene, R.D.; Marinca, V.; Negrea, R. Optimal Homotopy Asymptotic Method for viscous boundary layer flow in unbounded domain. In Proceedings of the 16th International Symposium on Symbolic and Nnumeric Algoritms for Scientific Computing (SYNASC 2014), Timisoara, Romania, 22–25 September 2014.
50. Ene, R.D.; Petrisor, C. Some mathematical approaches on the viscous flow problem on a continuous stretching surface: Nonlinear stability and dual approximate analytic solutions. *AIP Conf. Proc.* **2020**, *2293*, 350004.
51. Daftardar-Gejji, V.; Jafari, H. An iterative method for solving nonlinear functional equations. *J. Math. Anal. Appl.* **2006**, *316*, 753–763. [CrossRef]
52. Giles, R.V. *Theory and Problems of the Hydraulics*, 2nd ed.; Schaum's Outline Series; McGraw Hill Book Company: New York, NY, USA, 1977.
53. Greenshields, C.; Weller, H.G. *Notes on Computational Fluid Dynamics:General Principles*; CFD Direct Limited: Reading, UK, 2022.

mathematics

MDPI

Article

Turbulent Heat Transfer Augmentation in a Square Channel by Augmenting the Flow Pattern with Novel Arc-Shaped Ribs

Basma Souayeh [1,2,*] and Suvanjan Bhattacharyya [3]

1 Department of Physics, College of Science, King Faisal University, P.O. Box 400, Al-Ahsa 31982, Saudi Arabia
2 Laboratory of Fluid Mechanics, Physics Department, Faculty of Sciences of Tunis, University of Tunis El Manar, Tunis 2092, Tunisia
3 Department of Mechanical Engineering, Birla Institute of Technology and Science Pilani, Pilani Campus, Vidya Vihar, Pilani 333031, Rajasthan, India
* Correspondence: bsouayeh@kfu.edu.sa or basma.souayeh@gmail.com

Abstract: Solar water heaters (SWHs) are widely used in HVAC industries as well as in households for different heating purposes. The present numerical simulation focuses on the investigation of the thermo-hydraulic performance of novel semi-arc-rib SWHs. Semi-arc-shaped ribs in the square channel of the absorber plates with different pitch and height ratios are investigated in this study. The present novel modification disturbs the boundary layers by generating vortices, and thus, enhanced fluid mixing takes place. Water with a Reynolds number (Re) ranging from 4000 to 25,000 is used as a working fluid, and a 1.0 kW/m^2 heat flux is imposed on the tube wall. The results demonstrate a significant increase in the Nusselt number (Nu) as the fluid layers localize behind each rib near the absorber plates, and at the same time, the number of swirls generated inside the tube and the frictional losses both increased noticeably. To ensure the effectivity of the present novel SWH geometry, the thermo-hydraulic performance (η) for each case was calculated, and it was found that in all the cases, it was greater than unity, which signifies that the present semi-arc-rib SWH is promising and can be used in HVAC industrial and household applications.

Keywords: solar water heater; semi-circular arc; heat transfer; enhancement; swirl flow; thermal performance

MSC: 76-10

Citation: Souayeh, B.; Bhattacharyya, S. Turbulent Heat Transfer Augmentation in a Square Channel by Augmenting the Flow Pattern with Novel Arc-Shaped Ribs. *Mathematics* **2023**, *11*, 1490. https://doi.org/10.3390/math11061490

Academic Editor: Ramoshweu Solomon Lebelo

Received: 8 January 2023
Revised: 28 February 2023
Accepted: 9 March 2023
Published: 18 March 2023

1. Introduction

Energy consumption has increased as a result of welfare development and ongoing population rise across the world. Energy efficiency and sustainability are becoming more significant in today's world as a result of a supply and demand imbalance. Therefore, a significant amount of effort is being made to develop energy-efficient devices without any major financial investment.

A solar water heater is an important device that uses solar energy to heat water for domestic or commercial use. It consists of a solar collector, which absorbs the sun's energy, and a storage tank for hot water. The solar collector can be flat plate or evacuated tube and is typically mounted on a roof or a wall facing the sun. The storage tank is typically located near the collector and is insulated to reduce heat loss. The hot water produced by the solar collector is transferred to the storage tank, where it can be used as needed. Solar water heaters are an environmentally friendly and cost-effective alternative to traditional water heating systems that rely on fossil fuels.

Heat transfer enhancement for a solar water heater refers to methods used to improve the heat transfer rate from the solar collector to the water in the storage tank. This can be achieved by increasing the heat transfer area, the heat transfer coefficient, or the temperature difference between the two fluids. The common method for heat transfer enhancement is to improve the design of the solar collector by incorporating fins or other structures to increase the heat transfer area and, at the same time, use more thermally conductive fluids, such as water, with a high thermal conductivity, which can be used to transfer heat more efficiently. Implementing these heat transfer enhancement methods can lead to an improved performance and increased efficiency of a solar water heater. Generally, the technique of heat transfer enhancement (HTE) is divided into three categories: active techniques [1]; passive techniques [2,3]; and combined or combinational techniques [4,5]. Conventional heat exchangers (HEs) are incapable of satisfying the demand of heating and cooling in modern industries with their limited capacity. Modifications are being applied to the heat exchangers to enhance their thermal and flow performance. This includes use of ribs, turbulators, ultrasound, vibrations, magnetic field, electric field, advance heat transfer fluids, etc. [6]. The objective is to increase the thermohydraulic performance of a thermal device, whether by increasing the surface area for fluid interaction, by increasing the swirling motion of the fluid, or by increasing the thermal conductivity of the fluid. The thermohydraulic performance of a heat exchanger refers to its ability to transfer heat from one fluid to another. This performance is influenced by various factors, such as the temperature difference between the two fluids, the flow rate of the fluids, the heat transfer area, and the heat transfer coefficient. The overall thermohydraulic performance can be evaluated by calculating the heat transfer rate, the thermal efficiency, and the pressure drop across the heat exchanger. Improving the thermohydraulic performance of a heat exchanger can be achieved by optimizing these factors and increasing the heat transfer area and the heat transfer coefficient.

Among the various techniques of HTE in SWHs, ribs are the simplest, and the enhancement is appreciable when compared with other methods [7–10]. Deo et al. [8] experimentally investigated the influence of staggered inclined ribs on the thermal and flow performance of solar air heaters and reported significant enhancements in the performance and efficiency of solar air heaters (SAHs) when compared with a plane channel. Zhang et al. [9] studied the effect of longitudinal intersecting ribs on the thermohydraulic performance of gas turbine blades and reported an increase in the overall efficiency. Singh and Ekkad [11] reported on the implementation of ribs, along with dimples, for improving the cooling performance of gas turbine blades. It has been noted that the presence of both ribs and dimples enhanced the heat transfer (HT) and improved thermal hydraulic performance. Yang et al. [12] carried out a simulation study to study the impacts of high-blockage ribs on the HT and pressure drop (PD) characteristics of gas turbine blades. It was reported that the HT coefficient increases with an increase in the Re, the rib space to height ratio, and area of the ribbed portion covered in the channel. Alfarawi et al. [13] experimentally evaluated the thermal and hydrodynamic characteristics of air inside a rectangular ribbed channel. Tanda [14] presented performance results of SAHs fitted with four sets of ribs and found significant enhancements in the HT as well as the PD. Kumar et al. [15] reported on multiple arc-shaped ribs for SAHs, and although changing the roughness parameters resulted in an increase in the solar air heater's pumping power, there was a notable improvement in the thermal performance of the device. Hans et al. [16] examined single discrete arc ribs and how they affected SAH performance with constant parameter values. In single arc (discrete) ribs, the Nu was improved by 2.63 times more than the smooth SAH. Promvonge [17] experimentally reported on the influences of quadruple-twisted tapes arranged in four different combinations and V-fins on the HT and PD performance of square HE ducts. The V-finned counter-twisted tape had a much better thermal performance than the quadruple-twisted tapes. Mokkapati and Lin [18] reported on the thermohydraulic performance of corrugated tube HE tubes fitted with twisted tapes. The findings indicate that, as compared to plain tubes and corrugated tube HEs without twisted tapes, annularly corrugated tube heat exchangers with twisted tape boost the rate of HT by about 235.3% and 67.26%, respectively. Abraham and Vedula [19] investigated straight, V-shaped, and W-shaped ribs with apex angles of 45°. The ribs' height was kept as constant, and the pitch was varied. It was reported that the variation in the Nu ratios for the V and W configurations was negligible for the same pumping capacity. Chung et al. [20] reported on simple square ribs placed at a 60° angle in line with the direction of flow. They compared the performance of inclined ribs with that of intersecting ribs. The aspect ratio of the channel was varied between 1 and 4. It was found that the presence of intersecting ribs aided in the generation of more vortices, which resulted in a higher heat transfer enhancement for all the aspect ratios. Gawande et al. [7] employed L-shaped ribs for HTE in SAHs. The dimensions of the ribs remained constant, and the pitch or gap between the ribs varied from 10 to 25 mm. Configuration with a higher pitch ratio resulted in higher augmentation. Liu et al. [21] numerically investigated perforated square ribs for cooling channel application. The ribs under investigation were perforated with square and circular holes. It was reported that the perforated ribs showed an enhanced thermal performance when compared with non-perforated ribs and were suitable for cooling application. Bhattacharyya et al. [22] investigated alternating inclined ribs for their thermohydraulic performance in a circular channel. Four angles of attack were investigated for a wide range of Re values, covering laminar, transition, and turbulent flow regimes.

2. Objective

The use of novel arc-shaped ribs as a means of enhancing heat transfer is a new technique in the field of HTE. Arc-shaped ribs are small devices that are attached to the surface of a heat exchanger wall and cause the fluid flow to become turbulent, which can enhance the convective heat transfer and improve the overall heat transfer coefficient. From the literature, it was found that many researchers are working on circular tubes to enhance heat transfer by inserting only unidirectional ribs, and it was also noticed that in many cases, the thermo-hydraulic performance of the system falls below unity. As per the authors' knowledge, thus far, no research has been conducted on square channels with semi-arc-shaped ribs on four sides of the channel to enhance HT. Overall, the use of ribs for heat transfer enhancement remains an active area of research and development, with a significant degree of novelty and potential for continued innovation. In this study, a turbulent HTE in the square channel with the augmentation of the flow pattern using novel arc-shaped ribs in SWHs is presented.

3. Computational Domain, Boundary Conditions and Meshing

The dimensions of the 3D square channel are 20 mm × 20 mm, and a length of 2500 mm is taken for the present computational investigation; these dimensions are constant throughout the study. To enhance the convective HT rate, semi-arc ribs are attached to the SWH walls. The computational domain is shown in Figure 1. The current study concerns the region of the turbulent flow regime with a high Re ranging from 4000 to 25,000. A fully developed velocity profile is incorporated at the SWH inlet for different Re values. All side walls are assumed to have iso-heat flux boundary conditions with 1.0 kW/m^2 throughout the test section. Table 1 presents the tested parameters. To make efficient use of the computer resources, simulations were conducted based on the assumptions below:

- The flow is incompressible, and steady-state equations are solved to predict the results.
- The air travels easily over the solid surface with a no-slip boundary condition.
- In ambient circumstances, water as the working fluid, enters in the computational domain.

Table 1. Computational Parameters.

Parameter	Range
Inner diameter of the tube, D	20 mm
Height ratio, d/D = H	0.5, 0.4, 0.25
Pitch ratio, p/D = s	1, 1.5, 2.0
Reynolds number (Re)	4000–25,000

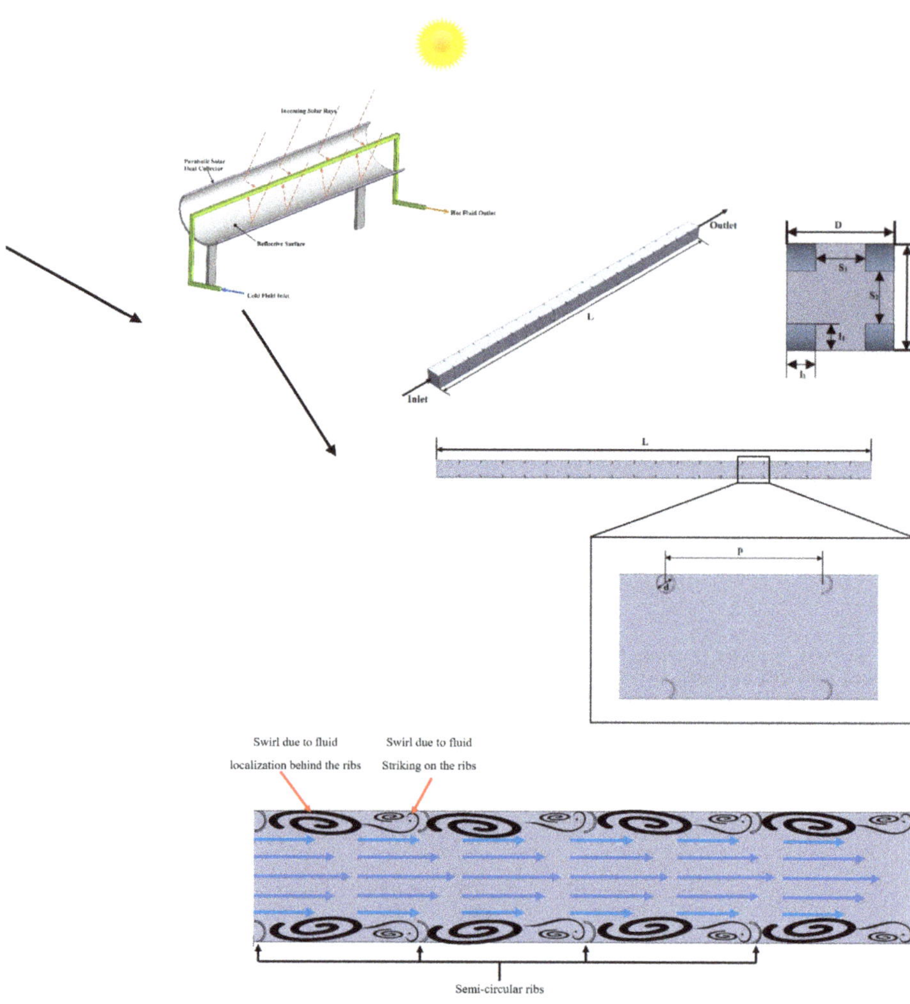

Figure 1. Computational domain.

The steady incompressible viscous flow is computationally simulated using a modified SWH with semi-arc cross-sectional roughness elements. The roughness elements are positioned on each corner of the square channel. The flow field and HT through the SWH are governed by the Navier–Stokes equation and the energy equation, which are written as follows [22]:

$$\frac{\partial}{\partial x_i}(\rho \bar{u}_i) = 0 \quad (1)$$

$$\frac{\partial}{\partial x_i}(\rho \bar{u}_i \bar{u}_j) + \frac{\partial}{\partial x_i}\bar{P}' \frac{\partial}{\partial x_i}\left[(\mu + \mu_t)\left(\frac{\partial}{\partial x_j}\bar{u}_i + \frac{\partial}{\partial x_i}\bar{u}_j\right)\right] \quad (2)$$

$$C_P\left[\bar{u}_i \frac{\partial}{\partial x_i}(rho\bar{T}) + \frac{\partial}{\partial x_i}\left(\frac{\mu_t}{\mathrm{Pr}_t} \times \frac{\partial}{\partial x_i}\bar{T}\right)\right] = \frac{\partial}{\partial x_i}\left(\lambda \times \frac{\partial}{\partial x_i}\bar{T}\right) \quad (3)$$

The variables *P*, *u*, and *T* in the equation above represent pressure, velocity, and temperature, respectively.

The important RANS turbulent model, using the k-ω RNG model, is implemented because this model includes an additional ω-equation that improves the accuracy of the projected results. This model includes the following equation:

$$\frac{\partial(\rho k u_i)}{\partial x_i} - \frac{\partial}{\partial x_j}\left(\alpha_k \mu_{eff} \frac{\partial k}{\partial x_j}\right) - G_k + \rho\varepsilon = 0 \tag{4}$$

$$\frac{\partial(\rho \varepsilon u_i)}{\partial x_i} - \frac{\partial}{\partial x_j}\left(\alpha_\varepsilon \mu_{eff} \frac{\partial \varepsilon}{\partial x_j}\right) - \frac{C_{1'\varepsilon}}{k} G_k + C_2 \rho \frac{\varepsilon^2}{k} = R_\varepsilon \tag{5}$$

$$\mu_{eff} = \mu_f + \mu_t = \mu_f + \rho C_\mu \frac{k^2}{\varepsilon} \tag{6}$$

The simulation software ANSYS Fluent 18.1 is used to simulate the HT and PD characteristics in the modified SWH with semi arc-shaped ribs. To discretize, the governing equations, second-order numeric techniques are used to improve the accuracy and reduce literation errors. In addition, the convergence threshold for the continuity and momentum components is established as 10^{-6}, and for the energy equation, it is 10^{-8} [22]. Steady-state calculations are performed on the quantities required to estimate the thermal–hydraulic performance.

In the test section, all the properties such as the velocity, pressure, temperature, etc., are measured under steady-state conditions. The tetrahedral element is formed in the semi-arc-shape-ribbed SWH. The Y+ value is kept as smaller than one for the rib roughness and the heated wall. ANSYS ICEM is used to construct the meshing, and the values of all the meshed elements are guaranteed to be more than 0.75. The procedure of constructing the mesh begins with 10 mm-sized coarse components, and the process is continued until the results have no or minimal influence on the mesh refinement (or element size). The meshing is shown in Figure 2, and Table 2 presents the specifics of the grid refinement along with the appropriate solution for the average Nu and friction factor (f). As a result, the grid of 2,987,675 nodes, which is grid 1 in Table 1, is used for all the simulations in order to save computational time and resources.

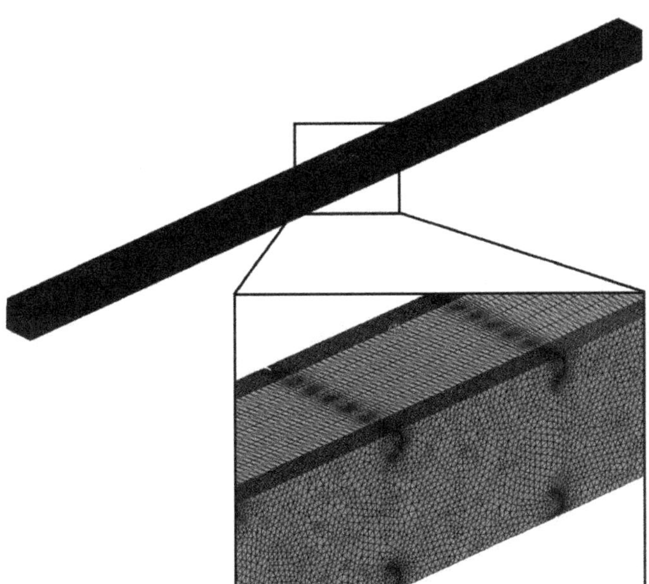

Figure 2. Meshing of the computational domain.

Table 2. Grid independence test.

	Nodes	Nu	f	η
	Re = 10,000, Semi-Arc-Shaped Rib, P = 1.0, H = 0.5			
Grid 1	2,987,675	132.370	0.037	1.455
Grid 2	3,345,789	132.373	0.037	1.455
Grid 3	3,876,444	132.391	0.038	1.456

Heat transport to the working fluid is quantified using the first law written below. It is known from the first law of thermodynamics that the rate of net HT to the working fluid is proportional to the temperature difference between the inlet and outlet fluids, as well as the mass flow rate and specific heat of the induced fluid [3,23].

$$Q = mc_p(T_o - T_i) \tag{7}$$

The heat transfer coefficient is calculated using the following equation, as stated by Bhattacharyya et al. [3]:

$$h = \frac{q_w}{T_w - T_b} \tag{8}$$

The Nusselt number measures the extent to which convection contributes to the total heat transfer relative to conduction. A comparison of the superiority of convective heat transfer vs. conductive heat transfer may be conducted using this dimensionless metric. In other words, a rising Nu value suggests that convection is a more important heat transfer mechanism than conduction. One possible expression of this is [23]:

$$Nu_{avg} = \frac{hD}{k} \tag{9}$$

where k is the thermal conductivity of the material.

One of the most critical parameters for assessment is known as Darcy's Friction Factor. This parameter is indicated by f and may be determined using the formula listed below [22]:

$$f = \frac{\Delta \mathrm{P}}{\frac{\mathrm{L}}{\mathrm{D}}\frac{1}{2}\rho \mathrm{V}^2} \tag{10}$$

Here, ρ, D, and L signify the density of the fluid, hydraulic diameter, and length of the SAH, respectively. Based on the free cross-section of an insert, the bulk fluid velocity and the static pressure drop are both designated by the ΔP. Reynolds number is used to compute the velocity of the working fluid (*Re*). The formula is as follows [3]:

$$Re = \frac{\rho V D}{\mu} \tag{11}$$

The increment in Nu_{avg} and f due to the semi-arc-shaped ribs' imposition on the SAHs, with respect to smooth SAHs, is determined by Nu_c and f_c, where $\mathrm{Nu_c}$ and $\mathrm{f_c}$ can be determined as [22]:

$$Nu_c = \frac{Nu_{\mathrm{avg}}\big|_{\text{with rib}}}{Nu_{\mathrm{avg}}\big|_{\text{smooth}}} \tag{12}$$

$$f_c = \frac{f\big|_{\text{with rib}}}{f\big|_{\text{smooth}}} \tag{13}$$

From Equations (12) and (13), the thermo-hydraulic performance (*THP*) can be determined as [22]:

$$THP = Nu_c \times f_c^{-\frac{1}{3}} \tag{14}$$

4. Validation

It is essential to validate the present simulations with the literature. The Nusselt numbers for plain channels obtained by simulation are compared in Figure 3a with those from Dittus–Boelter [23]. Figure 3b contrasts the friction factor with the historically reliable Blasius correlation [24]. The average absolute variance for the Nusselt number was found to be 4.0% and the average deviation for the friction factor was found to be 2.0% when comparing the simulated findings with the projected data. The current computational model can be used to calculate the flow and HT in a square channel with semi-arc-shaped ribs if the simulation accuracy is accepted.

$$Nu = 0.0023\mathrm{Re}^{0.8}\mathrm{Pr}^{0.4} \tag{15}$$

$$f = \frac{0.316}{Re^{\frac{1}{4}}} \tag{16}$$

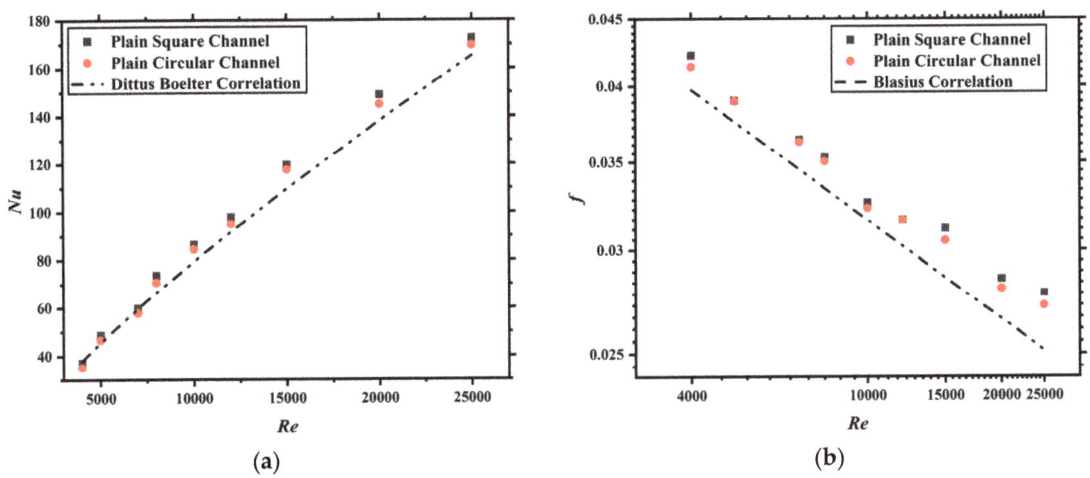

Figure 3. Validation of the computation study: (**a**) *Nu* and (**b**) *f*.

5. Results and Discussion

This important discussion section provides a detailed summary of the present computation investigation on the novel semi-arc-shaped ribs. Semi arc-shaped ribs are mainly responsible for disturbing the hydrodynamic and thermal boundary layers of the water flow, which leads to the formation of vortices near the heated walls and better mixing of the fluid layers. Due to the above-mentioned phenomenon, convective heat transfer is superior to conductive heat transfer. The current study concerns turbulent flow with Re ranging from 4000 to 25,000. Heat transfer, pressure drop, the j-factor, Bejan number, and thermo-hydraulic performance, as a function of the Re results, are discussed in this section.

Heat transfer in SWH was enhanced significantly by the incorporation of semi-arc-shaped ribs inside the square tube, as shown in Figure 4a. A decrease in the conduction HT and an increase in the convection HT occur when ribs are added to the SWH walls, and this is because the ribs increase the convective surface area. In addition, the development of the boundary layer acts as a fence for the HT, as this is a forced convection heat transfer scenario, and, as a result, the ribs installed in the flow field aid in the disruption of the boundary layer and boost the rate of HT.

Considering three different pitch ratios (P) of the semi-arc-shaped ribs, Figure 4a shows a plot of the average Nu as a function of Re for different height and pitch ratios. Convective heat transfer coefficients increase with increasing Re or flow velocities because this type of heat transfer is driven by forced convection. Increases in flow velocity raise the Re, which causes the flow to become more turbulent. This means that the fluid layers may be travelling over many other fluid layers, leading to greater mixing in the flow and a consequently greater heat transfer from the channel wall. The insertion of the novel-shaped ribs will increase the Nusselt number in heat transfer situations by creating turbulence in the fluid flow and enhancing the convective heat transfer. This will lead to an increase in the overall heat transfer coefficient and result in an increased Nusselt number. One can see from Figure 4a that the Nu increases uniformly with the Re for all the tested cases. In addition, when the roughness pitch ratio increases, the Nu drops for a certain Re. The ratio of the rib pitch to the rib height is the relative roughness pitch.

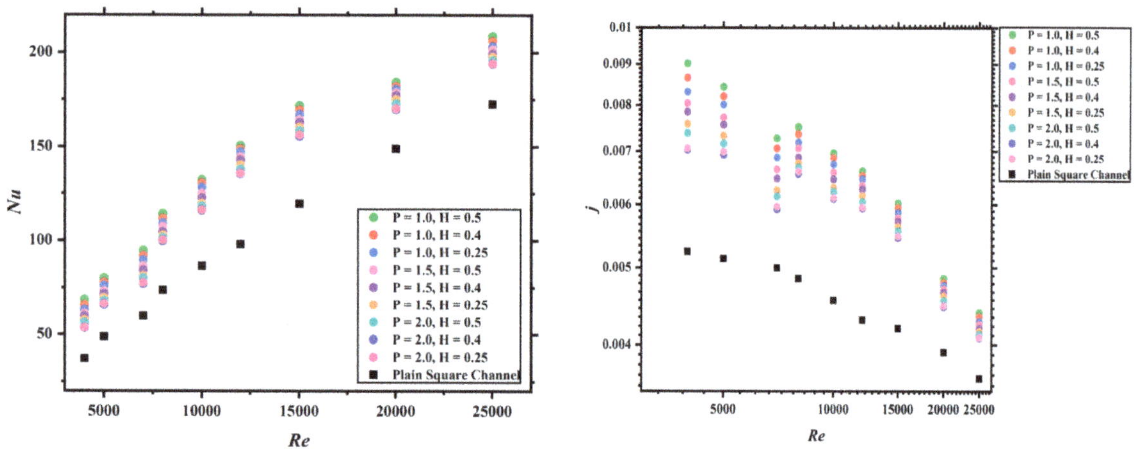

(a) (b)

Figure 4. (**a**) Nu as a function of Re at different pitch and height ratios, and (**b**) Colburn *j-factor* as a function of Re at different pitch and height ratios.

The Colburn j-factor is a dimensionless number used in the field of HT to predict the convective HT coefficient in forced convection. It is defined as the ratio of the convective HT coefficient to the product of the fluid's dynamic viscosity, the density of the fluid, and the square of the characteristic length (L) over which the HT occurs. It is important to note that the Colburn j-factor is an empirical coefficient, meaning that it is determined experimentally and may vary depending on the specific conditions of the HT process. In Figure 4b, it is apparent that the Colburn j-factor decreases with a rise in Re. This may be due to the novel rib creating turbulence in the fluid flow, which enhances the convective heat transfer and increases the overall heat transfer coefficient. Moreover, placing rib inserts in SWHs enhances the j-factor, meaning that higher heat transfer occurs due to the better mixing of the fluid. A higher j-factor can be observed when a small pitch ratio and high height ratio of the ribs are present. However, the effect of rib insertion on the Colburn j-factor will depend on the specific configuration and the thermal–fluid properties involved.

In the presence of semi-arc-shaped ribs, there is a flow separation that can be appreciated by examining Figure 5a,b and this can lead to a breakdown of the boundary layer, as well as the reattachment of the separated boundary layer, eddy generation at the rib tips and in the inter-rib regions, and enhanced mixing owing to the development of localized turbulence. The flow is obstructed (as one can see from Figure 5a,b), the boundary layer is stunted, and the local turbulence is increased as the number of ribs is increased. This leads to a greater rate of HT.

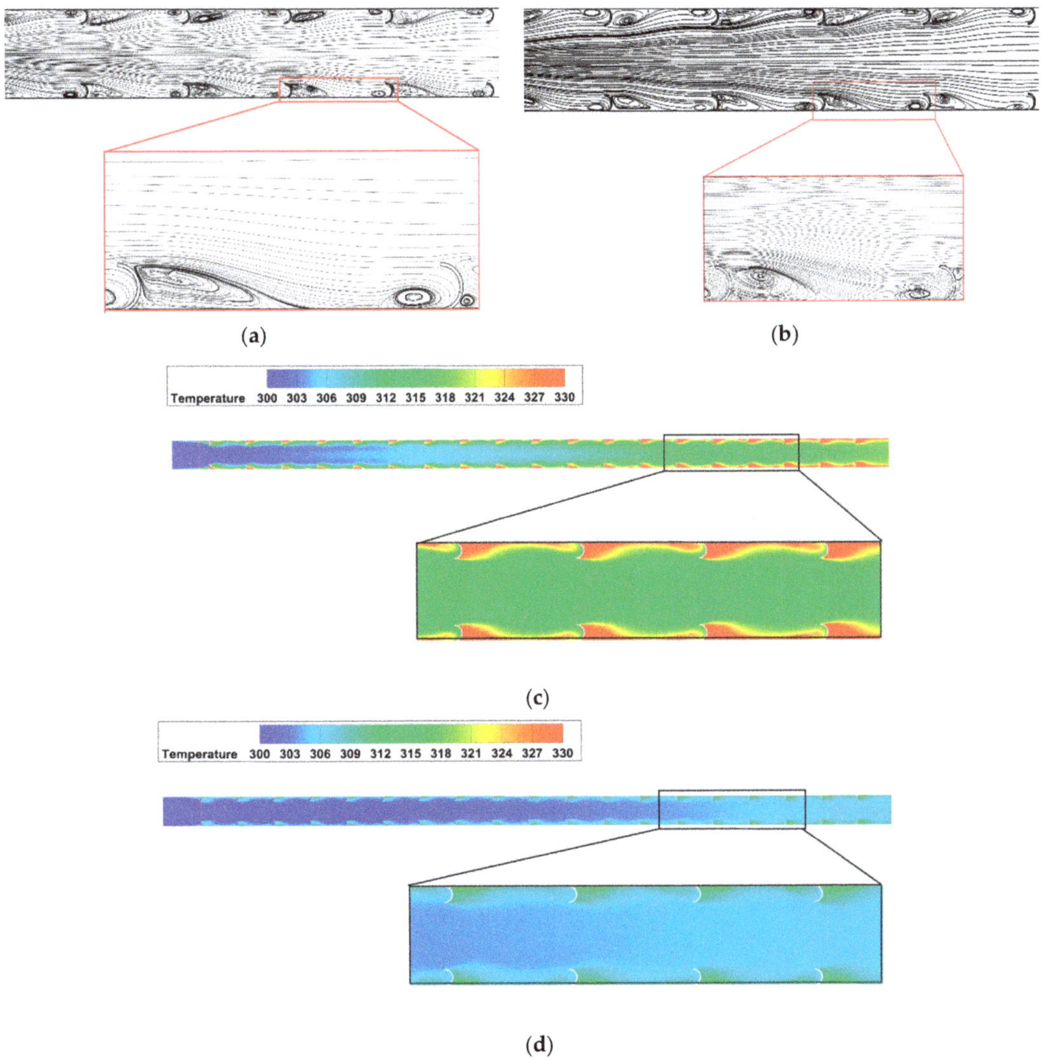

Figure 5. (**a**) Flow pattern around semi-arc-shaped ribs at Re 10,000, (**b**) flow pattern around semi-arc-shaped ribs at Re 25,000, (**c**) temperature contour around semi-arc-shaped ribs at Re 10,000, and (**d**) temperature contour around semi-arc-shaped ribs at Re 25,000.

In HT and fluid flow problems, the PD, in terms of f, is very important in order to calculate and, accordingly, understand the performance of the system. The f is a measure of the resistance to flow in a channel. It is an important factor for determining the pressure drop and flow rate in a system. In an SWH, the insertion of rib inserts can increase the friction factor by creating a more turbulent flow of water through the pipe. In the present study, it was observed that placing ribs inside the SWH increases the HT rate significantly. On the other hand, an increase in frictional loss is also observed after imposing the ribs inside the channel. Figure 6 shows that f decreases with a rise in Re. While modified SWHs are analyzed here, a noticeable increase in f is observed in comparison with smooth channel SWHs. The friction factor increases with a decrease in the pitch ratio and increase in the height ratio. A small pitch ratio signifies smaller gaps between two consecutive ribs, and the height ratio signifies the height of the ribs. From Figure 6, it is clearly apparent that when the gaps between the ribs decrease, the friction factor increases, because in this scenario, the number of vortices is high, and due to the localization of the fluid, the friction factor increases significantly. The height of the ribs plays a significant role in creating vortices. Increases in height mainly create larger vortices, and thus, f increases with an increase in the height of the ribs.

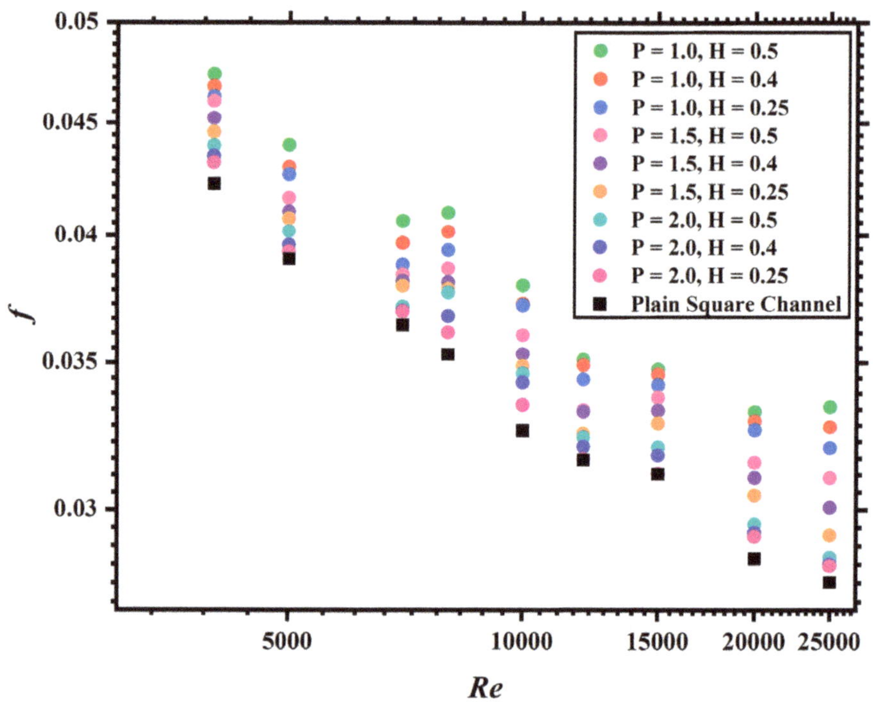

Figure 6. Friction factor as a function of Re at different pitch and height ratios.

Entropy analysis is an important aspect to study in thermal systems. The Bejan number is defined as the ratio of the system irreversibility to the total system irreversibility due to flow dynamics. It is often used to predict the flow patterns and HT characteristics of fluids in channels. The Bejan number can be used to design heat exchangers and predict the performance of heat transfer equipment. In general, a high Bejan number indicates that convective heat transfer is more important than conductive HT, while a low Bejan number indicates the opposite. The Bejan number is often used in conjunction with other dimensionless numbers, such as the Re and the Prandtl number, to predict the behavior of fluids in different flow regimes.

Generally, thermal entropy generation can be defined as [25,26]:

$$\dot{S}_{g,th} = \frac{Q_{avg}^2}{Nu\pi k T_i T_0 L} \tag{17}$$

Similarly, frictional entropy generation can be defined as [25,26]:

$$\dot{S}_{g.f} = \frac{8f\dot{m}^3 L}{\rho^2 \pi^2 D_i^5 (T_0 - T_i)} \ln\left(\frac{T_o}{T_i}\right) \tag{18}$$

Finally, the Bejan number is defined as [26]:

$$Be = \frac{\dot{S}_{g,th}}{\dot{S}_{g,th} + \dot{S}_{g,f}} \tag{19}$$

In Figure 7, one can see that the Bejan number decreases with an increase in Re. An enhancement of the Bejan number due to the placement of the ribs was clearly observed in the present study. This signifies that the HTE is due to the placement of ribs in SWHs.

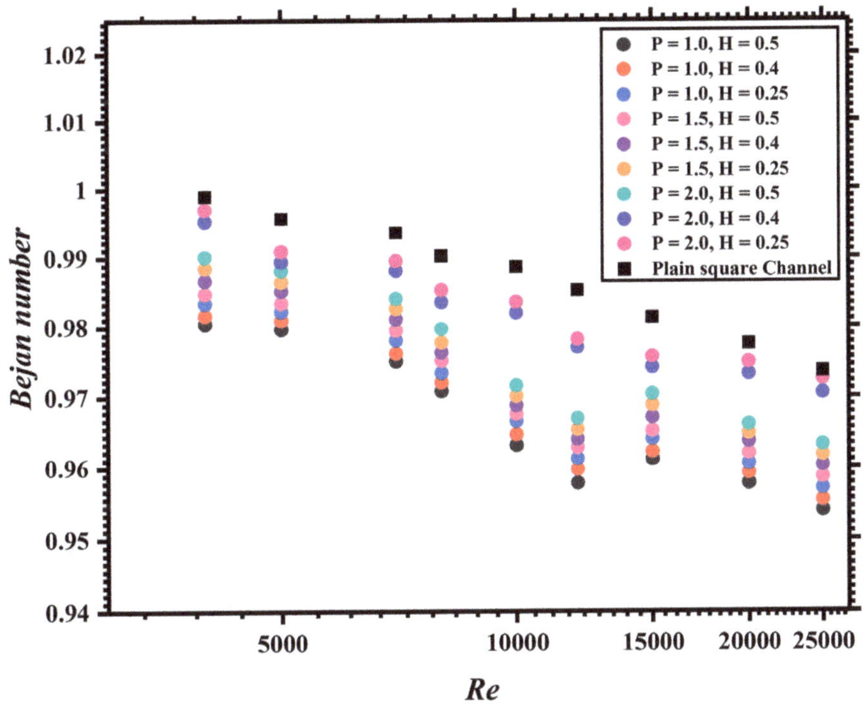

Figure 7. Bejan number (Be) as a function of Re at different pitch and height ratios.

Thermo-hydraulic performance (THP) is a parameter used to predict efficient modification in heat exchangers. THP is the ratio of Nu and f (Equation (11)). In the present study, thermo-hydraulic performance mainly signifies the effectivity of the rib inserts in SWHs. Figure 8 shows a decrease in THP with increasing Re. The placement of ribs improves the THP, which means that Nu increases higher than the increase in frictional loss. This signifies that ribs are an effective way to enhance the efficiency of SWHs. A lower pitch ratio and higher height ratio (P = 1.0, H = 0.5) perform better, as per Figure 8a. Keeping pitch ratio constant (P = 1.0), it is apparent that a higher height ratio (H = 0.5) results in higher performance increments. Similarly, when H = 0.5 is kept as constant, a lower pitch ratio (P = 1.0) results in high performance enhancement for each given height ratio. Moreover, it is important to note that for all the tested cases, the THP is greater than unity, which means that the present SWH system is promising and can be implemented.

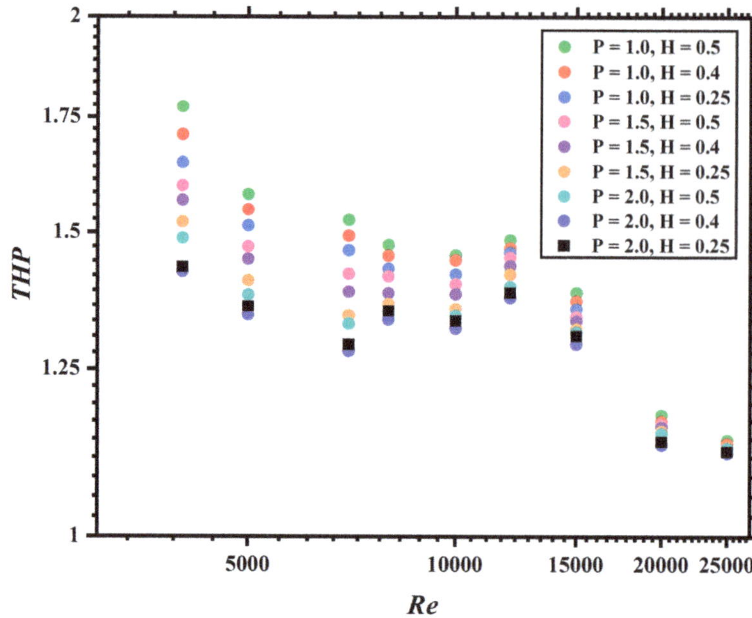

(a)

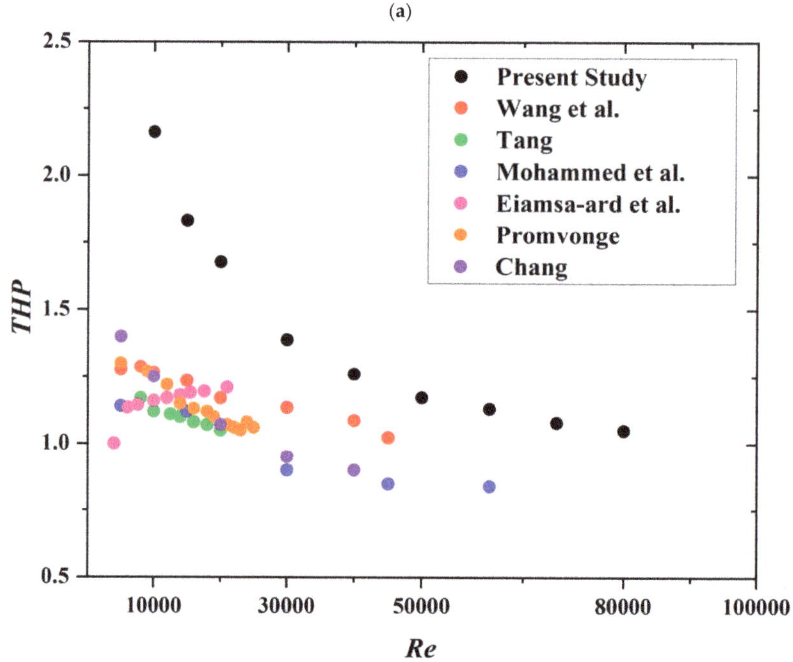

(b)

Figure 8. (**a**) Thermo-hydraulic performance varies with Re at different pitch and height ratios. (**b**) Comparison of the present study with the literature [27–32].

Figure 8b presents a comparison plot of the present study (best case: y = 0.5, h = 0.2, θ = 60°) with the literature (research in similar areas). It is clearly illustrated in Figure 8b that the outcomes of the present study are superior when compared with the literature on similar domains, such as that of Wang et al. [27], Tang [28], Mohammed et al. [29], Eiamsa-ard et al. [30], Promvonge [31], and Chang et al. [32].

6. Conclusions

A computational study on HT and PD in a square-channel SWH with the insertion of semi-arc-shaped ribs was presented. As the working fluid, water was utilized, and the Re ranged from 4000 to 25,000. The present geometry encompassed semi-arc-shaped ribs with pitch ratios of 1.0, 1.5, and 2.0 and rib height ratios of 0.5, 0.4, and 0.25.

From the above computational study, the following conclusions may be drawn:

1. The insertion of novel-shaped ribs increases the Nusselt number by creating swirl flow in the flow field and enhancing the convective HT, and eventually, this leads to an augmentation of the overall heat transfer coefficient and results in an increased Nusselt number.
2. With an increase in the semi-arc-shaped rib height, the heat transfer rate increases, and at the same time, the friction factor is also increased significantly.
3. A decrease in the semi-arc-shaped rib pitch leads to enhancements in the heat transfer. However, more enhancement is noted when the rib height ratio is highest and the pitch ratio is lowest.
4. The Colburn j-factor and Bejan number were also presented, and the outcome is promising.
5. The thermal performance factor remains higher than unity for all the configurations investigated in the present numerical investigation. A pitch ratio of 1.0 and height ratio of 0.5 show the highest performance. The enhanced geometry is promising and may be implemented in the HVAC sector.

Some practical guidelines for optimizing the performance of novel arc-shaped rib inserts in a solar water heater are as follows:

1. Arc-shaped rib pitch ratio: Choose a pitch ratio that is optimal for the specific flow conditions. A higher pitch ratio can increase turbulence, but a pitch ratio that is too high can also cause excessive pressure drop.
2. Arc-shaped rib height ratio: The height of the rib can affect the amount of turbulence generated, with large rib heights generally causing more turbulence. One should choose a rib height that balances the desired level of turbulence with the pressure drop acceptable for the given system.
3. Rib material: The material of the rib can affect the heat transfer and corrosion resistance of the system. One should consider using a material with good thermal conductivity and resistance to corrosion for a specific application.
4. Flow rate: The flow rate through the system can affect the heat transfer and turbulence generated by the novel arc-shaped rib insert. One should choose a flow rate that balances the desired level of heat transfer with the pressure drop acceptable for the given system.

Author Contributions: Conceptualization, B.S. and S.B.; methodology, B.S. and S.B.; software, B.S. and S.B.; validation, B.S. and S.B.; formal analysis, B.S. and S.B.; investigation, B.S. and S.B.; resources, B.S. and S.B.; data curation, B.S. and S.B.; writing—original draft preparation B.S. and S.B.; writing—review and editing, B.S. and S.B.; visualization, B.S. and S.B.; supervision, B.S. and S.B.; project administration, B.S.; funding acquisition, B.S. All authors have read and agreed to the published version of the manuscript.

Funding: This work was supported by the Deanship of Scientific Research, Vice Presidency for Graduate Studies and Scientific Research, King Faisal University, Saudi Arabia (Grant No. 2413).

Data Availability Statement: The data that support the findings of this study are available from the corresponding author upon reasonable request.

Acknowledgments: Authors appreciate support by the Deanship of Scientific Research, Vice Presidency for Graduate Studies and Scientific Research, King Faisal University, Saudi Arabia (Grant No. 2413).

Conflicts of Interest: The authors declare no conflict of interest.

References

1. Gui, N.G.J.; Stanley, C.; Nguyen, N.-T.; Rosengarten, G. Ferrofluids for heat transfer enhancement under an external magnetic field. *Int. J. Heat Mass Transf.* **2018**, *123*, 110–121. [CrossRef]
2. Bhattacharyya, S.; Vishwakarma, D.K.; Goel, V.; Chamoli, S.; Issakhov, A.; Meyer, J.P. Thermodynamics and heat transfer study of a circular tube embedded with novel perforated angular-cut alternate segmental baffles. *J. Therm. Anal. Calorim.* **2021**, *145*, 1445–1465. [CrossRef]
3. Bhattacharyya, S.; Vishwakarma, D.K.; Srinivasan, A.; Soni, M.K.; Goel, V.; Sharifpur, M.; Ahmadi, M.H.; Issakhov, A.; Meyer, J. Thermal performance enhancement in heat exchangers using active and passive techniques: A detailed review. *J. Therm. Anal. Calorim.* **2022**, *147*, 9229–9281. [CrossRef]
4. Bezaatpour, M.; Goharkhah, M. Effect of magnetic field on the hydrodynamic and heat transfer of magnetite ferrofluid flow in a porous fin heat sink. *J. Magn. Magn. Mater.* **2019**, *476*, 506–515. [CrossRef]
5. Bhattacharyya, S.; Sharma, A.K.; Vishwakarma, D.K.; Paul, A.R. Thermo-hydraulic performance of magnetic baffles for removal of concentrated heat fluxes in a heated mini channel. *Appl. Therm. Eng.* **2022**, *216*, 118992. [CrossRef]
6. Bhattacharyya, S.; Vishwakarma, D.K.; Roy, S.; Biswas, R.; Ardekani, M.M. Applications of Heat Transfer Enhancement Techniques: A State-of-the-Art Review. In *Inverse Heat Conduction and Heat Exchangers*; IntechOpen: London, UK, 2020. [CrossRef]
7. Gawande, V.B.; Dhoble, A.; Zodpe, D.; Chamoli, S. Experimental and CFD investigation of convection heat transfer in solar air heater with reverse L-shaped ribs. *Sol. Energy* **2016**, *131*, 275–295. [CrossRef]
8. Deo, N.S.; Chander, S.; Saini, J. Performance analysis of solar air heater duct roughened with multigap V-down ribs combined with staggered ribs. *Renew. Energy* **2016**, *91*, 484–500. [CrossRef]
9. Zhang, C.; Wang, Z.; Kang, J. Flow and Heat Transfer in a High-Aspect-Ratio Rib-Roughed Cooling Channel with Longitudinal Intersecting Ribs. *J. Appl. Mech. Tech. Phys.* **2018**, *59*, 679–686. [CrossRef]
10. Ngo, T.T.; Phu, N.M. Computational fluid dynamics analysis of the heat transfer and pressure drop of solar air heater with conic-curve profile ribs. *J. Therm. Anal. Calorim.* **2020**, *139*, 3235–3246. [CrossRef]
11. Singh, P.; Ekkad, S. Experimental study of heat transfer augmentation in a two-pass channel featuring V-shaped ribs and cylindrical dimples. *Appl. Therm. Eng.* **2017**, *116*, 205–216. [CrossRef]
12. Yang, W.; Xue, S.; He, Y.; Li, W. Experimental study on the heat transfer characteristics of high blockage ribs channel. *Exp. Therm. Fluid Sci.* **2017**, *83*, 248–259. [CrossRef]
13. Alfarawi, S.; Abdel-Moneim, S.; Bodalal, A. Experimental investigations of heat transfer enhancement from rectangular duct roughened by hybrid ribs. *Int. J. Therm. Sci.* **2017**, *118*, 123–138. [CrossRef]
14. Tanda, G. Performance of solar air heater ducts with different types of ribs on the absorber plate. *Energy* **2011**, *36*, 6651–6660. [CrossRef]
15. Kumar, R.; Goel, V.; Singh, P.; Saxena, A.; Kashyap, A.S.; Rai, A. Performance evaluation and optimization of solar assisted air heater with discrete multiple arc shaped ribs. *J. Energy Storage* **2019**, *26*, 100978. [CrossRef]
16. Hans, V.; Gill, R.; Singh, S. Heat transfer and friction factor correlations for a solar air heater duct roughened artificially with broken arc ribs. *Exp. Therm. Fluid Sci.* **2017**, *80*, 77–89. [CrossRef]
17. Promvonge, P. Thermal performance in square-duct heat exchanger with quadruple V-finned twisted tapes. *Appl. Therm. Eng.* **2015**, *91*, 298–307. [CrossRef]
18. Mokkapati, V.; Lin, C.-S. Numerical study of an exhaust heat recovery system using corrugated tube heat exchanger with twisted tape inserts. *Int. Commun. Heat Mass Transf.* **2014**, *57*, 53–64. [CrossRef]
19. Abraham, S.; Vedula, R.P. Heat transfer and pressure drop measurements in a square cross-section converging channel with V and W rib turbulators. *Exp. Therm. Fluid Sci.* **2016**, *70*, 208–219. [CrossRef]
20. Chung, H.; Park, J.S.; Park, S.; Choi, S.M.; Rhee, D.-H.; Cho, H.H. Augmented heat transfer with intersecting rib in rectangular channels having different aspect ratios. *Int. J. Heat Mass Transf.* **2015**, *88*, 357–367. [CrossRef]
21. Liu, J.; Hussain, S.; Wang, W.; Xie, G.; Sundén, B. Experimental and numerical investigations of heat transfer and fluid flow in a rectangular channel with perforated ribs. *Int. Commun. Heat Mass Transf.* **2021**, *121*, 105083. [CrossRef]
22. Bhattacharyya, S.; Chattopadhyay, H.; Benim, A.C. Computational investigation of heat transfer enhancement by alternating inclined ribs in tubular heat exchanger. *Prog. Comput. Fluid Dyn. Int. J.* **2017**, *17*, 390. [CrossRef]
23. Bhattacharyya, S.; Benim, A.C.; Bennacer, R.; Dey, K. Influence of Broken Twisted Tape on Heat Transfer Performance in Novel Axial Corrugated Tubes: Experimental and Numerical Study. *Heat Transf. Eng.* **2021**, *43*, 437–462. [CrossRef]
24. Bhattacharyya, S.; Bashir, A.I.; Dey, K.; Sarkar, R. Effect of novel short-length wavy-tape turbulators on fluid flow and heat transfer: Experimental study. *Exp. Heat Transf.* **2020**, *33*, 335–354. [CrossRef]
25. Huminic, G.; Huminic, A. The heat transfer performances and entropy generation analysis of hybrid nanofluids in a flattened tube. *Int. J. Heat Mass Transf.* **2018**, *119*, 813–827. [CrossRef]
26. Bejan, A. A Study of Entropy Generation in Fundamental Convective Heat Transfer. *J. Heat Transf.* **1979**, *101*, 718–725. [CrossRef]
27. Wang, W.; Zhang, Y.; Li, B.; Li, Y. Numerical investigation of tube-side fully developed turbulent flow and heat transfer in outward corrugated tubes. *Int. J. Heat Mass Transf.* **2018**, *116*, 115–126. [CrossRef]

28. Tang, X.; Dai, X.; Zhu, D. Experimental and numerical investigation of convective heat transfer and fluid flow in twisted spiral tube. *Int. J. Heat Mass Transf.* **2015**, *90*, 523–541. [CrossRef]
29. Mohammed, H.A.; Abbas, A.K.; Sheriff, J. Influence of geometrical parameters and forced convective heat transfer in transversely corrugated circular tubes. *Int. Commun. Heat Mass Transf.* **2013**, *44*, 116–126. [CrossRef]
30. Eiamsa-Ard, S.; Rattanawong, S.; Promvonge, P. Turbulent convection in round tube equipped with propeller type swirl generators. *Int. Commun. Heat Mass Transf.* **2009**, *36*, 357–364. [CrossRef]
31. Promvonge, P. Thermal performance in circular tube fitted with coiled square wires. *Energy Convers. Manag.* **2008**, *49*, 980–987. [CrossRef]
32. Chang, S.W.; Yang, T.L.; Liou, J.S. Heat transfer and pressure drop in tube with broken twisted tape insert. *Exp. Therm. Fluid Sci.* **2007**, *32*, 489–501. [CrossRef]

mathematics MDPI

Article

Optimization of MHD Flow of Radiative Micropolar Nanofluid in a Channel by RSM: Sensitivity Analysis

Reham A. Alahmadi [1], Jawad Raza [2,*], Tahir Mushtaq [2], Shaimaa A. M. Abdelmohsen [3], Mohammad R. Gorji [4] and Ahmed M. Hassan [5]

1 Basic Science Department, College of Science and Theoretical Studies, Saudi Electronic University, Riyadh 11673, Saudi Arabia
2 Department of Mathematics, COMSATS University Islamabad, Vehari Campus, Vehari 61100, Pakistan
3 Department of Physics, College of Science, Princess Nourah Bint Abdulrahman University, P.O. Box 84428, Riyadh 11671, Saudi Arabia
4 Faculty of Medicine and Health Sciences, Ghent University, 9000 Ghent, Belgium
5 Department of Mechanical Engineering, Future University in Egypt, New Cairo 11835, Egypt
* Correspondence: jawadraza@cuivehari.edu.pk

Abstract: These days, heat transfer plays a significant role in the fields of engineering and energy, particularly in the biological sciences. Ordinary fluid is inadequate to transfer heat in an efficient manner, therefore, several models were considered for the betterment of heat transfer. One of the most prominent models is a single-phase nanofluid model. The present study is devoted to solving the problem of micropolar fluid with a single-phase model in a channel numerically. The governing partial differential equations (PDEs) are converted into nonlinear ordinary differential equations (ODEs) by introducing similarity transformation and then solved numerically by the finite difference method. Response surface methodology (RSM) together with sensitivity analysis are implemented for the optimization analysis. The study reveals that sensitivity of the skin friction coefficient (Cf_x) to the Reynolds number (R) and magnetic parameter (M) is positive (directly proportional) and negative (inversely proportional) for the micropolar parameter.

Keywords: micropolar fluid; nanofluid; thermal radiation; response surface methodology; sensitivity analysis

MSC: 76D55

Citation: Alahmadi, R.A.; Raza, J.; Mushtaq, T.; Abdelmohsen, S.A.M.; R. Gorji, M.; Hassan, A.M. Optimization of MHD Flow of Radiative Micropolar Nanofluid in a Channel by RSM: Sensitivity Analysis. *Mathematics* **2023**, *11*, 939. https://doi.org/10.3390/math11040939

Academic Editor: Ramoshweu Solomon Lebelo

Received: 3 January 2023
Revised: 8 February 2023
Accepted: 10 February 2023
Published: 13 February 2023

1. Background

In previous decades, a demand to represent the fluid that depends on micro-components has concluded in the establishment of micropolar fluid. Eringen [1,2] was the first researcher to use the term micropolar. This term then became an area of dynamic exploration. A simple microfluid, by definition, is a fluent medium whose properties and actions in each of its volume elements are influenced by local movements of the material particles; such a fluid has local inertia. Eringen [2] presents a complete discussion of motion and micro-motions in the presentation of the theory, as well as evidence of the newly introduced micro-deformation rate tensors, which is a prerequisite for the creation of the constitutive equations used to characterise a simple microfluid. These classes of fluids identify many engineering and industrial applications physically and mathematically. On the other hand, a class of conventional Newtonian fluids cannot specifically identify the fluid flow for a range of applications in the area of engineering. The examples of such fluids are polymeric, colloidal solutions, paints, etc. In micropolar fluid, the micro-rotation vectors explain the rotational motion in microfluid. Therefore, the curl of the velocity vector in this case will be non-zero.

The control of magnetic induction past a plate in the existence of a micropolar fluid have mainly been evaluated by Gorla and Mohammeadein [3]. Pedieson [4] evaluated and studied boundary layer theory of micropolar fluids. Gupta [5] pursued this work in a study in which they examined the impact of the transmission of the heat of a fluid over a surface that was stretched. The flow of fluid across a stretched surface was then determined in order to do further research on Gupta's work [6].

1.1. Literature Review

The importance of heat transfer and heat exchangers cannot be overstated. For instance, raising the temperature will be necessary to increase the efficiency of the thermal processes for the production of heat and electricity [7,8]. The transfer of energy from one place (high concentration) to another place (low concentration) is known as heat transfer. Applications of heat transportation can be seen in our daily lives; for example, the human body emits heat continuously, and adjustment of the human body temperature is achieved by using clothing to adapt changing climatic conditions. Heat transportation is also utilised to manage temperature in our structures [9] and is required for cooking, refrigeration, and drying. It is also utilised for temperature regulation in automotive radiators [10] and mobile devices [11]. Solar thermal collectors [12,13] and spaceship thermal control elements [14] use heat conversion to turn solar energy into heat and power. Many of these systems require rapid heat dissipation to enable successful performance and optimum productivity inside the system [15]. As modern sciences progress, devices have become tiny, necessitating preferable temperature control. Basically, the smaller the scale, the more efficient cooling technology is required [16]. In thermal engineering, heat transfer enhancement is therefore a very important field.

Choi and Eastman [17] prepared nanofluids, which are colloidal suspensions of nanoscale metallic or non-metallic particles in a host fluid (HF). There are few fundamental conditions which met, low agglomeration of nanoparticles and steady-state suspension and the HF should be chemically constant. Nanofluids have two subcategories: non-metallic nanofluids (carbides: carbon, TiC, materials: SWCNT/MWCNT, graphene, diamond, etc.) and metallic nanofluids (metals: Cu, Fe, Al, Ag, Au; metal oxides: SiO_2, Al_2O_3, TiO_2, CuO). There are two methods to develop nanofluid: a one-step method, which involves developing the HF and nanoparticles at the same time; and the two-step method, in which it is generated separately and then mixed up [18]. For several applications, nanofluids have important properties, such as good stability, high heat conductivity, reduced erosion and friction coefficient, ultrafast heat transfer ability, and good lubrication.

Mathematical simulations by Rashid et al. [19] showed the combined effects of an angled magnetic field and a predetermined surface temperature (PST) on Cu-Al_2O_3-type nanoparticles in water. They found that the temperature rises by solid volume fraction ϕ, magnetic parameter M, and slip-parameter for both Cu-H_2O and Al_2O_3-H_2O. Haq and Aman [20] quantitatively evaluated the thermal performance of a water-based copper oxide (CuO) nanofluid in a trapezoidal cavity with the use of the finite element method (FEM). They concluded from this investigation that the velocity steadily decreases as the fluid thickens and becomes denser due to the presence of a solid volume fraction (=0–0.2). In a similar way, the rate of heat transmission is likewise decreasing as =0–0.2 increases, owing to convection. The characteristics of non-uniform melting heat transmission of a nanofluid over a sheet were investigated by Hayat et al. [21]. The base fluid (water, H_2O) was injected with copper (Cu) nanoparticles, and HAM was used to solve a governing self-similar system of differential equations. In this investigation, they found that when the volume fraction, Hartman number, and porosity parameter values increased, so did the skin friction coefficient and local Nusselt number. The effects of heat transmission on aluminium alloy nanoparticles suspended across a sheet under the influence of a magnetic field were studied by Sandeep et al. [22]. They took into consideration two distinct kinds of nanoparticles, AA 7072 (98% Al, 1% Zn, and 1% additives) and AA 7075 (90% Al, 5.6 Zn, 2.3 Mg, 1.2 Cu, and additives). Due to a larger proportion of copper used, the mathematical research revealed that AA 7075 had a substantially higher heat transfer rate than AA 7072. (Cu).

Shah et al. [23] conducted a mathematical investigation of aluminium and ethylene glycol nanoparticles on a sheet while taking the second law of thermodynamics into account.

1.2. Motivations

The great efforts and expertise of the researchers have succeeded in publishing the results of fluid flow between confined parallel plates. Suitable similar variables and numerical approaches were adopted in order to generate the results. The following constituents give the motivations of this research work.

- The problem of viscoelastic fluid in a confined space (channel) with extending walls was discussed by Misra et al. [24]. According to the study, reverse flow occurs close to the region's (channel's) centre and can be managed by applying an external magnetic field.
- Ashraf et al. [25] looked into the issue of micropolar fluid flow with heat transmission in a channel with stretching walls. Equations of fourth order coupled nonlinear ordinary differential type were solved using the quasi-linearization approach. The study exposed the fact that shear, coupled stresses and heat transfer rate at the walls are increased by stretching the channel walls. They also quantified that their investigation may be valuable for the flow and thermal control of polymeric processing.
- Researcher [26–28] examined the effect of heat transfer and nanoparticles on MHD water/kerosene-based nanofluid in a channel numerically. The studies revealed the fact that there exists a linear relationship between the thermal boundary layer thickness and the solid volume fraction.

The above-mentioned motivations of the study are either a problem of simple micropolar fluid or simple nanofluid in a channel with stretching walls. Therefore, without any doubt, it can be argued that there exists a potential research gap to investigate the problem related to micropolar nanofluid flow in a channel with stretching/shrinking walls.

1.3. Contributions

The following items are the main contribution of the current research.

- It proposes the single-phase nanofluid model of micropolar copper–blood nanoparticles in a channel with stretching and shrinking walls.
- Thermal radiations are also present in the channel to make the problem more appealing for heat transfer.
- To control the reversibility of the flow due to the stretching walls, we impose a transverse magnetic field.
- This research also investigates sensitivity analysis using response surface methodology (RSM).

2. Proposed Model

Consider two-dimensional steady laminar incompressible micropolar nanofluid in a channel with stretching and shrinking walls in the presence of a magnetic field and thermal radiation. In this study, copper nanoparticles are the solid dispersed phase while blood is the fluid continuum phase. The lower and upper walls of the channel stretch and shrink in the direction of the fluid (x-axis) with some constant rate $u = bx \; \forall \; b \in R$. If $b > 0$, then the case is known as stretching and $b < 0$ is for shrinking walls of the channel (see Figure 1).

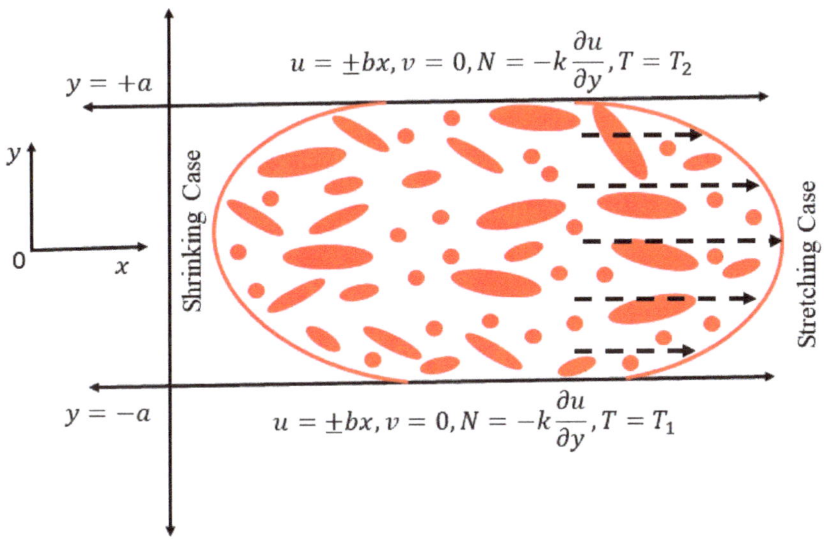

Figure 1. Physical sketch of the problem.

2.1. Governing Equations

The general equations of micropolar fluids as given by Eringen [1] can be represented in component form $\overline{V} = (u(x,y), v(x,y), 0), \overline{v} = (0,0,N(x,y))$, where $N = \nabla \times V \neq 0$ as:

$$\frac{\partial u}{\partial x} + \frac{\partial v}{\partial y} = 0 \tag{1}$$

$$u\frac{\partial u}{\partial x} + v\frac{\partial u}{\partial y} = \frac{-1}{\rho}\frac{\partial p}{\partial x} + \frac{\mu + \kappa}{\rho}\nabla^2 u + \frac{\kappa}{\rho}\frac{\partial N}{\partial y} - \frac{\sigma_f B_\circ^2}{\rho_{nf}}u \tag{2}$$

$$\frac{\partial v}{\partial x} + v\frac{\partial v}{\partial y} = \frac{-1}{\rho}\frac{\partial p}{\partial y} + \frac{\mu + \kappa}{\rho}\nabla^2 v - \frac{\kappa}{\rho}\frac{\partial N}{\partial x} \tag{3}$$

$$\rho_{nf}\bar{j}\left(u\frac{\partial N}{\partial x} + v\frac{\partial N}{\partial y}\right) = \gamma_{nf}\nabla^2 N + \kappa\left(\frac{\partial v}{\partial x} - \frac{\partial u}{\partial y}\right) - 2\kappa N \tag{4}$$

$$\left(u\frac{\partial T}{\partial x} + v\frac{\partial T}{\partial y}\right) = \frac{k_{nf}}{(\rho C_p)_{nf}}\left(\frac{\partial^2 T}{\partial x^2} + \frac{\partial^2 T}{\partial y^2}\right) - \frac{1}{(\rho C_p)_{nf}}\frac{\partial q_r}{\partial y} \tag{5}$$

Here, p is the pressure, N is the micro-rotational velocity, $\gamma_{nf} = j\left(\frac{k}{2} + \mu_{nf}\right)$ is the spine gradient viscosity, and u and v are the axial and transverse velocities components, respectively.

The appropriate boundary conditions for the current investigation are:

$$u = \pm bx,\ v = 0,\ N = -k\frac{\partial u}{\partial y},\ T = T_1 \text{ at } y = -a \tag{6}$$

$$u = \pm bx,\ v = 0,\ N = -k\frac{\partial u}{\partial y},\ T = T_2 \text{ at } y = +a \tag{7}$$

These physical quantities are described mathematically as:

$$\rho_{nf} = \rho_f(1-\varphi) + \varphi\rho_s \tag{8}$$

$$\mu_{nf} = \frac{\mu_f}{(1-\varphi)^{2.5}} \tag{9}$$

$$(\rho C_p)_{nf} = (\rho C_p)_f(1-\varphi) + (\rho C_p)_s\varphi \tag{10}$$

$$\frac{k_{nf}}{k_f} = \frac{k_s + 2k_f - 2\varphi\left(k_f - k_s\right)}{k_s + 2k_f + \varphi\left(k_f - k_s\right)} \tag{11}$$

Here, φ is the solid volume fraction, φ_s is for the nanosolid-particles, and φ_f is for the base fluid.

We apply a Rosseland approximation for radiation as:

$$q_r = -\frac{4\sigma^*}{3k^*}\frac{\partial T^4}{\partial y} \tag{12}$$

Here, the Stefan–Boltzmann constant is given by $\sigma^* = 5.6697 \times 10^{-8}\ \mathrm{Wm^{-2}K^{-4}}$ and the mean spectral absorption coefficient is denoted by k^*. Further, blackbody emission power, *eb* in terms of the Stefan–Boltzmann constant and absolute temperature, is given by $e_b = \sigma^* T^4$.

It is assumed that the temperature differences within the flow, such as the term T^4, may be expressed as a linear function of temperature. We obtain the Taylor series expansion for T^4 at a free stream temperature T_∞.

$$T^4 = T_\infty^4 + 4T_\infty^3(T - T_\infty) + 8T_\infty^2(T - T_\infty)^2 + \ldots$$

After neglecting higher-order terms as:

$$T^4 = 4T_\infty^4 T - 3T_\infty^4 \tag{13}$$

Using Equation (12) in (13), we obtain:

$$\frac{\partial q_r}{\partial y} = -\frac{16\sigma^* T_\infty^3}{3k^*}\frac{\partial^2 T}{\partial y^2} \tag{14}$$

Equation (5) is now converted in the light of (14) as:

$$(\rho C_p)_{nf}\left(u\frac{\partial T}{\partial x} + v\frac{\partial T}{\partial y}\right) = k_{nf}\left(\frac{\partial^2 T}{\partial x^2} + \frac{\partial^2 T}{\partial y^2}\right) + \frac{16\sigma^* T_\infty^3}{3k^*}\frac{\partial^2 T}{\partial y^2} \tag{15}$$

2.2. Similarity Solution

Now we introduce the similarity transformation as:

$$\eta = \frac{y}{a}, u = bxf'(\eta), v = -abf(\eta), \theta(\eta) = \frac{T - T_2}{T_1 - T_2} \tag{16}$$

Using Equation (16) in Equation (1), we see that Equation (1) is identically satisfied and eliminating the pressure term from (2) and (5), we obtain the required similarity coupled system of the ordinary differential equation as:

$$\left(1 + \frac{\kappa}{\Gamma_2}\right)f'''' - \frac{\Gamma_1}{\Gamma_2}R(f'f'' - ff''') - \frac{\kappa}{\Gamma_2}g' - \frac{M^2}{\Gamma_2}f'' = 0 \tag{17}$$

$$\left(1 + \frac{\kappa}{2\Gamma_2}\right)g'' + \frac{\kappa}{\Gamma_2}(f'' - 2g) + \frac{\Gamma_1}{\Gamma_2}R(fg' - f'g) = 0 \tag{18}$$

$$\frac{1}{\Gamma_3 Pr}\left(\Gamma_4 + \frac{4}{3}Rd\right)\theta'' + Rf\theta' = 0 \tag{19}$$

Subject to the boundary conditions:

$$f(-1) = 0,\ f'(-1) = \pm 1, g(-1) = 0, \theta(-1) = 1 \tag{20}$$

$$f(1) = 0,\ f'(1) = \pm 1, g(1) = 0, \theta(1) = 0 \tag{21}$$

Here, $R = \frac{a^2 b}{v_f}$ is the Reynolds number, $M^2 = \frac{\sigma_f B_0^2 a^2}{\mu_f}$ is the magnetic parameter, $K = \frac{k}{v_f}$ is the micropolar parameter, $Pr = \frac{v_f (\rho C_p)_f}{k_f}$ is the Prandtl number, and $Rd = \frac{\sigma^* T_\infty^3}{3k^* k_f}$ is the radiation parameter. Also,

$$\Gamma_1 = (1-\varphi) + \varphi \frac{\rho_s}{\rho_f}, \qquad \Gamma_2 = \frac{1}{(1-\varphi)^{2.5}},$$

$$\Gamma_3 = (1-\varphi) + \frac{(\rho C_p)_s}{(\rho C_p)_f}\varphi,\ \Gamma_4 = \frac{k_s + 2k_f - 2\varphi\left(k_f - k_s\right)}{k_s + 2k_f + \varphi\left(k_f - k_s\right)}$$

3. Results and Discussion

The numerical results have been handled in this section in the form of tables and graphs. Equations (17)–(19), subject to boundary conditions (20) and (21), are solved with the aid of a numerical scheme called Runge–Kutta 4th order and finite difference base scheme (bvp4c) [29]. Equations (17)–(19) are higher order ODEs, so we converted them into a system of first order ODEs, however three (03) initial conditions were missing $\left(\frac{d^2 f}{d\eta^2}\Big|_{\eta=0}, \frac{d^3 f}{d\eta^3}\Big|_{\eta=0}, \frac{d^2 g}{d\eta^2}\Big|_{\eta=0}, \frac{d\theta}{d\eta}\Big|_{\eta=0}\right)$. To find these missing initial conditions, we employed a shooting method. Once these missing conditions were found, then the solution computed and satisfied the boundary conditions (20) and (21). Thermophysical properties of blood and copper nanoparticles [27] are fetched from Table 1.

Table 1. Thermophysical properties of the blood and copper nanoparticles (see [27]).

Properties	Blood	Copper
Density $\left(\frac{\text{Kg}}{\text{m}^3}\right)$	1150	8933
Thermal conductivity $\left(\frac{\text{W}}{\text{mK}}\right)$	0.53	401
Specific Heat $\left(\text{JKg}^{-1}\text{K}^{-1}\right)$	3617	385

Figures 2–12 provide information on the hydrokinetic effects on velocity, angular velocity, temperature, and concentration. The comparison of the two different approaches of the numerical results is presented graphically in Figure 2 and it is depicted that the results coincide with each other. Figure 3 depicts how the micropolar parameter affects the velocity profile $f'(\eta)$ for stretching and contracting the wall. The velocity profile, which is parabolic in character, is shown to decline as the micropolar parameter rises. However, we deduced from Figure 4 that the micro-rotation $g(\eta)$ profile gradually increases as the micropolar parameter K increases after decreasing from the bottom wall to the channel centre. The micro-rotation profile displays an entirely different pattern in the case of diminishing walls. On other hand, the temperature profile $\theta(\eta)$ rises while the walls are extending and falls when the walls are contracting. As can be observed from Figure 8, the velocity profile decreases as the solid volume percentage increases for stretching walls and rises for contracting walls. The impact of the stretching Reynolds number R on the velocity profile is explained in Figure 9. This graphic demonstrates how the velocity profile for the stretched walls reduces towards the channel borders and increases near the channel centre. However, it can be seen in Figure 10 that the micro-rotation profile rises from the lower wall to the channel's centre, and then falls when the values of the Reynolds number for the

stretched walls are raised. Figure 11 also illustrates the effect of the Reynolds number on the temperature profile. We can observe that the temperature profile changes for the bottom and upper halves of the channel when the Reynolds number values are increased. Figure 12 shows the effect of the radiation parameter on the temperature profile. The temperature profile of the tube drops from the lower wall to the middle and climbs from the centre to the top wall as the radiation parameter rises. The reverse result, however, can be seen in the case of the diminishing walls.

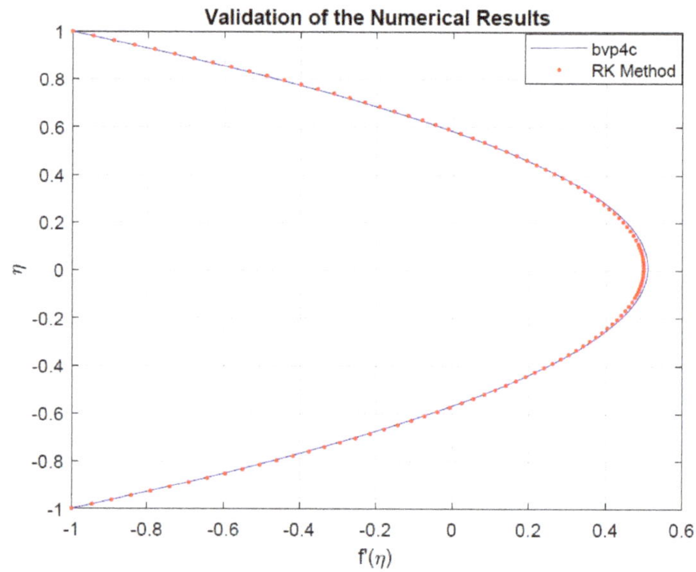

Figure 2. Code verification.

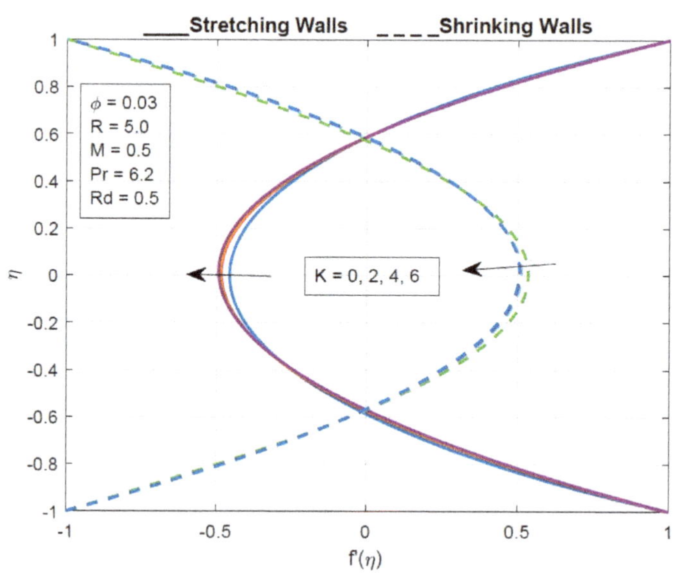

Figure 3. The effect of the micropolar parameter on the velocity profile.

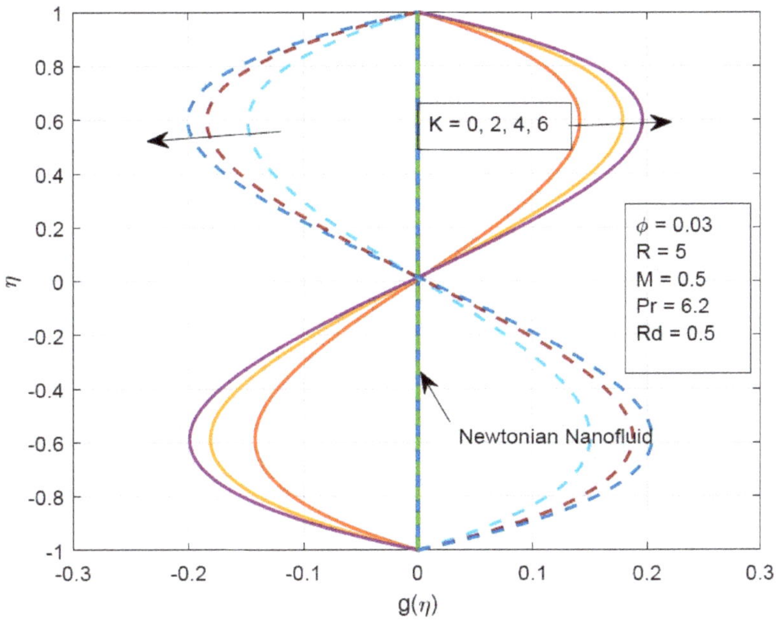

Figure 4. The effect of the micropolar parameter on the micro-rotation.

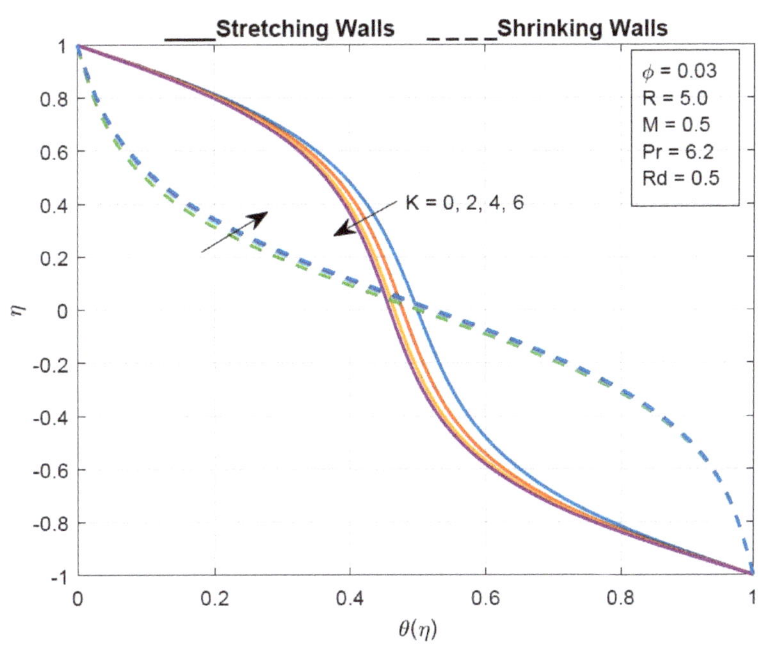

Figure 5. The effect of the micropolar parameter on the temperature profile.

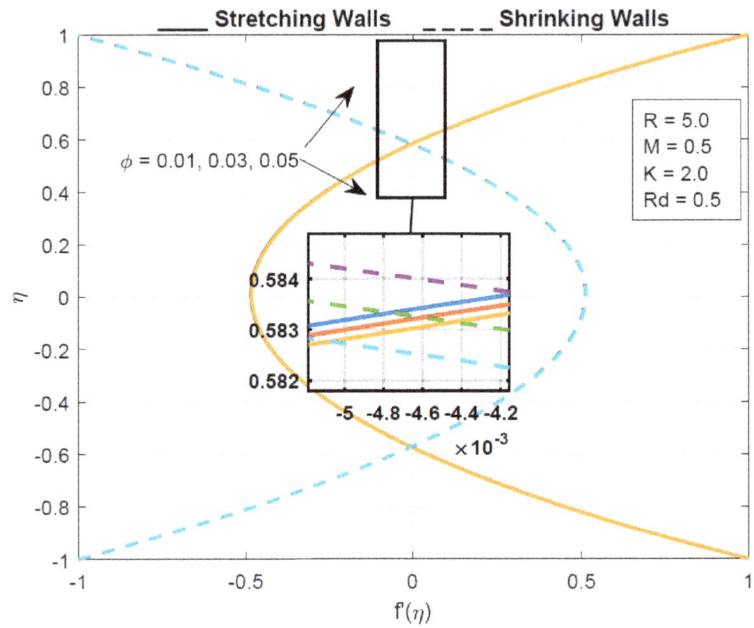

Figure 6. The effect of the solid volume fraction on the velocity profile.

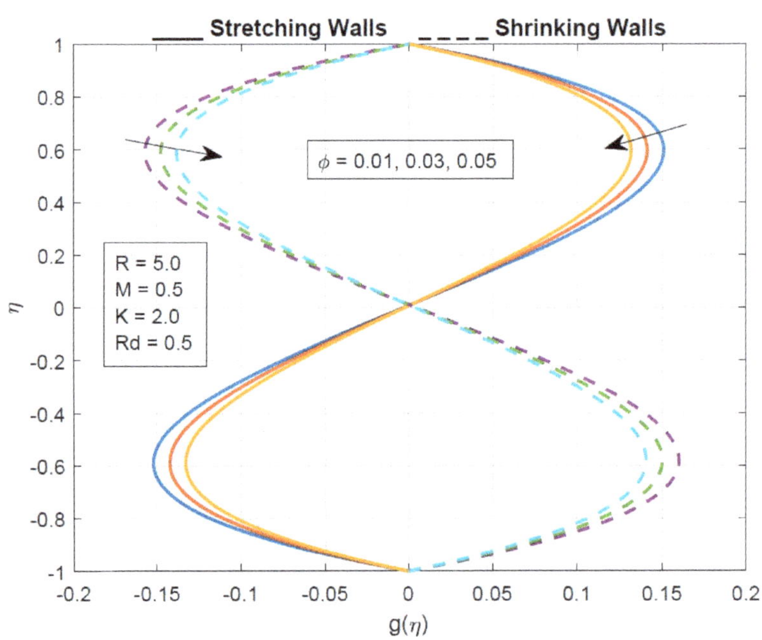

Figure 7. The effect of the solid volume fraction on the micro-rotation profile.

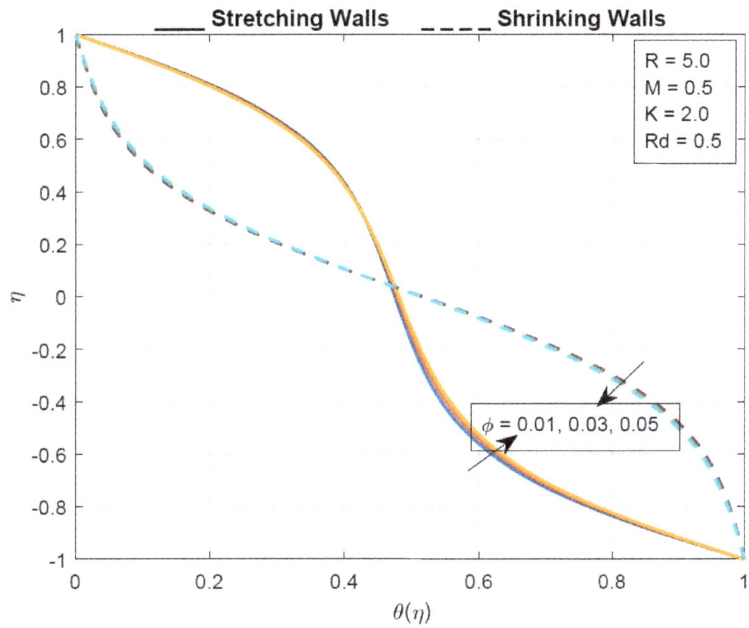

Figure 8. The effect of the solid volume fraction on the temperature profile.

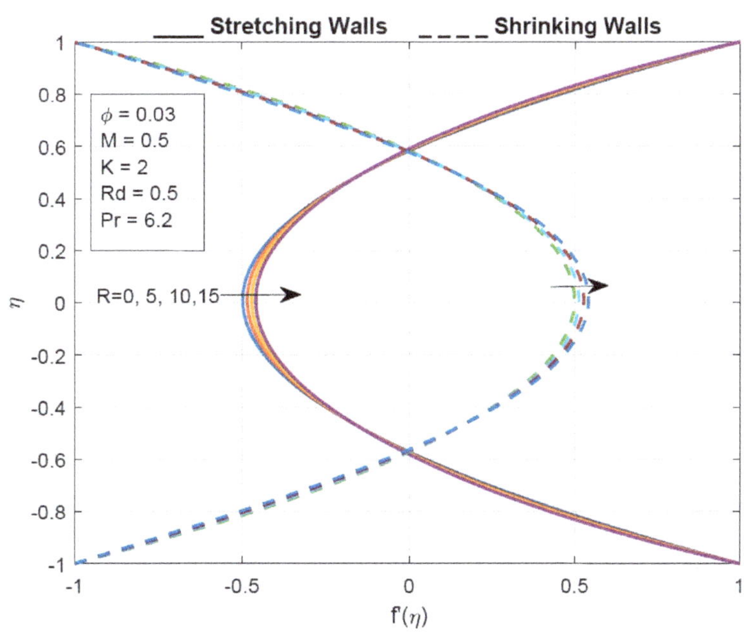

Figure 9. The effect of the Reynolds number on the velocity profile.

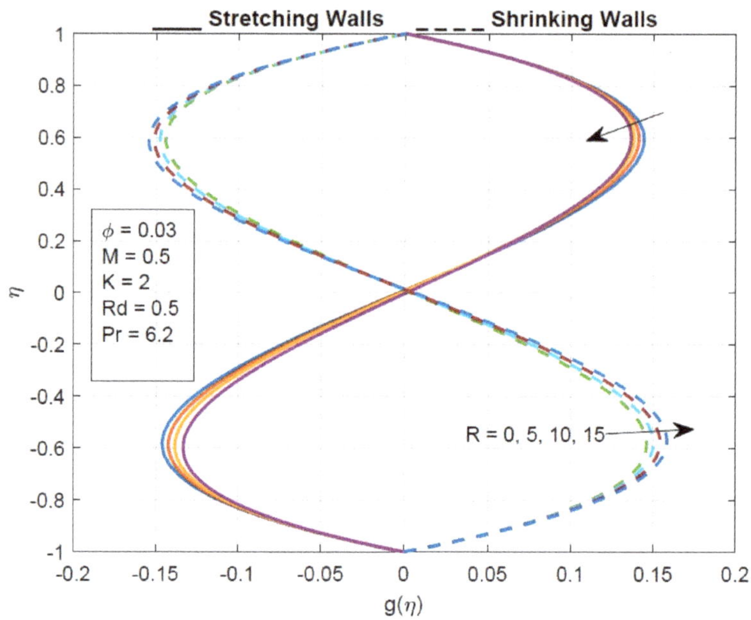

Figure 10. The effect of the Reynolds number on the micro-rotation profile.

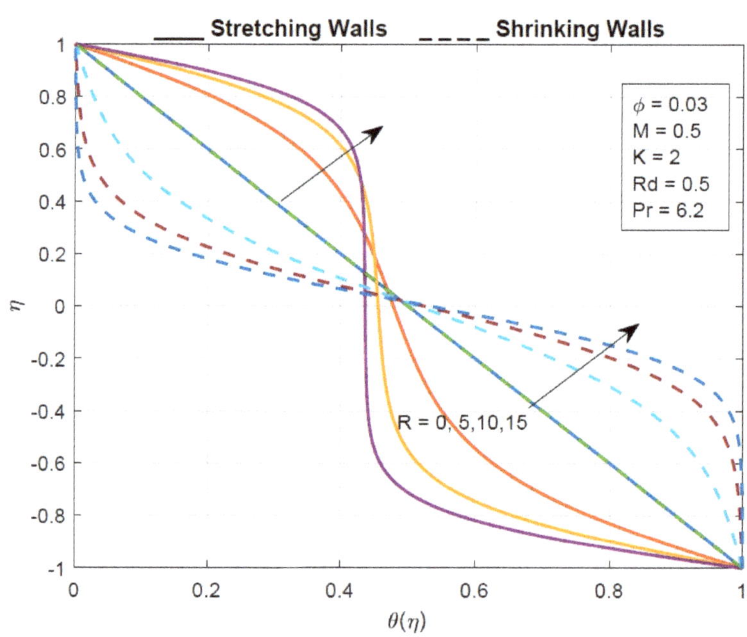

Figure 11. The effect of the Reynolds number on the temperature profile.

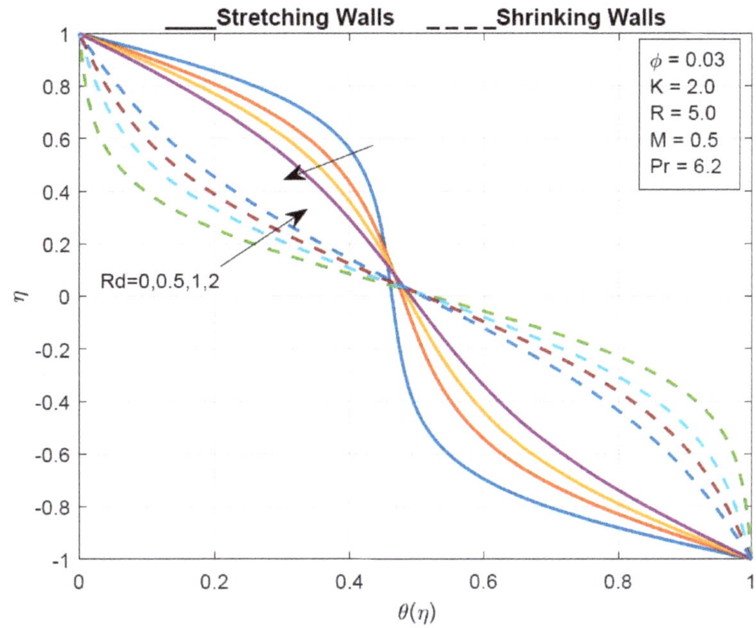

Figure 12. The effect of the radiation parameter on the temperature profile.

3.1. Application of Response Surface Methodology (RSM)

3.1.1. Optimization Process

RSM is one of many useful tools for describing a wide range of variables, together with limited resources, quantitative data, and the required test design (response surface methodology). The following steps were taken into account in this process:

1. To reach the suitable and believable requirements for the intended response, we planned and investigated the data values.
2. We outlined the most appropriate mathematical models for the response surface.
3. We described the mathematical models for the response surface that are most suited.
4. We used an analysis of the variance to examine the parametric direct and interaction impacts (ANOVA).

3.1.2. Optimization Analysis by RSM

The relation between the factor variables and the response variable (temperature gradient) was investigated using a face-centred central composite design. Tables 2 and 3 indicates the three factors and their levels. The quadratic model is presented in Equation (22), where three linear, square, and interactive terms are involved.

$$\text{Response} = \alpha_0 + \alpha_1 A + \alpha_2 B + \alpha_3 C + \alpha_{11} A^2 + \alpha_{22} B^2 + \alpha_{33} C^2 + \alpha_{12} AB + \alpha_{13} AC + \alpha_{23} BC \quad (22)$$

Table 2. Parameters with their levels for $Cf_x(-1)$.

Parameters	Symbols	Level		
		−1	0	1
R	A	−5	2	5
M	B	0	1	1.5
K	C	0.1	1	2

Table 3. Parameters with their levels $Nu_x(-1)$.

Parameters	Symbols	Level		
		−1	0	1
R	A	−5	2	5
φ	B	0	0.05	0.2
Rd	C	0.1	1	2

Here (Equation (22)), α_i and α_{ij} represent the regression coefficients. The statistical analysis was performed for 20 runs, as prescribed by the defined conditions.

$$Cf_x(-1) = 3.0015 + 0.08816R + 0.1126M - 0.1056K + 0.000686R^2 + 0.0843M^2 \\ +0.0479K^2 - 0.00175RM - 0.02615RK - 0.0731MK \quad (23)$$

$$Nu_x(-1) = 0.758 + 0.1441R - 2.00\varphi - 0.288Rd + 0.00412R^2 + 3.64\varphi^2 \\ +0.0665Rd^2 - 0.2273R\varphi - 0.03759Rd.R + 0.614\varphi Rd \quad (24)$$

The values of skin friction coefficient and Nusselt number for coded values are given in Table 4. The ANOVA Tables 5 and 6 provide a measure of accuracy for the approximate model. A parameter is important when the p-value is less than 0.05 (with 95 percent confidence). Since the p-value in the model is greater than 0.05, the linear, quadratic, and interaction terms may be omitted. Nonetheless, as seen in Table 5,6 the model proves to be superior since its coefficient of determination R^2 is higher. The correct regression equation is now as follows:

$$Cf_x(-1) = 3.0015 + 0.08816R + 0.1126M - 0.1056K + 0.0843M^2 + 0.0479K^2 \\ -0.02615RK - 0.0731MK \quad (25)$$

$$Nu_x(-1) = 0.758 + 0.1441R - 0.288Rd - 0.03759Rd.R \quad (26)$$

Table 4. Experimental design and responses.

Runs	Coded Values				
	A	B	C	Response $Cf_x(-1)$	Response $Nu_x(-1)$
1	−1	−1	−1	2.569977654	0.039218876
2	1	−1	−1	3.4412049	1.706219612
3	−1	1	−1	2.931369225	0.182650354
4	1	1	−1	3.797708428	0.975161305
5	−1	−1	1	2.808665994	0.256515544
6	1	−1	1	3.173524221	0.828662877
7	−1	1	1	2.954689881	0.315102036
8	1	1	1	3.315963627	0.717271607
9	−1	0	0	2.832551727	0.185583893
10	1	0	0	3.347006534	0.981815599
11	0	−1	0	3.092057592	0.701100814
12	0	1	0	3.302840544	0.617231054
13	0	0	−1	3.340485946	0.860674809
14	0	0	1	3.124384046	0.614658362
15	0	0	0	3.187362637	0.679561304
16	0	0	0	3.187362637	0.679561304
17	0	0	0	3.187362637	0.679561304
18	0	0	0	3.187362637	0.679561304
19	0	0	0	3.187362637	0.679561304
20	0	0	0	3.187362637	0.679561304

Table 5. ANOVA for $Cf_x(-1)$.

Source	DF	Adjusted Sum of Square	Adjusted Mean Square	F-Value	*p*-Value	Remarks
Model	9	1.27499	0.141666	211.26	0.000	Significant
Linear	3	1.05111	0.350369	522.48	0.000	Significant
R	1	0.87703	0.877027	1307.85	0.000	Significant
M	1	0.14695	0.146950	219.14	0.000	Significant
K	1	0.03201	0.032010	47.73	0.000	Significant
Square	3	0.03435	0.011451	17.08	0.000	Significant
R.R	1	0.00055	0.000546	0.81	0.388	Not Significant
M.M	1	0.00474	0.004738	7.07	0.024	Significant
K.K	1	0.00509	0.005093	7.59	0.020	Significant
2-Way Interaction	3	0.15276	0.050919	75.93	0.000	Significant
R.M	1	0.00036	0.000362	0.54	0.480	Not Significant
R.K	1	0.12739	0.127394	189.97	0.000	Significant
M.K	1	0.02220	0.022204	33.11	0.000	Significant
Error	10	0.00671	0.000671			
Lack-of-Fit	5	0.00671	0.001341	*	*	
Pure Error	5	0.00000	0.000000			
Total	19	1.28170				
	R^2 =99.48%			Adjusted R^2 = 99.01%		

Table 6. ANOVA for $Nu_x(-1)$.

Source	DF	Adj SS	Adj MS	F-Value	*p*-Value	Remarks
Model	9	2.39452	0.26606	26.11	0.000	Significant
Linear	3	1.76127	0.58709	57.62	0.000	Significant
R	1	1.65970	1.65970	162.90	0.000	Significant
Phi	1	0.03905	0.03905	3.83	0.079	Not Significant
Rd	1	0.06770	0.06770	6.64	0.028	Significant
Square	3	0.09999	0.03333	3.27	0.067	Not Significant
R.R	1	0.01970	0.01970	1.93	0.194	Not Significant
Phi.Phi	1	0.00191	0.00191	0.19	0.674	Not Significant
Rd.Rd	1	0.00982	0.00982	0.96	0.349	Not Significant
2-Way Interaction	3	0.41288	0.13763	13.51	0.001	Significant
R.Phi	1	0.11203	0.11203	11.00	0.008	Not Significant
R.Rd	1	0.26292	0.26292	25.81	0.000	Significant
Phi.Rd	1	0.02855	0.02855	2.80	0.125	Not Significant
Error	10	0.10188	0.01019			
Lack-of-Fit	5	0.10188	0.02038	*	*	
Pure Error	5	0.00000	0.00000			
Total	19	2.49640				
	R^2 =95.92%			Adjusted R^2 = 92.25%		

Tables 2 and 3 present the various levels of the parameters for $Cf_x(-1)$ and $Nu_x(-1)$, respectively. However, Table 4 represents the values of the response function for 20 different points. In Tables 5 and 6, the R^2 for $Cf_x(-1)$ and $Nu_x(-1)$ (99.48% and 95.92% respectively), which was obtained by the testing methods and statistical analysis of the model, is presented. However, the R^2-adj amounts for $Cf_x(-1)$ and $Nu_x(-1)$ (99.01% and 92.25%, respectively) are $\leq R^2$, but the model fits the data reasonably [30–33]. Moreover, the importance of the model for the response variables $Cf_x(-1)$ and $Nu_x(-1)$ is depicted from the F-value, which is equal to 211.26 and 26.11, respectively. According to Figure 13a,b, it is observed that the plots of normal probability are well-behaved and in good condition [33]. From these two figures, the residual histograms exhibit a skewed distribution. When the residual diagrams and fitted values were compared, the observed and fitted values showed a strong correlation.

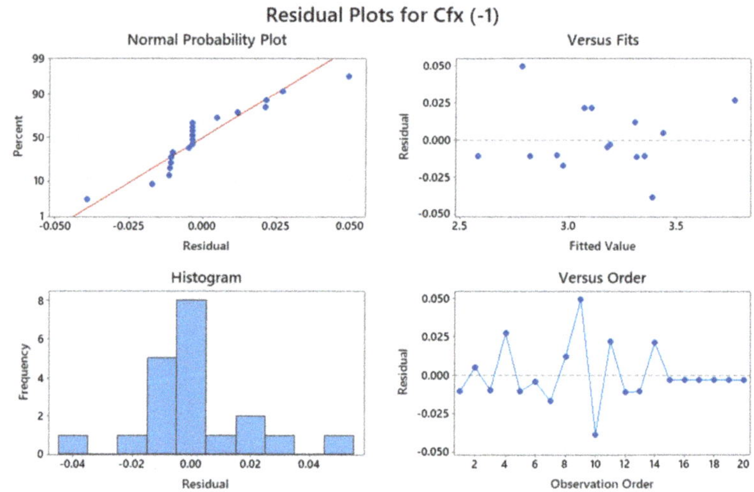

(a)

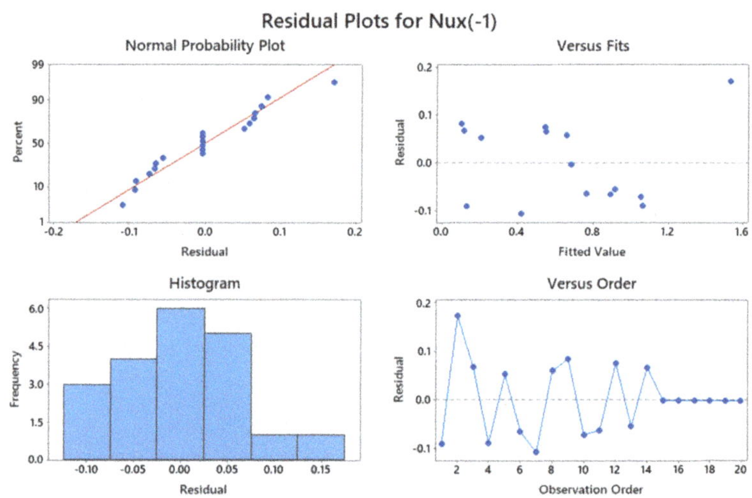

(b)

Figure 13. Residual plot for (**a**) Cf_x (**b**).

Figures 14 and 15 show the mean total skin friction coefficient and Nusselt number variations as effective parameters functions. The deviation of the skin friction coefficient in terms of M and R are shown in Figure 14a. The skin friction coefficient increases as M and R are increased, with the highest value $(+1)$ and lowest value (-1) for M and R, respectively. The variation in the response variable (skin friction) with respect to K and R are shown in Figure 14b. It is observed that reducing the values of K and increasing the value of R causes an increase in the skin friction coefficient. The highest value of skin friction is obtained in the level of (-1) and $(+1)$ and its lowest value is observed in the level of $(+1)$ and $(+1)$ for K and R, respectively. The variance of the skin friction coefficient in terms of K and M is shown in Figure 14c. The skin friction is increased when the value of K is reduced and the value of M is increased. Furthermore, for K and M, the highest and lowest values of the skin friction coefficient can be found at the levels of (-1) and (+1), respectively. The variance of the total Nusselt number in terms of the solid volume fraction of a nanofluid and the Reynolds number R is shown in Figure 15a. The Nusselt number is increased when the solid volume fraction is lower and the Reynolds number is higher. Moreover, the Nusselt number gains its minimum value in $(+1)$ and (-1) and its maximum in (-1) and $(+1)$ for φ and R, respectively. Figure 15b shows that the Nusselt number increases by increasing the values of R and decreasing the values of Rd. Nevertheless, the Nusselt number gains a maximum in level (-1) and $(+1)$ and a minimum in $(+1)$ and (-1) for Rd and R, respectively. In the same vein, Figure 15c shows that the Nusselt number reaches its highest value in (-1) and (-1) and its lowest value in $(+1)$ and $(+1)$ for Rd and the solid volume fraction.

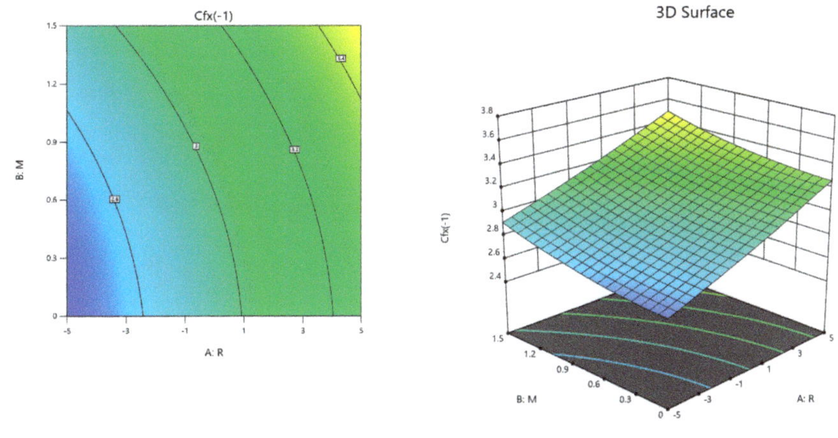

(**a**)

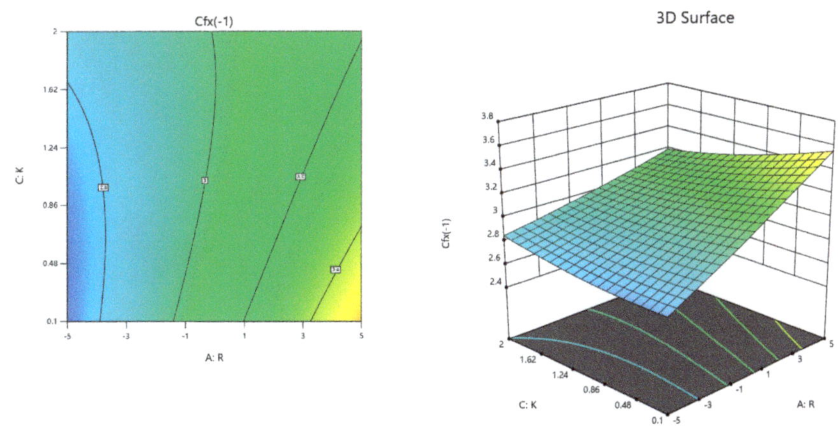

(**b**)

Figure 14. *Cont.*

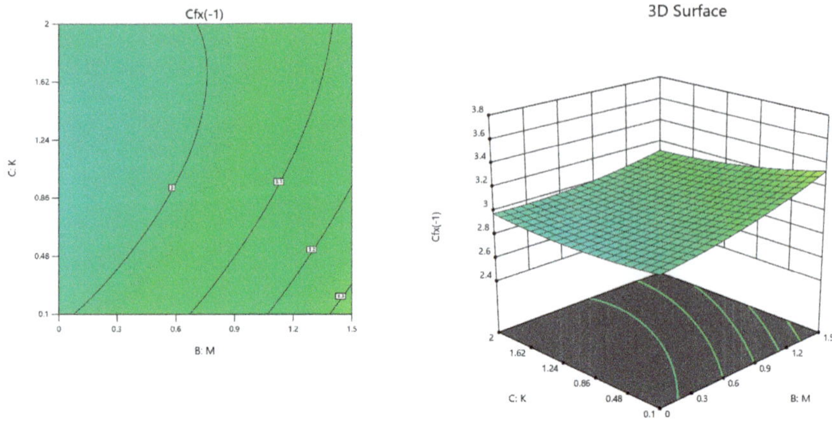

(c)

Figure 14. Three-dimensional surfaces and contour plots for all continuous parameters of M, R, and K on Cfx.

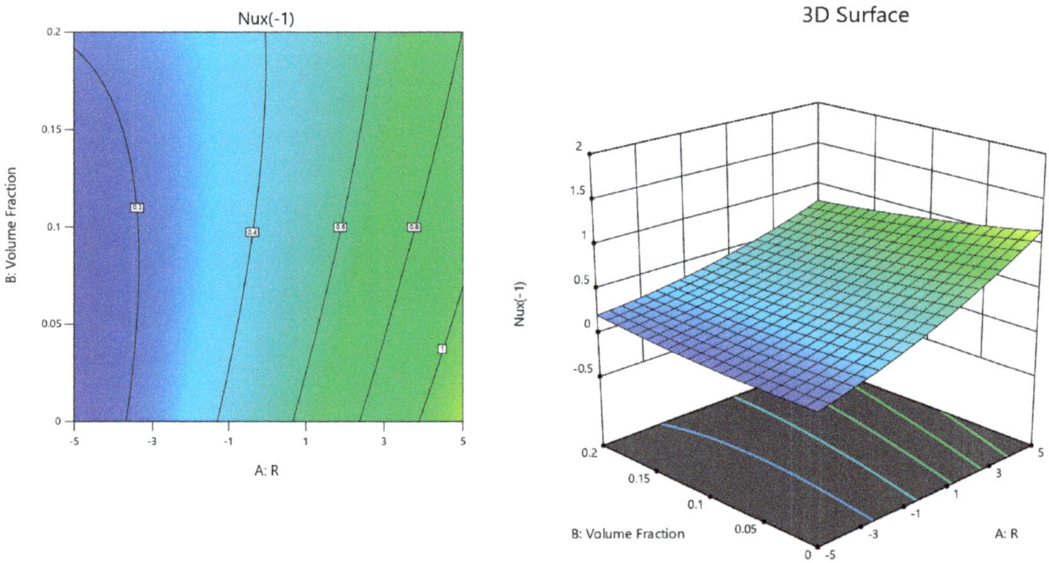

(a)

Figure 15. *Cont.*

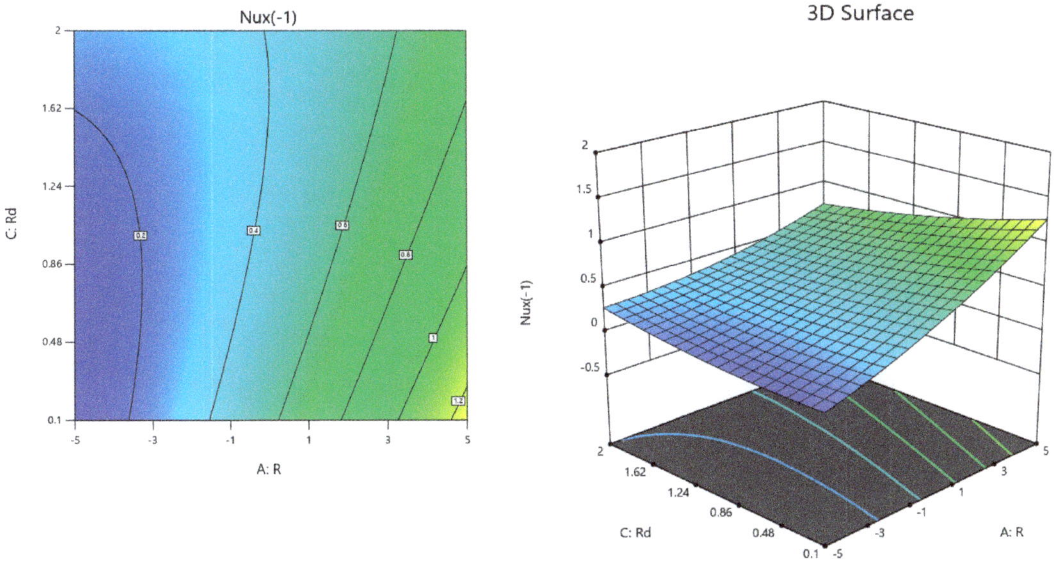

(b)

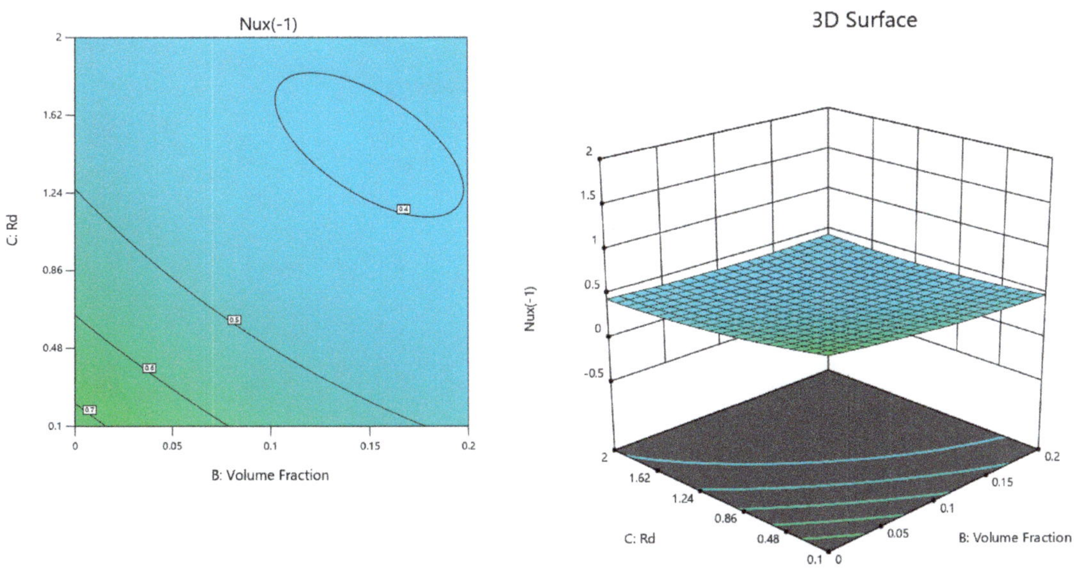

(c)

Figure 15. Three-dimensional surfaces and contour plots for all continuous parameters of R, Phi, and Rd on Nux.

The regression Equations (25) and (26) are used to calculate the sensitivity. The sensitivity functions are the partial derivatives of the response variables with respect to the factor variables, as shown below:

$$\frac{\partial Cf_x}{\partial A} = 0.08816 - 0.02615K \tag{27}$$

$$\frac{\partial Cf_x}{\partial B} = 0.1126 + 0.1686M - 0.0731K \tag{28}$$

$$\frac{\partial Cf_x}{\partial C} = -0.1056 + 0.0958K - 0.02615R - 0.0731M \tag{29}$$

$$\frac{\partial Nu_x}{\partial A} = 0.1441 - 0.03759Rd \tag{30}$$

$$\frac{\partial Nu_x}{\partial B} = 0 \tag{31}$$

$$\frac{\partial Nu_x}{\partial C} = -0.288 - 0.03759R \tag{32}$$

The positive sensitivity value indicates that the objective function has improved as a result of the improved input parameters. Its negative value, on the other hand, denotes a decline in the objective function due to the increased input parameters. From Table 7, it is seen that the sensitivity of Cf_x to A and B is positive and negative for C. Similarly, from Table 8, the sensitivity of Nu_x to A is positive and negative for C.

Table 7. Sensitivity analysis of the response Cf_x.

A	B	C	Sensitivity to A	Sensitivity to B	Sensitivity to C
0	−1	−1	0.11431	0.0171	−0.1283
0	−1	0	0.08816	−0.056	−0.0325
0	−1	1	0.06201	−0.1291	0.0633
0	0	−1	0.11431	0.1857	−0.2014
0	0	0	0.08816	0.1126	−0.1056
0	0	1	0.06201	0.0395	−0.0098
0	1	−1	0.11431	0.3543	−0.2745
0	1	0	0.08816	0.2812	−0.1787
0	1	1	0.06201	0.2081	−0.0829

Table 8. Sensitivity analysis of the response Nu_x.

A	B	C	Sensitivity to A	Sensitivity to B	Sensitivity to C
−1	0	−1	0.18169	0	−0.25041
−1	0	0	0.1441	0	−0.25041
−1	0	1	0.10651	0	−0.25041
0	0	−1	0.18169	0	−0.288
0	0	0	0.1441	0	−0.288
0	0	1	0.10651	0	−0.288
1	0	−1	0.18169	0	−0.32559
1	0	0	0.1441	0	−0.32559
1	0	1	0.10651	0	−0.32559

4. Conclusions

In this investigation, we considered two-dimensional steady laminar incompressible micropolar nanofluid in a channel with stretching and shrinking walls in the presence of a magnetic field and thermal radiation. The copper nanoparticles are the solid dispersed phase, while blood is the fluid continuum phase. The lower and upper walls of the channel stretch and shrink in the direction of the fluid (x-axis). The governing similar ODEs were solved and verified by two different numerical schemes. After in-depth discussions of the model, the main findings are drawn as below:

- The velocity of the fluid particles decreases by increasing the values of the micropolar parameter.
- As the radiation parameter increases, the temperature profile decreases from the lower wall to the middle and increases from the centre to the upper wall of the tube.
- The velocity profile declines as a solid volume fraction ϕ enhances for the stretching walls and increases for the shrinking walls.
- The skin friction coefficient increases as the magnetic and Reynolds number are increased.
- The Nusselt number is increased when the solid volume fraction is lower.
- The sensitivity of Cf_x to the Reynolds number and magnetic parameter is positive and negative for the micropolar parameter.
- Nu_x is optimized by taking higher values of the Reynolds number and lower values of the radiation parameter.

Author Contributions: Conceptualization, J.R. and T.M.; Methodology, R.A.A. and S.A.M.A.; Software, J.R., T.M., M.R.G. and A.M.H.; Validation, R.A.A. and M.R.G.; Formal analysis, S.A.M.A.; Investigation, M.R.G.; Resources, J.R. and A.M.H.; Data curation, T.M., S.A.M.A. and A.M.H.; Writing—original draft, J.R. and T.M.; Writing—review & editing, R.A.A. All authors have read and agreed to the published version of the manuscript.

Funding: Princess Nourah bint Abdulrahman University Researchers Supporting Project number (PNURSP2023R61), Princess Nourah bint Abdulrahman University, Riyadh, Saudi Arabia.

Data Availability Statement: Not applicable.

Acknowledgments: The authors express their gratitude to Princess Nourah bint Abdulrahman University Researchers Supporting Project number (PNURSP2023R61), Princess Nourah bint Abdulrahman University, Riyadh, Saudi Arabia.

Conflicts of Interest: The authors declare no conflict of interest.

References

1. Eringen, A.C. Theory of micropolar fluids. *J. Math. Mech.* **1966**, *16*, 1–18. [CrossRef]
2. Eringen, A.C. Theory of micropolar elasticity. In *Microcontinuum Field Theories*; Springer: New York, NY, USA, 1999; pp. 101–248.
3. Mohammadein, A.A.; Gorla, R.S.R. Effects of transverse magnetic field on mixed convection in a micropolar fluid on a horizontal plate with vectored mass transfer. *Acta Mech.* **1996**, *118*, 1–12. [CrossRef]
4. Peddieson, J. Boundary Layer Theory for a Micropolar Fluid. 1970. Available online: web (accessed on 1 January 2023).
5. Gupta, P.S.; Gupta, A.S. Heat and mass transfer on a stretching sheet with suction or blowing. *Can. J. Chem. Eng.* **1977**, *55*, 744–746. [CrossRef]
6. Chakrabarti, A.; Gupta, A.S. Hydromagnetic flow and heat transfer over a stretching sheet. *Q. Appl. Math.* **1979**, *37*, 73–78. [CrossRef]
7. Incropera, F.P.; Bergman, T.L.; Lavine, A.S.; DeWitt, D.P. *Fundamentals of Heat and Mass Transfer*; Springer: New York, NY, USA, 2011.
8. Cengel, Y.A.; Boles, M.A. *Thermodynamics: An Engineering Approach*, 8th ed.; McGraw-Hill: New York, NY, USA, 2015.
9. Cengel, Y.A. *Heat Transfer: A Practical Approach*, 2nd ed.; McGraw-Hill: New York, NY, USA, 2002.
10. Zhao, N.; Li, S.; Yang, J. A review on nanofluids: Data-driven modeling of thermalphysical properties and the application in automotive radiator. *Renew Sustain. Energy Rev.* **2016**, *66*, 596–616. [CrossRef]
11. Krishna, V.M.; Kumar, M.S. Numerical analysis of forced convective heat transfer of nanofluids in microchannel for cooling electronic equipment. *Mater. Today Proc.* **2019**, *17*, 295–302. [CrossRef]
12. Okonkwo, E.C.; Okwose, C.F.; Abid, M.; Ratlamwala, T.A.H. Second-law analysis and exergoeconomics optimization of a solar tower—Driven combined-cycle power plant using supercritical CO_2. *J. Energy Eng. ASCE* **2018**, *144*, 04018021. [CrossRef]
13. Okonkwo, E.C.; Abid, M.; Ratlamwala, T.A.H. Numerical analysis of heat transfer enhancement in a parabolic trough collector based on geometry modifications and working fluid usage. *J. Sol. Energy Eng.* **2018**, *140*, 0510091. [CrossRef]
14. Meseguer, J.; Pérez-Grande, I.; Sanz-Andrés, A. *Spacecraft Thermal Control*, 1st ed.; Elsevier: London, UK, 2012.
15. Sajid, M.U.; Ali, H.M. Thermal conductivity of hybrid nanofluids: A critical review. *Int. J. Heat Mass Transf.* **2018**, *126*, 211–234. [CrossRef]

16. Das, S.K.; Choi, S.U.S.; Patel, H.E. Heat transfer in nanofluids—A review heat transfer in nanofluids. *Heat Transf. Eng.* **2007**, *27*, 37–41.
17. Choi, S.U.; Eastman, J.A. *Enhancing Thermal Conductivity of Fluids with Nanoparticles*; No. ANL/MSD/CP-84938; CONF-951135-29; Argonne National Lab.: Lemont, IL, USA, 1995.
18. Liu, L.H.; Métivier, R.; Wang, S.; Wang, H. Advanced nanohybrid materials: Surface modification and applications. *J. Nanomater.* **2012**, *2012*, 536405. [CrossRef]
19. Rashid, I.; Haq, R.U.; Al-Mdallal, Q.M. Aligned magnetic field effects on water based metallic nanoparticles over a stretching sheet with PST and thermal radiation effects. *Phys. E Low-Dimens. Syst. Nanostruct.* **2017**, *89*, 33–42. [CrossRef]
20. ul Haq, R.; Aman, S. Water functionalized CuO nanoparticles filled in a partially heated trapezoidal cavity with inner heated obstacle: FEM approach. *Int. J. Heat Mass Transf.* **2019**, *128*, 401–417. [CrossRef]
21. Hayat, T.; Imtiaz, M.; Alsaedi, A. Melting heat transfer in the MHD flow of Cu–water nanofluid with viscous dissipation and Joule heating. *Adv. Powder Technol.* **2016**, *27*, 1301–1308. [CrossRef]
22. Sandeep, N.; Sharma, R.P.; Ferdows, M. Enhanced heat transfer in unsteady magnetohydrodynamic nanofluid flow embedded with aluminum alloy nanoparticles. *J. Mol. Liq.* **2017**, *234*, 437–443. [CrossRef]
23. Shah, F.; Khan, M.I.; Hayat, T.; Khan, M.I.; Alsaedi, A.; Khan, W.A. Theoretical and mathematical analysis of entropy generation in fluid flow subject to aluminum and ethylene glycol nanoparticles. *Comput. Methods Programs Biomed.* **2019**, *182*, 105057. [CrossRef] [PubMed]
24. Misra, J.C.; Shit, G.C.; Rath, H.J. Flow and heat transfer of a MHD viscoelastic fluid in a channel with stretching walls: Some applications to haemodynamics. *Comput. Fluids* **2008**, *37*, 1–11. [CrossRef]
25. Ashraf, M.; Jameel, N.; Ali, K. MHD non-Newtonian micropolar fluid flow and heat transfer in channel with stretching walls. *Appl. Math. Mech.* **2013**, *34*, 1263–1276. [CrossRef]
26. Raza, J.; Rohni, A.M.; Omar, Z. MHD flow and heat transfer of Cu–water nanofluid in a semi porous channel with stretching walls. *Int. J. Heat Mass Transf.* **2016**, *103*, 336–340. [CrossRef]
27. Reza, J.; Mebarek-Oudina, F.; Makinde, O.D. MHD slip flow of Cu-Kerosene nanofluid in a channel with stretching walls using 3-stage Lobatto IIIA formula. In *Defect and Diffusion Forum*; Trans Tech Publications Ltd.: Wollerau, Switzerland, 2018; Volume 387, pp. 51–62.
28. Raza, J.; Rohni, A.M.; Omar, Z. Numerical investigation of copper-water (Cu-water) nanofluid with different shapes of nanoparticles in a channel with stretching wall: Slip effects. *Math. Comput. Appl.* **2016**, *21*, 43. [CrossRef]
29. Lund, L.A.; Omar, Z.; Raza, J.; Khan, I. Magnetohydrodynamic flow of Cu–Fe 3 O 4/H 2 O hybrid nanofluid with effect of viscous dissipation: Dual similarity solutions. *J. Therm. Anal. Calorim.* **2021**, *143*, 915–927. [CrossRef]
30. Chan, S.Q.; Aman, F.; Mansur, S. Sensitivity analysis on thermal conductivity characteristics of a water-based bionanofluid flow past a wedge surface. *Math. Probl. Eng.* **2018**, *2018*, 9410167. [CrossRef]
31. Vahedi, S.M.; Pordanjani, A.H.; Raisi, A.; Chamkha, A.J. Sensitivity analysis and optimization of MHD forced convection of a Cu-water nanofluid flow past a wedge. *Eur. Phys. J. Plus* **2019**, *134*, 124. [CrossRef]
32. Khan, N.S.; Kumam, P.; Thounthong, P. Second law analysis with effects of Arrhenius activation energy and binary chemical reaction on nanofluid flow. *Sci. Rep.* **2020**, *10*, 19792. [CrossRef]
33. Thriveni, K.; Mahanthesh, B. Significance of variable fluid properties on hybrid nanoliquid flow in a micro-annulus with quadratic convection and quadratic thermal radiation: Response surface methodology. *Int. Commun. Heat Mass Transf.* **2021**, *124*, 105264. [CrossRef]

mathematics

Article

Artificial Neural Networking Magnification for Heat Transfer Coefficient in Convective Non-Newtonian Fluid with Thermal Radiations and Heat Generation Effects

Khalil Ur Rehman [1,2,*], Wasfi Shatanawi [1,3,*] and Andaç Batur Çolak [4]

[1] Department of Mathematics and Sciences, College of Humanities and Sciences, Prince Sultan University, Riyadh 11586, Saudi Arabia
[2] Department of Mathematics, Air University, PAF Complex E-9, Islamabad 44000, Pakistan
[3] Department of Mathematics, Faculty of Science, The Hashemite University, P.O. Box 330127, Zarqa 13133, Jordan
[4] Information Technologies Application and Research Center, Istanbul Commerce University, Istanbul 34445, Turkey
* Correspondence: kurrehman@psu.edu.sa (K.U.R.); wshatanawi@psu.edu.sa (W.S.)

Abstract: In this study, the Casson fluid flow through an inclined, stretching cylindrical surface is considered. The flow field is manifested with pertinent physical effects, namely heat generation, viscous dissipation, thermal radiations, stagnation point flow, variable thermal conductivity, a magnetic field, and mixed convection. In addition, the flow field is formulated mathematically. The shooting scheme is used to obtain the numerical data of the heat transfer coefficient at the cylindrical surface. Further, for comparative analysis, three different thermal flow regimes are considered. In order to obtain a better estimation of the heat transfer coefficient, three corresponding artificial neural networks (ANN) models were constructed by utilizing Tan-Sig and Purelin transfer functions. It was observed that the heat transfer rate exhibits an inciting nature for the Eckert and Prandtl numbers, curvature, and heat generation parameters, while the Casson fluid parameter, temperature-dependent thermal conductivity, and radiation parameter behave oppositely. The present ANN estimation will be helpful for studies related to thermal energy storage that have Nusselt number involvements.

Keywords: thermal energy; mixed convection; thermal radiation; nusselt number; artificial neural networking; casson fluid

MSC: 76R10; 76-10; 65K05

Citation: Rehman, K.U.; Shatanawi, W.; Çolak, A.B. Artificial Neural Networking Magnification for Heat Transfer Coefficient in Convective Non-Newtonian Fluid with Thermal Radiations and Heat Generation Effects. *Mathematics* **2023**, *11*, 342. https://doi.org/10.3390/math11020342

Academic Editor: Ramoshweu Solomon Lebelo

Received: 31 October 2022
Revised: 31 December 2022
Accepted: 4 January 2023
Published: 9 January 2023

1. Introduction

It is a well-known fact among researchers that the study of heat transfer has numerous applications, such as in combustion chambers, furnaces, individual nuclear reactors, heat exchangers with high temperatures, and recuperating thermal energy storage systems, to name just a few. In this regard, the heat transfer coefficient, namely the Nusselt number, contributed to a better heat exchange rate. Due to this motivation, various researchers investigate the heat transfer aspects of the Casson fluid model [1], such as Casson fluid flow in the vicinity of a stagnation point in the direction of a stretched sheet as described by Meraj et al. in [2]. The analysis is also done on the properties of heat transmission with viscous dissipation. In addition, through appropriate transformations, the equations describing heat transport in Casson fluid were reduced. The Casson fluid, velocity ratio parameters, Prandtl and Eckert numbers were the factors controlling the flow. The homotopy analysis method (HAM) was used to calculate the analytical solutions across the entire geographical domain. The Nusselt number and the skin friction coefficient were computed and analyzed. The heat transfer in Casson fluid flow across a nonlinearly extending surface was investigated by Swati [3]. The momentum and energy equations were transformed into reduced equations by utilizing the appropriate transformations. Further, with the aid of the shooting approach, numerical solutions were obtained. The velocity field was suppressed while temperature increased toward the Casson parameter. The heat transmission in a Casson fluid past a symmetric wedge with mixed convection was examined by Swati et al. [4]. The graphical representations of a representative collection of graphic outcomes were produced by the shooting method. It was discovered that while the temperature fell with a higher Falkner-Skan exponent, the velocity increased. Although the temperature was observed to drop in this situation. The temperature is found to decrease as the Prandtl number rises. Pramanik [5] looked into the boundary layer flow of Casson fluid combined with heat transfer in the presence of suction or blowing at the surface toward an exponentially extending surface. The equation for the temperature field included a factor for thermal radiation. The momentum and heat transmission equations were reduced by using suitable transformations. Then, numerical answers to these equations were discovered. Both velocity and temperature show an opposite nature toward the Casson fluid parameter. The temperature rises as a result of thermal radiation, which improves effective thermal diffusivity. Mahdy [6] offered numerical solutions for heat transfer in Casson fluid past a cylinder. Additionally, by using similarity transformations, the controlling partial differential equations were reduced to ordinary differential equations, and the resulting equations were then numerically solved using the shooting method. The primary goal was to look into how the governing variables affected the velocity, temperature profiles, skin friction coefficient, and temperature gradient at the surface.

In an unstable flow of a Casson fluid approaching a stagnation point across a stretching/shrinking sheet in the presence of thermal radiation, Abbas et al. [7] provided the heat and mass transfer study for Casson fluid. They took into account the linear Rosseland approximation for thermal radiation. In considering chemical reactions as a function of temperature, the influence of binary chemical reactions with Arrhenius activation energy was also taken into account. The bivariate spectral collocation quasi-linearization approach was used to produce the numerical solutions of the system of nonlinear PDEs that are constant throughout the entire domain and at all times. Subsequently, the numerical results for a number of relevant physical parameters were visually discussed as fields of velocity, temperature, and concentration. A moving wedge containing gyrotactic microorganisms was the subject of the study by Raju et al. [8] on the effects of thermophoresis and Brownian motion on two-dimensional magnetohydrodynamics (MHD) radiative Casson fluid. Using Runge-Kutta and Newton's methods, numerical results were presented graphically as well as in tabular form. In the two flow instances of suction and injection, the effects of pertinent parameters on the distributions of velocity, temperature, concentration, and density of motile organisms were given and addressed. Further, in comparing the obtained results to the existing prior studies, the results were validated and determined to be in good agreement. The temperature and concentration field are increased as the thermophoresis parameter values rise. The fact that gyrotactic microorganisms can speed up mass and heat transfer rates was a significant discovery of the present study. Reddy et al. [9] examined the consequences of conjugate heat transfer (CHT) on the idea of a heat function. The Casson fluid was physically represented as it passed through a thin, vertical cylinder. The hollow cylinder's inner wall was kept at a constant temperature. Additionally, by using an implicit methodology, the solutions to the linked, non-linear governing equations are discovered. All of the governing parameters were shown graphically in the flow charts. The Casson fluid parameters' steady-state times were prolonged. The heat function contours were concentrated near the leading edge at the cylinder's hotter wall. Furthermore, by increasing the values of all the regulating parameters, the heat lines' departures from the hot wall continue to decrease. In comparison to the Newtonian fluid, the Casson fluid is more important at the hot wall. The influence of nanoparticles suspended in the flow regime of Casson fluid towards an inclined plate was presented by Sulochana et al. [10]. The frictional heating, heat generation, and thermal radiation effects were all included in the energy and diffusion equations. TiO2-water and CuO-water were considered two different types of nanofluids to make the analysis more interesting. The analytical solutions to the transmuted governing partial differential equations (PDEs) were achieved by using the regular perturbation approach. In using graphical and tabular representations, the effects of relevant flow variables on thermal, momentum, mass transport, and mass and thermal transport rates were studied. According to the findings, heat radiation and limits on chemical reactions tend to increase the rates of thermal and mass transmission. Ali et al. [11] investigated the micropolar-Casson fluid flow in a restricted channel with MHD. The governing model of the issue was converted into a formulation based on the vorticity-stream function, and a finite difference method was used to solve it numerically. The effects of wall shear stress (WSS), axial velocity, and micro-rotation velocity on various flow regulating parameters, such as the Strouhal, Hartmann, porosity, micropolar, and Casson fluid parameters, were illustrated graphically and discussed. With increasing porosity parameter values, the WSS declines. It was discovered that the flow separation region was significantly influenced by the Hartman number as well. All of the axial locations had parabolic velocity profiles. The greatest velocity value was found near the throat of the constriction.

Gbadeyan et al. [12] looked at the impacts of nonlinear radiation, non-Darcian porous media, and variable thermal conductivity and viscosity on MHD Casson MHD nanofluid flow for vertical surfaces. The resulting flow equations were transformed into ordinary differential equations. The set of equations that resulted from this was then solved using the Galerkin weighted residual method (GWRM). The temperature, velocity, and nanoparticle volume percent were calculated using numbers (nanoparticle concentration). It is observed that as the nanoparticle volume fraction and temperature decrease, the viscosity and thermal conductivity increase. Alizadeh et al. [13] examined the impinging Casson fluid flow over a cylinder manifested with porous material, Soret, and Dufour effects. The flow equations were numerically solved, and Sherwood, Nusselt, and Bejan numbers were predicted. The results demonstrate that the Nusselt number decreased significantly, although the Sherwood number decreased less. It was also established that the fluid's improved non-Newtonian properties had a considerable impact on flow, temperature, and mass transfer irreversibilities. In terms of heat transport and entropy, Jamshed et al. [14] explored the Casson time-independent nanofluid. The impact of slip state and solar thermal transport on Casson nanofluid flow convection was comprehensively examined. The nanofluid was treated on a slippery surface with convective heat to evaluate the flow characteristics and thermal transport. The equations defining the flow problem were written using PDEs. After converting the equations to ODEs, their self-similar solution was discovered using a numerical approach known as the Keller box. The copper-water and titanium-water mixtures are two unique groups of nanofluids under consideration for the study. The numerical results for several flow parameters, such as skin friction, heat transfer, Nusselt number, and entropy, were visually depicted. Furthermore, increasing the Reynolds number enhanced the entropy in the system. In the case of the Casson phenomenon, rather than normal fluid, thermal conductivity increases. The recent developments on the subject enclosed above can be accessed in Refs. [15–18].

Additionally, on the basis of the literature reported above on non-Newtonian fluid, namely Casson fluid, we offer an estimation of the heat transfer coefficient at an inclined cylindrical surface. Further, mixed convection-casson fluid with a stagnation point is considered. The heat transfer aspects include heat generation, viscous dissipation, thermal radiation, and temperature-dependent variable thermal conductivity effects. The three different thermal flow fields and magnetic field assumptions are formulated mathematically. The obtained flow equations are reduced in terms of order and solved by using the shooting method. A Nusselt number as a heat transfer coefficient is predicted by using ANN models. The present article will help researchers obtain an accurate estimation of heat transfer coefficients from thermal engineering standpoints.

2. Mathematical Formulation

The heat transfer aspects of mixed convective magnetized Casson fluid flow over a stretching cylindrical surface are considered. Heat generation, viscous dissipation, thermal radiation, and temperature-dependent thermal conductivity are the key thermal effects held by energy equations. The mathematical formulation [9,10,13] concluded in this regard is as follows:

$$\frac{\partial(\widetilde{r}\widetilde{u})}{\partial\widetilde{x}}+\frac{\partial(\widetilde{r}\widetilde{v})}{\partial\widetilde{r}}=0, \tag{1}$$

$$\widetilde{u}\frac{\partial\widetilde{u}}{\partial\widetilde{x}}+\widetilde{v}\frac{\partial\widetilde{u}}{\partial\widetilde{r}}=v\left(\frac{\partial^2\widetilde{u}}{\partial\widetilde{r}^2}+\frac{1}{\widetilde{r}}\frac{\partial\widetilde{u}}{\partial\widetilde{r}}\right)\left(1+\frac{1}{\beta}\right)+g_0\beta_T(\widetilde{T}-\widetilde{T}_\infty)\cos(\alpha)+u_e\frac{\partial\widetilde{u}_e}{\partial\widetilde{x}}-\frac{\sigma B_0^2}{\rho}(\widetilde{u}-u_e), \tag{2}$$

$$\rho c_p\left(\widetilde{u}\frac{\partial\widetilde{T}}{\partial\widetilde{x}}+\widetilde{v}\frac{\partial\widetilde{T}}{\partial\widetilde{r}}\right)=\overline{\mu}\left(1+\frac{1}{\beta}\right)\left(\frac{\partial\widetilde{u}}{\partial\widetilde{r}}\right)^2+\frac{1}{\widetilde{r}}\frac{\partial}{\partial\widetilde{r}}\left(\kappa\frac{\partial\widetilde{T}}{\partial\widetilde{r}}\right)+Q_0\left(\widetilde{T}-\widetilde{T}_\infty\right)-\frac{1}{\widetilde{r}}\frac{\partial}{\partial\widetilde{r}}(\widetilde{r}\overline{q}), \tag{3}$$

with endpoint conditions:

$$\begin{aligned}\widetilde{u}(\widetilde{x},\widetilde{r})=\widetilde{U}_w=a\widetilde{x},\ \widetilde{v}(\widetilde{x},\widetilde{r})=0,\ \widetilde{T}=\widetilde{T}_w \text{ at } \widetilde{r}=c,\\ \widetilde{u}=\widetilde{u}_e=d\widetilde{x},\ \widetilde{T}\to\widetilde{T}_\infty \text{ as } \widetilde{r}\to\infty.\end{aligned} \tag{4}$$

The relation between thermal conductivity and radioactive heat flux is as follows:

$$\kappa\left(\widetilde{T}\right)=\kappa_\infty\left(1+\varepsilon\frac{\widetilde{T}-\widetilde{T}_\infty}{\Delta T}\right),\text{where } \Delta T=\widetilde{T}_w-\widetilde{T}_\infty. \tag{5}$$

$$\overline{q}=-\frac{16\sigma^* T_\infty^3}{3k^*}\frac{\partial\widetilde{T}}{\partial\widetilde{r}}. \tag{6}$$

For the solution of Equations (1)–(4), we have variables [13].

$$\begin{aligned}\widetilde{v}=-\tfrac{c}{\widetilde{r}}\sqrt{\tfrac{vU_0}{L}}f(\eta),\ \widetilde{u}=\widetilde{x}\tfrac{U_0}{L}f'(\eta),\\ \eta=\tfrac{\widetilde{r}^2-c^2}{2c}\sqrt{\tfrac{U_0}{vL}},\ \theta(\eta)=\tfrac{\widetilde{T}-\widetilde{T}_\infty}{\widetilde{T}_w-\widetilde{T}_\infty}.\end{aligned} \tag{7}$$

Equations (1)–(3) under Equation (7) take the form

$$(1+1/\beta)(f'''(1+2\gamma\eta)+2\gamma f'')-f'^2+ff''-M^2(f'-A)+A^2+G\theta\cos(\alpha)=0, \tag{8}$$

$$\begin{aligned}&(\theta''(1+2\eta\gamma)+2\gamma\theta')\left(1+\tfrac{4}{3}R\right)+\varepsilon\left((\theta\theta''+\theta'^2)(1+2\eta\gamma)+2\gamma\theta\theta'\right)\\ &+\Pr Ec(1+2\eta\gamma)\left(1+\tfrac{1}{\beta}\right)f''^2+\Pr f\theta'+\Pr H\theta=0,\end{aligned} \tag{9}$$

while the reduced boundary conditions are:

$$\begin{aligned}f'=1, f=0,\ \theta=1 \text{ at } \eta=0,\\ f'=A,\ \theta=0 \text{ as } \eta\to\infty.\end{aligned} \tag{10}$$

The mathematical relation for the Nusselt number is as follows:

$$\mathrm{Nu}=\frac{\widetilde{x}q_w}{\kappa(\widetilde{T}_w-\widetilde{T}_\infty)},q_w=-\kappa\left(1+\frac{16\sigma^*\widetilde{T}_\infty^3}{3k^*\kappa}\right)\left(\frac{\partial\widetilde{T}}{\partial\widetilde{r}}\right)_{\widetilde{r}=c}. \tag{11}$$

$$\frac{\mathrm{Nu}}{\sqrt{\mathrm{Re}_{\widetilde{x}}}}=-\left(1+\frac{4}{3}R\right)\theta'(0) \tag{12}$$

2.1. Formulation without Thermal Radiation

In the present case, the heat transfer is examined in the absence of thermal radiation [7,14]. The flow field includes physical effects such as viscous dissipation, heat generation, temperature-dependent thermal conductivity, mixed convection, an externally applied magnetic field, and a stagnation point. The flow field is mathematically concluded as follows:

$$\frac{\partial(\tilde{r}\tilde{u})}{\partial\tilde{x}} + \frac{\partial(\tilde{r}\tilde{v})}{\partial\tilde{r}} = 0, \tag{13}$$

$$\tilde{u}\frac{\partial\tilde{u}}{\partial\tilde{x}} + \tilde{v}\frac{\partial\tilde{u}}{\partial\tilde{r}} = v\left(\frac{\partial^2\tilde{u}}{\partial\tilde{r}^2} + \frac{1}{\tilde{r}}\frac{\partial\tilde{u}}{\partial\tilde{r}}\right)\left(1+\frac{1}{\beta}\right) + g_0\beta_T(\tilde{T}-\tilde{T}_\infty)\cos(\alpha) + u_e\frac{\partial\tilde{u}_e}{\partial\tilde{x}} - \frac{\sigma B_0^2}{\rho}(\tilde{u}-u_e), \tag{14}$$

$$\rho c_\rho\left(\tilde{u}\frac{\partial\tilde{T}}{\partial\tilde{x}} + \tilde{v}\frac{\partial\tilde{T}}{\partial\tilde{r}}\right) = \overline{\mu}\left(1+\frac{1}{\beta}\right)\left(\frac{\partial\tilde{u}}{\partial\tilde{r}}\right)^2 + \frac{1}{\tilde{r}}\frac{\partial}{\partial\tilde{r}}\left(\kappa\frac{\partial\tilde{T}}{\partial\tilde{r}}\right) + Q_0\left(\tilde{T}-\tilde{T}_\infty\right). \tag{15}$$

while endpoint conditions for the present case are:

$$\begin{aligned} \tilde{u}(\tilde{x},\tilde{r}) = \tilde{U}_w = a\tilde{x},\ \tilde{v}(\tilde{x},\tilde{r}) = 0,\ \tilde{T} = \tilde{T}_w \text{ at } \tilde{r} = c, \\ \tilde{u} = \tilde{u}_e = d\tilde{x},\ \tilde{T} \to \tilde{T}_\infty \text{ as } \tilde{r} \to \infty. \end{aligned} \tag{16}$$

The relation for thermal conductivity is given as follows:

$$\kappa\left(\tilde{T}\right) = \kappa_\infty\left(1+\varepsilon\frac{\tilde{T}-\tilde{T}_\infty}{\Delta T}\right), \text{where } \Delta T = \tilde{T}_w - \tilde{T}_\infty. \tag{17}$$

To obtain the solutions of Equations (1)–(3), we have used the variables given in Equation (7).

The reduced set of equations to describe the heat transfer in Casson fluid flow over the inclined surface are concluded as follows:

$$(1+1/\beta)(f'''(1+2\gamma\eta)+2\gamma f'') - f'^2 + ff'' - M^2(f'-A) + A^2 + G\theta\cos(\alpha) = 0, \tag{18}$$

$$\begin{aligned} &(\theta''(1+2\eta\gamma)+2\gamma\theta') + \varepsilon\left((\theta\theta''+\theta'^2)(1+2\eta\gamma)+2\gamma\theta\theta'\right) + \mathrm{Pr}Ec(1+2\eta\gamma)\left(1+\tfrac{1}{\beta}\right)f''^2 \\ &+\mathrm{Pr}f\theta' + \mathrm{Pr}H\theta = 0, \end{aligned} \tag{19}$$

while the respective boundary conditions are:

$$\begin{aligned} f' = 1, f = 0,\ \theta = 1 \text{ at } \eta = 0, \\ f' = A,\ \theta = 0 \text{ as } \eta \to \infty. \end{aligned} \tag{20}$$

Since the thermal radiations are not considered, the mathematical relation for Nusselt number reduces to

$$\mathrm{Nu} = \frac{\tilde{x}q_w}{\kappa(\tilde{T}_w - \tilde{T}_\infty)}, q_w = -\kappa\left(\frac{\partial\tilde{T}}{\partial\tilde{r}}\right)_{\tilde{r}=c}. \tag{21}$$

$$\frac{\mathrm{Nu}}{\sqrt{\mathrm{Re}_{\tilde{x}}}} = -\theta'(0) \tag{22}$$

2.2. Formulation without Heat Generation

In this case, we have considered the thermal flow regime without the heat generation effect. The energy equation is carried in the presence of temperature-dependent thermal conductivity and viscous dissipation, while the momentum equation makes assumptions about stagnation point flow, mixed convection, and the magnetic field [15,16]. The concluding mathematical equations in this regard are as follows:

$$\frac{\partial(\tilde{r}\tilde{u})}{\partial\tilde{x}} + \frac{\partial(\tilde{r}\tilde{v})}{\partial\tilde{r}} = 0, \tag{23}$$

$$\widetilde{u}\frac{\partial \widetilde{u}}{\partial \widetilde{x}} + \widetilde{v}\frac{\partial \widetilde{u}}{\partial \widetilde{r}} = v\left(\frac{\partial^2 \widetilde{u}}{\partial \widetilde{r}^2} + \frac{1}{\widetilde{r}}\frac{\partial \widetilde{u}}{\partial \widetilde{r}}\right)\left(1+\frac{1}{\beta}\right) + g_0\beta_T(\widetilde{T}-\widetilde{T}_\infty)\cos(\alpha) + u_e\frac{\partial \widetilde{u}_e}{\partial \widetilde{x}} - \frac{\sigma B_0^2}{\rho}(\widetilde{u}-u_e), \tag{24}$$

$$\rho c_\rho\left(\widetilde{u}\frac{\partial \widetilde{T}}{\partial \widetilde{x}} + \widetilde{v}\frac{\partial \widetilde{T}}{\partial \widetilde{r}}\right) = \overline{\mu}\left(1+\frac{1}{\beta}\right)\left(\frac{\partial \widetilde{u}}{\partial \widetilde{r}}\right)^2 + \frac{1}{\widetilde{r}}\frac{\partial}{\partial \widetilde{r}}\left(\kappa\frac{\partial \widetilde{T}}{\partial \widetilde{r}}\right) - \frac{1}{\widetilde{r}}\frac{\partial}{\partial \widetilde{r}}(\widetilde{r}\widetilde{q}), \tag{25}$$

with boundary conditions:

$$\begin{gathered}\widetilde{u}(\widetilde{x},\widetilde{r}) = \widetilde{U}_w = a\widetilde{x},\ \widetilde{v}(\widetilde{x},\widetilde{r}) = 0,\ \widetilde{T} = \widetilde{T}_w \text{ at } \widetilde{r} = c,\\ \widetilde{u} = \widetilde{u}_e = d\widetilde{x},\ \widetilde{T} \to \widetilde{T}_\infty \text{ as } \widetilde{r} \to \infty.\end{gathered} \tag{26}$$

The relation for thermal conductivity and radioactive heat flux is the same as in Equations (5) and (6), respectively. In changing the order of Equations (24) and (25), we used variables given in Equation (7). The ultimate outcome in this regard is as follows:

$$(1+1/\beta)(f'''(1+2\gamma\eta)+2\gamma f'') - f'^2 + ff'' - M^2(f'-A) + A^2 + G\theta\cos(\alpha) = 0, \tag{27}$$

$$(\theta''(1+2\eta\gamma)+2\gamma\theta')\left(1+\frac{4}{3}R\right) + \varepsilon\Big((\theta\theta''+\theta'^2)(1+2\eta\gamma)+2\gamma\theta\theta'\Big) + \mathrm{Pr}Ec(1+2\eta\gamma)\left(1+\frac{1}{\beta}\right)f''^2 + \mathrm{Pr}f\theta' = 0, \tag{28}$$

The reduced boundary conditions are as follows:

$$\begin{gathered}f' = 1, f = 0,\ \theta = 1 \text{ at } \eta = 0,\\ f' = A,\ \theta = 0 \text{ as } \eta \to \infty.\end{gathered} \tag{29}$$

In following all thermal flow fields, it should be noted that heat generation coefficient, free stream velocity, dynamics viscosity, temperature, electrical conductivity, fluid density, uniform magnetic field, inclination, gravitational acceleration, thermal exponential coefficient, mean absorption coefficient, kinematic viscosity, Stefan-Boltzmann constant, Eckert number, heat generation parameter, Prandtl number, mixed convection, magnetic field, ratio of free stream to stretching velocity, and radiation parameters are denoted as Q_0, $\widetilde{u}_e$, $\overline{\mu}$, $\widetilde{T}$, ρ, σ, B_0, α, g_0, β_T, k^*, v, σ^*, Ec, H, Pr, G, M, A, and R, respectively. Further, the involved flow parameters are defined as follows:

$$\begin{gathered}M = \sqrt{\frac{\sigma B_0^2 L}{\rho U_0}},\ G = \frac{g_0\beta_T(\widetilde{T}_w-\widetilde{T}_\infty)L^2}{U_0\widetilde{x}},\ Pr = \frac{\overline{\mu}c_p}{\kappa},\\ \beta = \frac{\overline{\mu}\sqrt{2\pi_c}}{\tau_r},\ R = \frac{4\sigma^*\widetilde{T}_\infty^3}{\kappa k^*},\ \gamma = \sqrt{\frac{vL}{c^2U_0}},\ A = \frac{d}{a},\\ H = \frac{LQ_0}{U_0\rho c_p},\ Ec = \frac{U_0^2(\widetilde{x}/L)^2}{c_p(\widetilde{T}_w-\widetilde{T}_\infty)}.\end{gathered} \tag{30}$$

3. Numerical Method

In the ANN Model-I, the characteristics of heat transfer for mixed convective magnetized Casson fluid flow are considered. The main thermal effects held by energy equations include heat generation, viscous dissipation, thermal radiation, and temperature-dependent thermal conductivity. Subject to these physical effects, Equations (1)–(6) are the ultimate flow-narrating differential equations. The reduced system obtained by means of Equation (7) is given in Equations (8)–(10). The dimensionless relation for the Nusselt number is given in Equation (12). In the ANN Model-II, the heat transfer aspects without thermal radiation are addressed. The Equations (13)–(17) represent the mathematical formulation for ANN Model-II, which is heat transfer aspects without thermal radiations.

In addition, using Equation (7), the dimensionless differential equations for the ANN Model-II are summarized as Equations (18)–(20). In the absence of thermal radiations, the dimensionless form of the Nusselt number is offered in Equation (22). In ANN Model-III, we considered heat transfer aspects in the absence of a heat generation effect for the Casson fluid flow over a stretched surface. The originating partial differential equation for ANN

Model-III is concluded in Equations (23)–(26). The reduced differential system for ANN Model-III is summarized as Equations (27)–(29). In the absence of the heat generation effect, the Nusselt number relation holds as it does for Model-I. Our key interest is to obtain the numerical data of the Nusselt number for each case, namely ANN Model-I, ANN Model-II, and ANN Model-III. Firstly, we deal with the major case, which is ANN Model-I. Various schemes [19–22] exist to narrate the fluid flow problems, but to execute the shooting method [23,24] along with the Runge-Kutta scheme, the following necessary procedure is carried out:

$$Y_1 = f(\eta), Y_2 = f'(\eta), Y_3 = f''(\eta), Y_4 = \theta(\eta), Y_5 = \theta'(\eta). \tag{31}$$

Owning Equation (31) in Equations (8) and (9), one has

$$\begin{gathered} {Y_1}' = Y_2, \\ {Y_2}' = Y_3, \\ {Y_3}' = \frac{1}{\left(1+\frac{1}{\beta}\right)(1+2\eta\gamma)}\left[-2\gamma y_3\left(1+\frac{1}{\beta}\right) + Y_2^2 - Y_1Y_3 - GY_4\cos\alpha + M^2(Y_2 - A) - A^2\right], \\ {Y_4}' = Y_5, \\ {Y_5}' = -\frac{1}{\left(1+\frac{4}{3}R\right)(1+2\eta\gamma)+\varepsilon(1+2\eta\gamma)Y_4}\left[\begin{array}{c} (1+\frac{4}{3}R)(2\gamma Y_5) + \varepsilon((1+2\eta\gamma)Y_5^2 + 2\gamma Y_4Y_5) + \\ \Pr Y_1Y_5 + \Pr Ec(1+2\eta\gamma)(1+\frac{1}{\beta})Y_2^2 + \Pr HY_4 \end{array}\right]. \end{gathered} \tag{32}$$

and

$$\begin{gathered} Y_1 = 0,\ Y_2 = 1,\ Y_4 = 1 \text{ at } \eta = 0, \\ Y_2 \to A,\ Y_4 \to 0 \text{ as } \eta = \infty. \end{gathered} \tag{33}$$

Then the self-coding is implemented in Matlab, and outcomes are reported for ANN Model-I in terms of graphs and tables. Similarly, we find numerical solutions for ANN Model-II and ANN Model-III.

4. Development of ANN Models (I,II,III)

The ANN models were created using the multilayer perceptron (MLP) method, one of the models that researchers frequently adopt due to its strong learning capabilities. In terms of structure, MLP networks consist of three linked primary layers.

The prediction data is derived from the input layer, the hidden layer, and the output layer, which are the first, second, and third layers, respectively. In the developed ANN model, MLP structures with a single hidden layer are preferred. In the performance analysis of the designed ANN model, it has been shown that, in order to achieve ideal results, the number of hidden layers is sufficient and that it is not necessary to experiment on MLP models with multiple hidden layer structures. An MLP network model's symbolic architecture is depicted in Figure 1. In each of the three distinct MLP network models, different input parameters were defined in order to estimate Nu values. Table 1 displays the input and output parameters of three distinct ANN models that were created. Moreover, *R1* and *R2* represent the thermal radiation parameter values 0 and 0.5, respectively. The same is the case for heat generation parameters H_1 and H_2. The performance of forecasts is impacted by the best data optimization during the building of ANN models. According to the grouping technique frequently employed in the literature, the data utilized in ANN models, each of which was produced with a different number of data sets, were segmented. A total of 15% of the data is set aside for validation, 15% for testing, and 70% is set aside for training. Table 2 provides details about the data set used to create three distinct ANN models. The optimization of the computational component known as the neuron in the hidden layer of MLP models is one of the challenges. Further, there is no model or guideline for calculating the number of neurons, which is the fundamental cause of this challenge. In the hidden layer, the number of neurons between 5 and 25 was tested. The MLP network model with 10 neurons in the hidden layer was chosen after the performances of other MLP networks with various numbers of hidden layer neurons were assessed. In determining the optimal number of neurons, parameters such as deviation rates, mean squared error (MSE) values, and coefficient of determination (Rm) values were taken into account. The Levenberg-Marquardt method, a popular training technique with excellent estimate performance, served as the ANN model's training procedure [25]. In the hidden and output layers, respectively, there are Tan-Sig and Purelin functions acting as transfer functions. The transfer function mathematical expressions are shown below [26]:

$$f(x) = \frac{1}{1+e^{-x}}, \tag{34}$$

$$\text{Pureline}(x) = x. \tag{35}$$

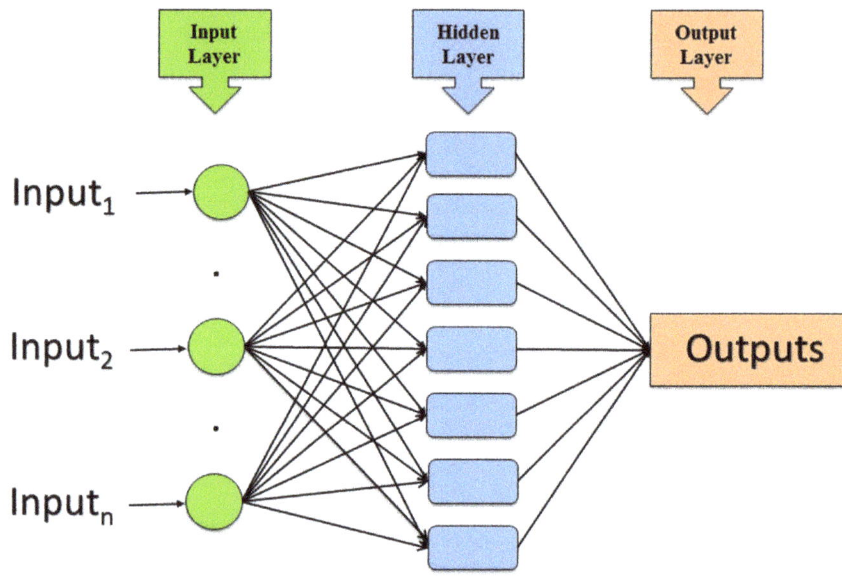

Figure 1. The symbolic architecture of an MLP network model.

Table 1. Output and input values of three different ANN models.

	Inputs								Output
ANN Model-I	β	γ	Ec	Pr	ε	R	H		Nu
ANN Model-II	β	γ	Ec	Pr	ε	H	R_1	R_2	Nu
ANN Model-III	β	γ	Ec	Pr	ε	R	H_1	H_2	Nu

Table 2. Information about the data set used in the development of three different ANN models.

	ANN Model-I	ANN Model-II	ANN Model-III
Training Data	48	84	84
Validation Data	11	18	18
Test Data	11	18	18
Total Data	70	120	120

The MSE, Rm, and margin of deviation (MoD) metrics, which are often used in the literature, were chosen to examine the estimated performance of three ANN models.

The following lists the mathematical formulas [27–30] used to calculate the performance parameters:

$$\mathrm{MSE} = \frac{1}{N}\sum_{i=1}^{N}\left(X_{\mathrm{num}(i)} - X_{\mathrm{ANN}(i)}\right)^2. \tag{36}$$

$$Rm = \sqrt{1 - \frac{\sum_{i=1}^{N}\left(X_{\mathrm{num}(i)} - X_{\mathrm{ANN}(i)}\right)^2}{\sum_{i=1}^{N}\left(X_{\mathrm{num}(i)}\right)^2}}, \tag{37}$$

$$\mathrm{MoD}(\%) = \left[\frac{X_{\mathrm{num}} - X_{\mathrm{ANN}}}{X_{\mathrm{num}}}\right]. \tag{38}$$

5. Comparative Analysis

The aim of this study is to predict the values of the heat transfer coefficient at a cylindrical surface when a non-Newtonian fluid passes over the surface. A total of three different flow regimes have been considered when constructing the corresponding ANN models. The ANN Model-I is developed by considering the Casson fluid flow over an inclined stretching cylinder along with the involved physical effects, namely an externally applied magnetic field, stagnation point flow, mixed convection, heat generation, viscous dissipation, thermal radiations, and variable thermal conductivity. In this model, we consider seven inputs and the Nusselt number as an output. The ANN Model-II is used to predict the Nusselt number values for two different thermal regimes, namely, the thermal regime with radiations and the thermal regime without radiations. The ANN Model-III offers the prediction of the Nusselt number for two different thermal regimes, namely, thermal regimes with and without heat generation. In using the shooting method, we obtained the numerical values of the Nusselt number for three different models (see Tables 3–21). The impact of the Casson fluid parameter on the Nusselt number is presented in detail in Table 3. The numerical information for the Nusselt number for a positive iteration of the curvature parameter is presented in Table 4. Table 5 demonstrates the influence of Eckert number on Nusselt number, and as can be observed, Nusselt number exhibits a direct relationship with larger Eckert number values, i.e., Nusselt number grows in magnitude as Eckert number rises. Table 6 shows how Pr affects the Nusselt number.

Table 3. Effect of Casson fluid parameter on Nusselt number.

β	$\theta'(0)$	$-[1 + 4/3R]\,\theta'(0)$, $R = 0.2$
0.2	−2.0221	2.5613
0.3	−1.9371	2.4536
0.4	−1.8933	2.3982
0.5	−1.8664	2.3641
0.6	−1.8481	2.3409
0.7	−1.8347	2.3239
0.8	−1.8246	2.3112
0.9	−1.8165	2.3009
01	−1.8100	2.2927
02	−1.7794	2.2539

Table 4. Impact of curvature parameter on Nusselt number.

γ	$\theta'(0)$	$-[1+4/3R]\,\theta'(0)$, $R=0.2$
0.2	−0.9511	1.2047
0.3	−1.0468	1.3259
0.4	−1.1443	1.4494
0.5	−1.2441	1.5759
0.6	−1.3466	1.7057
0.7	−1.4515	1.8386
0.8	−1.5587	1.9744
0.9	−1.6681	2.1129
01	−1.7794	2.2539
02	−2.9651	3.7559

Table 5. Impact of Eckert number on Nusselt number.

Ec	$\theta'(0)$	$-[1+4/3R]\,\theta'(0)$, $R=0.2$
0.2	−1.9015	2.4086
0.3	−2.0216	2.5608
0.4	−2.1397	2.7104
0.5	−2.2560	2.8577
0.6	−2.3705	3.0027
0.7	−2.4832	3.1455
0.8	−2.5943	3.2862
0.9	−2.7037	3.4247
01	−2.8116	3.5615
02	−3.2282	4.0892

Table 6. Impact of Prandtl number on Nusselt number.

Pr	$\theta'(0)$	$-[1+4/3R]\,\theta'(0)$, $R=0.2$
0.2	−1.3536	1.7146
0.3	−1.3813	1.7496
0.4	−1.4084	1.7840
0.5	−1.4348	1.8175
0.6	−1.4607	1.8502
0.7	−1.4861	1.8824
0.8	−1.5109	1.9139
0.9	−1.5353	1.9448
01	−1.5593	1.9752
02	−1.7794	2.2540

Table 7. Impact of variable thermal conductivity on Nusselt number.

ε	$\theta'(0)$	$-[1 + 4/3R]\,\theta'(0)$, $R = 0.2$
0.2	−3.0653	3.8828
0.3	−2.8466	3.6057
0.4	−2.6458	3.3514
0.5	−2.4588	3.1146
0.6	−2.2821	2.8907
0.7	−2.1121	2.6754
0.8	−1.9457	2.4646
0.9	−1.7794	2.2540
01	−1.6097	2.0390
02	−1.0271	1.3010

Table 8. Impact of radiation parameter on Nusselt number.

R	$\theta'(0)$	$-[1 + 4/3R]\,\theta'(0)$
0.2	−1.7794	2.2539
0.3	−1.6752	2.3452
0.4	−1.5591	2.3906
0.5	−1.4325	2.3875
0.6	−1.2956	2.3320
0.7	−1.1483	2.2200
0.8	−0.9895	2.0449
0.9	−0.8162	1.7956
01	−0.6202	1.4471
02	−0.3629	1.3306

Table 9. Influence of heat generation parameter on Nusselt number.

H	$\theta'(0)$	$-[1 + 4/3R]\,\theta'(0)$, $R = 0.2$
0.2	−1.8322	2.3208
0.3	−1.8836	2.3859
0.4	−1.9338	2.4495
0.5	−1.9827	2.5115
0.6	−2.0305	2.5720
0.7	−2.0772	2.6312
0.8	−2.1230	2.6892
0.9	−2.1678	2.7459
01	−2.2118	2.8017
02	−2.2421	2.8401

Table 10. Impact of curvature parameter on Nusselt number for non-radiative and radiative flow fields.

γ	$\theta'(0)$		$-[1 + 4/3R]\,\theta'(0)$	
	$R = 0$	$R = 0.5$	$R = 0$	$R = 0.5$
0.2	−1.0677	−0.9493	1.0677	1.5822
0.3	−1.3171	−1.2061	1.3171	2.0105
0.4	−1.5766	−1.4739	1.5766	2.4569
0.5	−1.8443	−1.7494	1.8443	2.9163
0.6	−2.1182	−2.0306	2.1182	3.3851
0.7	−2.3970	−2.3160	2.3970	3.8607
0.8	−2.6796	−2.6045	2.6796	4.3417
0.9	−2.9653	−2.8953	2.9653	4.8264
1.0	−3.2535	−3.1881	3.2535	5.3146
2.0	−6.2049	−6.1651	6.2049	10.277

Table 11. Impact of Casson fluid parameter on Nusselt number for radiative and non-radiative flow fields.

β	$\theta'(0)$		$-[1 + 4/3R]\,\theta'(0)$	
	$R = 0$	$R = 0.5$	$R = 0$	$R = 0.5$
0.2	−3.3407	−3.2765	3.3407	5.4619
0.3	−3.3106	−3.2453	3.3106	5.4099
0.4	−3.2950	−3.2294	3.2950	5.3834
0.5	−3.2854	−3.2196	3.2854	5.3660
0.6	−3.2788	−3.2130	3.2788	5.3561
0.7	−3.2740	−3.2081	3.2740	5.3479
0.8	−3.2703	−3.2044	3.2703	5.3417
0.9	−3.2674	−3.2015	3.2674	5.3369
1.0	−3.2650	−3.1992	3.2650	5.3331
2.0	−3.2535	−3.1881	3.2535	5.3145

Table 12. Impact of Eckert number on Nusselt number for radiative and non-radiative flow fields.

E	$\theta'(0)$		$-[1 + 4/3R]\,\theta'(0)$	
	$R = 0$	$R = 0.5$	$R = 0$	$R = 0.5$
0.2	−3.2969	−3.2327	3.2969	5.3889
0.3	−3.3403	−3.2773	3.3403	5.4632
0.4	−3.3836	−3.3219	3.3836	5.5376
0.5	−3.4270	−3.3665	3.4270	5.6119
0.6	−3.4704	−3.4111	3.4704	5.6863
0.7	−3.5138	−3.4558	3.5138	5.7608
0.8	−3.5572	−3.5004	3.5572	5.8352
0.9	−3.6006	−3.5451	3.6006	5.9096
1.0	−3.6440	−3.5898	3.6440	5.9841
2.0	−4.0787	−4.0372	4.0787	6.7301

Table 13. Impact of Prandtl number on Nusselt number for non-radiative and radiative flow fields.

Pr	$\theta'(0)$		$-[1 + 4/3R]\,\theta'(0)$	
	$R = 0$	$R = 0.5$	$R = 0$	$R = 0.5$
0.2	−3.0932	−3.0717	3.0932	5.1205
0.3	−3.1265	−3.0956	3.1265	5.1604
0.4	−3.1592	−3.1191	3.1592	5.1995
0.5	−3.1912	−3.1424	3.1912	5.2384
0.6	−3.2227	−3.1654	3.2227	5.2767
0.7	−3.2535	−3.1881	3.2535	5.3145
0.8	−3.2839	−3.2105	3.2839	5.3519
0.9	−3.3137	−3.2326	3.3137	5.3887
1.0	−3.3431	−3.2546	3.3431	5.4254
2.0	−3.6149	−3.4617	3.6149	5.7706

Table 14. Impact of variable thermal conductivity on Nusselt numbers for radiative and non-radiative flow fields.

ε	$\theta'(0)$		$-[1 + 4/3R]\,\theta'(0)$	
	$R = 0$	$R = 0.5$	$R = 0$	$R = 0.5$
0.2	−2.7937	−2.7799	2.7937	4.6341
0.3	−2.5894	−2.5450	2.5894	4.2425
0.4	−2.4047	−2.3224	2.4047	3.8714
0.5	−2.2353	−2.1076	2.2353	3.5134
0.6	−2.0777	−1.8965	2.0777	3.1614
0.7	−1.9285	−1.6847	1.9285	2.8084
0.8	−1.7848	−1.4671	1.7848	2.4456
0.9	−1.6436	−1.2373	1.6436	2.0625
1.0	−1.5720	−0.9852	1.5720	1.6423
2.0	−1.2825	−0.6896	1.2825	1.1496

Table 15. Impact of heat generation parameter on Nusselt number for radiative and non-radiative flow fields.

H	$\theta'(0)$		$-[1 + 4/3R]\,\theta'(0)$	
	$R = 0$	$R = 0.5$	$R = 0$	$R = 0.5$
0.2	−3.2670	−3.2020	3.2670	5.3377
0.3	−3.2805	−3.2159	3.2805	5.3609
0.4	−3.2938	−3.2298	3.2938	5.3841
0.5	−3.3072	−3.2436	3.3072	5.4071
0.6	−3.3204	−3.2573	3.3204	5.4299
0.7	−3.3337	−3.2709	3.3337	5.4526
0.8	−3.3468	−3.2845	3.3468	5.4753
0.9	−3.3599	−3.2981	3.3599	5.4979
1.0	−3.3730	−3.3116	3.3730	5.5204
2.0	−3.5007	−3.4432	3.5007	5.7398

Table 16. Impact of Casson fluid parameter on Nusselt number with and without heat generation.

β	$\theta'(0)$		$-[1 + 4/3R]\ \theta'(0),\ R = 0.2$	
	$H = 0$	$H = 0.5$	$H = 0$	$H = 0.5$
0.2	−3.2951	−3.3603	4.1739	4.2565
0.3	−3.2642	−3.3305	4.1347	4.2187
0.4	−3.2483	−3.3152	4.1146	4.1994
0.5	−3.2386	−3.3058	4.1023	4.1875
0.6	−3.2319	−3.2994	4.0938	4.1794
0.7	−3.2271	−3.2947	4.0877	4.1734
0.8	−3.2233	−3.2911	4.0829	4.1688
0.9	−3.2204	−3.2883	4.0793	4.1653
1.0	−3.2180	−3.2860	4.0762	4.1624
2.0	−3.2066	−3.2750	4.0618	4.1484

Table 17. Impact of curvature fluid parameter on Nusselt number with and without heat generation.

γ	$\theta'(0)$		$-[1 + 4/3R]\ \theta'(0),\ R = 0.2$	
	$H = 0$	$H = 0.5$	$H = 0$	$H = 0.5$
0.2	−0.9679	−1.1551	1.2256	1.4632
0.3	−1.2281	−1.3848	1.5556	1.7542
0.4	−1.4969	−1.6307	1.8961	2.0656
0.5	−1.7722	−1.8885	2.2448	2.3922
0.6	−2.0528	−2.1552	2.6003	2.7299
0.7	−2.3372	−2.4286	2.9605	3.0763
0.8	−2.6248	−2.7070	3.3248	3.4289
0.9	−2.9147	−2.9895	3.6921	3.7867
1.0	−3.2066	−3.2750	4.0618	4.1484
2.0	−6.1786	−6.2150	7.8264	7.8725

Table 18. Impact of Eckert number on Nusselt number with and without heat generation.

E	$\theta'(0)$		$-[1 + 4/3R]\ \theta'(0),\ R = 0.2$	
	$H = 0$	$H = 0.5$	$H = 0$	$H = 0.5$
0.2	−3.2508	−3.3182	4.1177	4.2032
0.3	−3.2950	−3.3613	4.1737	4.2577
0.4	−3.3392	−3.4045	4.2297	4.3125
0.5	−3.3835	−3.4477	4.2858	4.3672
0.6	−3.4277	−3.4909	4.3418	4.4219
0.7	−3.4720	−3.5341	4.3979	4.4766
0.8	−3.5162	−3.5773	4.4539	4.5313
0.9	−3.5605	−3.6206	4.4501	4.5862
1.0	−3.6048	−3.6638	4.5662	4.6409
2.0	−4.0481	−4.0967	5.1277	5.1893

Table 19. Impact of Prandtl number on Nusselt number with and without heat generation.

Pr	$\theta'(0)$		$-[1 + 4/3R]\,\theta'(0)$, $R = 0.2$	
	$H = 0$	$H = 0.5$	$H = 0$	$H = 0.5$
0.2	−3.0786	−3.0996	3.8996	3.9263
0.3	−3.1050	−3.1360	3.9331	3.9724
0.4	−3.1310	−3.1717	3.9661	4.0176
0.5	−3.1566	−3.2068	3.9984	4.0621
0.6	−3.1818	−3.2412	4.0304	4.1058
0.7	−3.2066	−3.2750	4.0618	4.1484
0.8	−3.2311	−3.3083	4.0928	4.1906
0.9	−3.2553	−3.3410	4.1234	4.2321
1.0	−3.2791	−3.3732	4.1536	4.2728
2.0	−3.5026	−3.6719	4.4367	4.6512

Table 20. Impact of variable thermal conductivity on Nusselt number with and without heat generation.

ε	$\theta'(0)$		$-[1 + 4/3R]\,\theta'(0)$, $R = 0.2$	
	$H = 0$	$H = 0.5$	$H = 0$	$H = 0.5$
0.2	−2.7477	−2.8195	3.4805	3.5715
0.3	−2.5369	−2.6115	3.2135	3.3079
0.4	−2.3414	−2.4194	2.9658	3.0646
0.5	−2.1571	−2.2393	2.7324	2.8365
0.6	−1.9804	−2.0679	2.5086	2.6194
0.7	−1.8079	−1.9017	2.2901	2.4088
0.8	−1.6357	−1.7376	2.0719	2.2011
0.9	−1.4601	−1.5721	1.8495	1.9914
1.0	−1.2761	−1.4015	1.1664	1.7751
2.0	−0.8518	−1.0240	1.0789	1.2971

Table 21. Impact of thermal radiation parameter on Nusselt number with and without heat generation.

R	$\theta'(0)$		$-[1 + 4/3R]\,\theta'(0)$	
	$H = 0$	$H = 0.5$	$H = 0$	$H = 0.5$
0.2	−3.2066	−3.2750	4.0618	4.1484
0.3	−3.1941	−3.2629	4.4717	4.5681
0.4	−3.1833	−3.2526	4.8799	4.9862
0.5	−3.1740	−3.2436	5.2911	5.4071
0.6	−3.1658	−3.2356	5.6984	5.8241
0.7	−3.1585	−3.2286	6.1054	6.2408
0.8	−3.1519	−3.2222	6.5141	6.6593
0.9	−3.1459	−3.2164	6.9209	7.0761
1.0	−3.1404	−3.2111	7.3265	7.4915
2.0	−3.1012	−3.1731	11.372	11.635

The impact of the temperature-dependent variable viscosity parameter on the Nusselt number is inspected and given in Tables 7–9 shows the impact of heat generation and thermal radiation parameters on the Nusselt number. In detail, larger values of the thermal radiation parameter cause a decline in the Nusselt number while for large heat generation parameter, the Nusselt number shows inciting values.

Collectively, for Tables 3–9, it has been observed that the heat transfer normal to the cylindrical surface enhances for curvature parameter, Prandtl number and heat generation parameter while for Casson fluid, thermal conductivity, radiation parameters and Eckert number. In addition, it behaved in opposition to the impact of the curvature parameter on the Nusselt number is observed for two different values of the thermal radiation parameter that is $R = 0$ and $R = 0.5$ see Table 10. Further, for the two alternative values of the thermal radiation parameter, $R = 0$ and $R = 0.5$, are used to examine the impact of the Casson fluid parameter on the Nusselt number. Table 11 is provided in this context. It was observed that the Nusselt number dramatically decreases for positive Casson fluid parameter fluctuation. In both the presence and non-existence scenarios of thermal radiations, the influence of the Eckert number on the Nusselt number is seen see Table 12. Table 13 offers the impact of Pr on the Nusselt number for both radiative and non-radiative cases. Furthermore, in both cases, the Nusselt number is an increasing function of positive variation in Pr. Table 14 examines and provides information on the impact of a temperature-dependent variable viscosity parameter on the Nusselt number. The effects of heat generation on the Nusselt number are shown in Table 15. In this case, there were higher values of the heat production parameter reveal increasing levels for the Nusselt number. Collectively for Tables 10–15, the magnitude of heat transfer normal to the cylindrical surface is higher for the case of presence of the thermal radiation effect. Table 16 offered the influence of the Casson fluid parameter on the Nusselt number is noticed for two different values namely $H = 0$ and $H = 0.5$. Further, $H = 0$ corresponds to the non-existence of the heat generation effect while $H = 0.5$ implied the existence of heat generation effect. In both cases, it is seen that the Nusselt number shows an inverse relation towards Casson fluid parameter.

The effects of the curvature fluid parameter on the Nusselt number are shown in Table 17 for two distinct values, $H = 0$ and $H = 0.5$. In addition, the Nusselt number is stronger in the case of the heat generating effect.

For two distinct scenarios, a thermal flow field with heat generation and a thermal flow field without heat generation, the impact of the Eckert number on Nusselt is explored see Table 18. The finding on the Nusselt number toward a positive fluctuation in the Prandtl number is presented in Table 19. As seen by past events, the Nusselt number rises sharply when provoked. Both the presence and absence of the heat generating effect are noted by such measurements. Additionally, it is noted that the Nusselt number's magnitude is greater for thermal flow fields with heat generating effects. For both thermal fields, namely thermal flow regimes with and without heat generating effect, the effect of changing thermal conductivity parameter on Nusselt number is perceived. To that end, Table 20 is provided. When there is a thermal flow regime and a heat generating impact, the Nusselt number is larger. The observation of the Nusselt number toward a positive fluctuation in the thermal radiation parameter is shown in Table 21. Both the presence and absence of the heat generating effect are observed and recorded. Furthermore, it is shown that the Nusselt number magnitude is a little bit bigger when the influence of heat generation is present. While the heat transfer normal to the cylindrical surface exhibits encouraging values for the Prandtl number, curvature parameter, Eckert number, and thermal radiations, the Casson fluid parameter and the temperature dependent variable viscosity parameter exhibit the opposite behavior. Furthermore, we have shown that the magnitude of the Nusselt number is larger when thermal radiations are present. The estimated MSE and R values for every ANN model for the training, validation, and testing phases are displayed in Table 22. The fact that the R-value is extremely near to 1 and the MSE value is low demonstrates the great accuracy with which the generated ANN models can predict the Nusselt number. For each flow regime, we have constructed ANN models and the procedure is supported graphically.

In particular, the validation that the training period, which began with the entrance of the data into the system, is ideally finished, is the first stage in the construction of ANN models. In Figure 2a, the ANN Model-I training performance graphs are created while the training performance of the ANN Model-II is developed in Figure 2b. Figure 2c provides the training performance of the ANN model-III. In networks with MLP design, the training cycle is repeated until there is the minimal error between the target data and the prediction data acquired in the output layer. With each epoch, the MSE values, which are large at the start of the training phase, go smaller. The findings shown in Figure 2a–c demonstrate that the constructed ANN models for predicting the Nusselt number have successfully completed their training phases. The examination of error histograms are a further step in evaluating the training performance of ANN models to forecast the Nusselt number. The error histograms for ANN models I, II, and III are shown in Figure 3a–c, respectively. The error histograms display the discrepancies between the goal values attained during the training phase and the anticipated values. The errors obtained for each ANN model are shown to cluster around the zero-error line, according to our observations. The numerical quantities of the inaccuracies are also relatively modest, which should be emphasized.

Table 22. Performance results for ANN models.

	MSE			*Rm*		
	Training	**Validation**	**Test**	**Training**	**Validation**	**Test**
ANN Model-I	9.88×10^{-3}	3.87×10^{-2}	3.24×10^{-3}	0.98624	0.96049	0.96282
ANN Model-II	3.34×10^{-4}	1.31×10^{-3}	2.11×10^{-3}	0.99993	0.99972	0.99201
ANN Model-III	3.59×10^{-4}	1.73×10^{-2}	1.78×10^{-3}	0.99993	0.99576	0.99861

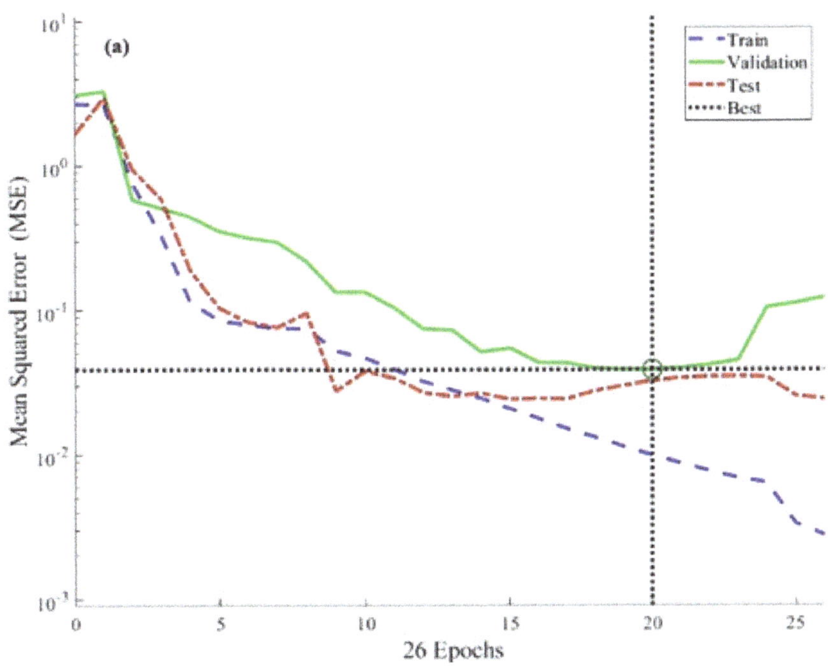

Figure 2. *Cont.*

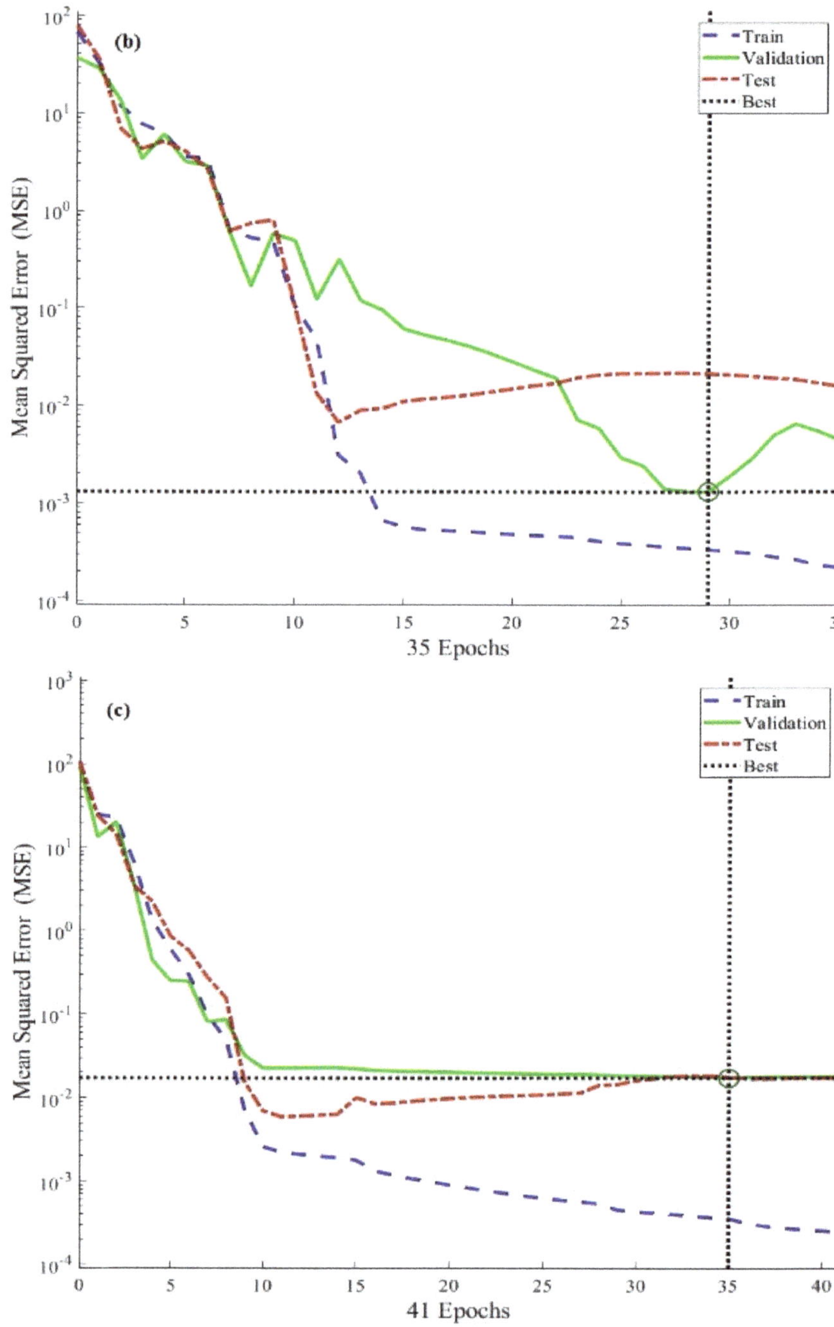

Figure 2. (**a**) Training performance of ANN model-I. (**b**) Training performance of ANN model-II. (**c**) Training performance of ANN model-III.

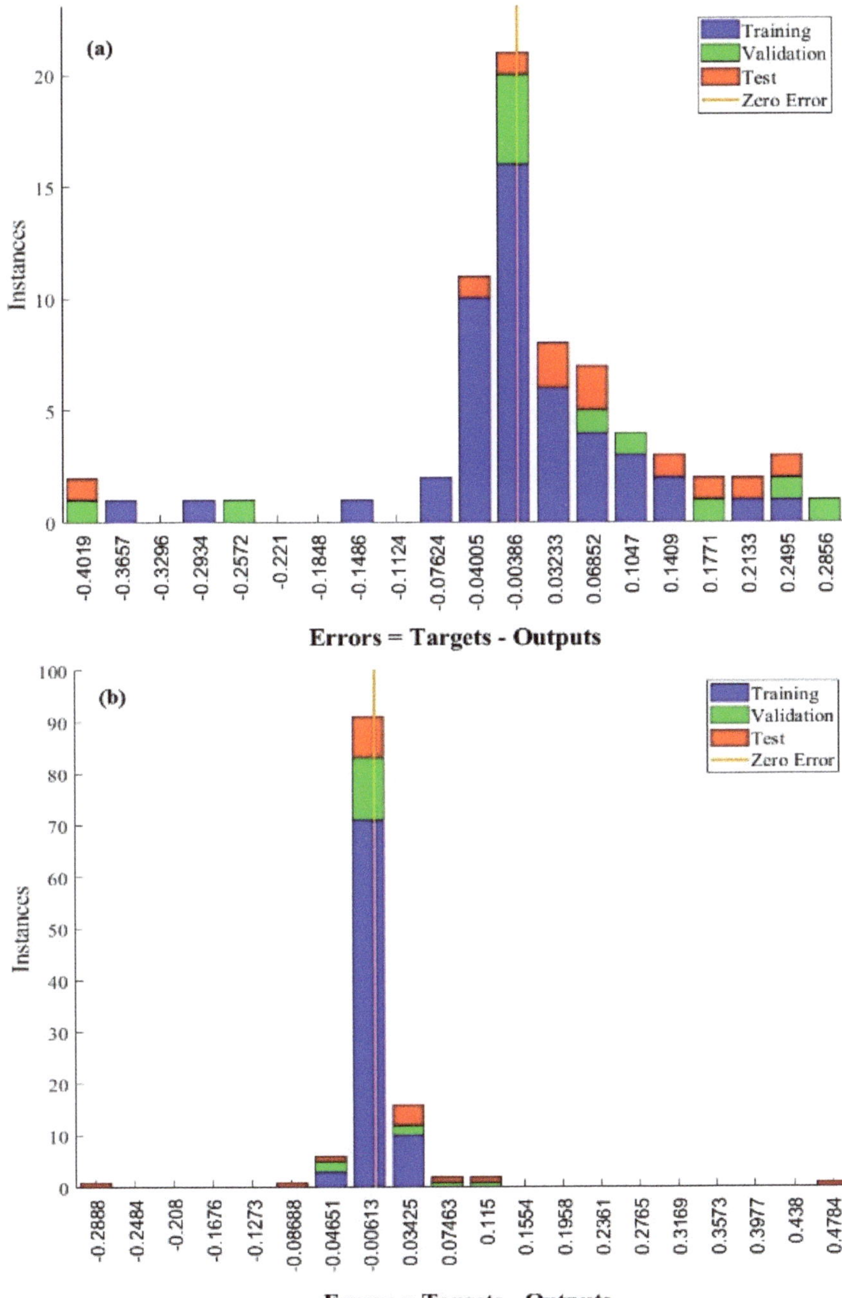

Figure 3. *Cont.*

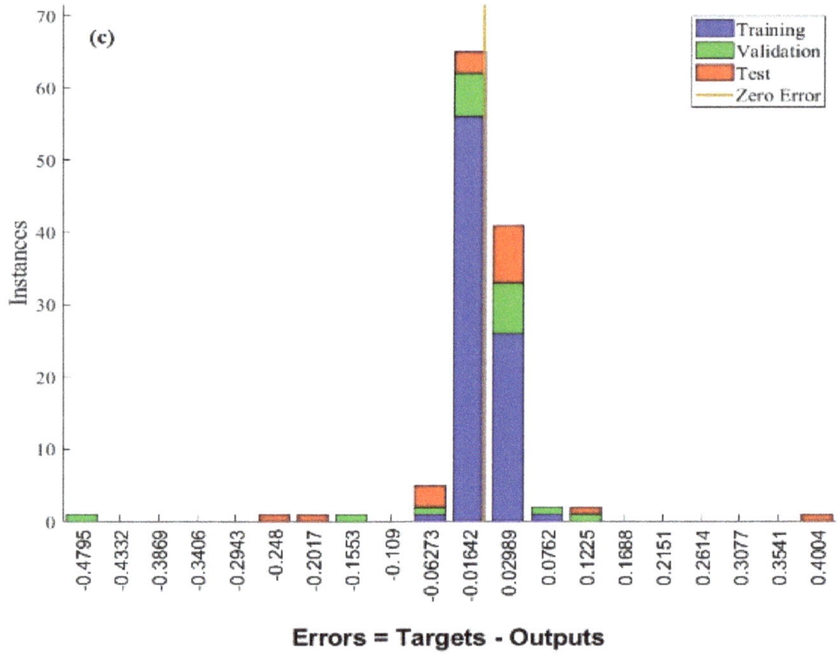

Figure 3. (**a**) Error histograms for of ANN model-I. (**b**) Error histograms for of ANN model-II. (**c**) Error histograms for of ANN model-III.

The findings from the error histograms demonstrate that a little error is carried out throughout the training stages of the three distinct ANN models that were created to predict the Nusselt number. Figure 4a–c depict the output values and target values obtained from the ANN models, designated as Model-I, Model-II and Model-III, each of which was developed using data sets with different data numbers.

In the analysis of each data point, it is evident that the goal data and the data from the ANN models (I,II,III) are in perfect harmony. The generated ANN models can estimate the Nusselt number with great accuracy, as demonstrated by the perfect match of the outputs derived from the ANN estimations with the target data. Figure 5a. Figure 5b,c display the MoD values that represent the proportional deviation between the target data and the outputs from three distinct ANN models created for predicting Nusselt number parameters based on various input parameters.

It can be noted that the data points are typically close to the zero-deviation line and have low values when the data points reflecting the MoD values for ANN Models I, II, and III are inspected. The average MoD values calculated for Model-I, Model-II and Model-III are obtained as 0.01%, 0.01% and 0.06%, respectively. The low MoD values show that there is relatively little variation between the goal values and the projected values derived from the created ANN models. In addition to the MoD values, the disparities between the target values and the ANN models' outputs are examined for each output value in Figure 6a–c. Each ANN model has, in general, relatively modest differences when the different values obtained for each data point utilized in ANN model training are taken into account. The findings from the analysis of MoD and difference values show that both ANN models (I,II,III) developed can predict Nusselt number with very low errors. Figure 7a–c titled Model-I, Model-II, and Model-III, respectively, illustrate the targeted and ANN outputs for each of the three ANN models. The data for each ANN model is often found on the zero-error line when the positions of the data points are taken into account. Additionally, it should be mentioned that the data points fall within a 10% error range. It is noticed that in the absence of magnetic field and heat generation effects, our problems reduced to Hayat et al. [31]. Additionally, for comparison, the Nusselt number is taken into consideration. In this direction, Table 23 is offered in this regard. A perfect match that yields the surety of the present study was found.

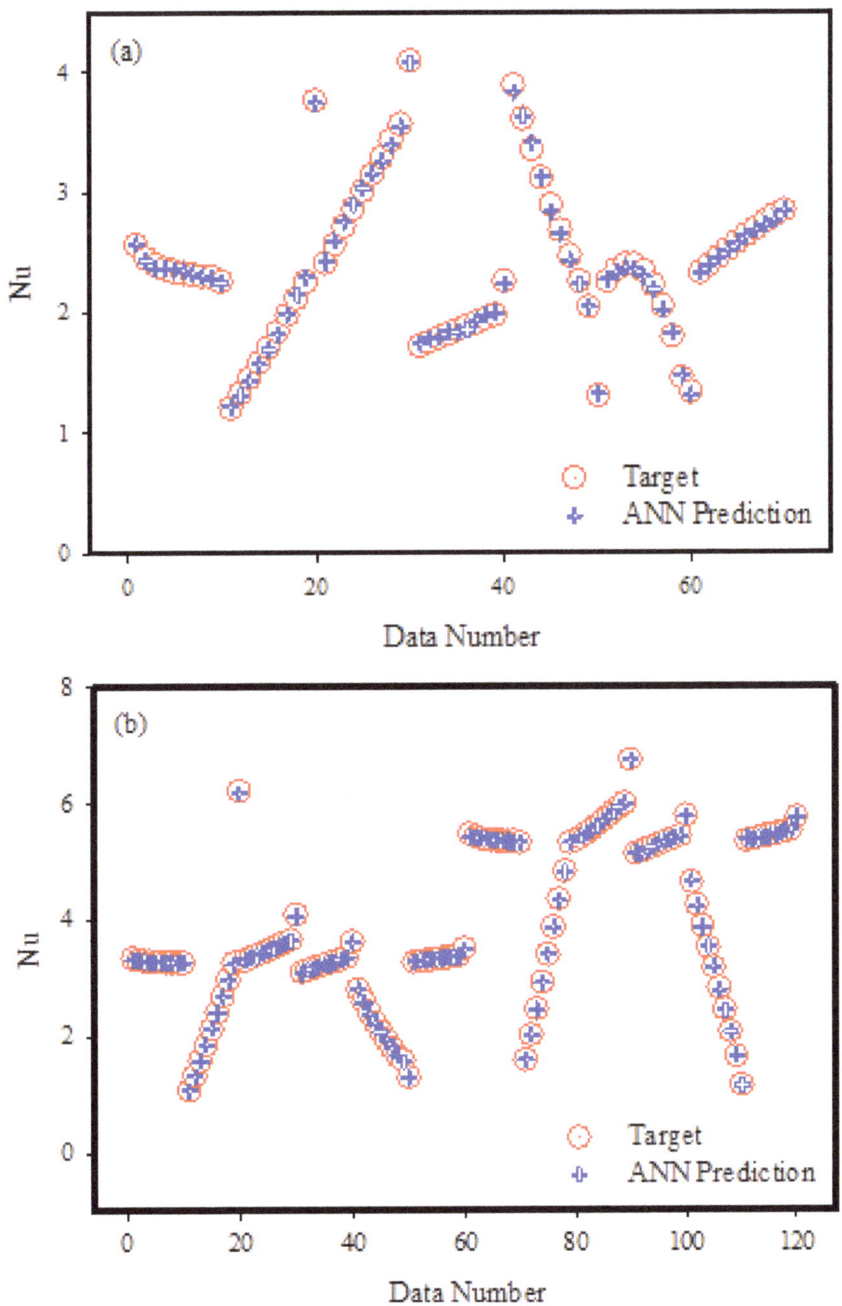

Figure 4. *Cont.*

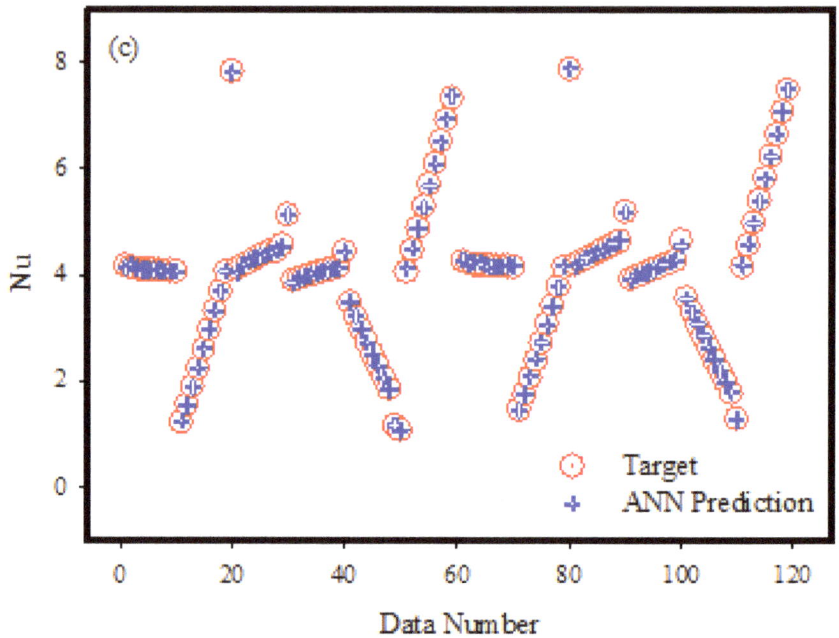

Figure 4. (**a**) The output values and target values obtained from ANN model-I. (**b**) The output values and target values obtained from ANN model-II. (**c**) The output values and target values obtained from ANN model-III.

Table 23. Comparison of Nusselt number with Hayat et al. [31].

ε	β	γ	**Hayat et al. [31]**	**Present Study**
0.0	1.0	0.2	0.5276	0.5054
0.0	1.4	0.2	0.5316	0.5203
0.0	1.8	0.2	0.5336	0.5124
0.0	2.0	0.0	0.5442	0.5220
0.0	2.0	0.12	0.5336	0.5213
0.0	2.0	0.19	0.5279	0.5016
0.0	2.0	0.19	0.5739	0.5216
0.2	2.0	0.19	0.5308	0.5124
0.3	2.0	0.19	0.5123	0.5061

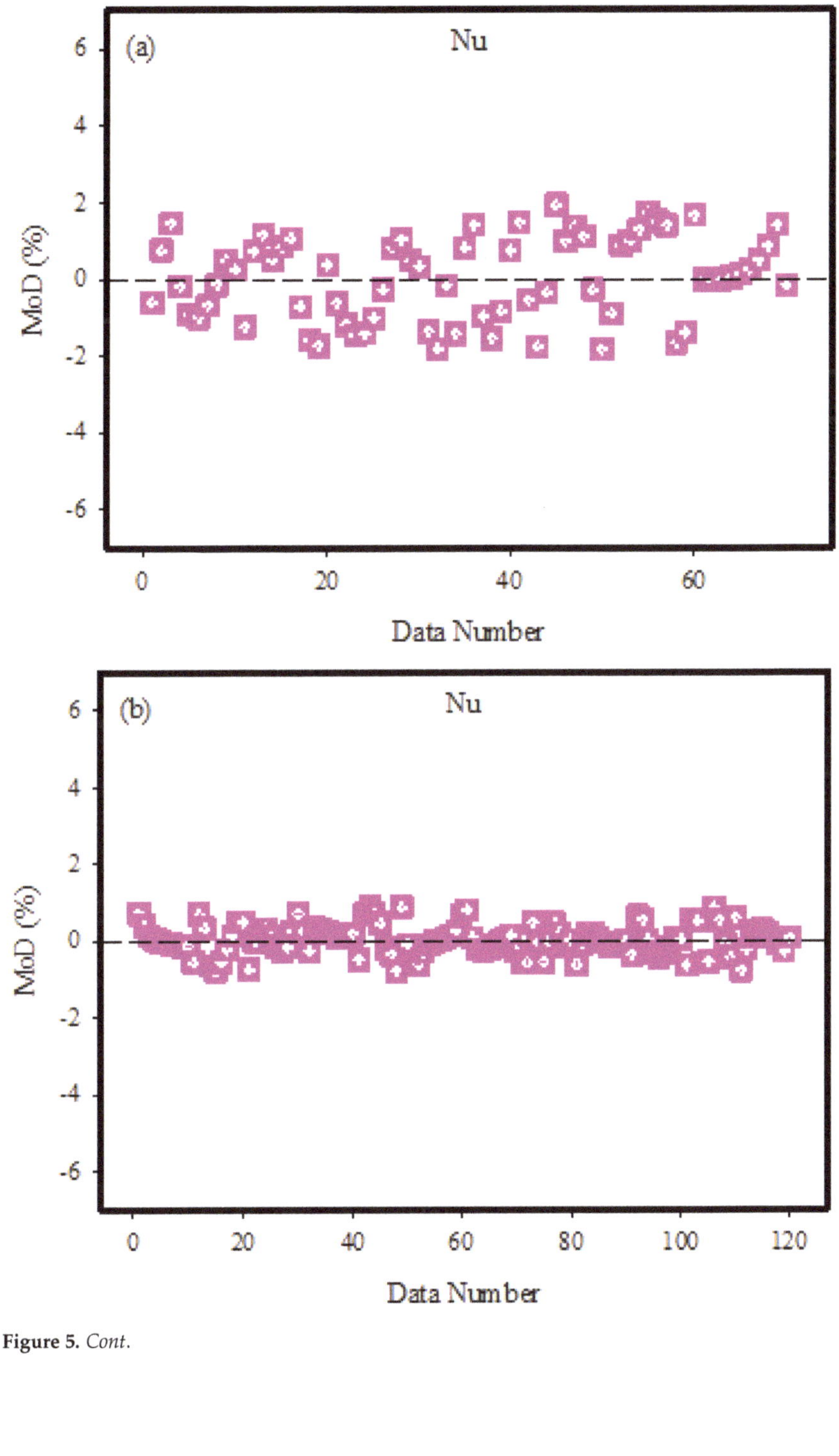

Figure 5. *Cont.*

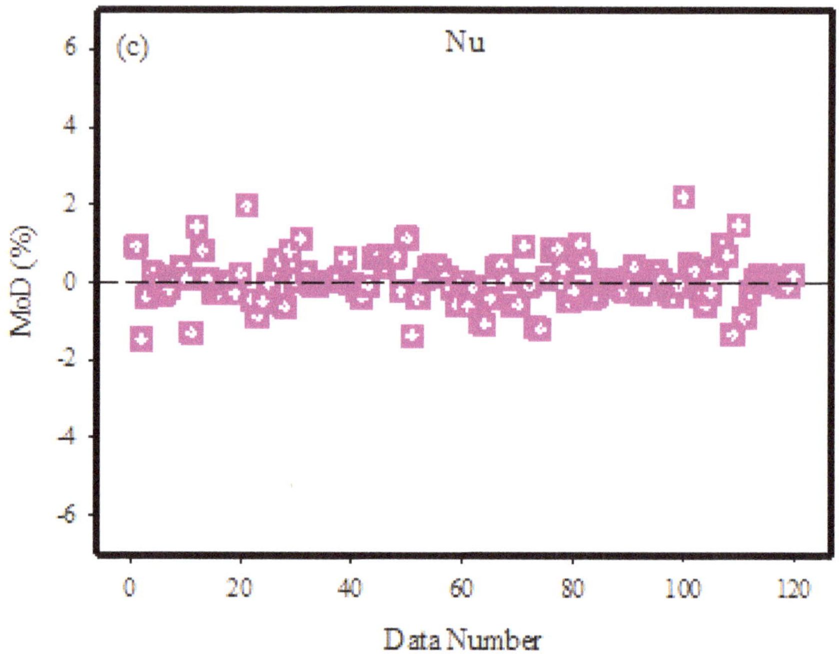

Figure 5. (**a**) The MoD values for ANN model-I. (**b**) The MoD values for ANN model-II. (**c**) The MoD values for ANN model-III.

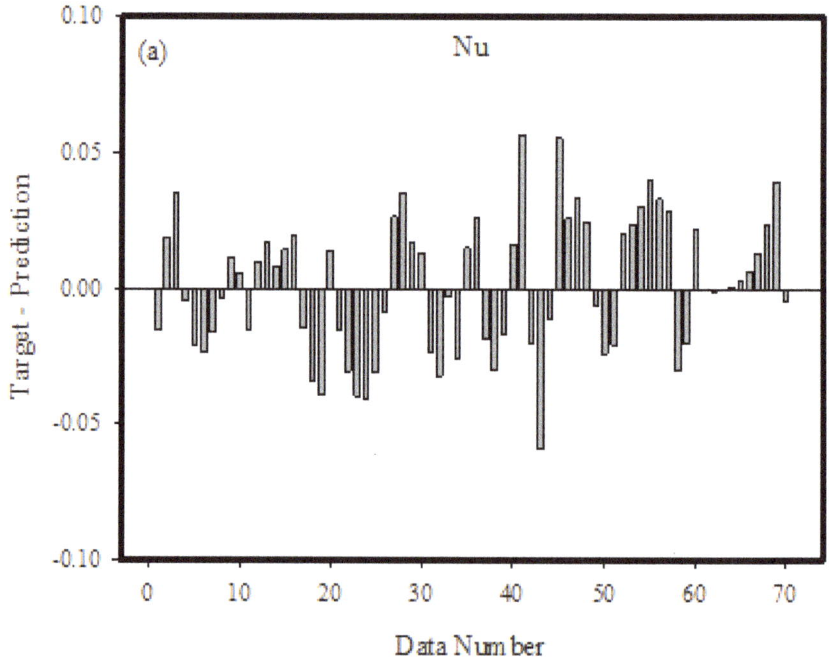

Figure 6. *Cont.*

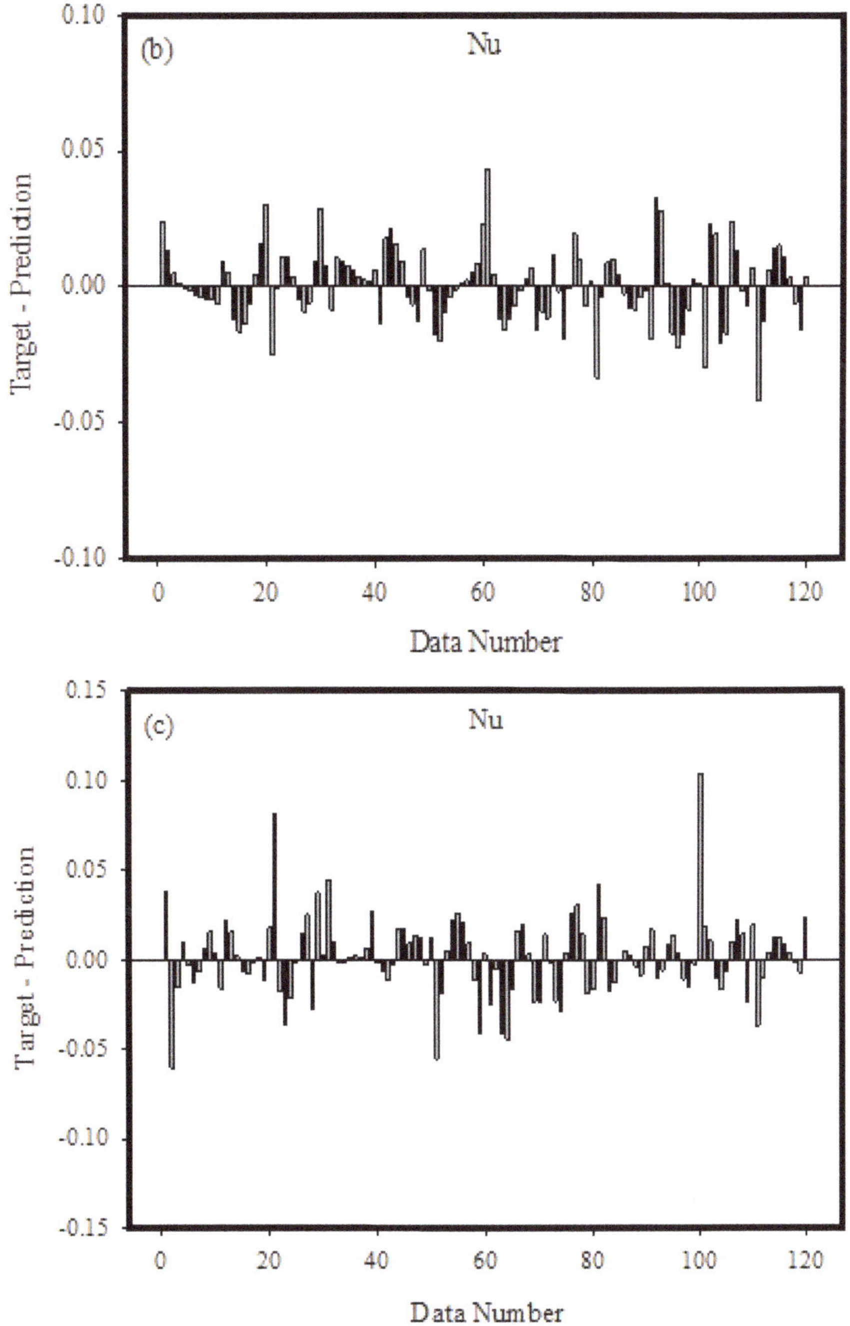

Figure 6. (**a**) The variations between the targeted values and the outputs of the ANN model-I. (**b**) The variations between the goal values and the outputs of the ANN model-II. (**c**) The variations between the goal values and the outputs of the ANN model-III.

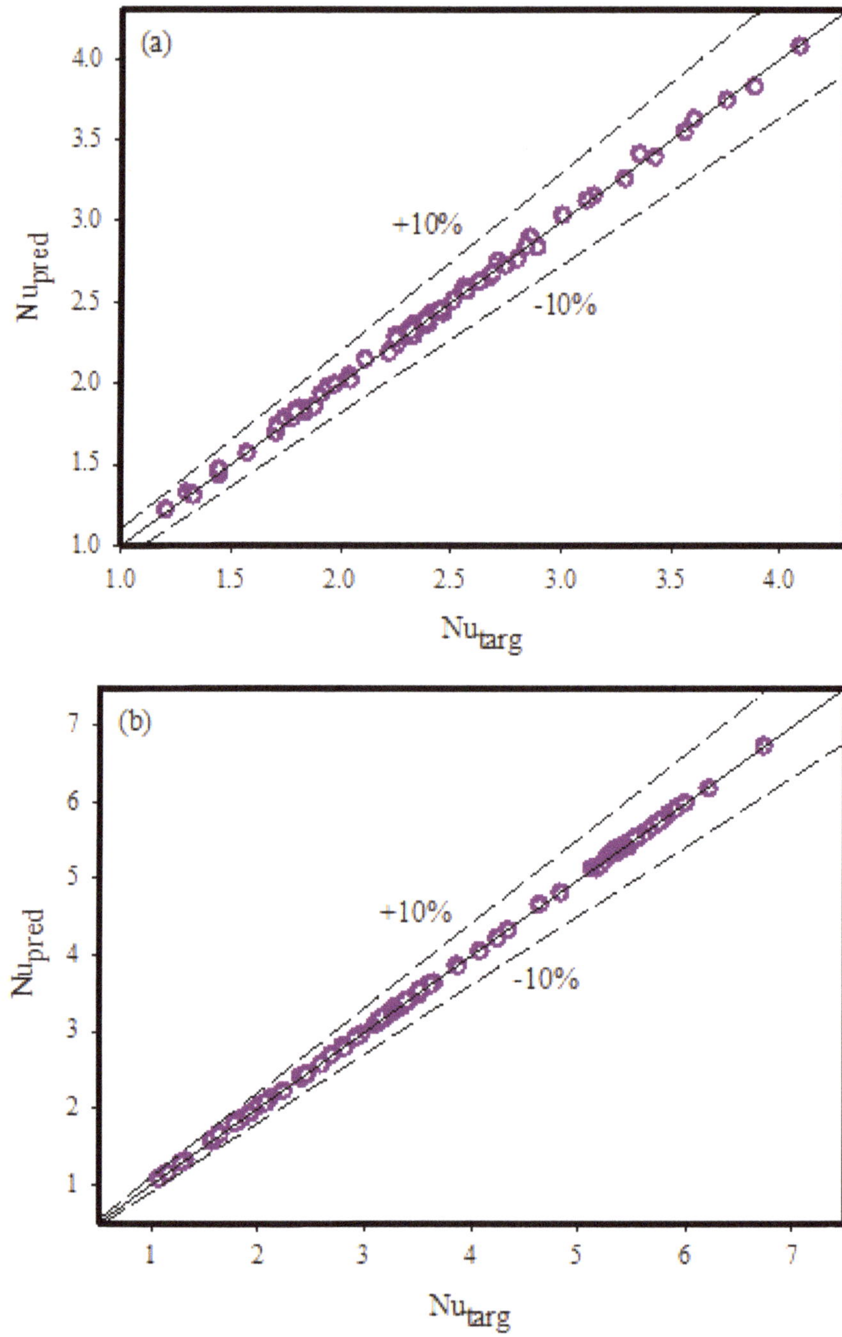

Figure 7. *Cont.*

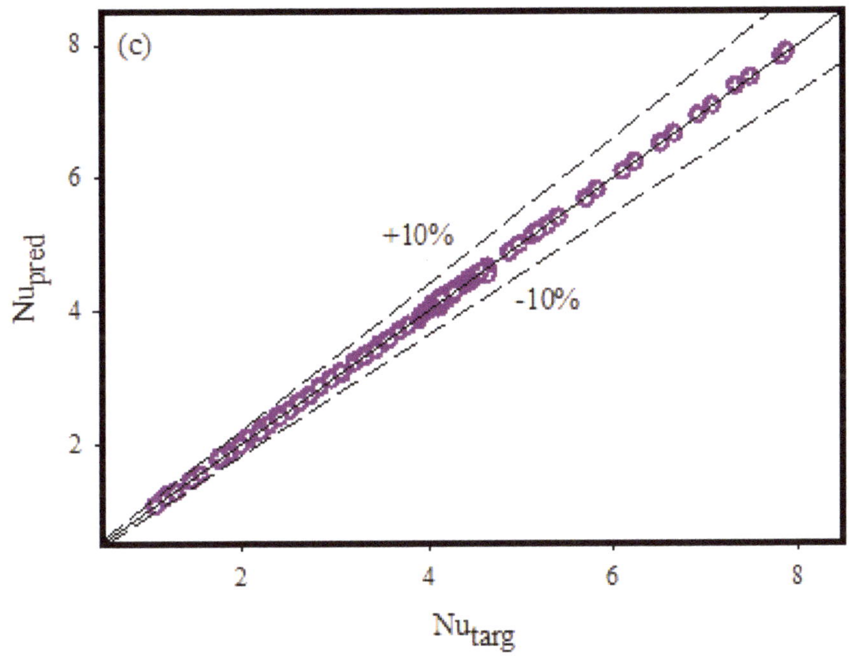

Figure 7. (**a**) The target and prediction values for ANN model-I. (**b**) The target and prediction values for ANN model-II. (**c**) The target and prediction values for ANN model-III.

6. Conclusions

An artificial neural networking models are developed to predict the heat transfer normal to the cylindrical surface for the incompressible flow of a two-dimensional mixed convection Casson fluid. The magnetic field is generated outside. In addition, it is presumed that the surface temperature is stronger than the surrounding fluid temperature. Further, the energy equation is carried with viscous dissipation, variable thermal conductivity, heat production, and thermal radiations. The following are the main results:

- Nusselt number shows inciting nature towards the Eckert number, curvature parameter, Prandtl number, and heat generation parameter
- Nusselt number admits declining trends toward the Casson fluid parameter, temperature-dependent thermal conductivity, and radiation parameters.
- The MSE and R values for Models I, II, and III are low and hence the developed ANN models can predict the Nusselt number with good accuracy.
- MoD outcomes show that there is not much of a discrepancy between the predicted and targeted values of the Nusselt number produced by the ANN models I, II, and III.
- The data points are often positioned on the zero-error line and fall within the 10% error region for ANN models I, II, and III to forecast the Nusselt number.
- Obtaining future data by using ANN models can provide many advantages in terms of both time and finance. In particular, obtaining specific parameters that can be obtained as a result of experimental studies by using ANN models can be considered an important advantage in both industrial applications and scientific studies.

Author Contributions: Conceptualization, K.U.R. and A.B.Ç.; Data curation, A.B.Ç.; Formal analysis, K.U.R. and W.S.; Investigation, K.U.R. and A.B.Ç.; Methodology, K.U.R.; Software, W.S.; Supervision, W.S. All authors have read and agreed to the published version of the manuscript.

Funding: This research received no external funding.

Institutional Review Board Statement: Not applicable.

Informed Consent Statement: Not applicable.

Data Availability Statement: The adopted methodology can be offered upon request by readers.

Acknowledgments: The authors would like to thank Prince Sultan University, Saudi Arabia, for the technical support through the TAS research lab.

Conflicts of Interest: The authors declare no conflict of interest.

Nomenclature

$\widetilde{u}, \widetilde{v}$	Velocity components	M	Magnetic field parameter
v	Kinematic viscosity	R	Radiation parameter
$\widetilde{x}, \widetilde{r}$	Cylindrical coordinates	G	Mixed convection parameter
β	Casson fluid parameter	Pr	Prandtl number
β_T	Thermal expansion coefficient	τ_r	Yield stress
g_0	Gravitational acceleration	A	Velocities ratio parameter
α	Angle of inclination	γ	Curvature parameter
$\widetilde{T}_\infty$	Ambient temperature	R	Radiation parameter
$\widetilde{T}$	Temperature of fluid	Ec	Eckert number
B_0	Magnetic field constant	Nu	Nusselt number
$\widetilde{u}_e$	Free stream velocity	σ^*	Stefan-Boltzmann constant
σ	Fluid electrical conductivity	Q_0	Heat generation
c_p	Specific heat at constant pressure	L	Characteristic length
ρ	Fluid density	ε	Small parameter
$\widetilde{q}$	Radiative heat flux	c	Radius of cylinder
κ	Variable thermal conductivity	U_0	Reference velocity
$\overline{\mu}$	Dynamic viscosity	$\widetilde{T}_w$	Surface temperature
$\theta(\eta)$	Fluid temperature		

References

1. Casson, N. A flow equation for pigment-oil suspensions of the printing ink type. In *Rheology of Disperse Systems;* Mill, C.C., Ed.; Pergamon Press: Oxford, UK, 1959; pp. 84–104.
2. Mustafa, M.; Hayat, T.; Ioan, P.; Hendi, A. Stagnation-point flow and heat transfer of a Casson fluid towards a stretching sheet. *Z. Für Nat. A* **2012**, *67*, 70–76. [CrossRef]
3. Mukhopadhyay, S. Casson fluid flow and heat transfer over a nonlinearly stretching surface. *Chin. Phys. B* **2013**, *22*, 074701. [CrossRef]
4. Mukhopadhyay, S.; Mondal, I.C.; Chamkha, A.J. Casson fluid flow and heat transfer past a symmetric wedge. *Heat Transf. Asian Res.* **2013**, *42*, 665–675. [CrossRef]
5. Pramanik, S. Casson fluid flow and heat transfer past an exponentially porous stretching surface in presence of thermal radiation. *Ain Shams Eng. J.* **2014**, *5*, 205–212. [CrossRef]
6. Mahdy, A. Heat transfer and flow of a Casson fluid due to a stretching cylinder with the Soret and Dufour effects. *J. Eng. Phys. Thermophys.* **2015**, *88*, 928–936. [CrossRef]
7. Abbas, Z.; Sheikh, M.; Motsa, S.S. Numerical solution of binary chemical reaction on stagnation point flow of Casson fluid over a stretching/shrinking sheet with thermal radiation. *Energy* **2016**, *95*, 12–20. [CrossRef]
8. Raju, C.S.K.; Hoque, M.M.; Sivasankar, T. Radiative flow of Casson fluid over a moving wedge filled with gyrotactic microorganisms. *Adv. Powder Technol.* **2017**, *28*, 575–583. [CrossRef]
9. Reddy, G.J.; Kethireddy, B.; Umavathi, J.C.; Sheremet, M.A. Heat flow visualization for unsteady Casson fluid past a vertical slender hollow cylinder. *Therm. Sci. Eng. Prog.* **2018**, *5*, 172–181. [CrossRef]
10. Sulochana, C.; Ashwinkumar, G.P.; Sandeep, N. Effect of frictional heating on mixed convection flow of chemically reacting radiative Casson nanofluid over an inclined porous plate. *Alex. Eng. J.* **2018**, *57*, 2573–2584. [CrossRef]
11. Ali, A.; Umar, M.; Bukhari, Z.; Abbas, Z. Pulsating flow of a micropolar-Casson fluid through a constricted channel influenced by a magnetic field and Darcian porous medium: A numerical study. *Results Phys.* **2020**, *19*, 103544. [CrossRef]
12. Gbadeyan, J.A.; Titiloye, E.O.; Adeosun, A.T. Effect of variable thermal conductivity and viscosity on Casson nanofluid flow with convective heating and velocity slip. *Heliyon* **2020**, *6*, e03076. [CrossRef]
13. Alizadeh, R.; Gomari, S.R.; Alizadeh, A.; Karimi, N.; Li, L.K. Combined heat and mass transfer and thermodynamic irreversibilities in the stagnation-point flow of Casson rheological fluid over a cylinder with catalytic reactions and inside a porous medium under local thermal nonequilibrium. *Comput. Math. Appl.* **2021**, *81*, 786–810. [CrossRef]

14. Jamshed, W.; Goodarzi, M.; Prakash, M.; Nisar, K.S.; Zakarya, M.; Abdel-Aty, A.H. Evaluating the unsteady Casson nanofluid over a stretching sheet with solar thermal radiation: An optimal case study. *Case Stud. Therm. Eng.* **2021**, *26*, 101160. [CrossRef]
15. Khan, M.R.; Al-Johani, A.S.; Elsiddieg, A.M.A.; Saeed, T.; Mousa, A.A.A. The computational study of heat transfer and friction drag in an unsteady MHD radiated Casson fluid flow across a stretching/shrinking surface. *Int. Commun. Heat Mass Transf.* **2022**, *130*, 105832. [CrossRef]
16. Ali, G.; Ali, F.; Khan, A.; Ganie, A.H.; Khan, I. A generalized magnetohydrodynamic two-phase free convection flow of dusty Casson fluid between parallel plates. *Case Stud. Therm. Eng.* **2022**, *29*, 101657. [CrossRef]
17. Ahmad, S.; Haq, S.U.; Ali, F.; Khan, I.; Nisar, K.S. Time fractional analysis of channel flow of couple stress Casson fluid using Fick's and Fourier's Laws. *Sci. Rep.* **2022**, *12*, 2956. [CrossRef]
18. Shahzad, H.; Wang, X.; Ghaffari, A.; Iqbal, K.; Hafeez, M.B.; Krawczuk, M.; Wojnicz, W. Fluid structure interaction study of non-Newtonian Casson fluid in a bifurcated channel having stenosis with elastic walls. *Sci. Rep.* **2011**, *12*, 12219. [CrossRef]
19. Xiao, B.; Wang, W.; Zhang, X.; Long, G.; Fan, J.; Chen, H.; Deng, L. A novel fractal solution for permeability and Kozeny-Carman constant of fibrous porous media made up of solid particles and porous fibers. *Powder Technol.* **2019**, *349*, 92–98. [CrossRef]
20. Liang, M.; Fu, C.; Xiao, B.; Luo, L.; Wang, Z. A fractal study for the effective electrolyte diffusion through charged porous media. *Int. J. Heat Mass Transf.* **2019**, *137*, 365–371. [CrossRef]
21. Hosseini, N.; Khoei, A.R. Modeling Fluid Flow in Fractured Porous Media with the Interfacial Conditions Between Porous Medium and Fracture. *Transp. Porous Media* **2021**, *139*, 109–129. [CrossRef]
22. Xiao, B.; Fang, J.; Long, G.; Tao, Y.; Huang, Z. Analysis of Thermal Conductivity of Damaged Tree-Like Bifurcation Network with Fractal Roughened Surfaces. *Fractals* **2022**, *30*, 2250104. [CrossRef]
23. Rehman, K.U.; Malik, M.Y.; Zehra, I.; Alqarni, M.S. Group theoretical analysis for MHD flow fields: A numerical result. *J. Braz. Soc. Mech. Sci. Eng.* **2019**, *41*, 156. [CrossRef]
24. Khan, I.; Rehman, K.U.; Malik, M.Y.; Aly, S. On magnetized non-Newtonian rotatory fluid flow field. *Adv. Mech. Eng.* **2019**, *11*, 1687814019878914. [CrossRef]
25. Çolak, A.B. An experimental study on the comparative analysis of the effect of the number of data on the error rates of artificial neural networks. *Int. J. Energy Res.* **2021**, *45*, 478–500. [CrossRef]
26. Öcal, S.; Gökçek, M.; Çolak, A.B.; Korkanç, M. A comprehensive and comparative experimental analysis on thermal conductivity of TiO_2-$CaCO_3$/Water hybrid nanofluid: Proposing new correlation and artificial neural network optimization. *Heat Transf. Res.* **2021**, *52*, 55–79. [CrossRef]
27. Çolak, A.; Güzel, T.; Yıldız, O.; Özer, M. An experimental study on determination of the shottky diode current-voltage characteristic depending on temperature with artificial neural network. *Phys. B Condens. Matter* **2021**, *608*, 412852. [CrossRef]
28. Rehman, K.U.; Shatanawi, W.; Çolak, A.B. Thermal analysis of flowing stream in partially heated double forward-facing step by using artificial neural network. *Case Stud. Therm. Eng.* **2022**, *37*, 102221. [CrossRef]
29. Çolak, A.B.; Karakoyun, Y.; Acikgoz, O.; Yumurtaci, Z.; Dalkilic, A.S. A numerical study aimed at finding optimal artificial neural network model covering experimentally obtained heat transfer characteristics of hydronic underfloor radiant heating systems running various nanofluids. *Heat Transf. Res.* **2022**, *53*, 51–71. [CrossRef]
30. Cao, Y.; Kamrani, E.; Mirzaei, S.; Khandakar, A.; Vaferi, B. Electrical efficiency of the photovoltaic/thermal collectors cooled by nanofluids: Machine learning simulation and optimization by evolutionary algorithm. *Energy Rep.* **2022**, *8*, 24–36. [CrossRef]
31. Hayat, T.; Asad, S.; Alsaedi, A. Flow of variable thermal conductivity fluid due to inclined stretching cylinder with viscous dissipation and thermal radiation. *Appl. Math. Mech.* **2014**, *35*, 717–728. [CrossRef]

mathematics

MDPI

Article

Computational Analysis on Magnetized and Non-Magnetized Boundary Layer Flow of Casson Fluid Past a Cylindrical Surface by Using Artificial Neural Networking

Khalil Ur Rehman [1,2,*], Wasfi Shatanawi [1,3] and Andaç Batur Çolak [4]

1 Department of Mathematics and Sciences, College of Humanities and Sciences, Prince Sultan University, Riyadh 11586, Saudi Arabia
2 Department of Mathematics, Air University, PAF Complex E-9, Islamabad 44000, Pakistan
3 Department of Mathematics, Faculty of Science, The Hashemite University, P.O. Box 330127, Zarqa 13133, Jordan
4 Information Technologies Application and Research Center, Istanbul Commerce University, Istanbul 34445, Turkey
* Correspondence: kurrehman@psu.edu.sa

Abstract: In this article, we constructed an artificial neural networking model for the stagnation point flow of Casson fluid towards an inclined stretching cylindrical surface. The Levenberg–Marquardt training technique is used in multilayer perceptron network models. Tan–Sig and purelin transfer functions are carried in the layers. For better novelty, heat and mass transfer aspects are taken into account. The viscous dissipation, thermal radiations, variable thermal conductivity, and heat generation effects are considered by way of an energy equation while the chemical reaction effect is calculated by use of the concentration equation. The flow is mathematically modelled for magnetic and non-magnetic flow fields. The flow equations are solved by the shooting method and the outcomes are concluded by means of line graphs and tables. The skin friction coefficient is evaluated at the cylindrical surface for two different flow regimes and the corresponding artificial neural networking estimations are presented. The coefficient of determination values' proximity to one and the low mean squared error values demonstrate that each artificial neural networking model predicts the skin friction coefficient with high accuracy.

Keywords: Casson fluid; mixed convection; thermal radiations; shooting method; artificial neural networking; Levenberg–Marquardt technique

MSC: 76R10; 76-10; 65K05

Citation: Rehman, K.U.; Shatanawi, W.; Çolak, A.B. Computational Analysis on Magnetized and Non-Magnetized Boundary Layer Flow of Casson Fluid Past a Cylindrical Surface by Using Artificial Neural Networking. *Mathematics* **2023**, *11*, 326. https://doi.org/10.3390/math11020326

Academic Editor: Ramoshweu Solomon Lebelo

Received: 28 October 2022
Revised: 3 January 2023
Accepted: 4 January 2023
Published: 8 January 2023

1. Introduction

Alfven [1] was the pioneer of the field of magnetohydrodynamics (MHD) and since the study of MHD is still a subject that researchers are quite interested in Due to its extensive applications in daily life, for example blood flow control during surgery, magnetic endoscopy, cell separation, magnetic devices, tumor treatment, and drug targeting to mention just a few. Collectively, MHD plays a key role in industrial and biomedical sciences [2]. Owing to such importance, various recent studies performed by researchers such as Mustafa [3] have studied magnetized viscous flow by way of nonlinear surfaces. It has been demonstrated that temperature and flow fields have a straightforward analytical expression. He offered precise formulations for wall shear stress. He concludes that strong magnetic fields thin both the momentum and temperature layers. Additionally, as opposed to lower branches, upper branch solutions were more thoroughly chilled, resulting in increased heat transfer rates. The magnetized fluid by way of a porous channel with a radiation assumption was investigated by Akinbowale [4]. Heat and mass transfer are examined in relation to important rheological parameters such as the magnetic and pressure gradient, the radiation parameter, and Prandtl and Reynolds numbers. It was found that increasing the pressure results in an increase in velocity, with the greatest effect occurring toward the center of the flow channel, whereas increasing the radiation parameter causes the temperature distribution to decrease, with the greatest effect occurring toward the electrically conducting wall. Hanumesh et al. [5] studied MHD peristaltic flow through an asymmetric tapered tube. Through a porous material, fluid with varied transport characteristics is transported. A low Reynolds number and a long wavelength were the fundamental assumptions used to formulate the problem. The momentum and energy equations' solutions were obtained using the perturbation method. The graphed answers show that a key factor in controlling the fluid velocity in the channel's center is the varying viscosity. The MHD fluid caused by an unstable stretched sheet with an expanded heat flux was into consideration by Ahmed et al. [6]. It was presumed that thermal conductivity and viscosity would change with the temperature. The flow equations were solved by an efficient shooting method and the Runge–Kutta algorithm. Graphical representations and in-depth analysis were carried out to examine the flow field. Liaqat et al. [7] studied heat transfer using self-propelled bioconvective microorganisms submerged in a water-based MHD nanofluid that included Cattaneo–Christov characteristics. Through Matlab programming, a finite element method was used to establish the numerical outcomes of the collection of non-linear equations. An important finding was that the density of the liquid was enhanced toward the melting factor. MHD micropolar tangent hyperbolic fluid flow toward the stretched surface was investigated by Pardeep et al. [8]. The collection of partial differential equations was transformed using similarity transformations to obtain the theoretically specified ordinary differential system. The issue was mathematically resolved using the bvp4c method. The major goal of this extensive investigation was to enhance heat transformation under the influence of numerous parameters. A number of physical factors were used to depict the heat transfer, skin friction, temperature, and velocity. It was discovered that changes in velocity and temperature profiles drove changes in the parameters that affected the size of the nanoparticles and the rate of heat transfer.

The Casson fluid model [9] has received a lot of attention from researchers due to its unique characteristics. Compared to conventional viscoplastic models, the Casson fluid model more closely matches rheological data for a variety of materials. Casson fluid, a shear-thinning fluid, is predicted to have a yield stress below which there is no flow, an infinite viscosity at a zero shear rate, and zero viscosity at an infinite shear rate. Casson fluids include intense fruit liquids, tomato sauce, soup, honey, and jellies. Furthermore, it is an approximate rheological model for chocolate and blood. Additionally, Casson fluid exhibits yield stress and is crucial in the biomechanics and polymer processing sectors. Owing to such importance, various researchers have considered the examination of the Casson flow field in various configurations such as Reddy et al. [10] who investigated the importance of the Soret and Hall effects on Casson fluid toward a vertical surface. The dimensional equations that control flow were converted into dimensionless equations by dimensionless variables, leading to the discovery of the analytical solution via the homotopy analysis method (HAM), which was then contrasted with the Adomain decomposition method (ADM) solution. With a particular focus on the physical factors involved in the current investigation, the heat and mass transfer rates against the Casson fluid parameter were visually illustrated. When the upper disk is assumed to be impermeable and the bottom one is assumed to be porous, Mohyud-Din and Khan [11] explored the time-dependent Casson fluid flow. The controlling equations were transformed by using transformations. The formulas for the temperature and velocity were obtained using HAM. The effects of several physical parameters were explored towards Eckert, squeeze numbers, and dimensionless length. The system's overall inaccuracy was calculated for both the suction and injection situations using Mathematica Package BVPh2.0. For emerging parameters, surface quantities were reported. For both the presence and absence of a magnetic field, Casson fluid was studied by Abro and Khan [12]. The Fabrizio–Caputo fractional derivative was used to obtain the flow formulation. Analytical solutions were identified. The Fox-H and Mittag–Leffler functions were used to express the generic solutions for the flow field. Finally, a graphical representation was provided using the relevant parameters, and it was seen that the behavior of the Caputo–Fabrizio and ordinary fractional fluid models for the fluid flow was reciprocal. The analysis of time independent naturally convective flow was identified by Kataria and Patel [13]. A vertical plate was passed over by a Casson fluid flow. The flow equations were resolved numerically in Matlab and resolved analytically using the Laplace transform method. Sherwood, Nusselt numbers, and skin friction expressions were discovered. By creating graphs, the properties of the flow field were examined, and the physical elements were thoroughly explained. The examination of the Casson liquid over a disk as a semi-infinite zone was presented by Rehman et al. [14]. The Casson nano-liquid flow was achieved by rotating a rigid disk at a fixed angular frequency. By creating a homogenous magnetic field normal to the axial direction, magnetic interaction was taken into consideration. The chemical reaction, heat generation, heat absorption, and Navier's slip condition were manifested during disk rotation. In order to create an ordinary differential system, the obtained flow narrating differential equations was reduced. The Von Karman method of the scheme was used to achieve this. Instead of continuing with the standard built-in system, a computational approach was developed to produce correct trends. By using graphical and tabular structures, the effects of the flow parameters were studied. It was found that the Casson fluid parameter caused both the tangential and radial velocities to decrease. Neeraja et al. [15] explored convective and viscous dissipation effects on magnetized Casson fluid. Using the gunshot method, the flow equations were resolved. The governing parameters affected the temperature, solid displacement, liquid velocity, and concentration. For the Casson parameter, the liquid velocity and consequently the solid displacement were reduced. When compared to the previous results, the current results showed a logical agreement. In the context of emerging mass and heat transfer technologies, Rasool et al. [16] examined the properties of Casson nanofluid flow via porous media across a non-linear stretching surface. The Darcy–Forchheimer relation allows for an incompressible viscous nanofluid of the Casson type to pass through the specified porous material. For the nanoparticles' velocity, temperature, and concentration, slip boundary conditions were applied. Attendance was made to Brownian diffusion and thermophoresis. To numerically solve the problem, a Runge–Kutta (RK) scheme of fourth order was used. Graphs were created for a range of progressive non-dimensionalized parameter values, and numerical data were used to examine changes in wall drag factor, heat transfer rates, and mass transfer rates. The results show that the porous media offer resistance to fluid flow and the strength of the inertial impact decreases the momentum boundary layer. The thermophoresis and Brownian motion were discovered to have a progressive relationship with temperature. For increasing values of the slip parameters, there is a decrease in the magnitude of the rate of heat and mass transfers. Over a horizontal plate the Casson nanofluid flow through use of the non-Darcy porous medium, Farooq et al. [17] reported their findings. By utilizing the proper non-similar transformations, the equations were converted into a dimensionless model. Through the use of bvp4c, local non-similarity was used to solve the dimensionless partial differential system. In-depth research was carried out on the effects of the newly discovered non-dimensional characteristics on the flow field. Additionally, the influences of variables on the skin friction and the rate of heat transfer were investigated. Finally, using publicly available data, comparisons between locally similar solutions and non-similar solutions were completed. Ramesh et al. [18] investigated the time-dependent and incompressible Casson squeezing flow in between disks. In the flow phenomena, the nanofluid theory (Buongiorno model) was realized. For the lower disks, concentration, temperature, and velocity slip were also included. The similarity functions were completed first to ultimate flow equations and they were solved by the RK-5 scheme. The results were presented in relation to the various physical quantities. A higher Reynolds number caused a decline in radial velocity. The vortex viscosity parameter first increases and subsequently decreases the microrotational field.

Owing to the importance of the Casson fluid model, artificial neural networking models are constructed for two different boundary layer flow regimes, namely non-magnetized and magnetized flow fields. The Casson fluid flow towards the stretching cylinder is mathematically formulated in the presence of a heat generation effect, viscous dissipation, mixed convection, temperature-dependent variable thermal conductivity, thermal radiations, and first-order chemical reaction effects. The ultimate flow equations are solved by the use of the shooting method. The skin friction coefficient (SFC) is estimated at the cylindrical surface. We constructed artificial neural networking models for better estimation of the skin friction coefficient. We believe that by following the present outcomes of the Casson fluid flow regime, one can extend the idea to investigate the time-independent shear rate and shear stress characteristics of molten chocolate, yogurt, blood, and many other culinary and biological materials. The present research contributes to answering the following concerns:

- Formulation of Casson fluid flow towards cylindrical surfaces with pertinent physical effects.
- Comparative examination of Casson velocity for magnetized and non-magnetized flow fields.
- Examination of Casson concentration for chemically reactive and non-reactive flow fields.
- Evaluation of the SFC at the cylindrical surface for non-magnetized and magnetized flow fields.
- Estimation of the SFC by using an artificial neural networking model.

2. Mathematical Formulation

The Casson fluid flow field is considered towards the inclined surface in the presence of mixed convection, a magnetic field, and stagnation point flow. Both heat and mass transfer aspects are considered for better novelty. Thermal effects, namely thermal radiations, viscous dissipation, variable thermal conductivity, and heat generation, are considered by way of an energy equation while the first-order chemical reaction effect is calculated by the use of a concentration equation. Both the concentration and temperature at the cylindrical surface are presumed higher in strength as compared to the field far away from the surface.

The geometry of the problem is given in Figure 1a. The ultimate mathematical equations [19,20] for the present problem are stated as follows:

$$\frac{\partial(\widetilde{R}\widetilde{U})}{\partial\widetilde{X}}+\frac{\partial(\widetilde{R}\widetilde{V})}{\partial\widetilde{R}}=0\,, \tag{1}$$

$$\begin{aligned}\widetilde{U}\tfrac{\partial\widetilde{U}}{\partial\widetilde{X}}+\widetilde{V}\tfrac{\partial\widetilde{U}}{\partial\widetilde{R}}&=\nu\left(1+\tfrac{1}{\beta}\right)\left(\tfrac{\partial^2\widetilde{U}}{\partial\widetilde{R}^2}+\tfrac{1}{\widetilde{R}}\tfrac{\partial\widetilde{U}}{\partial\widetilde{R}}\right)+g_0\beta_T(\widetilde{T}-\widetilde{T}_\infty)cos(\alpha)\\&+g_0\beta_C(\widetilde{C}-\widetilde{C}_\infty)cos(\alpha)+\widetilde{U}_e\tfrac{\partial\widetilde{U}_e}{\partial\widetilde{X}}-\tfrac{\sigma B_0^2}{\rho}(\widetilde{U}-\widetilde{U}_e)\,,\end{aligned} \tag{2}$$

$$\rho c_p\left(\widetilde{U}\frac{\partial\widetilde{T}}{\partial\widetilde{X}}+\widetilde{V}\frac{\partial\widetilde{T}}{\partial\widetilde{R}}\right)=\frac{1}{\widetilde{R}}\frac{\partial}{\partial\widetilde{R}}\left(\kappa\frac{\partial\widetilde{T}}{\partial\widetilde{R}}\right)-\frac{1}{\widetilde{R}}\frac{\partial}{\partial\widetilde{R}}\left(\widetilde{R}\overline{q}\right)+\overline{\mu}\left(1+\frac{1}{\beta}\right)\left(\frac{\partial\widetilde{U}}{\partial\widetilde{R}}\right)^2+Q_0\left(\widetilde{T}-\widetilde{T}_\infty\right). \tag{3}$$

$$\widetilde{U}\frac{\partial\widetilde{C}}{\partial\widetilde{X}}+\widetilde{V}\frac{\partial\widetilde{C}}{\partial\widetilde{R}}=D_m\frac{\partial^2\widetilde{C}}{\partial\widetilde{R}^2}-k_c(\widetilde{C}-\widetilde{C}_\infty). \tag{4}$$

In Equations (1)–(4), B_0 is the uniform magnetic field, D_m is the mass diffusivity, $\widetilde{C}$ is the concentration, β_C denotes the solutal expansion coefficient, β_T is the thermal expansion coefficient, and k_c is the rate of the chemical reaction,. The relation for radioactive heat flux is given as:

$$\overline{q}=-\frac{\partial\widetilde{T}}{\partial\widetilde{R}}\,\frac{16\sigma^*T_\infty^3}{3k^*}\,. \tag{5}$$

The variable thermal conductivity relation is given as:

$$\kappa\left(\widetilde{T}\right) = \left(\varepsilon\frac{\widetilde{T} - \widetilde{T}_\infty}{\Delta T} + 1\right)\kappa_\infty, \tag{6}$$

with

$$\Delta T = \widetilde{T}_w - \widetilde{T}_\infty. \tag{7}$$

The flow endpoint conditions are particularized as:

$$\widetilde{U}(\widetilde{X},\widetilde{R}) = \widetilde{U}_w = a\widetilde{X},\ \widetilde{V}(\widetilde{X},\widetilde{R}) = 0\ ,\widetilde{C} = \widetilde{C}_w,\ \widetilde{T} = \widetilde{T}_w,\ \text{at}\ \widetilde{R} = R_1\ , \tag{8}$$

$$\widetilde{U} = \widetilde{U}_e = d\widetilde{X},\ \widetilde{C} \to \widetilde{C}_\infty,\ \widetilde{T} \to \widetilde{T}_\infty,\ \text{ as }\ \widetilde{R} \to \infty\ . \tag{9}$$

To obtain a reduced differential system, we have:

$$\begin{gathered}\widetilde{U} = \widetilde{X}\tfrac{U_0}{L}F'_C(\eta),\ \widetilde{V} = -\tfrac{R_1}{R}\sqrt{\tfrac{\nu U_0}{L}}F_C(\eta),\\ \theta_C(\eta) = \tfrac{\widetilde{T}-\widetilde{T}_\infty}{\widetilde{T}_w-\widetilde{T}_\infty},\ \phi_C(\eta) = \tfrac{\widetilde{C}-\widetilde{C}_\infty}{\widetilde{C}_w-\widetilde{C}_\infty},\ \eta = \tfrac{\widetilde{r}^2-R_1{}^2}{2R_1}\sqrt{\tfrac{U_0}{\nu L}}\ .\end{gathered} \tag{10}$$

With Equation (10) in Equations (2)–(4), we obtain:

$$\begin{gathered}(1+1/\beta)(F_C'''(1+2\gamma\eta)+2\gamma F_C'') - F_C'^2 + F_C F_C'' + G_T\theta_C\cos(\alpha) + G_C\phi_C\cos(\alpha)\\ -M^2(F_C' - A) + A^2 = 0\ ,\end{gathered} \tag{11}$$

$$\begin{gathered}\left(1+\tfrac{4}{3}R\right)(\theta_C''(1+2\eta\gamma)+2\gamma\theta_C') + \varepsilon\left((\theta_C\theta_C'' + \theta_C'^2)(1+2\eta\gamma)+2\gamma\theta_C\theta_C'\right)\\ +\mathrm{Pr}E(1+2\eta\gamma)\left(1+\tfrac{1}{\beta}\right)F_C''^2 + \mathrm{Pr}H\theta_C + \mathrm{Pr}F_C\theta_C' = 0\ ,\end{gathered} \tag{12}$$

$$\phi_C''(1+2\eta\gamma) + 2\gamma\phi_C' + Scf\phi_C' - ScRc\phi_C = 0\ , \tag{13}$$

The reduced endpoint conditions are:

$$F_C = 0,\ F_C' = 1\ ,\theta_C = 1, \phi_C = 1, \text{at}\ \eta = 0 \tag{14}$$

$$F_C' = A,\ \theta_C = 0,\ \phi_C = 0, \text{as}\ \eta \to \infty\ . \tag{15}$$

The flow parameters are identified as:

$$\begin{gathered}\beta = \tfrac{\overline{\mu}\sqrt{2\pi_c}}{\tau_r},\ R = \tfrac{4\sigma^*\widetilde{T}_\infty^3}{\kappa k^*},\ A = \tfrac{d}{a}, \gamma = \sqrt{\tfrac{\nu L}{c^2 U_0}},\\ G_C = \tfrac{g_0\beta_C(\widetilde{C}_w-\widetilde{C}_\infty)L^2}{U_0\widetilde{x}},\ G_T = \tfrac{g_0\beta_T(\widetilde{T}_w-\widetilde{T}_\infty)L^2}{U_0\widetilde{x}}, M = \sqrt{\tfrac{\sigma B_0^2 L}{\rho U_0}},\ Pr = \tfrac{\overline{\mu}c_p}{\kappa},\\ E = \tfrac{U_0^2(\widetilde{x}/L)^2}{c_p(\widetilde{T}_w-\widetilde{T}_\infty)},\ H = \tfrac{LQ_0}{U_0\rho c_p}\ ,\ Sc = \tfrac{\nu}{D_m},\ \ Rc = \tfrac{k_c L}{U_0}.\end{gathered} \tag{16}$$

The chemical reaction parameter, radiation, magnetic field, heat generation parameters, Prandtl, Schmidt, Eckert, concentration Grashof, temperature Grashof numbers, Casson fluid parameter, curvature parameter, and velocities ratio parameter are symbolized as Rc, R, M, H, Pr, Sc, E, G_C, G_T, β, γ and A, respectively. For the present case, our interest lies in evaluating the SFC at the cylindrical surface. Therefore, the mathematical relationship for SFC is as followes:

$$\left.\begin{aligned} C_f &= \tfrac{2\tau_w}{\rho U_w^2}\ ,\\ \tau_w &= \overline{\mu}\left(\tfrac{\partial\widetilde{u}}{\partial\widetilde{r}}\right)_{\widetilde{r}=c},\\ \sqrt{\mathrm{Re}_{\widetilde{x}}}C_f &= \left(1+\tfrac{1}{\beta}\right)f''\ (0)\end{aligned}\right\}. \tag{17}$$

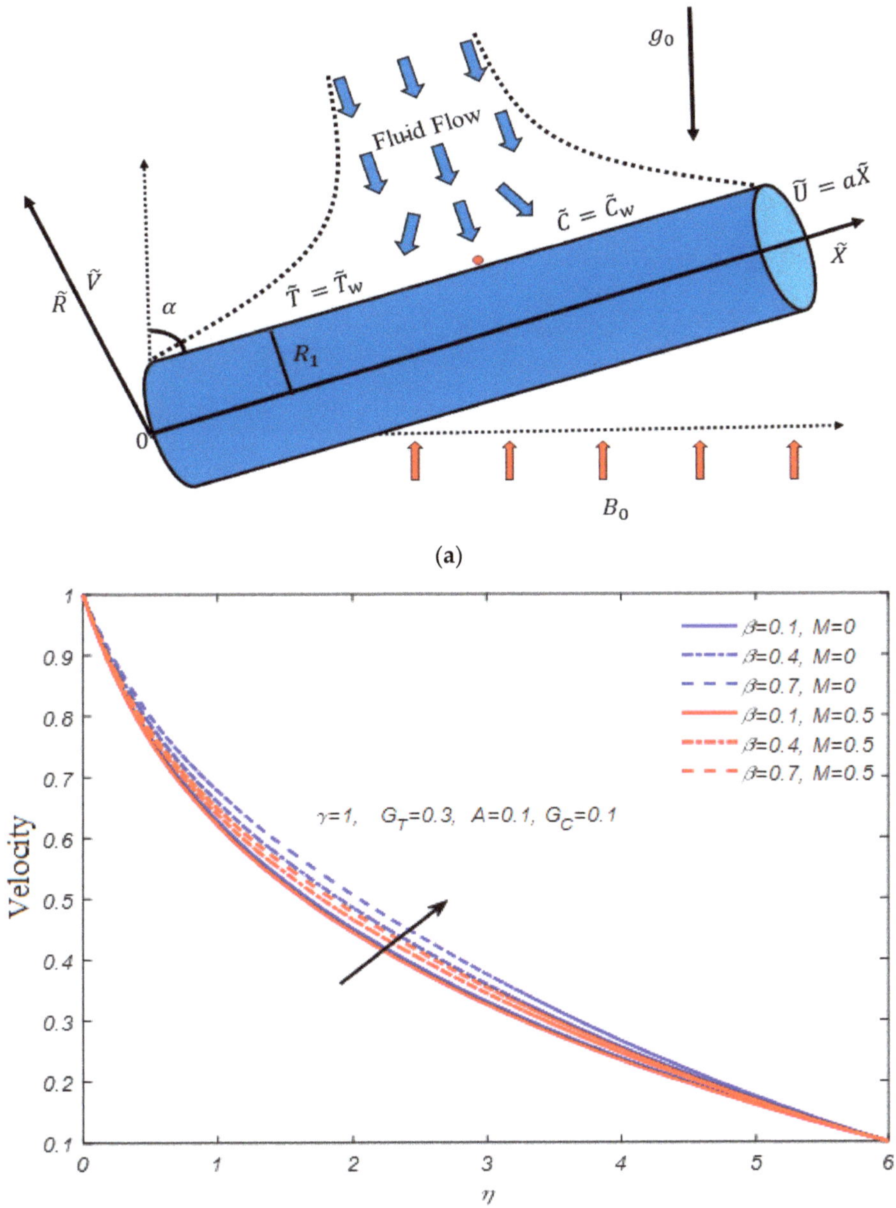

(a)

(b)

Figure 1. *Cont.*

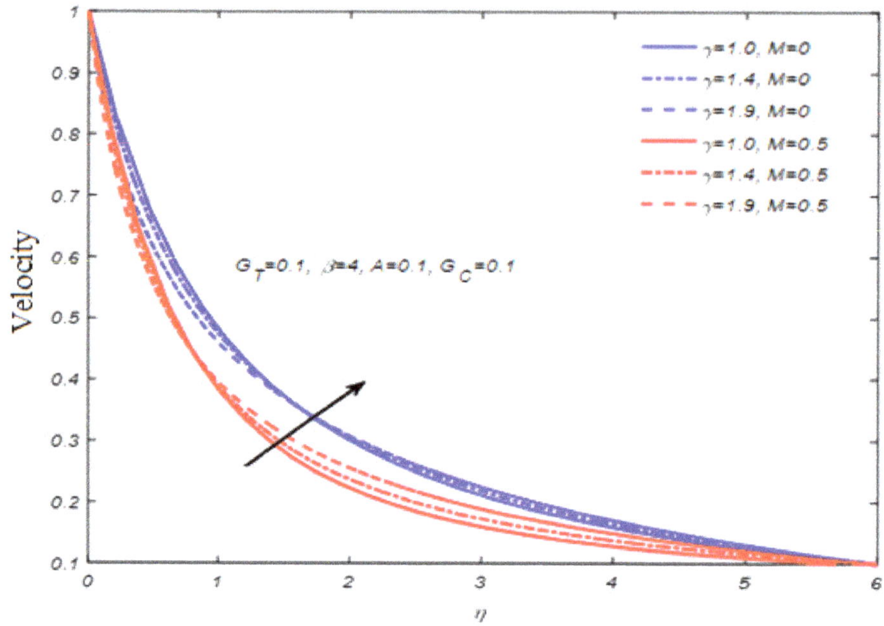

(c)

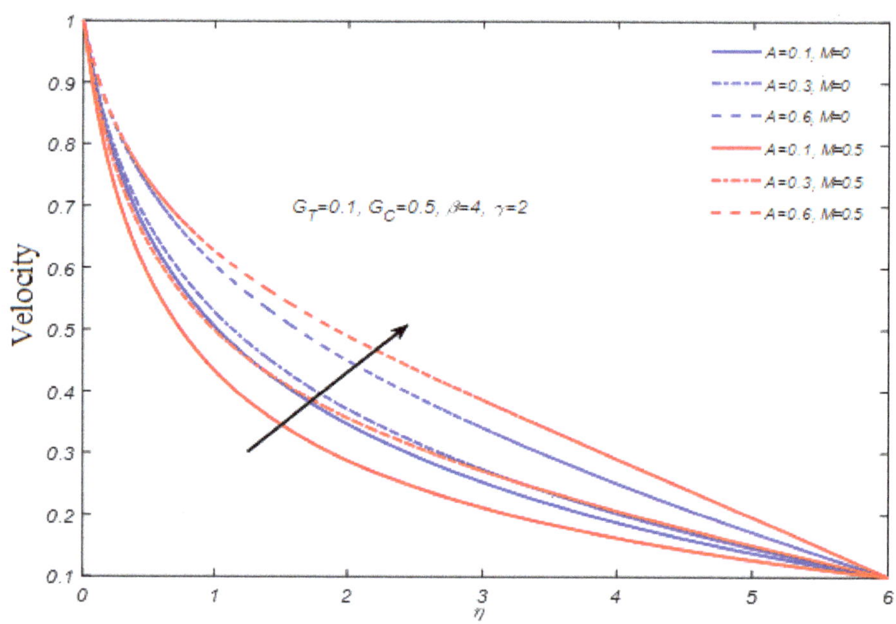

(d)

Figure 1. (**a**) Geometry of the problem. (**b**) Impact of β on $f'(\eta)$. (**c**) Impact of γ on $f'(\eta)$. (**d**) Impact of A on $f'(\eta)$.

3. Non-Magnetized Mathematical Model

The heat transfer characteristics in the magnetized flow field of Casson fluid at the cylindrical surface are considered. Mixed convection, stagnation point flow, viscous dissipation, heat generation, variable thermal conductivity, thermal radiation effects, and first-order chemical reaction effects are combined for better novelty. The flow narrating the differential system for a non-magnetized flow regime can be obtained by using $M = 0$ in Equations (11)–(15) and we have:

$$(1+1/\beta)(F_C'''(1+2\gamma\eta)+2\gamma F_C'')+F_C F_C''-F_C'^2+G_T\theta_C\cos(\alpha)+G_C\phi_C\cos(\alpha)+A^2=0\,, \tag{18}$$

$$\begin{aligned}&\left(1+\tfrac{4}{3}R\right)(\theta_C''(1+2\eta\gamma)+2\gamma\theta_C')+\varepsilon\left((\theta_C\theta_C''+\theta_C'^2)(1+2\eta\gamma)+2\gamma\theta_C\theta_C'\right)\\&+\mathrm{Pr}E(1+2\eta\gamma)\left(1+\tfrac{1}{\beta}\right)F_C''^2+\mathrm{Pr}H\theta_C+\mathrm{Pr}F_C\theta_C'=0\,,\end{aligned} \tag{19}$$

$$\phi_C''(1+2\eta\gamma)+2\gamma\phi_C'+Scf\phi_C'-ScRc\phi_C=0\,, \tag{20}$$

The boundary conditions for the present cases remain the same:

$$F_C=0,\ F_C'=1\,,\theta_C=1,\phi_C=1,\ \text{at}\ \eta=0 \tag{21}$$

$$F_C'=A,\ \theta_C=0,\ \phi_C=0,\ \text{as}\ \eta\to\infty\,. \tag{22}$$

It is important to note that for the evaluation of SFC, see Equation (17) to see that the mathematical relationship remains the same.

4. Solution Methodology

The flow equations are highly non-linear therefore one cannot obtain an exact solution. For the numerical solution, we choose the shooting scheme along with the Runge–Kutta algorithm due to higher convergence. To implement the shooting method [21,22], we need to transform Equations (11)–(15) into a set of first-order initial value systems (IVSs). To achieve such IVSs, we considered:

$$\begin{aligned}&Y_1=F_C(\eta),\ Y_2=F_C'(\eta),\quad Y_3=F_C''(\eta),\quad Y_4=\theta_C(\eta),\\&Y_5=\theta_C'(\eta),Y_6=\phi_C(\eta),\quad Y_7=\phi_C'(\eta),\end{aligned} \tag{23}$$

Using Equation (23) into Equations (11)–(15), we have:

$$Y_1'=Y_2, \tag{24}$$

$$Y_2'=Y_3, \tag{25}$$

$$Y'_3=\frac{1}{\left(1+\frac{1}{\beta}\right)(1+2\eta\gamma)}\left[\begin{array}{l}-2\gamma Y_3\left(1+\frac{1}{\beta}\right)+Y_2^2-Y_1Y_3-G_TY_4\cos\alpha-G_CY_6\cos\alpha\\-M^2(Y_2-A)-A^2\end{array}\right], \tag{26}$$

$$Y_4'=Y_5, \tag{27}$$

$$Y_5'=-\frac{1}{\left(1+\frac{4}{3}R\right)(1+2\eta\gamma)+\varepsilon(1+2\eta\gamma)Y_4}\left[\begin{array}{c}(1+\frac{4}{3}R)(2\gamma Y_5)+\varepsilon((1+2\eta\gamma)Y_5^2+2\gamma Y_4Y_5)+\\\mathrm{Pr}Y_1Y_5+\mathrm{Pr}E(1+2\eta\gamma)(1+\frac{1}{\beta})Y_3^2+\mathrm{Pr}HY_4\end{array}\right], \tag{28}$$

$$Y_6'=Y_7, \tag{29}$$

$$Y_7'=\frac{ScRcY_6-ScY_1Y_6-2\gamma Y_7}{(1+2\eta\gamma)}. \tag{30}$$

Meanwhile, the conditions are transformed as:

$$Y_6 = 1, Y_4 = 1, Y_2 = 1,\ Y_1 = 0, \quad \text{at} \quad \eta = 0, \qquad Y_6 \to 0, Y_4 \to 0\,,\ Y_2 \to A, \quad \text{as} \quad \eta = \infty. \tag{31}$$

5. Numerical Outcomes

The numerical outcomes are detailed as line graphs and tables. Figure 1b–d is plotted for the examination of velocity while Figure 2a,b offers the concentration outcomes. In detail, for the velocity analysis, two separate flow fields, namely the magnetic flow field and the non-magnetic flow field, are taken into consideration. We possess a non-magnetic flow field by selecting $M = 0$ and $M = 0.5$ for the magnetic regime. For velocity dependency, we focused on the following inputs, namely the curvature, Casson fluid [23], and velocities ratio parameters for these two regimes.

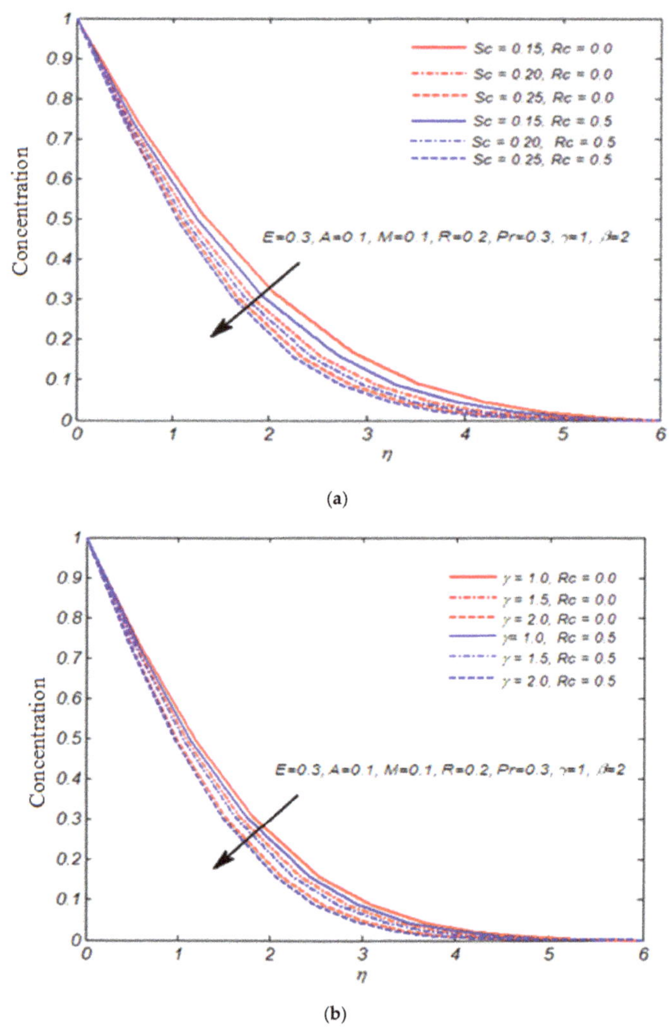

Figure 2. (**a**) Impact of Sc on $\phi(\eta)$. (**b**) Impact of γ on $\phi(\eta)$.

Figure 1b displays the influence of the Casson fluid parameter on the velocity in scenarios involving magnetic and non-magnetic fields. For higher Casson parameters, the velocity profile significantly increases. Both magnetic and non-magnetic fluid flows have the same effect. It is important to remember that greater Casson fluid parameter values cause the viscosity to decrease and as a result the velocity increases. In addition, it should be noted that the velocity is greater for a non-magnetic flow regime than it is in a magnetic flow field. When an external magnetic field was taken into consideration, the Lorentz force existed as a resistive force. The effect of the curvature parameter on the Casson fluid velocity is shown in Figure 1c for the flow regimes. The Casson fluid velocity noticeably increases for positive variations in the curvature parameter, and this effect is shared by both flow fields. This is due to the fact that the radius of the curvature decreases when the curvature parameter is iterated positively. As a result, there is less of a surface area in contact with the fluid flow, resulting in reduced resistance for the fluid particles. It is important to keep in mind that the strength of the velocity is higher for non-magnetic flow fields than for magnetic flow fields. This is because of Lorentz force's existence when a magnetic field is applied externally.

The effect of the velocities ratio is seen in Figure 1d. We have seen that the Casson fluid velocity is directly dependent on the velocities ratio parameter, and that larger values in the velocities ratio parameter led to higher magnitudes of fluid velocity. It is key to remember that the velocities ratio parameter is the ratio of free stream velocity to stretching velocity. When the ratio parameter has a value lower than one, it is inferred that the role of the free stream is less important than the stretching velocity. As a result, the inclined stretched cylindrical surface causes a considerable disturbance in fluid flow. It is necessary to keep in mind that the effect of the velocities ratio parameter on the Casson fluid flow is the same for the magnetic and non-magnetic flow fields for the stagnation point flow. The study of the impact of flow parameters on Casson fluid concentration is seen in Figure 2a,b. For the purpose of examining concentrations, we specifically took into account the reactive and non-reactive flows. We take $Rc = 0$ for a non-reactive flow field and $Rc = 0.5$ for a chemically reactive environment.

We took into account differences in the curvature parameter and Schmidt number for these two regimes. The combined effect of the Sc on the concentration for reactive and non-reactive scenarios is depicted in Figure 2a. We noticed that the concentration profile greatly declines for higher values of the Schmidt number. Both reactive and non-reactive fluid flow scenarios have the same effect. In addition, it is important to remember that the strength of the concentration profile is higher in the case of a reactive flow regime than it is in a non-reactive flow field. The influence of the curvature parameter on the concentration of Casson fluid is shown in Figure 2b for both reactive and non-reactive flow patterns.

The Casson fluid concentration decreases noticeably for positive variations In the curvature parameter, as we have seen. The effects are the same in both flow fields. This is because the radius of curvature decreases when we iterate the curvature parameter affirmatively. As a result, there is a reduced surface area in contact with the fluid flow, which lowers the resistance that the fluid particles must overcome. In the current flow problem, SFC is physical quantity at cylindrical surfaces; see Tables 1–8. The deviation in skin friction is noticed when chemically reactive Casson fluid flow at cylindrical surfaces is assumed along with magnetic, mixed convection, stagnation point flow, heat generation, viscous dissipation, and variable thermal conductivity effects. Tables 1–8 offer an investigation into the impact of flow parameters on skin friction for two distinct values of the magnetic field parameter, $M = 0$, and $M = 0.2$. In detail, Tables 1 and 2 provide an analysis of the effect of the Casson fluid parameter [24] on the SFC.

Here, $M = 0$ denotes the flow of non-magnetized Casson fluid, whereas $M = 0.2$ denotes the flow of magnetized Casson fluid. Tables 1 and 2 demonstrate that the SFC exhibits a decreasing trend as the Casson fluid parameter values increase. These findings hold true for both flow fields. It is noticeable that the magnetic flow field has stronger skin friction coefficients than the non-magnetic flow field. The fluctuation in the SFC for positive variations in the velocities ratio parameter is shown in Tables 3 and 4. Such observations are evaluated for both flow fields. We have seen that the SFC decreases noticeably for higher values of the velocities ratio parameter. Additionally, in the case of non-magnetized Casson fluid, the SFC is lower in magnitude. For two distinct values of the magnetic field parameter, Tables 5 and 6 display the impact of the curvature parameter on the SFC. We have seen that the SFC considerably rises with an increasing curvature parameter.

It should be noted that for magnetic flow fields, the skin friction coefficient values are higher. The fluctuation in the SFC for positive values of the thermal Grashof number is seen in Tables 7 and 8. We have seen that as for initial values of the thermal Grashof number, the skin friction coefficient tends to decrease but for positive values of the thermal Grashof number, the SFC increases. It can be seen that the strength of skin friction is substantially higher for non-magnetic flow fields than for magnetic flows.

Table 1. Impact of the Casson fluid parameter on the SFC for the nonmagnetic flow field.

β	$f''(0)$	$(1+1/\beta)f''(0)$	ANN Values
	$M = 0$	$M = 0$	$M = 0$
0.2	−0.8224	−4.9344	5.205703
0.3	−0.8633	−3.7409	3.64113
0.4	−0.8965	−3.1377	3.207106
0.5	−0.9241	−2.7723	2.685965
0.6	−0.9474	−2.5264	2.468694
0.7	−0.9673	−2.3492	2.21511
0.8	−0.9846	−2.2154	2.107406
0.9	−0.9998	−2.1106	2.041828
1.0	−1.0131	−2.0262	1.926957
2.0	−1.0928	−1.6392	1.526076

Table 2. Impact of the Casson fluid parameter on the SFC for the magnetic flow field.

β	$f''(0)$	$(1+1/\beta)f''(0)$	ANN Values
	$M = 0.2$	$M = 0.2$	$M = 0.2$
0.2	−0.8517	−5.1102	5.127272
0.3	−0.9017	−3.9074	3.906281
0.4	−0.9420	−3.1377	3.142552
0.5	−0.9752	−2.9256	2.913623
0.6	−1.0039	−2.6771	2.689821
0.7	−1.0271	−2.4944	2.491419
0.8	−1.0478	−2.3575	2.352806
0.9	−1.0659	−2.2503	2.252857
1.0	−1.0818	−2.1639	2.16358
2.0	−1.1763	−2.3526	2.370091

Table 3. Impact of the velocities ratio parameter on the SFC for the nonmagnetic flow field.

A	$f''(0)$	$(1+1/\beta)f''(0)$	ANN Values
	$M = 0$	$M = 0$	$M = 0$
0.2	−0.4365	−1.8913	1.798471
0.3	−0.4153	−1.7995	1.775878
0.4	−0.3859	−1.6721	1.761253
0.5	−0.3486	−1.5105	1.47786
0.6	−0.3036	−1.3154	1.397436
0.7	−0.2511	−1.0881	1.114926
0.8	−0.1914	−0.8293	0.84088
0.9	−0.1245	−0.5395	0.560701
1.0	−0.0497	−0.2154	0.221825
2.0	−0.0097	−0.0421	0.044358

Table 4. Impact of the velocities ratio parameter on the SFC for the nonmagnetic flow field.

A	$f''(0)$	$(1+1/\beta)f''(0)$	ANN Values
	$M = 0.2$	$M = 0.2$	$M = 0.2$
0.2	−0.4757	−2.0612	2.061184
0.3	−0.4457	−1.9312	1.937046
0.4	−0.4081	−1.7683	1.776405
0.5	−0.3633	−1.5742	1.580536
0.6	−0.3114	−1.3494	1.351205
0.7	−0.2526	−1.0945	1.090686
0.8	−0.1871	−0.8107	0.81182
0.9	−0.1146	−0.4965	0.498124
1.0	−0.0344	−0.1491	0.150031
2.0	−0.0067	−0.0291	0.028967

Table 5. Impact of the curvature parameter on the SFC for the nonmagnetic flow field.

γ	$f''(0)$	$(1+1/\beta)f''(0)$	ANN Values
	$M = 0$	$M = 0$	$M = 0$
0.2	−0.5089	−2.2052	2.054256
0.3	−0.5595	−2.4243	2.433774
0.4	−0.6068	−2.6293	2.688039
0.5	−0.6523	−2.8264	2.789083
0.6	−0.6965	−3.0179	3.168692
0.7	−0.7396	−3.2046	3.04553
0.8	−0.7820	−3.3884	3.265665
0.9	−0.8236	−3.5686	3.535665
1.0	−0.8647	−3.7467	3.75129
2.0	−1.2520	−5.4249	5.194255

Table 6. Impact of the curvature parameter on the SFC for the magnetic flow field.

γ	$f''(0)$	$(1+1/\beta)f''(0)$	ANN Values
	$M = 0.2$	$M = 0.2$	$M = 0.2$
0.2	−0.5547	−2.4035	2.397728
0.3	−0.6038	−2.6163	2.618119
0.4	−0.6499	−2.8161	2.820131
0.5	−0.6944	−3.0088	3.010963
0.6	−0.7377	−3.1964	3.195347
0.7	−0.7800	−3.3797	3.376525
0.8	−0.8216	−3.5599	3.55692
0.9	−0.8626	−3.7376	3.738581
1.0	−0.9030	−3.9126	3.923491
2.0	−1.2857	−5.5709	5.600951

Table 7. Impact of the temperature Grashof number on the SFC for the nonmagnetic flow field.

G_T	$f''(0)$	$(1+1/\beta)f''(0)$	ANN Values
	$M = 0$	$M = 0$	$M = 0$
0.2	−0.3822	−1.6561	1.545852
0.3	−0.3107	−1.3463	1.296924
0.4	−0.2319	−1.0048	1.027428
0.5	−0.1363	−1.0016	0.938263
0.6	−1.1515	−4.9894	5.305239
0.7	−1.1315	−4.9027	4.950683
0.8	−1.1115	−4.8162	4.683114
0.9	−1.0715	−4.6428	4.742156
1.0	−1.0915	−4.7295	4.449012
2.0	−2.0515	−8.8892	8.987435

Table 8. Impact of the temperature Grashof number on the SFC for the magnetic flow field.

G_T	$f''(0)$	$(1+1/\beta)f''(0)$	ANN Values
	$M = 0.2$	$M = 0.2$	$M = 0.2$
0.2	−0.8802	−3.8139	3.786153
0.3	−0.8563	−3.7103	3.708539
0.4	−0.8324	−3.6067	3.609377
0.5	−0.8084	−3.5027	3.498244
0.6	−0.7845	−3.3993	3.403159
0.7	−0.7606	−3.2956	3.298737
0.8	−0.7367	−3.1921	3.187463
0.9	−0.7128	−3.0885	3.082916
1.0	−0.6889	−2.9850	2.986601
2.0	−0.4509	−1.9537	1.955074

6. Artificial Neural Networking Outcomes

The Casson flow field was mathematically modelled and solved by use of the shooting method. At the surface, skin friction is a quantity of interest. We evaluated the values of the skin friction coefficient at the cylindrical surface by assuming two different flow regimes.

In the first case, we considered the flow regime without an externally applied magnetic field while in another case the flow regime was assumed in the presence of the externally applied magnetic field. For both cases, we have developed an artificial neural networking (ANN) model. For a non-magnetic flow field, we selected $M = 0$, and for the magnetic flow field, we selected $M = 0.2$. Multilayer perceptron (MLP) is used because of its strong structural characteristics. MLP networks are made up of layers, and each layer is coupled to the next. Because the input factors influencing each estimated skin friction coefficient

(SFC) value vary, two distinct ANN models were created, with distinct input and output values produced for each. Figure 3a depicts the MLP architecture, which depicts the layered structure of the produced ANN models. Table 9 shows the input and output parameters for each generated model. It is critical to optimize the data set used in the construction of ANN models [25]. Seventy percent of the data set used to create the two separate models was set aside for training, 15% for validation, and 15% for testing [26]. One of the challenges in developing ANN models is the lack of a rule for identifying the neurons in the hidden layer [27]. Figure 3b,c depicts the structural topologies of two different ANN models. From both figures, one can see that four input values are defined in the input layer of each of the given structural topologies, while the number of neurons in the hidden layers is 20 and 10, respectively. In the output layer, it is seen that the SFC parameter, which is the only value, is obtained. For ANN models, it is important to ideally optimize the data set being used [28]. Seventy percent of the data set used for the two different models developed was reserved for training the model, 15% for testing, and 15% for validation. The lack of a set formula for calculating the number of neurons in the hidden layer is one of the challenges in designing ANN models. For this reason, the performance of ANN models with different numbers of neurons in the hidden layer has been examined and the number of neurons in the hidden layer has been ideally optimized.

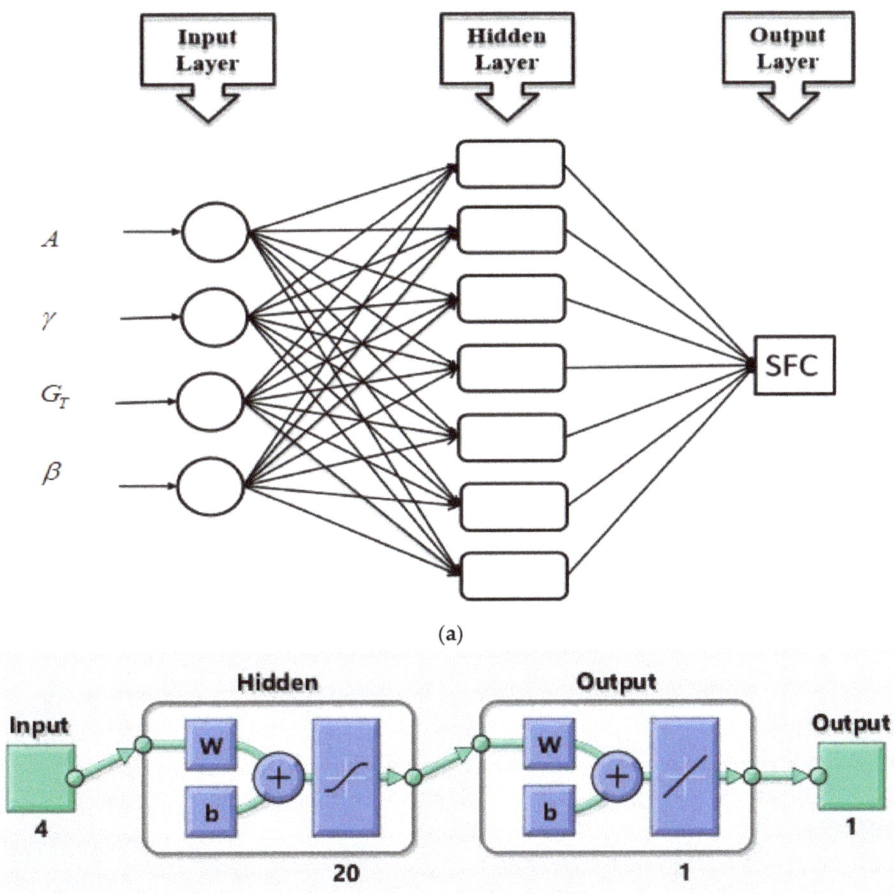

(a)

(b)

Figure 3. *Cont.*

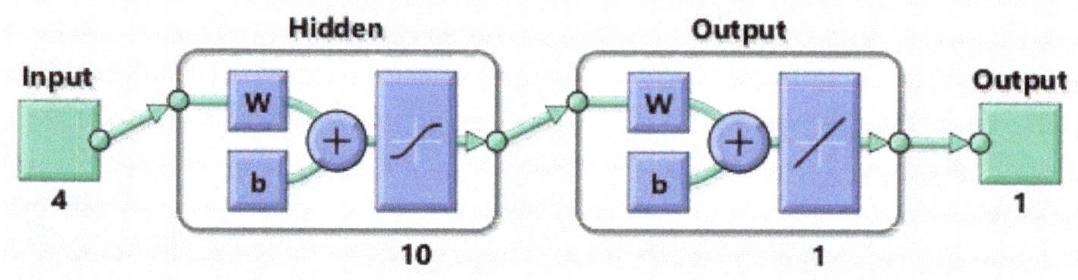

(c)

Figure 3. (**a**) The layered structure of the developed ANN models for SFC. (**b**) Structural topology of the ANN model for the non-magnetized flow field. (**c**) Structural topology of the ANN model for the magnetized flow field.

Table 9. The output and input parameters for each developed model.

Model	Inputs				Output
Model 1 (M = 0)	A	γ	G_T	β	SFC
Model 2 (M = 0.2)	A	γ	G_T	β	SFC

Table 10 contains information about the data set utilized in each ANN model as well as the number of neurons in each models hidden layer. The Levenberg–Marquardt training technique, which is extensively utilized due to its great learning performance, is used in MLP network models. The hidden and output layers both have access to purelin and tan–sig transfer functions. The following are the mathematical expressions for the transfer functions utilized:

$$f(x) = \frac{1}{1 + \exp(-x)}, \tag{32}$$

$$\text{purelin}(x) = x. \tag{33}$$

Table 10. Information about the data set used in each ANN model and the number of neurons.

Model	Number of Neuron	Training	Validation	Test	Total
Model 1 (M = 0)	20	28	6	6	40
Model 2 (M = 0.2)	10	28	6	6	40

Mean squared error (MSE), coefficient of determination (R), and margin of deviation (MoD) parameters, which are extensively used in the literature, were chosen to evaluate the performance of two different ANN models [29,30]. The mathematical equations used to calculate the performance parameters are as follows:

$$\text{MSE} = \frac{1}{N}\sum_{i=1}^{N}\left(X_{\text{targ(i)}} - X_{\text{pred(i)}}\right)^2, \tag{34}$$

$$\text{R} = \sqrt{1 - \frac{\sum_{i=1}^{N}\left(X_{\text{targ(i)}} - X_{\text{pred(i)}}\right)^2}{\sum_{i=1}^{N}\left(X_{\text{targ(i)}}\right)^2,}}, \tag{35}$$

$$\text{MoD}(\%) = \left[\frac{X_{\text{targ}} - X_{\text{pred}}}{X_{\text{targ}}}\right] \times 100. \tag{36}$$

The training accuracy of the ANN models designed to predict SFC values in two different situations has been thoroughly studied. Training performance graphs for two ANN models are shown in Figure 4a,b for $M = 0$ and $M = 0.2$, respectively. The graphs in both figures depict the training cycle (epoch) that occurs in an MLP network. Furthermore, the MSE values of each ANN model are greater at the start of the training period and drop as the model progresses.

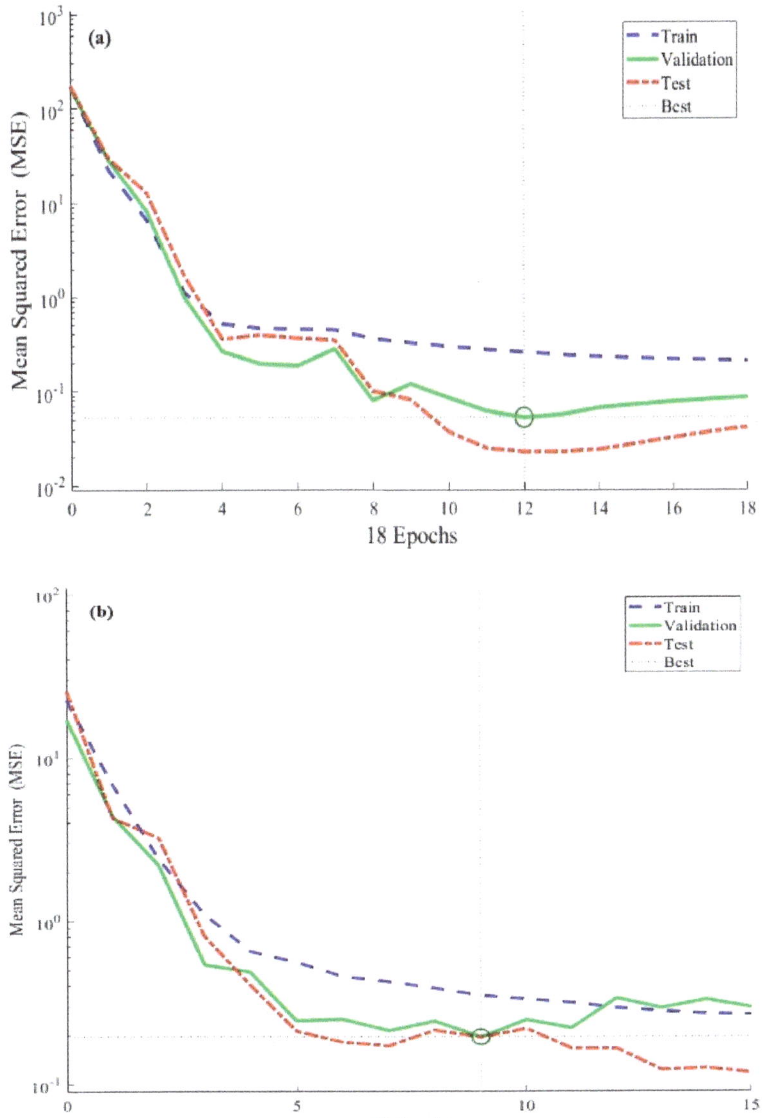

Figure 4. (**a**,**b**). The training performance graph of the ANN model for $M = 0$ and $M = 0.2$.

It is seen that the ANN model's training phases ended with the MSE values obtained for each data set meeting the most optimal position. The error histograms for the ANN model of both magnetized ($M = 0.2$) and non-magnetized ($M = 0$) flow fields are shown in Figure 5a,b. It should be observed that the calculated error rates for three different data sets are positioned quite close to the zero-error line in the error histograms. The error levels in the error histograms are also quite low. The results of the performance and error histograms reveal that the training stages of the ANN models designed for estimating SFC values have been completed optimally.

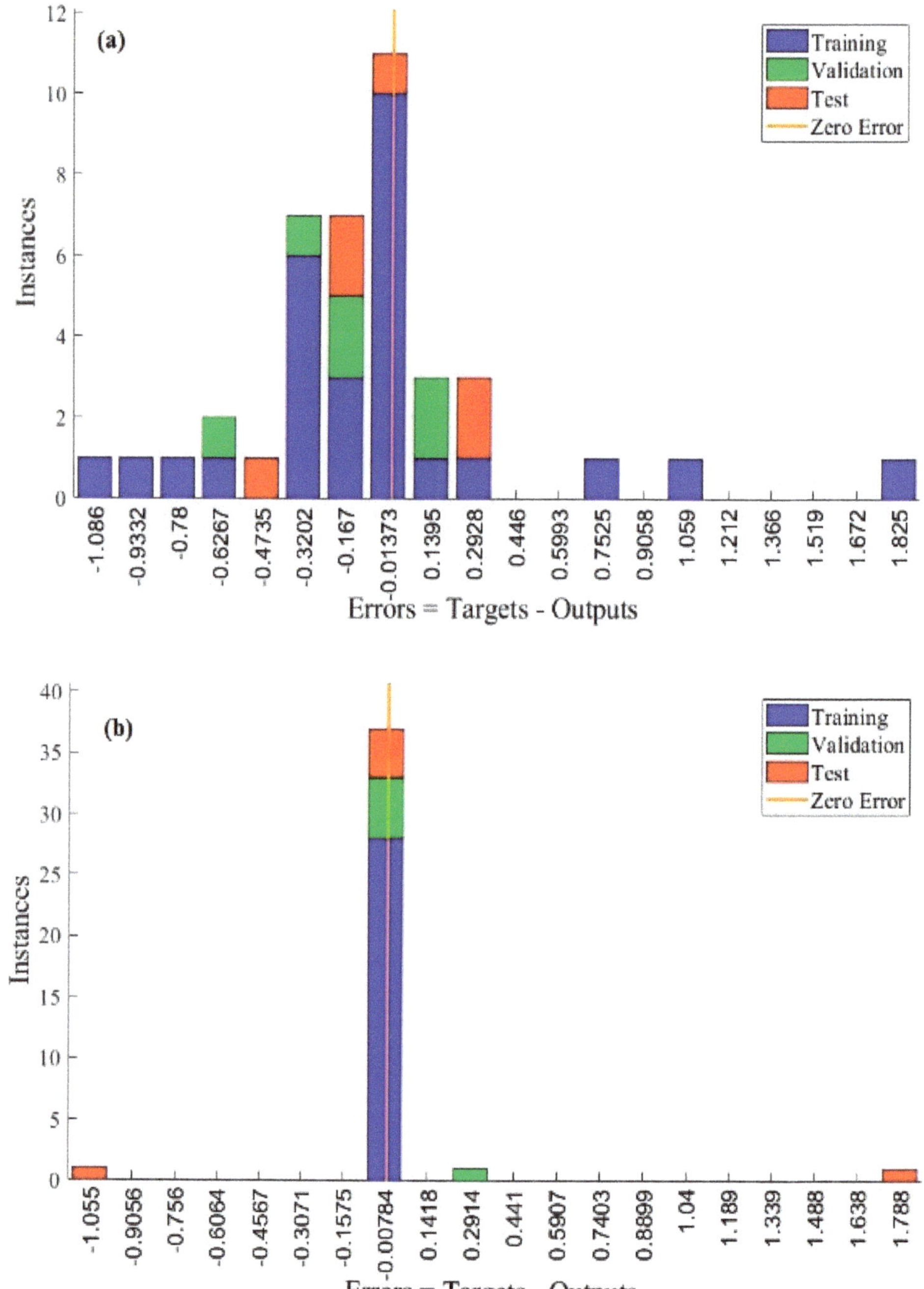

Figure 5. (**a**,**b**). Error histogram of the ANN model for $M = 0$ and $M = 0.2$.

Figure 6a,b shows the SFC values for both magnetic ($M = 0.5$) and non-magnetized ($M = 0$) flow fields, as well as the values produced using the ANN models (b).

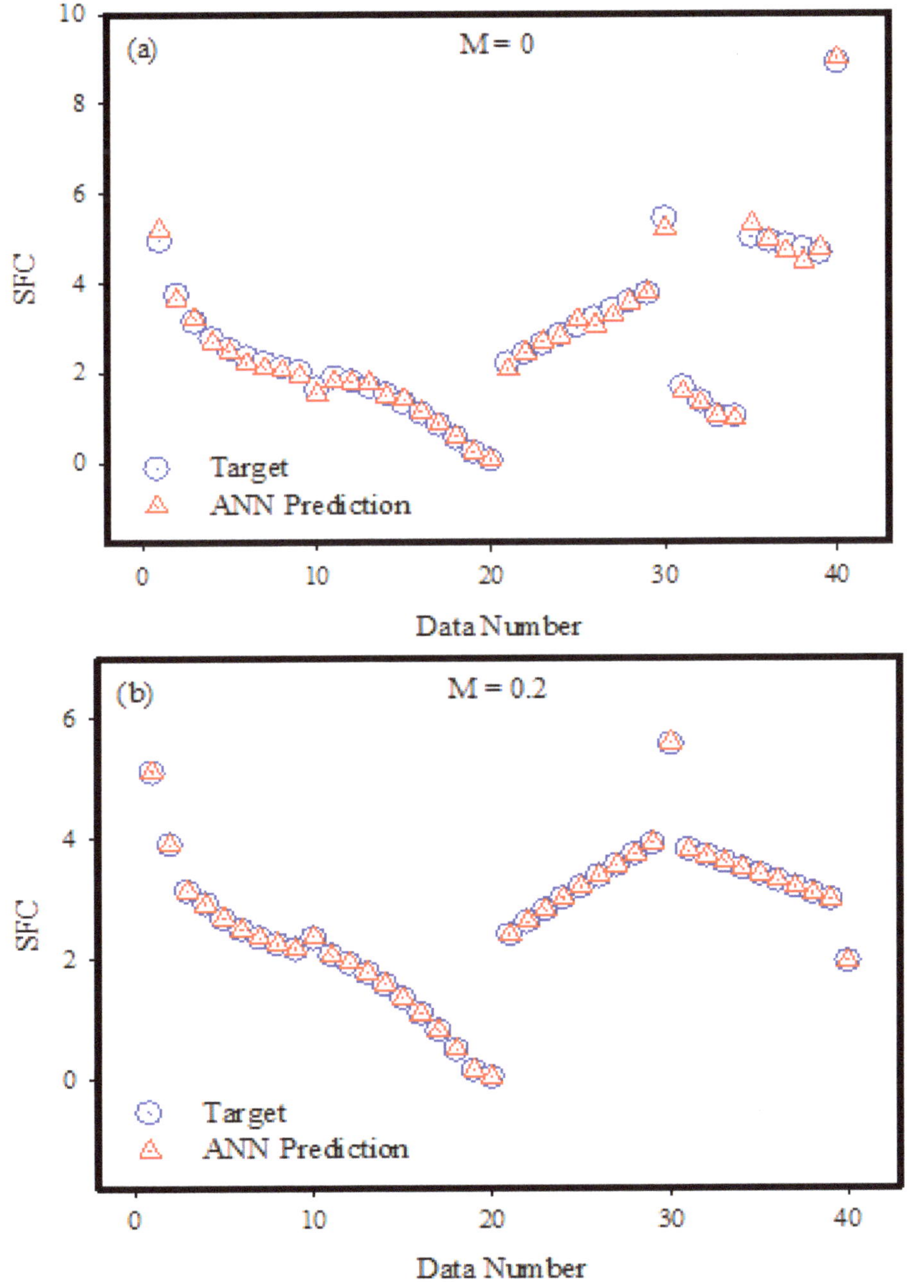

Figure 6. (**a**,**b**). The predicted and target values for $M = 0$ and $M = 0.2$.

When the outcomes for each data point are analyzed, the graphs clearly show that the ANN model results are in very good agreement with the goal values. This excellent fit of the goal and ANN outputs demonstrates that the created ANN models can accurately anticipate SFC values.

Figure 7a,b demonstrates the deviation ratio between the SFC values derived from two different ANN models with the goal data. Such observations are completed for $M = 0$ and $M = 0.2$, respectively. Examining the MoD values obtained for each data point utilized in the creation of the ANN models reveals that the data are often concentrated around the zero-error line. The MoD data to the zero-error line indicate that the ANN outputs have low deviation rates. For a close error study of the ANN models ($M = 0$, $M = 02$), the discrepancies between the ANN model outputs and the target data are designed for the individual data points and are shown in Figure 8a,b. When the difference values for the two different ANN models are considered, it is apparent that for all of the data points, the calculated differences are quite low. The examination of the MoD and difference values reveals that both ANN models can predict SFC values with relatively low error values. The target values are on the x-axis in Figure 9a,b, and the ANN outputs are on the y-axis. When the data for two distinct ANN models ($M = 0$, $M = 02$) were reviewed, the data points were found to be quite close to the zero-error line. The numbers also show that the data points are inside the +10% error band range. Table 11 shows the performance parameters derived for two different ANN models created to estimate SFC values under different scenarios. The MoD values calculated for the ANN models are quite low, as can be observed. The low MoD values suggest that the deviation rates of the ANN model outputs are quite low. The R values' proximity to one and the low MSE values demonstrate that each ANN model can make predictions with high accuracy. The collected results suggest that each ANN model can accurately calculate the output parameter that is the SFC.

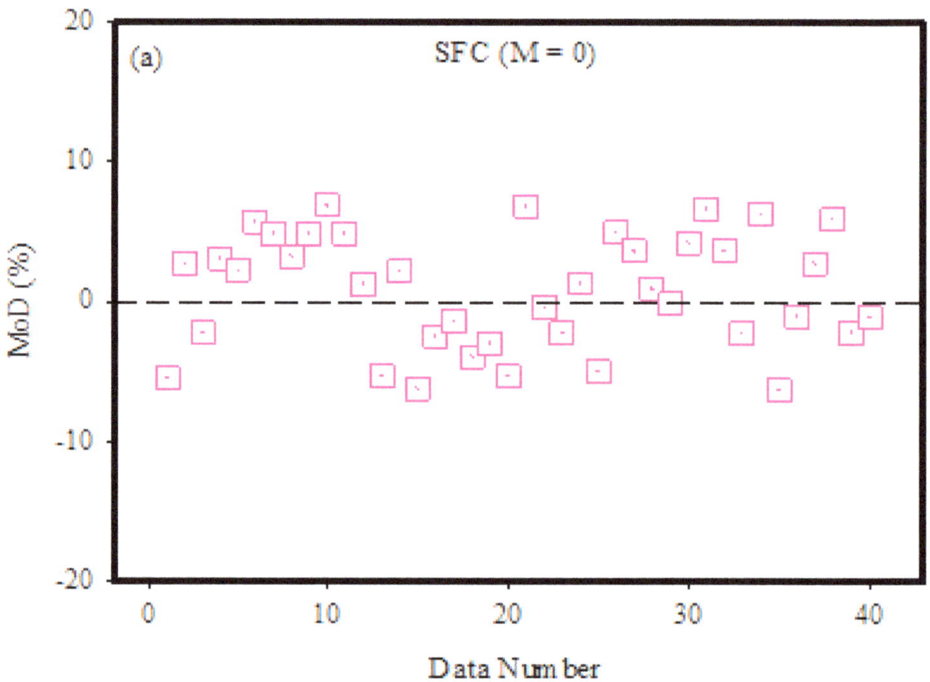

Figure 7. *Cont.*

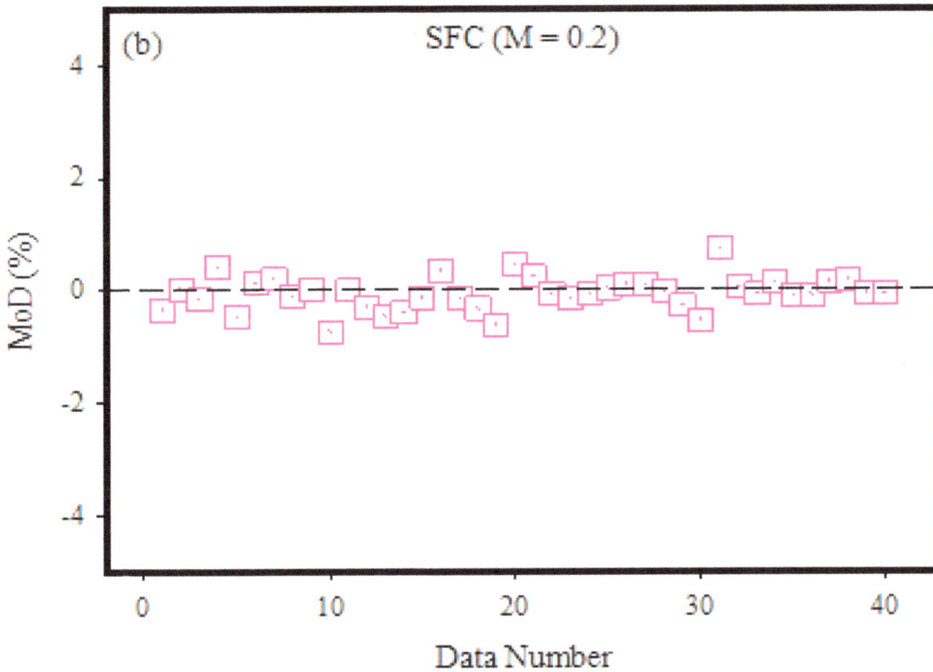

Figure 7. (**a**,**b**). MoD values for each data point for $M = 0$ and $M = 0.2$.

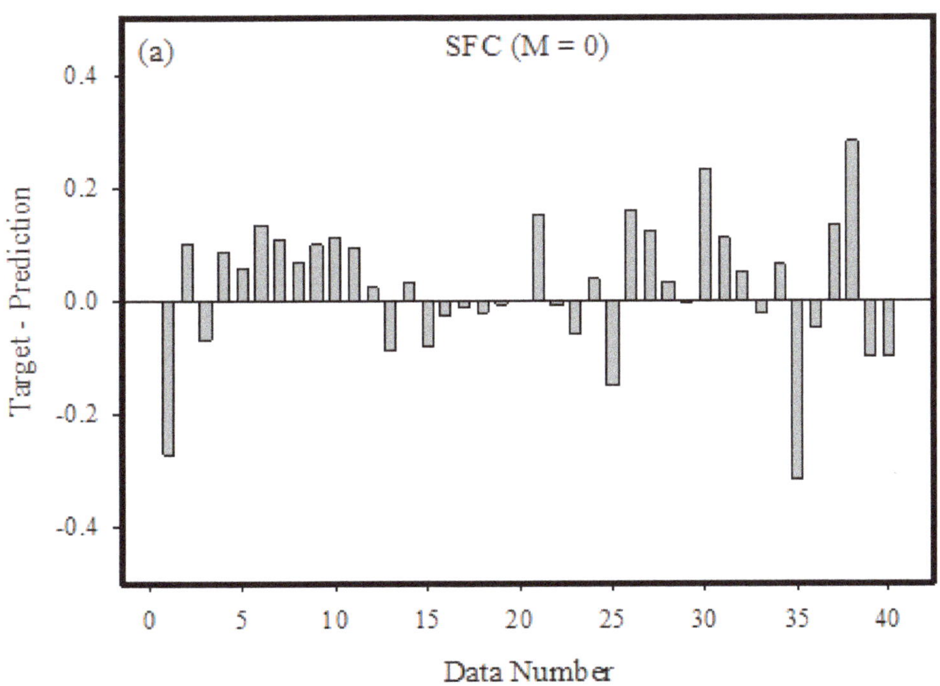

Figure 8. *Cont.*

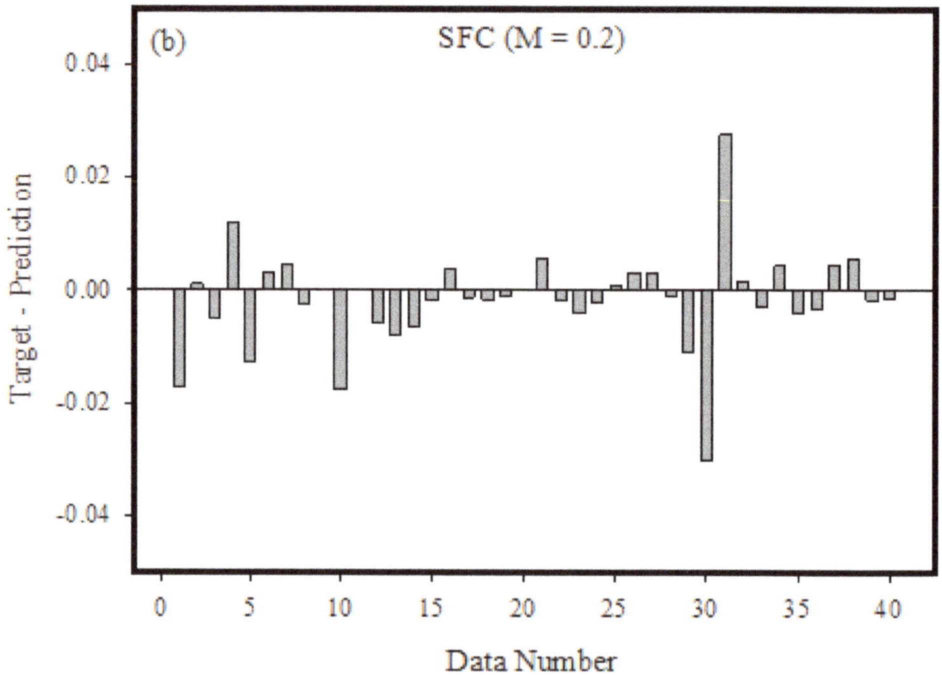

Figure 8. (**a**,**b**). The differences between the target values and the ANN outputs for each set of data when $M = 0$ and $M = 0.2$.

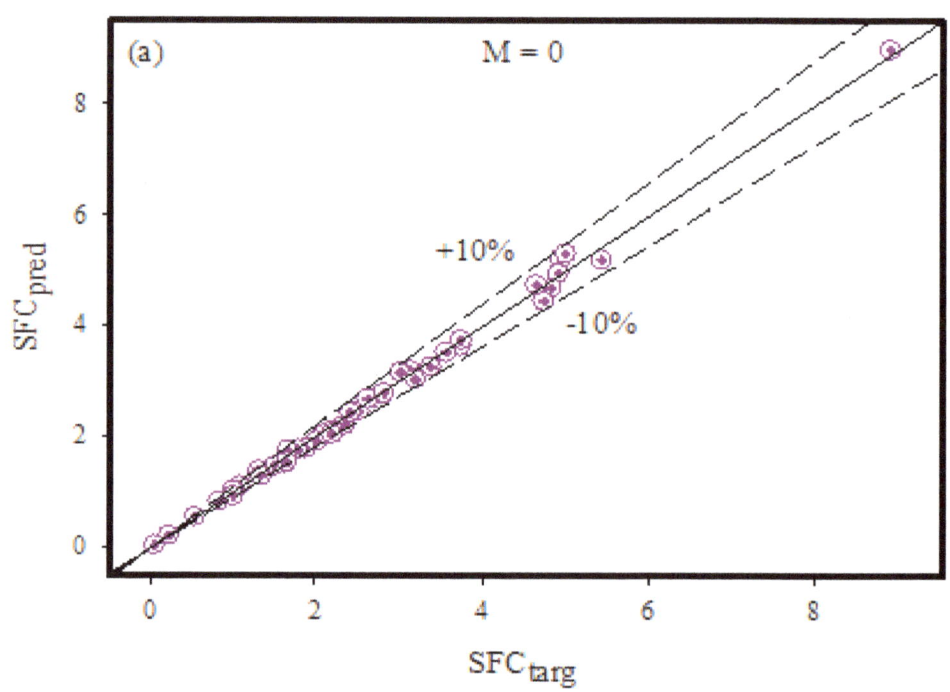

Figure 9. *Cont.*

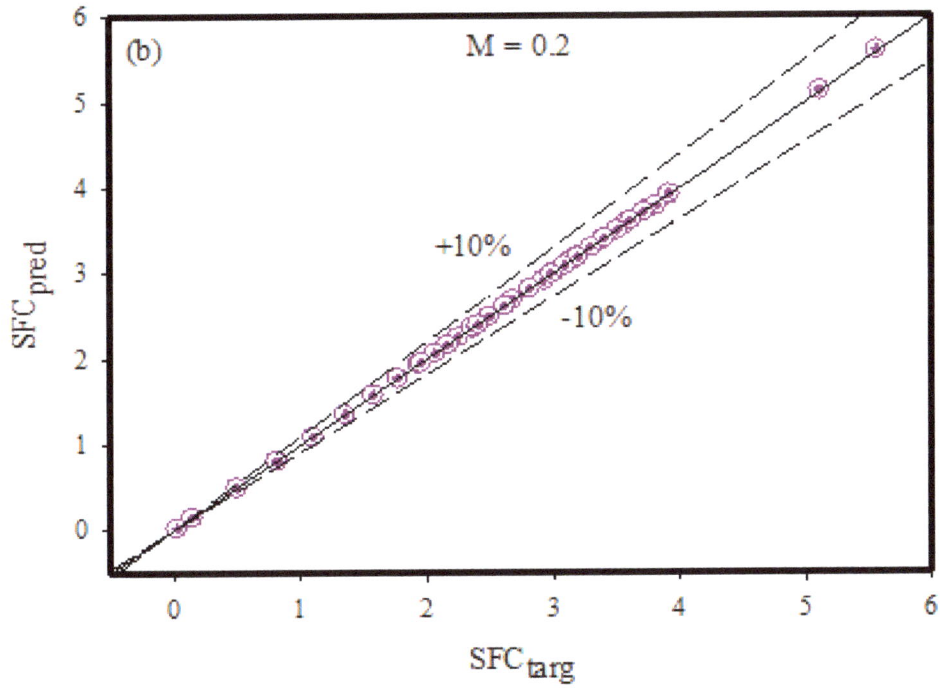

Figure 9. (**a**,**b**). Target and prediction values for $M = 0$ and $M = 0.2$.

Table 11. The performance parameters calculated for the two different ANN models.

Model	MSE	R	MoD_{min} (%)	MoD_{max} (%)
Model 1 ($M = 0$)	5.28×10^{-2}	0.96081	−0.12	6.9
Model 2 ($M = 0.2$)	2.39×10^{-5}	0.99998	0.0008	−0.74

7. Conclusions

The stagnation point flow of Casson fluid has been mathematically modelled for magnetized and non-magnetized flow fields. The novelty was enhanced by considering heat generation, viscous dissipation, mixed convection, temperature-dependent variable thermal conductivity, thermal radiations, and chemical reaction effects. The ultimate flow equations were solved by using the shooting method. The SFC values were evaluated at the cylindrical surface and the corresponding ANN models were constructed. The key outcomes are as follows:

- The margin of deviation and difference values reveals that both ANN models can predict SFC values with relatively low error values.
- The error levels in the error histograms are also quite low. Furthermore, for both ANN models, we noticed that the data points were inside the +10% error band range.
- The coefficient of determination values' proximity to one and the low mean squared error values demonstrate that each ANN model can carry out predictions with high accuracy.
- The magnitude of velocity is higher for the case of non-magnetized Casson fluid flow as compared to non-magnetic flow.
- For both chemically reactive and non-reactive flows, the concentration profiles show a declining nature towards the Schmidt number and curvature parameter.
- The SFC is found to be the decreasing function of the Casson fluid parameter and the velocities ratio parameter while the opposite is the case for the curvature parameter.

- For variation in the Casson fluid parameter, thermal Grashof number, and curvature parameter, the magnitude of SFC is higher for the case of magnetized flow as compared to the non-magnetized flow regime.

Author Contributions: Conceptualization, K.U.R. and A.B.Ç.; data curation, A.B.Ç.; formal analysis, K.U.R. and W.S.; investigation, K.U.R. and A.B.Ç.; methodology, K.U.R.; software, W.S.; Supervision, W.S. All authors have read and agreed to the published version of the manuscript.

Funding: This research received no external funding.

Institutional Review Board Statement: Not applicable.

Informed Consent Statement: Not applicable.

Data Availability Statement: The adopted methodology can be offered upon request by the readers.

Acknowledgments: The authors would like to thank Prince Sultan University, Saudi Arabia, for the technical support offered through the TAS research lab.

Conflicts of Interest: The authors declare no conflict of interest.

Nomenclature

$\widetilde{X}, \widetilde{R}$	Cylindrical coordinates
$\widetilde{U}, \widetilde{V}$	Velocity components
ν	Kinematic viscosity
β	Casson fluid parameter
β_T	Thermal expansion coefficient
g_0	Gravitational acceleration
α	Angle of inclination
β_C	Solutal expansion coefficient
$\widetilde{T}_\infty$	Ambient temperature
$\widetilde{T}$	Temperature of fluid
B_0	Magnetic field constant
$\widetilde{C}$	Concentration of fluid
$\widetilde{C}_\infty$	Ambient concentration
$\widetilde{U}_e$	Free stream velocity
σ	Fluid electrical conductivity
c_p	Specific heat at constant pressure
ρ	Fluid density
$\overline{q}$	Radiative heat flux
κ	Variable thermal conductivity
$\overline{\mu}$	Dynamic viscosity
Q_0	Heat generation coefficient
L	Characteristic length
ε	Small parameter
R_1	Radius of cylinder
k_c	Chemical reaction rate
$\widetilde{C}_w$	Surface concentration
U_0	Reference velocity
$\widetilde{T}_w$	Surface temperature
D_m	Mass diffusivity
$F_C{}'(\eta)$	Fluid velocity
$\theta_C(\eta)$	Fluid temperature
$\phi_C(\eta)$	Fluid concentration
G_T	Temperature Grashof number
G_C	Concentration Grashof number
Pr	Prandtl number
A	Velocities ratio parameter
R	Radiation parameter

γ	Curvature parameter
k^*	Coefficient of mean absorption
E	Eckert number
M	Magnetic field parameter
Rc	Chemical reaction parameter
Sc	Schmidt number
σ^*	Stefan–Boltzmann constant
H	Heat generation parameter

References

1. Alfvén, H. Existence of Electromagnetic-Hydrodynamic Waves. *Nature* **1942**, *150*, 405–406. [CrossRef]
2. Rashidi, S.; Esfahani, J.A.; Maskaniyan, M. Applications of magnetohydrodynamics in biological systems-a review on the numerical studies. *J. Magn. Magn. Mater.* **2017**, *439*, 358–372. [CrossRef]
3. Turkyilmazoglu, M. Analytical solutions to mixed convection MHD fluid flow induced by a nonlinearly deforming permeable surface. *Commun. Nonlinear Sci. Numer. Simul.* **2018**, *63*, 373–379. [CrossRef]
4. Akinshilo, A.T. Mixed convective heat transfer analysis of MHD fluid flowing through an electrically conducting and non-conducting walls of a vertical micro-channel considering radiation effect. *Appl. Therm. Eng.* **2019**, *156*, 506–513. [CrossRef]
5. Vaidya, H.; Rajashekhar, C.; Divya, B.; Manjunatha, G.; Prasad, K.; Animasaun, I. Influence of transport properties on the peristaltic MHD Jeffrey fluid flow through a porous asymmetric tapered channel. *Results Phys.* **2020**, *18*, 103295. [CrossRef]
6. Megahed, A.M.; Reddy, M.G.; Abbas, W. Modeling of MHD fluid flow over an unsteady stretching sheet with thermal radiation, variable fluid properties and heat flux. *Math. Comput. Simul.* **2021**, *185*, 583–593. [CrossRef]
7. Ali, L.; Ali, B.; Ghori, M.B. Melting effect on Cattaneo–Christov and thermal radiation features for aligned MHD nanofluid flow comprising microorganisms to leading edge: FEM approach. *Comput. Math. Appl.* **2022**, *109*, 260–269. [CrossRef]
8. Kumar, P.; Poonia, H.; Ali, L.; Areekara, S. The numerical simulation of nanoparticle size and thermal radiation with the magnetic field effect based on tangent hyperbolic nanofluid flow. *Case Stud. Therm. Eng.* **2022**, *37*, 102247. [CrossRef]
9. Casson, N. A flow equation for pigment-oil suspensions of the printing ink type. In *Rheology of Disperse Systems*; Mill, C.C., Ed.; Pergamon Press: Oxford, UK, 1959; pp. 84–104.
10. Reddy, C.; Rao, C.V.; Surender, O. Soret, Joule heating and Hall effects on free convection in a Casson fluid saturated porous medium in a vertical channel in the presence of viscous dissipation. *Procedia Eng.* **2015**, *127*, 1219–1226. [CrossRef]
11. Mohyud-Din, S.T.; Khan, S.I. Nonlinear radiation effects on squeezing flow of a Casson fluid between parallel disks. *Aerosp. Sci. Technol.* **2016**, *48*, 186–192. [CrossRef]
12. Abro, K.A.; Khan, I. Analysis of the heat and mass transfer in the MHD flow of a generalized Casson fluid in a porous space via non-integer order derivatives without a singular kernel. *Chin. J. Phys.* **2017**, *55*, 1583–1595. [CrossRef]
13. Kataria, H.; Patel, H. Heat and mass transfer in magnetohydrodynamic (MHD) Casson fluid flow past over an oscillating vertical plate embedded in porous medium with ramped wall temperature. *Propuls. Power Res.* **2018**, *7*, 257–267. [CrossRef]
14. Rehman, K.U.; Malik, M.; Zahri, M.; Tahir, M. Numerical analysis of MHD Casson Navier's slip nanofluid flow yield by rigid rotating disk. *Results Phys.* **2018**, *8*, 744–751. [CrossRef]
15. Neeraja, A.; Devi, R.R.; Devika, B.; Radhika, V.N.; Murthy, M.K. Effects of viscous dissipation and convective boundary conditions on magnetohydrodynamics flow of casson liquid over a deformable porous channel. *Results Eng.* **2019**, *4*, 100040. [CrossRef]
16. Rasool, G.; Chamkha, A.J.; Muhammad, T.; Shafiq, A.; Khan, I. Darcy-Forchheimer relation in Casson type MHD nanofluid flow over non-linear stretching surface. *Propuls. Power Res.* **2020**, *9*, 159–168. [CrossRef]
17. Farooq, U.; Hussain, M.; Ijaz, M.A.; Khan, W.A.; Farooq, F.B. Impact of non-similar modeling on Darcy-Forchheimer-Brinkman model for forced convection of Casson nano-fluid in non-Darcy porous media. *Int. Commun. Heat Mass Transf.* **2021**, *125*, 105312. [CrossRef]
18. Ramesh, G.; Roopa, G.; Rauf, A.; Shehzad, S.; Abbasi, F. Time-dependent squeezing flow of Casson-micropolar nanofluid with injection/suction and slip effects. *Int. Commun. Heat Mass Transf.* **2021**, *126*, 105470. [CrossRef]
19. Zeeshan, A.; Mehmood, O.U.; Mabood, F.; Alzahrani, F. Numerical analysis of hydromagnetic transport of Casson nanofluid over permeable linearly stretched cylinder with Arrhenius activation energy. *Int. Commun. Heat Mass Transf.* **2021**, *130*, 105736. [CrossRef]
20. Awan, A.U.; Ahammad, N.A.; Shatanawi, W.; Allahyani, S.A.; Tag-ElDin, E.M.; Abbas, N.; Ali, B. Significance of magnetic field and Darcy–Forchheimer law on dynamics of Casson-Sutterby nanofluid subject to a stretching circular cylinder. *Int. Commun. Heat Mass Transf.* **2022**, *139*, 106399. [CrossRef]
21. Gireesha, B.J.; Shankaralingappa, B.M.; Prasannakumar, B.C.; Nagaraja, B. MHD flow and melting heat transfer of dusty Casson fluid over a stretching sheet with Cattaneo–Christov heat flux model. *Int. J. Ambient. Energy* **2022**, *43*, 2931–2939. [CrossRef]
22. Rehman, K.U.; Shatanawi, W. Thermal analysis on mutual interaction of temperature stratification and solutal stratification in the presence of non-linear thermal radiations. *Case Stud. Therm. Eng.* **2022**, *35*, 102080. [CrossRef]
23. Abbas, Z.; Khaliq, S. Numerical study of non-isothermal analysis of exiting sheet thickness in the calendering of micropolar-Casson fluid. *J. Plast. Film Sheeting* **2022**, *38*, 105–129. [CrossRef]

24. Shahzad, H.; Wang, X.; Ghaffari, A.; Iqbal, K.; Hafeez, M.B.; Krawczuk, M.; Wojnicz, W. Fluid structure interaction study of non-Newtonian Casson fluid in a bifurcated channel having stenosis with elastic walls. *Sci. Rep.* **2022**, *12*, 12219. [CrossRef] [PubMed]
25. Çolak, A.B. An experimental study on the comparative analysis of the effect of the number of data on the error rates of artificial neural networks. *Int. J. Energy Res.* **2021**, *45*, 478–500. [CrossRef]
26. Güzel, T.; Çolak, A.B. Artificial intelligence approach on predicting current values of polymer interface Schottky diode based on temperature and voltage: An experimental study. *Superlattices Microstruct.* **2021**, *153*, 106864. [CrossRef]
27. Rehman, K.U.; Çolak, A.B.; Shatanawi, W. Artificial Neural Networking (ANN) Model for Drag Coefficient Optimization for Various Obstacles. *Mathematics* **2022**, *10*, 2450. [CrossRef]
28. Gedik, E.; Kurt, H.; Pala, M.; Alakour, A. An experimental and artificial neural network investigation on the laminar flow of magnetorheological fluids through circular pipes. *J. Magn. Magn. Mater.* **2022**, *546*, 168893. [CrossRef]
29. Rehman, K.U.; Çolak, A.B.; Shatanawi, W. Artificial Neural Networking (ANN) Model for Convective Heat Transfer in Thermally Magnetized Multiple Flow Regimes with Temperature Stratification Effects. *Mathematics* **2022**, *10*, 2394. [CrossRef]
30. Srinivasacharya, D.; Kumar, R.S. Artificial neural network modeling of the Casson fluid flow over unsteady radially stretching sheet with Soret and Dufour effects. *J. Therm. Anal. Calorim.* **2022**, *147*, 14891–14903. [CrossRef]

mathematics

MDPI

Article

Radiative MHD Nanofluid Flow Due to a Linearly Stretching Sheet with Convective Heating and Viscous Dissipation

Haifaa Alrihieli [1,*], Mohammed Alrehili [2] and Ahmed M. Megahed [3]

1 Department of Mathematics, Faculty of Science, University of Tabuk, Tabuk 71491, Saudi Arabia
2 Department of Mechanical Engineering, Faculty of Engineering, University of Tabuk, Tabuk 71491, Saudi Arabia
3 Department of Mathematics, Faculty of Science, Benha University, Benha 13518, Egypt
* Correspondence: halrehaili@ut.edu.sa

Abstract: This article describes a two-dimensional steady laminar boundary layer flow and heat mass transfer caused by a non-Newtonian nanofluid due to a horizontally stretching sheet. The non-dimensional parameters take into consideration and regulate the effects of convective boundary condition, slip velocity, Brownian motion, thermophoresis and viscous dissipation. The thermal radiation, which affects the flow's thermal conductivity and the nanofluid's variable viscosity are also taken into consideration. We propose that a hot fluid could exist beneath the stretching sheet's bottom surface, which could aid in warming the surface via convection. The physical boundary conditions are non-dimensionalized, as are the governing transport set of nonlinear partial differential equations. By using the shooting approach, numerical values for dimensionless velocity, temperature and nanoparticle concentration are achieved. Distributions of velocity, temperature and concentration are plotted against a number of newly important governing factors, and the outcomes are then provided in accordance with those graphs. Additionally, the local skin-friction coefficient, the local Sherwood number and the local Nusselt number are discussed in order to further clarify and thoroughly explain the current problem. In order to validate the numerical results, comparisons are made with previously published data in the literature. There is a really good accord. Additionally, the current work has implications in the nanofluid applications.

Keywords: Maxwell nanofluid; thermal radiation; convective boundary condition; variable conductivity; viscous dissipation

MSC: 76A05; 76D10; 76W05

Citation: Alrihieli, H.; Alrehili, M.; Megahed, A.M. Radiative MHD Nanofluid Flow Due to a Linearly Stretching Sheet with Convective Heating and Viscous Dissipation. *Mathematics* **2022**, *10*, 4743. https://doi.org/10.3390/math10244743

Academic Editor: Ramoshweu Solomon Lebelo

Received: 11 November 2022
Accepted: 9 December 2022
Published: 14 December 2022

1. Introduction

Nanofluid is a crucial fluid type for energy conveyance since it includes both base fluid and nanoparticles. In light of the demands of applications across decades, nanofluid subjects are hence sustainable. There have been a number of general hypotheses put forth on the thermophysical properties and heat transport of changed base fluid nanoliquids up until this point. Choi [1] created the word "nanofluids" to describe the investigation and examination of nanoparticles. Additionally, he looked into how adding nanoparticles to the basic fluid improves the thermal characteristics of fluids. When a nanoparticle has at least one of its major dimensions smaller than 100 nm, it is said to be suspended in a thin liquid, or a nanofluid. Due to their amazing ability to increase heat conductivity, nanofluids have proven beneficial in a variety of technical and industrial applications. Because common heat transfer fluids have poorer thermal conductivities, it is impossible to meet cooling rate requirements with them. The nanoparticles can be dispersed to improve the thermal conductivity and the total thermal performance of common heat transfer fluids. Nanofluids have unique features that make them potentially useful in a variety of heat transfer processes, including microelectronics, fuel cells, hybrid engines, etc.

Engine oils, radiators, engines, coolants, automatic transmission fluids, lubricants and other synthetic high-temperature heat transfer fluids are all common components of conventional truck thermal systems. These might profit from the increased heat conductivity provided by nanofluids as a result of the addition of nanoparticles [2]. Due to its numerous manufacturing applications, certain numerical and experimental investigations on nanofluids that focus on the thermal conductivity under different physical conditions have been carefully examined [3–9].

The mechanism by which the work done by a nanofluid on adjacent layers as a result of shear stresses is irreversibly transformed into heat is known as the viscous dissipation phenomena. The viscous dissipation phenomenon, which shows as an increase in fluid temperature, is caused by the irreversible work done by the fluid motion to resist the layers of shear stresses in the flow. The viscous dissipation phenomenon is crucial in heat transport research, particularly in boundary layer flows, because of the greater velocity gradients inside the boundary layer's region. The impact of viscous dissipation in nanofluids may vary due to the influence of particle migration, which significantly alters the distribution of temperature and nanoparticle concentration. Heat transfer is significantly impacted by the viscous dissipation phenomena, especially in high-velocity flows and very viscous flows at low velocities. Many scientists [10–15] have already looked at a range of real-world issues connected to the phenomenon of viscous dissipation in nanofluid flow under varied circumstances.

Most physical models are controlled by a system of differential equations, some of which cannot be solved analytically. In order to solve this problem, we must use some numerical methods that are connected to numerical analyses. Numerical approximation is sufficient in many situations, including most practical engineering applications, chemistry, economics, physics and biology. As a result, numerous models are addressed numerically using different techniques [16,17]. According to the research described above, many researchers have investigated the Newtonian and non-Newtonian nanofluid flow problems using a variety of numerical methodologies. The current study, which is inspired by the aforementioned literature and applications, investigates numerically the boundary layer flow and heat transfer of a Maxwell nanofluid model that is exposed to a magnetic field and thermal radiation. Along with convective boundary conditions, slip velocity, viscous dissipation and the variable properties of nanofluids are also examined. Utilizing the shooting technique, numerical solutions are obtained for the domains of velocity, temperature and concentration.

2. Problem Formulation

Consider an incompressible nanofluid flowing through a permeable stretched sheet in a boundary layer with effects from radiation, viscous dissipation and convective heating. The x-axis and y-axis are perpendicular to each other in the problem geometry, with the surface lying along the x-axis. Here, heat and mass transmission mechanisms are explained in terms of the Brownian and thermophoresis characteristics with diffusion coefficient D_B and thermophoretic diffusion coefficient D_T, respectively. Additionally, the phenomenon of convective heat transfer is also taken into account. In the x-direction, the sheet is stretched with velocity $u_w = ax$, where a is a positive constant with dimension s^{-1}. Likewise, the nanofluid flow is assumed to have the velocity vector $\vec{U}$ with two components u and v, which can take the form:

$$\vec{U} = u\,\vec{i} + v\,\vec{j}. \tag{1}$$

In this investigation, we assume that a hot fluid exists beneath the stretching sheet's bottom surface. By using a convection phenomenon, this hot fluid, which has a temperature of T_f, significantly contributes to warming the stretching sheet's surface. So, the heat transfer coefficient h_f is created as a result. This temperature T_f is thought to be in the following form:

$$T_f = T_\infty + Ax^2, \tag{2}$$

where T_∞ refers to a constant ambient cold fluid temperature, A is constant. Here, we have that $T_f > T_\infty$ is the highest temperature in the system. Furthermore, the nanofluid concentration $C_w(x)$ is assumed to take the form:

$$C_w(x) = C_\infty + cx^2, \tag{3}$$

where C_∞ stands for the nanofluid concentration away from the sheet and c is a constant. Additionally, it is anticipated that the vector of the applied magnetic field would permeate the Maxwell nanofluid with electrical conductivity σ, which can be considered to be as follows:

$$\vec{B}(x) = B_0 \vec{j}, \tag{4}$$

where B_0, as shown in Figure 1, is a factor that indicates the intensity of the magnetic field acting in the y-positive axis's direction.

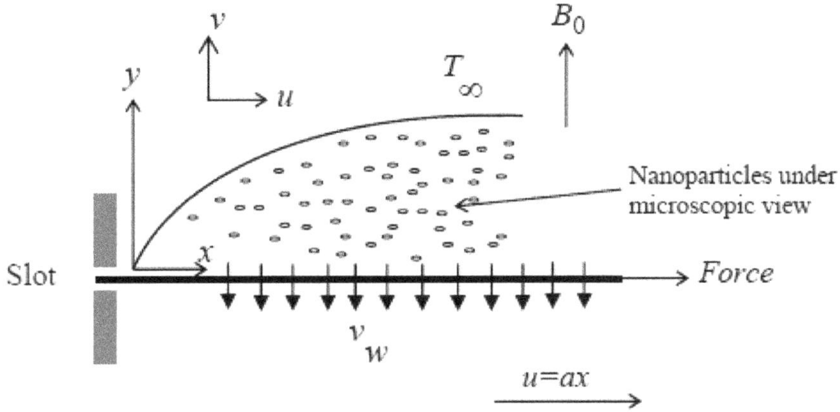

Figure 1. Schematic diagram for the nanofluid flow.

Further, we assume that the sheet is porous and the nanofluid moves through the holes at a constant speed v_w. In Cartesian coordinates, x and y, the fundamental steady equations for the conservation of mass, momentum, thermal energy and nanoparticles for nanofluids can be expressed as [18]:

$$\frac{\partial u}{\partial x} + \frac{\partial v}{\partial y} = 0, \tag{5}$$

$$u\frac{\partial u}{\partial x} + v\frac{\partial u}{\partial y} = \frac{1}{\rho_\infty}\frac{\partial}{\partial y}\left(\mu(T)\frac{\partial u}{\partial y}\right) - \beta_1\left(u^2\frac{\partial^2 u}{\partial x^2} + v^2\frac{\partial^2 u}{\partial y^2} + 2uv\frac{\partial^2 u}{\partial x \partial y}\right) - \frac{\sigma B_0^2}{\rho_\infty}u, \tag{6}$$

$$\begin{aligned} u\frac{\partial T}{\partial x} + v\frac{\partial T}{\partial y} = \frac{1}{\rho_\infty c_p}\frac{\partial}{\partial y}\left(\kappa(T)\frac{\partial T}{\partial y}\right) + \frac{1}{\rho_\infty c_p}\left(\mu(T)(\frac{\partial u}{\partial y})^2 - \rho_\infty\beta_1\left(2uv\frac{\partial u}{\partial x}\frac{\partial u}{\partial y} + v^2(\frac{\partial u}{\partial y})^2\right)\right) + \\ \tau\left[D_B\frac{\partial C}{\partial y}\frac{\partial T}{\partial y} + \frac{D_T}{T_\infty}\left(\frac{\partial T}{\partial y}\right)^2\right] - \frac{1}{\rho_\infty c_p}\frac{\partial q_r}{\partial y}, \end{aligned} \tag{7}$$

$$u\frac{\partial C}{\partial x} + v\frac{\partial C}{\partial y} = D_B\frac{\partial^2 C}{\partial y^2} + \frac{D_T}{T_\infty}\frac{\partial^2 T}{\partial y^2}, \tag{8}$$

where, β_1 is the Maxwell coefficient, ρ_∞ is the ambient nanofluid density, $\mu(T)$ is the nanofluid viscosity, T is the nanofluid's Maxwell temperature and κ is the nanofluid thermal conductivity. Here, we must remember that the Maxwell fluid class, characterized by the Maxwell coefficient β_1, is the most basic category of non-Newtonian fluids. The properties of the relaxation time can be accurately described by this model. Furthermore, we must point out that if $\beta_1 = 0$, our model can be reduced to a Newtonian model. Further, according to the Rosseland approximation, the radiative heat transfer q_r is represented by the expression [19]:

$$q_r = -\frac{4\sigma^*}{3k^*}\frac{\partial T^4}{\partial y}. \tag{9}$$

The term T^4 in the last equation is used to denote the slight temperature variation in the fluid. The Taylor's series about T_∞ is used to expand the variable T^4 as a linear function. Consequently, disregarding the higher order terms produces the following [20]:

$$T^4 \cong 4T_\infty^3 T - 3T_\infty^4. \tag{10}$$

In order to fully formulate the suggested problem, following is an introduction to the boundary conditions for the distributions of velocity, temperature and concentration [21]:

$$\begin{gathered} u = ax + (\frac{\lambda_1}{\mu_\infty})\left(\mu(T)\frac{\partial u}{\partial y} - \rho_\infty\beta_1(2uv\frac{\partial u}{\partial x} + v^2\frac{\partial u}{\partial y})\right), \quad v = -v_w, \\ -\kappa(T)\left(\frac{\partial T}{\partial y}\right)_w = h_f\left(T_f - T_w\right), \quad C = C_w(x) = C_\infty + cx^2, \quad at \quad y = 0, \end{gathered} \tag{11}$$

$$u \to 0, \quad T \to T_\infty, \quad C \to C_\infty \text{ as } y \to \infty, \tag{12}$$

where μ_∞ is the ambient viscosity of the nanofluid and λ_1 is the slip velocity factor. We now begin with dimensionless variables that can transform partial differential equations into ordinary differential equations before creating the solution procedure [21]:

$$\eta = \left(\frac{a}{\nu_\infty}\right)^{\frac{1}{2}} y, \quad u = axf'(\eta), \quad v = -(a\nu_\infty)^{\frac{1}{2}} f(\eta), \tag{13}$$

$$\theta(\eta) = \frac{T - T_\infty}{T_f - T_\infty}, \quad \phi(\eta) = \frac{C - C_\infty}{C_w - C_\infty}, \tag{14}$$

where f is the non-dimensional stream function, θ is the non-dimensional fluid temperature, ϕ is the dimensionless concentration and η is the dimensionless similarity variable. Furthermore, we assume in this study that the dimensionless temperature impacts the nanofluid thermal conductivity $\kappa(T)$ as well as the nanofluid viscosity $\mu(T)$ according to these laws [22]:

$$\mu = \mu_\infty e^{-\alpha\theta}, \kappa = \kappa_\infty(1 + \varepsilon\theta), \tag{15}$$

where κ_∞ is the thermal conductivity away from the sheet, ε is the factor of the thermal conductivity, α is the viscosity parameter, μ_∞ is a constant viscosity of the nanofluid at the ambient. Here, we must observe that $\kappa = \kappa_\infty$ when the nanofluid temperature T is equal to the ambient temperature T_∞. Therefore, the thermal conductivity of the nanofluid varies with temperature along the thermal boundary layer before being constant at ambient. The governing ordinary differential equations with boundary conditions are written as follows when the aforementioned Equations (13) and (14) are introduced into the momentum, energy and concentration equations:

$$(f''' - \alpha f''\theta')e^{-\alpha\theta} + ff'' - f'^2 + \beta\left(2ff'f'' - f^2 f'''\right) - Mf' = 0, \tag{16}$$

$$\frac{1}{\Pr}\left(\varepsilon\theta'^2 + (1 + R + \varepsilon\theta)\theta''\right) + f\theta' - 2\theta f' + Nt\theta'^2 + Nb\theta'\phi' + Ecf''\left(e^{-\alpha\theta}f'' + \beta\left(2ff'^2 - f^2 f''\right)\right) = 0, \tag{17}$$

$$\phi'' + \Pr\text{Le}(f\phi' - 2\phi f') + \frac{Nt}{Nb}\theta'' = 0. \tag{18}$$

According to the following modified boundary condition:

$$f = f_w, \quad f' = 1+\lambda\left(e^{-\alpha\theta} f'' + \beta(2ff'^2 - f^2 f'')\right), \quad \phi = 1, \\ \theta' = -\delta\left(\frac{1-\theta}{1+R+\varepsilon\theta}\right), \quad at \quad \eta = 0, \tag{19}$$

$$f' \to 0, \quad \theta \to 0, \quad \phi \to 0, \quad \text{as } \eta \to \infty. \tag{20}$$

Nevertheless, it is crucial to remember that the current non-Newtonian model can be converted into a Newtonian model if β is missing from the previous system. In addition, the equations above include the following dimensionless quantities and parameters:

$$\beta = a\beta_1, M = \frac{\sigma B_0^2}{a\rho_\infty}, \lambda = \lambda_1\sqrt{\frac{a}{\nu_\infty}}, Nb = \frac{\tau D_B(C_w - C_\infty)}{\nu_\infty}, Ec = \frac{a^2}{Ac_p}, \tag{21}$$

$$Nt = \frac{\tau D_T\left(T_f - T_\infty\right)}{\nu_\infty T_\infty}, R = \frac{16\sigma^* T_\infty^3}{3\kappa_\infty k^*}, \delta = \frac{h_f}{\kappa_\infty}\sqrt{\frac{\nu_\infty}{a}}, Le = \frac{\kappa_\infty}{\rho_\infty c_p D_B}, \Pr = \frac{\mu_\infty c_p}{\kappa_\infty}, \tag{22}$$

where β is the Maxwell parameter and measures the relaxation time and M, λ, Nb, Ec are the magnetic number, slip velocity parameter, Brownian motion parameter and Eckert number, which denote the viscous dissipation phenomenon. Nt is the thermophoresis parameter, R is the thermal radiation parameter and $\delta, Le, \Pr$ are the surface-convection parameter, Lewis parameter and Prandtl number, respectively.

Skin friction Cf_x, heat transfer rate in term of Nux and mass transfer rate in terms of Shx are the physical characteristics of Maxwell nanofluid flow. These terms are denoted as follows:

$$\frac{Cf_x}{2}Re_x^{\frac{1}{2}} = -\left[e^{-\alpha\theta(0)} f''(0) - \beta(f''(0)f^2(0) - 2f(0)f'^2(0))\right], \\ \frac{Nu_x}{\sqrt{Re_x}} = -\theta'(0), \qquad \frac{Sh_x}{\sqrt{Re_x}} = -\phi'(0), \tag{23}$$

where $Re_x = \frac{u_w x}{\nu_\infty}$ is the local Reynolds number.

3. Physical And Graphical Interpretation of Results

Here, a comprehensive investigation of radiative, MHD Maxwell nanofluid flow is described in this research using the shooting method, under the impact of slip velocity, viscous dissipation and convective heating phenomenon. Firstly, a comparison of numerical values representing the rate of heat transfer $(-\theta'(0))$ for various suction parameters f_w and the Prandtl number Pr with results previously published (Ishak et al. [23]) is presented in Table 1 as evidence of the reliability of the existing solutions.

Figure 2 shows three different values of the Maxwell parameter side-by-side comparisons of the fields of velocity, temperature and concentration. As can be seen, Figure 2 depicts a significant resistance to flow velocity with increasing Maxwell parameter values due to the development of shear stress. Additionally, the temperature of the sheet $\theta(0)$ as well as the temperature $\theta(\eta)$ and concentration $\phi(\eta)$ distributions were all dramatically increased by the same parameter.

Table 1. Comparison of Nusselt number $-\theta'(0)$ for different values of f_w and Pr with the results of Ishak et al. [23] when $\alpha = \beta = M = Ec = \lambda = R = \varepsilon = Nt = Nb = 0$.

Pr	f_w,	Ishak et al. [23]	Present Work
0.72	−1.5	0.4570	0.457001520
1.0	−1.5	0.5000	0.500000000
10	−1.5	0.6542	0.654211910
0.72	0.0	0.8086	0.808589088
1.0	0.0	1.0000	1.000000000
3.0	0.0	1.9237	1.923689985
10.0	0.0	3.7207	3.720699510
0.72	1.5	1.4944	1.494389791
1.0	1.5	2.0000	2.000002010
10	1.5	16.0842	16.08419892

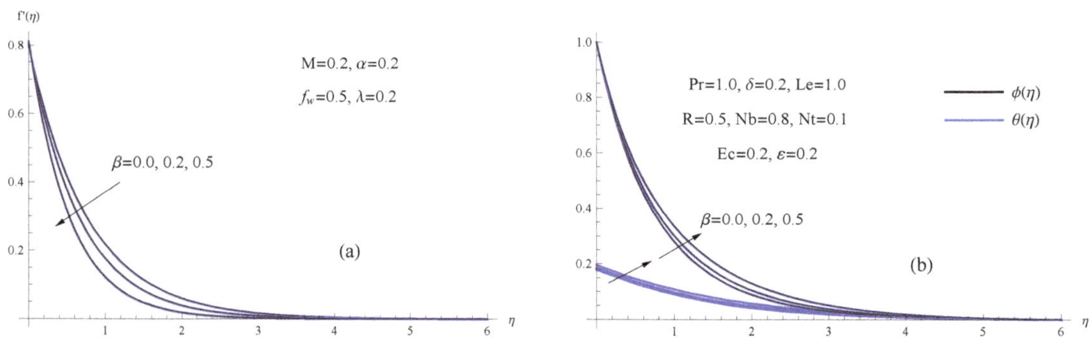

Figure 2. (**a**) The $f'(\eta)$ for chosen β, (**b**) $\theta(\eta)$ and $\phi(\eta)$ for chosen β.

According to the change of the suction parameter f_w, Figure 3 shows the analysis of the flow and heat mass transportation performance of Maxwell nanofluid. An intriguing finding is that the classical model specifies the lowest fluid flow for high values of the suction parameter while having the fastest velocity profile for low values. Higher suction parameter values lessen the mass distribution while enhancing the cooling of the nanofluid and the sheet temperature $\theta(0)$ since the fluid displays the fastest heat and mass transfer in the absence of the suction parameter (impermeable sheet).

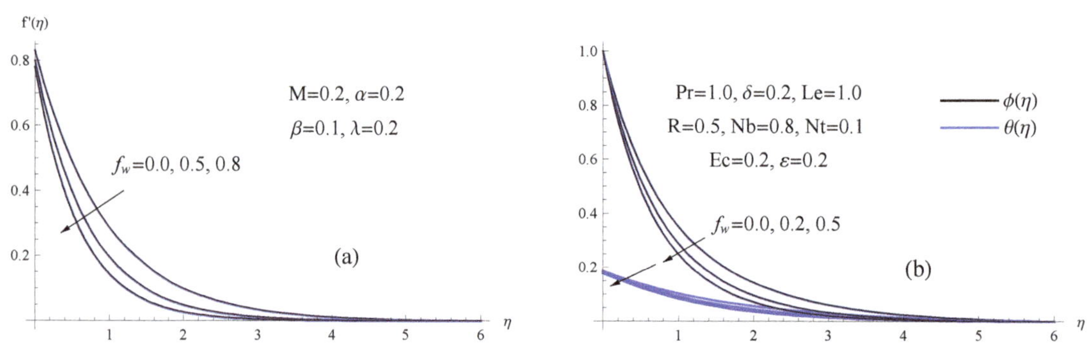

Figure 3. (**a**) The $f'(\eta)$ for chosen f_w, (**b**) $\theta(\eta)$ and $\phi(\eta)$ for chosen f_w.

Figure 4 depicts variations in velocity $f'(\eta)$, temperature $\theta(\eta)$ and concentration $\phi(\eta)$ caused by the magnetic field's M impact on the Maxwell nanofluid. Various inputs of the non-dimensional magnetic parameter M are used in this investigation while holding the inputs of other related physical parameters constant. The velocity of the nanofluid is observed to be restricted by an increase in the magnetic parameter. The existence of the Lorentz force is the physical component that causes this result. One of the viscous forces of this type, the Lorentz force, works in the opposite direction of nanofluid flow and slows down the fluid velocity. As a result, the creation of these viscous forces has a profound impact and causes the fluid to be warmed and concentrated to enhance the magnetic parameter.

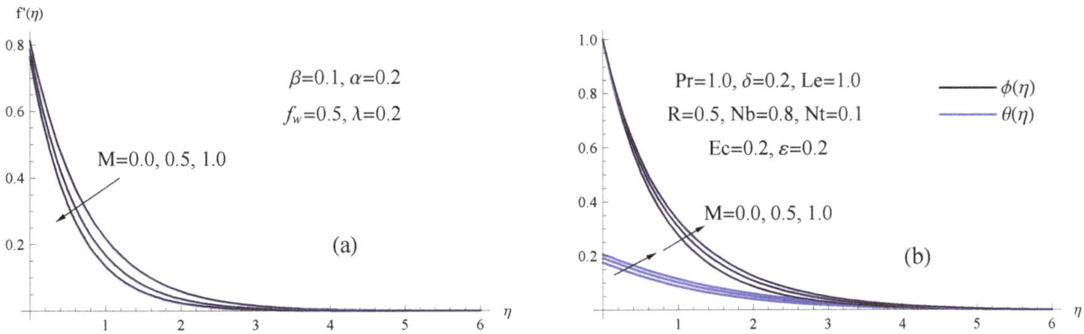

Figure 4. (**a**) The $f'(\eta)$ for chosen M, (**b**) $\theta(\eta)$ and $\phi(\eta)$ for chosen M.

For various values of the viscosity parameter α, Figure 5 explains the demeanor of the velocity $f'(\eta)$, temperature $\theta(\eta)$ and concentration $\phi(\eta)$ fields. The graphic shows that as the viscosity parameter climbs, the momentum boundary layer thickness and velocity field slow down. The essential duty of nanofluid viscosity, which mostly depends on temperature, is to promote mass and heat transfer rates within the boundary layer. As a result, it is evident that as the viscosity parameter improves, the concentration and temperature of nanofluid as well as the sheet temperature $\theta(0)$ and the thickness of the thermal boundary layer rise. Clearly the velocity distribution through the boundary layer was greatly impacted by the viscosity parameter α since the viscosity parameter directly influences the velocity field, as seen from Equation (16). While the concentration and temperature fields are both indirectly influenced by the viscosity parameter, and as a result, both are only marginally impacted.

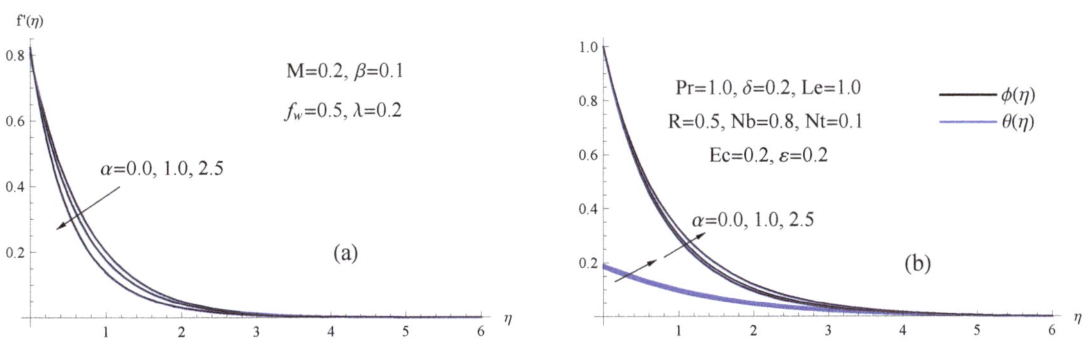

Figure 5. (**a**) The $f'(\eta)$ for chosen α, (**b**) $\theta(\eta)$ and $\phi(\eta)$ for chosen α.

Figure 6 examines the impact of the slip velocity parameter λ on the velocity $f'(\eta)$, temperature $\theta(\eta)$ and concentration $\phi(\eta)$ fields while the other physical governing parameters are unchanged. We notice that anytime the slip velocity λ increases, both the sheet velocity $f'(0)$ and the nanofluid velocity $f'(\eta)$ dramatically decrease with the dimensionless variable η. Additionally, we see that the velocity distribution changes as η is increased in the interval $0 \leq \lambda < 2$. Further, when the slip velocity parameter λ improves, the same drop tendency is shown for both the temperature distribution $\theta(\eta)$ and the sheet temperature $\theta(0)$. Moreover, we can see from the following graphic that with rising values of λ, the boundary layer thickness and the concentration field both get better.

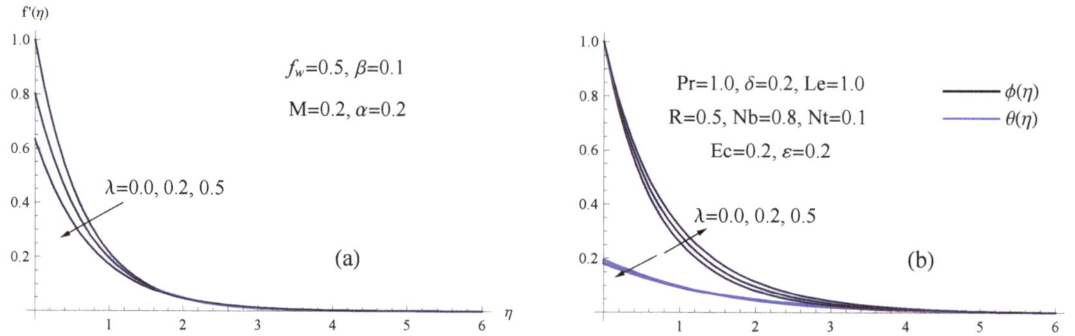

Figure 6. (**a**) The $f'(\eta)$ for chosen λ, (**b**) $\theta(\eta)$ and $\phi(\eta)$ for chosen λ.

According to the influence of the Eckert number Ec, the temperature distribution $\theta(\eta)$ shows modification in Figure 7a. In the nanofluid heat transfer mechanism, the principal objective of the viscous dissipation phenomena is to alter the thermal performance with sources of energy. Larger Eckert number values indicate that heat is dissipating and traveling in the direction of the fluid as a result of viscous force. Fluid particles travel quickly as a result, causing more collisions between them. This greater collision produces thermal energy by converting kinetic energy. Furthermore, the graph of the temperature field $\theta(\eta)$ for miscellaneous values of thermal radiation parameter R is designed in Figure 7b. With the higher thermal radiation parameter along the sheet, as opposed to away from it, a significant drop in both the sheet temperature $\theta(0)$ and the nanofluid temperature is seen.

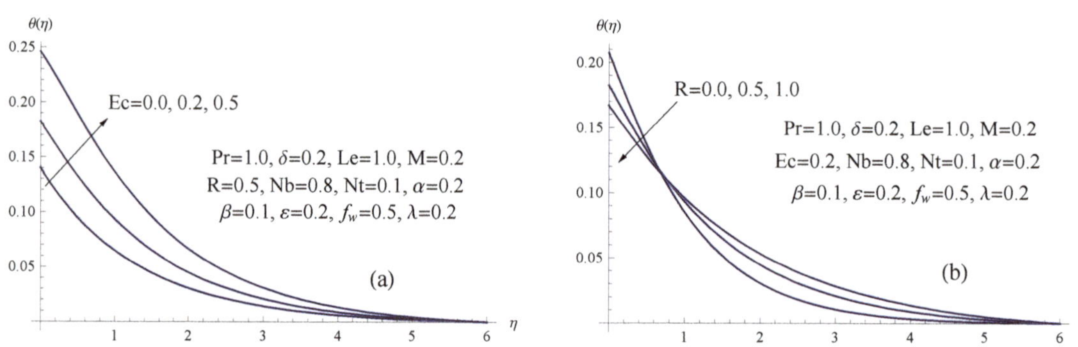

Figure 7. (**a**) $\theta(\eta)$ for chosen Ec (**b**) $\theta(\eta)$ for chosen R.

Figure 8 shows how the surface-convection parameter δ affects the temperature distributions $\theta(\eta)$ in the region of the thermal boundary layer. The nanofluid temperature $\theta(\eta)$ is seen to dramatically increase along the sheet wall but only modestly rise away from the sheet, especially when η is greater than 4.0. As a result, the nanofluid along the sheet warmed due to higher values of the surface-convection parameter, which thickened the thermal boundary layer. Figure 8b illustrates the graphical behavior of the nanofluid's dimensionless temperature for various values of the thermal conductivity parameter ε. It is evident that as ε increases, the Maxwell nanofluid temperature and the thermal boundary layer thickness grow away from the sheet wall, whereas the reverse trend is noted beside the sheet. In contrast to nanofluids with constant thermal conductivity, this causes the thermal boundary layer thickness of the nanofluid with variable thermal conductivity to be greater.

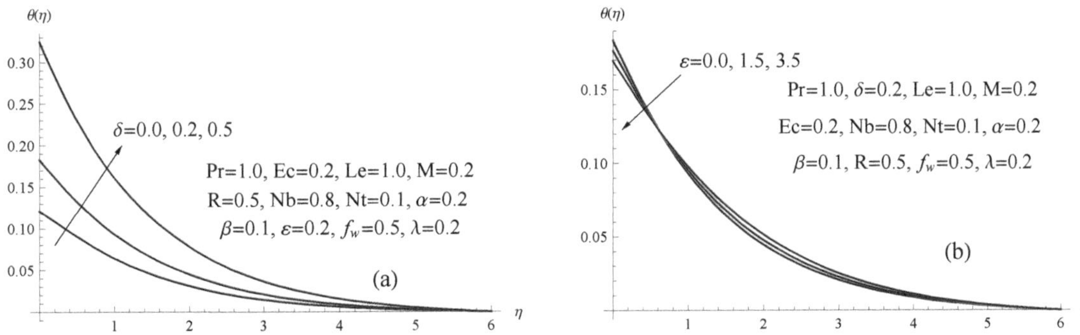

Figure 8. (**a**) The $\theta(\eta)$ for chosen δ, (**b**) $\theta(\eta)$ for chosen ε.

Additionally, Figure 9 shows the temperature field $\theta(\eta)$ outcomes for the Brownian motion parameter Nb and the thermophoresis parameter Nt. The particles' erratic motion generates extra kinetic energy, which boosts the thermal energy that is already there. As a result, a rise in Nb causes the fluid temperature in the boundary layer's thermal domain to increase faster. The Maxwell nanofluid temperature and the sheet temperature $\theta(0)$ both accelerated similarly as a result of the increase in the thermophoresis parameter Nt.

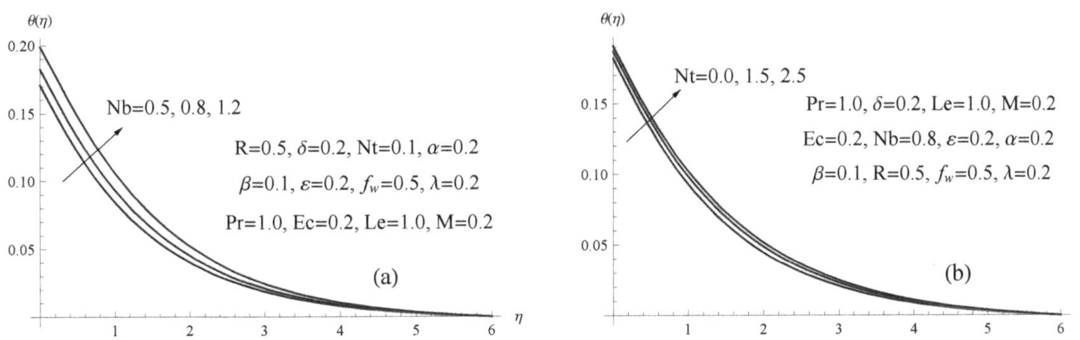

Figure 9. (**a**) The $\theta(\eta)$ for chosen Nb, (**b**) $\theta(\eta)$ for chosen Nt.

Before we have finished our analysis, we must concentrate on the drag force, which can be calculated using the skin-friction coefficient $\frac{Cf_x}{2} Re_x^{\frac{1}{2}}$, rate of heat transfer (which can be evaluated using the local Nusselt number $\frac{Nu_x}{\sqrt{Re_x}}$), and the rate of mass transfer, which can be determined using the local Sherwood number $\frac{Sh_x}{\sqrt{Re_x}}$. Therefore, we are interested to examine the key physical characteristics that can influence how these quantities behave. In order to obtain these values for this work, we constructed Table 2. It is evident from Table 2 that as the Maxwell number, viscosity parameter, slip velocity parameter and surface-convection parameter grow, the local skin-friction coefficient changes inversely. Additionally, a reduction in the Sherwood number is brought about by rising values of the Maxwell number, magnetic number, viscosity parameter and slip velocity parameter, whereas the remaining parameters affect the Sherwood number in the opposite direction.

In addition, increasing the suction, slip and surface-convection parameters elevates the values of the local Nusselt number, but the opposite trend is shown for the remaining parameters. Last but not least, the thermal conductivity parameter can increase the local Sherwood number and the skin friction coefficient values while having an opposite impact on the Nusselt number.

Table 2. Values of $\frac{Cf_x}{2}Re_x^{\frac{1}{2}}$, $\frac{Nu_x}{\sqrt{Re_x}}$ and $\frac{Sh_x}{\sqrt{Re_x}}$ for various values of $\beta, f_w, M, \alpha, \lambda, Ec, R, \delta$ and ε with $Nb = 0.8, Le = 1.0, Pr = 1.0$ and $Nt = 0.1$.

β	f_w	M	α	λ	Ec	R	δ	ε	$\frac{Cf_x}{2}Re_x^{\frac{1}{2}}$	$\frac{Nu_x}{\sqrt{Re_x}}$	$\frac{Sh_x}{\sqrt{Re_x}}$
0.0	0.5	0.2	0.2	0.2	0.2	0.5	0.2	0.2	1.010101	0.106808	1.404140
0.2	0.5	0.2	0.2	0.2	0.2	0.5	0.2	0.2	0.980241	0.105987	1.361981
0.5	0.5	0.2	0.2	0.2	0.2	0.5	0.2	0.2	0.937128	0.104581	1.291542
0.1	0.0	0.2	0.2	0.2	0.2	0.5	0.2	0.2	0.845444	0.105775	1.168950
0.1	0.5	0.2	0.2	0.2	0.2	0.5	0.2	0.2	0.995036	0.106410	1.383612
0.1	0.8	0.2	0.2	0.2	0.2	0.5	0.2	0.2	1.095980	0.106979	1.536251
0.1	0.5	0.0	0.2	0.2	0.2	0.5	0.2	0.2	0.937012	0.107398	1.414660
0.1	0.5	0.5	0.2	0.2	0.2	0.5	0.2	0.2	1.071621	0.105056	1.342411
0.1	0.5	1.0	0.2	0.2	0.2	0.5	0.2	0.2	1.179310	0.103068	1.284550
0.1	0.5	0.2	0.0	0.2	0.2	0.5	0.2	0.2	1.003950	0.106514	1.388790
0.1	0.5	0.2	1.0	0.2	0.2	0.5	0.2	0.2	0.958801	0.105961	1.360981
0.1	0.5	0.2	2.5	0.2	0.2	0.5	0.2	0.2	0.888192	0.104945	1.308652
0.1	0.5	0.2	0.2	0.0	0.2	0.5	0.2	0.2	1.329681	0.104874	1.515341
0.1	0.5	0.2	0.2	0.2	0.2	0.5	0.2	0.2	0.995036	0.106410	1.383612
0.1	0.5	0.2	0.2	0.5	0.2	0.5	0.2	0.2	0.734743	0.106767	1.258561
0.1	0.5	0.2	0.2	0.2	0.0	0.5	0.2	0.2	0.997361	0.112547	1.383421
0.1	0.5	0.2	0.2	0.2	0.2	0.5	0.2	0.2	0.995036	0.106410	1.383612
0.1	0.5	0.2	0.2	0.2	0.5	0.5	0.2	0.2	0.991547	0.097316	1.383845
0.1	0.5	0.2	0.2	0.2	0.2	0.0	0.2	0.2	0.994306	0.152219	1.379111
0.1	0.5	0.2	0.2	0.2	0.2	0.5	0.2	0.2	0.995036	0.106410	1.383612
0.1	0.5	0.2	0.2	0.2	0.2	1.0	0.2	0.2	0.995541	0.081922	1.386090
0.1	0.5	0.2	0.2	0.2	0.2	0.5	0.0	0.2	0.997938	0.057677	1.389350
0.1	0.5	0.2	0.2	0.2	0.2	0.5	0.2	0.2	0.995036	0.106410	1.383612
0.1	0.5	0.2	0.2	0.2	0.2	0.5	0.5	0.2	0.988322	0.215834	1.370541
0.1	0.5	0.2	0.2	0.2	0.2	0.5	0.2	0.0	0.995008	0.108859	1.383361
0.1	0.5	0.2	0.2	0.2	0.2	0.5	0.2	1.5	0.995196	0.093315	1.384910
0.1	0.5	0.2	0.2	0.2	0.2	0.5	0.2	3.5	0.995396	0.079346	1.386282

4. Conclusions

Maxwell nanofluid flow caused by stretching surfaces has presented numerous challenges to the fluid mechanics research community as a result of widespread applications in the commercial and industrial sectors. As a result, the main goal of this work is to elucidate how the convective heating and viscous dissipation phenomena, which are connected to the variable thermo-physical properties, affect the slippery flow of the Maxwell MHD nanofluid toward a stretching horizontal sheet. The simplified reduced core governing equations are numerically solved using the shooting method. Graphs and tables are used to investigate how physical parameters affect fluctuations in velocity, temperature, concentration, skin-friction coefficient, Sherwood number and the Nusselt number. The following findings are achieved after computation and observation.

1. The increased Maxwell parameter, slip velocity parameter, viscosity parameter, magnetic number and suction parameter diminishes the nanofluid velocity.
2. Eckert number and surface-convection parameter values that are larger result in magnifying values for the temperature field.

3. The suction parameter, thermal conductivity parameter and magnetic parameter all raise the skin-friction coefficient.
4. The results showed that the existence of thermophoresis and Brownian motion makes the heat transmission phenomena more effective.
5. Higher radiation and suction parameter values result in a larger Sherwood number, while Maxwell and slip velocity parameter values result in a smaller Sherwood number.
6. A larger magnetic number, Brownian motion parameter, viscosity parameter and Maxwell parameter will result in a temperature rise whereas a higher suction parameter and slip velocity parameter will reduce the temperature.
7. The concentration of the nanofluid is severely degraded as the viscosity, magnetic number, Maxwell and slip velocity parameters drop.

Author Contributions: Data curation, H.A.; Formal analysis, M.A.; Funding acquisition, H.A.; Methodology, A.M.M. All authors have read and agreed to the published version of the manuscript.

Funding: This research received no external funding.

Institutional Review Board Statement: Not applicable.

Informed Consent Statement: Not applicable.

Data Availability Statement: Not applicable.

Acknowledgments: The authors wish to express their sincere thanks to the honorable referees for their valuable comments and suggestions to improve the quality of the paper.

Conflicts of Interest: The authors declare no conflict of interest.

Greek Symbols

ρ	nanofluid density (kg m^{-3})
ρ_∞	the ambient nanofluid density (kg m^{-3})
β	the dimensionless Maxwell parameter
β_1	the Maxwell coefficient (S)
μ	coefficient of viscosity (kg m^{-1}s^{-1})
μ_∞	the ambient nanofluid viscosity (kg m^{-1}s^{-1})
ν	kinematic viscosity (m^2 s^{-1})
ν_∞	the ambient kinematic viscosity (m^2 s^{-1})
θ	dimensionless temperature
ϕ	dimensionless concentration
λ_1	slip velocity factor (m)
λ	slip velocity parameter
σ	electrical conductivity (S m^{-1})
σ^*	Stefan–Boltzmann constant (W m^{-2} K^{-4})
δ	the surface convection parameter
η	similarity variable
κ	thermal conductivity (W m^{-1} K^{-1})
κ_∞	the ambient nanofluid thermal conductivity (W m^{-1} K^{-1})
ε	thermal conductivity parameter

Superscripts

$'$	differentiation with respect to η
∞	free stream condition
w	wall condition

Nomenclature

a	velocity coefficient (s^{-1})
A	is a constant ($K\ m^{-2}$)
B_0	strength of a uniform magnetic field (T)
c	is a constant ($mol\ L^{-1}\ m^{-2}$)
c_p	specific heat at constant pressure ($J\ kg^{-1}\ K^{-1}$)
C	nanoparticles concentration ($mol\ L^{-1}$)
Cf_x	skin friction coefficient
C_w	surface nanoparticles concentration ($mol\ L^{-1}$)
C_∞	ambient nanoparticles concentration ($mol\ L^{-1}$)
D_B	Brownian diffusion coefficient ($m^2\ s^{-1}$)
D_T	thermophoresis diffusion coefficient ($m^2\ s^{-1}$)
Ec	Eckret number
f	dimensionless stream function
f_w	suction parameter
h_f	the heat transfer coefficient ($W\ m^{-2}\ K^{-1}$)
k^*	mean absorption coefficient (m^{-1})
Le	Lewis parameter
M	magnetic parameter
Nb	Brownian motion parameter
Nt	thermophoresis parameter
Nu_x	local Nusselt number
Pr	Prandtl number
R	radiation parameter
Re_x	local Reynolds number
Sh_x	local Sherwood number
T	nanofluid temperature (K)
T_f	convection temperature (K)
T_∞	ambient temperature (K)
u	velocity component in the x-direction ($m\ s^{-1}$)
v	velocity component in the y-direction ($m\ s^{-1}$)
v_w	suction velocity ($m\ s^{-1}$)
x, y	Cartesian coordinates (m)

References

1. Choi, S.U.S. Enhancing thermal conductivity of fluids with nanoparticles. *ASME FED* **1995**, **231**, 99–103.
2. Yu, W.; France, D.M.; Routbort, J.L.; Choi, S.U.S. Review and comparison of nanofluid thermal conductivity and heat transfer enhancements. *Heat Transf. Eng.* **2008**, **29**, 432–460. [CrossRef]
3. Buongiorno, J. Convective transport in nanofluids. *ASME J. Heat Mass Transf.* **2006**, *128*, 240–250. [CrossRef]
4. Awais, M.; Hayat, T.; Irum, S.; Alsaedi, A. Heat generation/absorption effects in a boundary layer stretched flow of Maxwell nanofluid; Analytic and Numeric solutions. *PLoS ONE* **2015**, *10*, e0129814. [CrossRef]
5. Shafique, Z.; Mustafa, M.; Mushtaq, A. Boundary layer flow of Maxwell fluid in rotating frame with binary chemical reaction and activation energy. *Res. Phys.* **2016**, *6*, 627–633. [CrossRef]
6. Hayat, T.; Khan, M.I.; Waqas, M.; Alsaedi, A.; Khan, M.I. Radiative flow of micropolar nanofluid accounting thermophoresis and Brownian moment. *Int. J. Hydrogen Energy* **2017**, *42*, 16821–16833. [CrossRef]
7. Sharma, K.; Gupta, S. Viscous dissipation and thermal radiation effects in MHD flow of Jeffrey nanofluid through impermeable surface with heat generation/absorption. *Nonlinear Eng.* **2017**, *6*, 153–166. [CrossRef]
8. Patel, H.R.; Mittal, A.S.; Darji, R.R. MHD flow of micropolar nanofluid over a stretching/shrinking sheet considering radiation. *Int. Commun. Heat Mass Transf.* **2019**, *108*, 104322. [CrossRef]
9. Noor, N.A.M.; Shafie, S.; Admon, M.A. Slip effects on MHD squeezing flow of Jeffrey nanofluid in horizontal channel with chemical reaction. *Mathematics* **2021**, *9*, 1215. [CrossRef]
10. Mangathai, P.; Reddy, B.R.; Sidhartha, C. Unsteady MHD Williamson and Casson nanofluid flow in the presence of radiation and viscous dissipation. *Turk. J. Comput. Math. Educ.* **2021**, *12*, 1036–1051.
11. Alotaibi, H.; Althubiti, S.; Eid, M.R.; Mahny, K.L. Numerical treatment of MHD flow of Casson nanofluid via convectively heated non-linear extending surface with viscous dissipation and suction/injection effects. *Comput. Mater. Contin.* **2021**, *66*, 229–245. [CrossRef]

12. Sheikholeslami, M.; Abelman, S.; Ganji, D.D. Numerical simulation of MHD nanofluid flow and heat transfer considering viscous dissipation. *Int. J. Heat Mass Transf.* **2014**, *79*, 212–222. [CrossRef]
13. Hussain, S. Finite element solution for MHD flow of nanofluids with heat and mass transfer through a porous media with thermal radiation, viscous dissipation and chemical reaction effects. *Adv. Appl. Math. Mech.* **2017**, *9*, 904–923. [CrossRef]
14. Ibrahim, W.; Negera, M. Viscous dissipation effect on Williamson nanofluid over stretching/shrinking wedge with thermal radiation and chemical reaction. *J. Phys. Commun.* **2020**, *4*, 045015. [CrossRef]
15. Lund, L.A.; Omar, Z.; Khan, I.; Raza, J.; Sherif, E.M.; Seikh, A.H. Magnetohydrodynamic (MHD) flow of micropolar fluid with effects of viscous dissipation and joule heating over an exponential shrinking sheet: Triple solutions and stability analysis. *Symmetry* **2020**, *12*, 142. [CrossRef]
16. Saadi, F.A.; Worthy, A.; Alrihieli, H.; Nelson, M. Localised spatial structures in the Thomas model. *Math. Comput. Simul.* **2022**, *194*, 141–158. [CrossRef]
17. Li, Xi.; Alrihieli, H.; Algehyne, E.A.; Khan, M.A.; Alshahrani, M.Y.; Alraey, Y.; BilalRiaze, M. Application of piecewise fractional differential equation to COVID-19 infection dynamics. *Results Phys.* **2022**, *39*, 105685. [CrossRef]
18. Ramzan, M.; Bilal, M.; Chung, J.D.; Farooq, U. Mixed convective flow of Maxwell nanofluid past a porous vertical stretched surface—An optimal solution. *Results Phys.* **2016**, *6*, 1072–1079. [CrossRef]
19. Bardos, C.; Golse, F.; Perthame, B. The Rosseland approximation for the radiative transfer equations. *Commun. Pure Appl. Math.* **1987**, *40*, 691–721. [CrossRef]
20. Algehyne, E.A.; Alrihieli, H.; Saeed, A.; Alduais, F.S.; Hayat, A.U.; Kumam, P. Numerical simulation of 3D Darcy & Forchheimer fluid flow with the energy and mass transfer over an irregular permeable surface. *Sci. Rep.* **2022**, *12*, 14629.
21. Megahed, A.M. Improvement of heat transfer mechanism through a Maxwell fluid flow over a stretching sheet embedded in a porous medium and convectively heated. *Math. Comput. Simul.* **2021**, *187*, 97–109. [CrossRef]
22. Megahed, A.M.; Reddy, M.G.; Abbas, W. Modeling of MHD fluid flow over an unsteady stretching sheet with thermal radiation, variable fluid properties and heat flux. *Math. Comput. Simul.* **2021**, *185*, 583–593. [CrossRef]
23. Ishak, A.; Nazar, R.; Pop, I. Heat transfer over an unsteady stretching permeable surface with prescribed wall temperature. *Nonlinear Anal. Real World Appl.* **2009**, *10*, 2909–2913. [CrossRef]

mathematics

MDPI

Article

Numerical Investigation of the Magnetized Reactive Viscous Couple Stress Fluid Flow Down an Inclined Riga Plate with Variable Viscosity

Samuel Olumide Adesanya [1], Tunde Abdulkadir Yusuf [2,*] and Ramoshweu Solomon Lebelo [3]

[1] Department of Mathematical Sciences, Redeemer's University, Ede 232101, Nigeria
[2] Department of Mathematics, Faculty of Science, Adeleke University, Ede 232104, Nigeria
[3] Education Department, Vaal University of Technology, Private Bag X021, Vanderbijlpark 1911, South Africa
* Correspondence: tunde.yusuf@adelekeuniversity.edu.ng

Abstract: Accurate determination of optimum flow and heat transfer condition is one of the major challenges faced in the application of magnetic fluid in the field of medicine and engineering, especially when applied as ferrofluids for targeted drug deliveries, treatment of hyperthermia, sealants in computer hard drives, lubricants in car shafts. In view of these important applications, a mathematical investigation of the flow and heat transfer behavior of reactive magnetic fluids containing nanostructures is presented based on a couple of stress constitutive models. The reactive fluid is assumed to flow through inclined magnetized solid boundaries for energy conversion. The formulation leads to nonlinear coupled equations. The dimensionless equations are numerically solved using the spectral Chebyshev assumed solution for the weighted residual technique, and the correctness of the solution is confirmed using the shooting Runge–Kutta method. The effects of various fluid parameters on velocity, temperature, skin friction, and heat transfer rates are described in tabular and graphical form, along with suitable physical explanations. Thermal analysis computations are also presented. According to the findings, an enhanced couple of stress fluid and variable viscosity parameters reduced the skin drag and heat transfer rate at the bottom wall. Furthermore, the thermal stability of the flow can be achieved with increasing values modified Hartman number while increasing couple stress parameter encourages thermal instability in the flow domain.

Keywords: reactive magnetic fluid; couple stress; Riga surface; Chebyshev spectral method

MSC: 76D99; 76W99

Citation: Adesanya, S.O.; Yusuf, T.A.; Lebelo, R.S. Numerical Investigation of the Magnetized Reactive Viscous Couple Stress Fluid Flow Down an Inclined Riga Plate with Variable Viscosity. *Mathematics* **2022**, *10*, 4713. https://doi.org/10.3390/math10244713

Academic Editors: Efstratios Tzirtzilakis and Mario Versaci

Received: 11 November 2022
Accepted: 9 December 2022
Published: 12 December 2022

1. Introduction

The biomedical and rheological properties of ferrofluids have recently been widely studied. This is due to their applications in some areas of medicine and engineering [1–3]. Regarding real-world applications, the bulk of ferrofluids has non-Newtonian properties. The couple stress fluid is one of the most well-known non-Newtonian fluid models. The theory of couple stress is a broad notion of viscous fluid that allows for polar effects like the presence of couple stresses and body couples in classical theory. Couple stress fluids have a high viscosity and polar effects (Abbas et al. [4]). This is first considered by Stokes [5] and has been an area of potential interest in many physical and engineering processes. Hence, the study of flow and heat transfer analysis of couple stress fluid has captured the mind of several researchers [6–9]. The work in [4] examines the mixed convection flow of an electrically conducting couple stress fluid in an inclined channel. They noticed a reduction in the thermal field as the channel was tilted. The concept of channel flow having non-zero inclination has been instrumental in diverse industrial processes such as chemical processing, iron removal, and electrical system. Because of its applications, Sui et al. [10] examined the behavior of physical components on non-Newtonian fluid flow past an inclined surface. RamReddy et al. [11] reported the nonlinear convective flow of non-Newtonian liquid over an inclined porous regime. Significant change is observed as the angle of inclination is improved. Yusuf et al. [12] also presented a semi-analytical solution on a micropolar fluid flow over an inclined geometry with velocity slip. Very recently, Ahmad et al. [13] analytically examined Stoke's theory of couple stress fluid over an inclined channel under variable viscosity.

To improve the fluid thermophysical characteristics of some industrial liquids that are poor conductors of electricity, an external device known as a Riga plate is introduced to enhance the heat transfer attributes. Riga plate is an array of a magnetic bar supported by permanent magnets and alternating electrons. It could also act as an external device to boost fluid electricity. Gailitis and Lielausis [14] built a Riga plate under the influence of magnetic and electrical fields that generate a Lorentz force along the wall, which helps to control fluid flow. Given this, several researchers have discussed laminar flow under the influence of magnetic force along the Riga surface under various geometries afterward [15–18]. Nadeem et al. [19] studied the slip effect on a rotation fluid over a Riga channel. They discovered that the momentum boundary layer thickness declined due to the Riga surface. Naseem et al. [20] examined a grade three fluid flow with the influence of Cattaneo-Cristov heat flux over a Riga surface. Recently, stagnation point fluid flow and heat transfer over a vertical surface consisting of magnetic electrodes have been investigated by Khashi'ie et al. [21]. They indicated a diminution in the heat transfer rate due to an upsurge in the magnet and electrode width. Furthermore, fluid physical properties are affected by temperature changes. Previous heat conduction research relied heavily on the ambient fluid's unchanging physical properties. However, certain qualities, remarkably fluid viscosity, are known to alter with temperature. It is vital to account for the temperature-dependent fluctuation of viscosity in order to effectively describe flow and heat transfer rates. In this regard, the influence of temperature-dependent viscosity on a transient flow along a slanted surface is examined by Chinyoka et al. [22]. Further, Nadeem et al. [23] discussed nanofluid flow with magnetic and variable viscosity effects over a curved surface. They further established that thermal distribution declined for augmenting variable viscosity. Megahed et al. [24] considered the slip effect on a viscoelastic fluid flow over a surface with variable viscosity.

Except for a few problems, numerical algorithms have proven to be useful in solving continuum mechanics problems in which flow and heat transport are approximated by highly nonlinear differential equations. This typically results in the issue of convergence and computational time. Several approximate methods have become enormously popular in solving nonlinear models. However, of interest is the Chebyshev spectral collocation method. The main advantage of these methods lies in their accuracy for a given number of unknowns. For problems whose solutions are sufficiently smooth, they exhibit exponential convergence/spectral accuracy rates as in Mai-Duy and Tanner [25]. For instance, the method's efficiency is described in some different kinds of nonlinear partial differential equations by Khater et al. [26]. Comprehensive discussions on Chebyshev spectral collocation methods are available in different review articles [27–29]. The investigations mentioned above have demonstrated reactive flow along the Riga surface, but they have yet to examine how reactive fluid flow is across a Riga channel. This investigation will address this gap and offer a thorough understanding of fluid flow through heated, magnetic boundaries. As a result, three areas are the focus of our attention: first, the flow of a reactive couple stress fluid in incline walls with temperature-dependent viscosity; second, the impact of the fluid reactiveness through a magnetized heated channel; and third, the analysis of the numerical solution using a reliable solution technique. The Solutions to the nonlinear ODEs are generated via the Chebyshev spectral collocation technique, and results are displayed through graphs. Further, the solution compared with a different numerical scheme is also presented to corroborate our results.

2. Model Formulation

This paper establishes a continuous incompressible flow and heat transfer of a temperature-dependent reactive couple stress fluid in a non-porous inclined Riga plate. As demonstrated by the flow geometry (see Figure 1), the (x, y) Cartesian coordinate is chosen along the plates so that the x-axis is parallel to the flow and the y-axis is perpendicular to it. The plates are at an angle M to the horizontal. Furthermore, when considering a gravity-driven steady flow, the fully evolved incompressible fluid implies $V = (u(y),0,0)$, $T = T(y)$, as indicated in the flow geometry. The governing equations for this problem are based on the Cartesian system for continuity, momentum, and energy equations as follows:

$$\nabla.V = 0 \tag{1}$$

$$\rho\frac{DV}{Dt} = \nabla\tau - \eta\nabla^4 V + \rho f + \mathrm{J}\times\mathrm{B} \tag{2}$$

$$\rho C_P\frac{DT}{Dt} = k\nabla^2 T + \tau L + QAC_0 e^{-\frac{E}{RT}} \tag{3}$$

In Equations (1)–(3), V represents the velocity vector, ∇ is the grad operator, (ρ, C_P, t, T) —represents the fluid density, specific heat, time, and fluid temperature, respectively, $\left(k, \tau, L, \frac{D}{Dt}\right)$ are the thermal conductivity of the fluid, Cauchy stress tensor, the gradient of the velocity vector and material derivative. $(\eta, f, \mathrm{J}, \mathrm{B})$—couple stress coefficient, body force, the current density of electric field and a sum of the magnetic field, (Q, A, C_0), (E, R), Q—the heat of reaction, A—rate constant, C_0 initial reactant concentration, E—activation energy, R—universal gas constant,
where

$$\tau = -PI + \mu A_1,\ A_1 = L + L^T,\ \frac{D(*)}{Dt} = \frac{\partial(*)}{\partial t} + v.\nabla(*) \tag{4}$$

are the formal definitions of the Cauchy tensor, the First Rilivin–Ericksen tensor, and the material derivative.

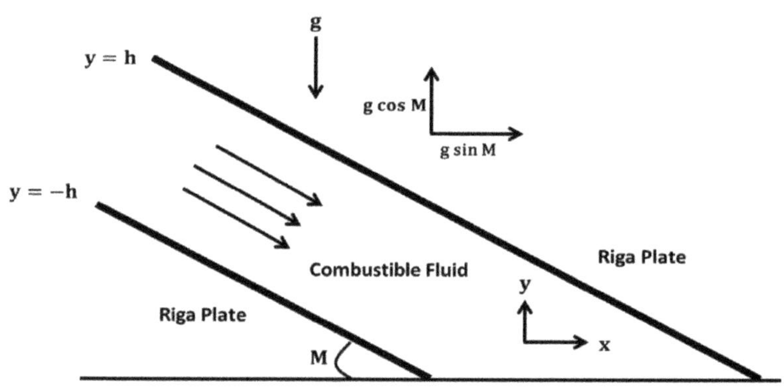

Figure 1. Flow Channel.

Using the Ginsberg approximation, then the Lorentz force is expressed as

$$\mathrm{J}\times\mathrm{B} \approx \sigma(E\times\mathrm{B}) = \frac{\pi}{8}J_0 M_0 e^{-\frac{\pi}{l}y'} \tag{5}$$

Given the combustible fluid assumptions, Equations (1)–(3) can be written as

$$\frac{\partial u'}{\partial x'} = 0 \tag{6}$$

$$0 = -\frac{dp}{dx} + \frac{d}{dy'}\left(\mu\prime\frac{du'}{dy'}\right) - \eta\frac{d^4u'}{dy'^4} + \rho g\cos M + \frac{\pi}{8}J_0 M_0 e^{-\frac{\pi}{l}y\prime} \tag{7}$$

$$0 = k\frac{d^2T}{dy'^2} + \mu'\left(\frac{du'}{dy'}\right)^2 + \eta\left(\frac{d^2u'}{dy'^2}\right)^2 + QAC_0 e^{-\frac{E}{RT}} \tag{8}$$

Together with the boundary conditions

$$u'(\pm h) = 0 = \frac{d^2u'}{dy'^2}(\pm h),\ T(\pm h) = T_0 \tag{9}$$

The inverse relationship between fluid temperature and temperature is a well-known phenomenon in the literature

$$\mu' = \mu_0 e^{-\alpha'(T-T_0)} \cong \mu_0(1 - \alpha'(T - T_0)) \tag{10}$$

Since viscosity variation temperature is usually, i.e., $0 < \alpha' << 1$. To make Equations (7)–(9) dimensionless, we introduce the following dimensionless variables and parameters

$$\begin{gathered} y = \tfrac{y'}{h},\ u = \tfrac{\mu_0 u'}{\rho g h^2 \cos M},\ H^2 = \tfrac{\pi J_0 M_0}{8\rho g\cos M}, \beta^2 = \tfrac{\mu_0 h^2}{\eta},\ \gamma = \tfrac{\pi h}{l},\ \theta = \tfrac{E(T-T_0)}{RT_0^2}, Nu = \tfrac{hEq_w}{kRT_0^2} \\ \alpha = \tfrac{\alpha' RT_0^2}{E}, \lambda = \tfrac{QEAh^2C_0 e^{-\frac{E}{RT_0}}}{T_0^2 RK},\ \varepsilon = \tfrac{RT_0}{E},\ \delta = \left(\tfrac{\rho g\cos M}{h}\right)^2 \tfrac{e^{\frac{E}{RT_0}}}{QC_0Ah^2}, C_F = \tfrac{\mu_0\tau_w}{h\rho g\cos M}. \end{gathered} \tag{11}$$

So that the dimensionless form of (7)–(9) becomes:

$$0 = 1 + \frac{d}{dy}\left((1-\alpha\theta)\frac{du}{dy}\right) - \frac{1}{\beta^2}\frac{d^4u}{dy^4} + H^2 e^{-\gamma y};\ u(\pm 1) = 0 = \frac{d^2u}{dy^2}(\pm 1) \tag{12}$$

$$0 = \frac{d^2\theta}{dy^2} + \lambda\left(e^{\frac{\theta}{1+\varepsilon\theta}} + \delta(1-\alpha\theta)\left(\frac{du}{dy}\right)^2 + \frac{\delta}{\beta^2}\left(\frac{d^2u}{dy^2}\right)^2\right);\quad \theta(\pm 1) \tag{13}$$

The shear stress at the wall τ_w and heat flux q_w are respectively expressed as

$$\tau_w = \mu'\frac{du'}{dy'} - \eta\frac{d^3u'}{dy'^3}\bigg|_{y'=-h} \quad\text{and}\quad q_w = -k\frac{dT}{dy'}\bigg|_{y'=-h} \tag{14}$$

Using Equation (11) in (14), we get

$$S_F = (1-\alpha\theta)\frac{du}{dy} - \frac{1}{\beta^2}\frac{d^3u}{dy^3}\bigg|_{y=-1},\qquad Nu = -\frac{d\theta}{dy}\bigg|_{y=-1} \tag{15}$$

3. Spectral Chebyshev Collocation Method of Solution

In view of the nonlinear nature of the model, Equations (12) and (13) are numerically solved. A collocation method based on the Chebyshev polynomial is remarkable and instrumental in addressing the solution to the boundary value problem. The numerical solutions obtained in [−1, 1] are expanded as a finite series in Chebyshev polynomial expressed as

$$u(y) \approx u^{Np}(y) = \sum_{i=0}^{Np} a_n T_n(y),\ \text{and}\ \theta(y) \approx \theta^{Np}(y) = \sum_{i=0}^{Np} b_n T_n(y) \tag{16}$$

where $\{a_n\}_{n=0}^{Np}$ and $\{b_n\}_{n=0}^{Np}$ are sets of expansion coefficients to be determined. Now, substituting $u^n(y)$ and $\theta^n(y)$ in Equations (12) and (13) yields the residual approximation of the form:

$$R_1(y) = 1 + \left(\left(1-\alpha\theta^{Np}\right)u_y^{Np}\right)' - \frac{1}{\beta^2}u_{y,y,y,y}^{Np} + Ha^2 e^{-\gamma y}, \tag{17}$$

$$R_2(y) = \theta_{y,y}^{Np} + \lambda\left(e^{\frac{\theta^{Np}}{1+\varepsilon\theta^{Np}}} + \delta\left(1-\alpha\theta^{Np}\right)\left(u_y^{Np}\right)^2 + \frac{\delta}{\beta^2}\left(u_{y,y}^{Np}\right)^2\right) \tag{18}$$

the approximation formula must be exact at y equal to y_i by the transformed zeros of the Chebyshev Gauss- Lobato collocation nodes to $[-1, 1]$.

$$y_i = \left(-\cos\left(\frac{i\pi}{Np}\right)\right), \quad i = 0, 1, 2, \ldots, Np \tag{19}$$

The relation

$$R_1(y_i) = 0, \text{ and } R_2(y_i) = 0, \quad i = 0, 1, 2, \ldots, Np \tag{20}$$

must be established. The rth-order derivative of the variables is then yielded through differentiation as

$$\frac{d^r u}{dy^r} = \sum_{n=0}^{Np} a_n \frac{d^r T_n}{dy^r} \text{ and } \frac{d^r \theta}{dy^r} = \sum_{n=0}^{Np} b_n \frac{d^r T_n}{dy^r} \tag{21}$$

The values of the derivative with $n = 1, 2, \ldots, r$ at the Gauss–Lobatto can be evaluated by

$$\begin{array}{lcl} \frac{d\widehat{u}}{dy} = D^{(1)}\widehat{u} = D\widehat{u} & & \frac{d\widehat{\theta}}{dy} = D^{(1)}\widehat{\theta} = D\widehat{\theta} \\ \frac{d^2\widehat{u}}{d^2y} = D^{(2)}\widehat{u} = D^2\widehat{u} & \text{and} & \frac{d^2\widehat{\theta}}{d^2y} = D^{(2)}\widehat{\theta} = D^2\widehat{\theta} \\ \cdots\ \cdots\ \cdots\ \cdots & & \cdots\ \cdots\ \cdots\ \cdots \\ \frac{d^r\widehat{u}}{d^ry} = D^{(r)}\widehat{u} = D^r\widehat{u} & & \frac{d^r\widehat{\theta}}{d^ry} = D^{(r)}\widehat{\theta} = D^r\widehat{\theta} \end{array} \tag{22}$$

where $\widehat{u} = (u(y_0), u(y_1), \ldots, u(y_{Np}))^T$ and $\widehat{\theta} = (\theta(y_0), \theta(y_1), \ldots, \theta(y_{Np}))^T$ are the vectors, $D^{(\bullet)}$ are the differential matrices. Using (16), the ordinary differential equations are reduced to systems of algebraic equations. For the sake of validation, Equations (12) and (13) are also solved iteratively via the Runge–Kutta–Fehlberg (RKF) integration technique. This computation is carried out on a computer symbolic package. The RKF is introduced to guarantee the accuracy of the present method.

4. Results and Discussion

In this section, the solutions obtained using Chebyshev polynomial as an admissible trial function for the nonlinear problem are displayed graphically and explained for the profile of velocity, the temperature distributions, the coefficient of skin friction, and the local Nusselt number are presented for various emerging parameters. Tabular results are also presented for comparison in Tables 1–3. In Tables 4 and 5, stability analysis and convergence results are displayed. Table 1 establishes the correctness of the spectral collocation method when compared with its exact solution. The uniqueness of the solution is confirmed since $\alpha << 1$ is infinitely small. Therefore, the Chebyshev trial function provides a very good approximation for the uncoupled problem. Moreover, for the Coupled system when $\alpha \neq 0$, tables for the comparison of the numerical solutions by spectral collocation and the Runge–Kutta–Fehlberg (RKF) integration technique are presented for Equations (12) and (13). In Tables 2 and 3, the error ranges from $10^{-17}$and 10^{-8} for Equation (12) which means that the two results converge to one. The error range for the solution of (13) is $10^{-18}$and 10^{-8}. Therefore, there is a perfect agreement between the two numerical solutions; as a result, the approximation is okay. Table 4 represents the influence of flow parameters on the wall friction and heat transfer rate. The result revealed that increasing values of the viscous dissipation parameter reduces the wall skin friction while it increases the wall Nusselt number. This is true due to the irreversible work done in converting kinetic energy to heat energy. Furthermore, the Nusselt number increases due to decreasing heat transfer by conduction across the channel width.

Moreover, magnets and electrodes at the solid boundaries increase the skin friction and the wall Nusselt number. This shows that flow resistance and rate of heat transfer at the wall become higher with the increasing values of the modified Hartmann number. Similar behavior is noticed with increasing couple stress inverse parameters. Finally, increasing values of the viscosity variation parameter show that momentum transfer from the laminar flow to the wall increases with the variation of the viscosity parameter. Table 5 shows the flow parameters' effect on the flow's thermal stability. The result shows that the activation energy parameter increases the critical values, thus stabilizing the flow. The contribution of the modified Hartman number is seen to encourage thermal instability in the flow. This is correct physically since the magnetic term increases the fluid temperature. It is expected to encourage thermal runaway in the flow domain. A similar explanation holds for the viscosity variation parameter and the inverse of the couple stress parameter in terms of encouraging thermal instability in the flow domain. In Table 6, the convergence of the Chebyshev series is obtained when the number of terms in the approximating polynomial is 10 for the given parameter values.

Table 1. Comparison of Exact solution and Numerical solution when $\alpha = 0$, $\beta = 0.1$, $H = 1$, $\gamma = 1$.

y	Exact	SCM	Absolute Error
−1	$-5.03992583264522 \times 10^{14}$	0.000000000000000000	$5.039925832645223 \times 10^{14}$
−0.75	0.0017454436901435553	0.001745454926416422	$1.123627286700457 \times 10^{-8}$
−0.5	0.003173227582038684	0.003173231306211247	$3.724172563217276 \times 10^{-9}$
−0.25	0.004076000485437167	0.004076004401859833	$3.916422666207231 \times 10^{-9}$
0	0.0043502281691282264	0.004350231164298437	$2.995170210809417 \times 10^{-9}$
0.25	0.003981732876595737	0.003981735066616083	$2.190020346494459 \times 10^{-9}$
0.5	0.0030344730267644108	0.003034472001333134	$1.025431276783367 \times 10^{-9}$
0.75	0.0016418517123099419	0.001641848839780530	$2.872529411600613 \times 10^{-9}$
1	$4.46751760135205 \times 10^{-14}$	$-4.910861206864707 \times 10^{-9}$	$4.910905882040721 \times 10^{-9}$

Table 2. Validation of result of Equation (12) when $\delta = \beta = \gamma = H = 1$, $\lambda = \varepsilon = \alpha = 0.1$.

y	$u(y)_{SCM}$	$u(y)_{RK4}$	$\lvert u(y)_{SCM} - u(y)_{RK4} \rvert$
−1	1.35477×10^{-17}	0.000000	1.35477×10^{-17}
−0.75	0.126368	0.126368	3.73265×10^{-9}
−0.50	0.228617	0.228617	7.89402×10^{-10}
−0.25	0.292212	0.292212	1.68506×10^{-9}
0.0	0.310659	0.310659	1.45745×10^{-9}
0.25	0.283667	0.283667	1.01772×10^{-10}
0.50	0.216004	0.216004	3.25084×10^{-9}
0.75	0.116914	0.116914	8.42008×10^{-9}
1	-7.58235×10^{-19}	1.41621×10^{-8}	1.41621×10^{-8}

Table 3. Validation of result of Equation (13) when $\delta = \beta = \gamma = H = 1$, $\lambda = \varepsilon = \alpha = 0.1$.

y	$\theta(y)_{WRM}$	$\theta(y)_{RK4}$	$\mid \theta(y)_{WRM} - \theta(y)_{RK4} \mid$
−1.0	-2.32225×10^{-18}	0.0000000	2.32225×10^{-18}
−0.75	0.0333263	0.0333263	6.36403×10^{-9}
−0.50	0.0578145	0.0578142	6.82415×10^{-9}
−0.25	0.0725453	0.0725453	6.93838×10^{-9}
0.0	0.0770979	0.0770979	1.05644×10^{-8}
0.25	0.0715674	0.0715674	1.21143×10^{-8}
0.50	0.0564163	0.0564163	1.43481×10^{-8}
0.75	0.0323197	0.0323197	1.81587×10^{-8}
1.0	-2.63179×10^{-19}	-2.03698×10^{-8}	2.03698×10^{-8}

Table 4. Skin Friction and Nusselt number for various fluid parameters.

H	δ	ε	λ	α	β	γ	S_F	Nu
1	1	0.3	0.5	0.1	1	1	2.54320	0.89053
1	2	0.3	0.5	0.1	1	1	2.54319	1.17403
1	3	0.3	0.5	0.1	1	1	2.54318	1.46102
2	1	0.3	0.5	0.1	1	1	7.17259	2.58254
3	1	0.3	0.5	0.1	1	1	14.8866	9.30193
1	1	0.3	0.5	0.1	2	1	2.54350	1.20465
1	1	0.3	0.5	0.1	3	1	2.54385	1.36500
1	1	0.3	0.5	0.2	1	1	2.54333	0.89233
1	1	0.3	0.5	0.3	1	1	2.54345	0.89417

Table 5. Computation of Critical values of Frank–Kameneskii parameter.

δ	ε	α	β	γ	H	λ_C
1	0.1	0.1	1	1	1	0.861553
1	0.2	0.1	1	1	1	1.046770
1	0.3	0.1	1	1	1	1.838140
1	0.1	0.1	1	1	2	0.560669
1	0.1	0.1	1	1	3	0.305436
1	0.1	0.2	1	1	1	0.858432
1	0.1	0.3	1	1	1	0.855150
1	0.1	0.1	2	1	1	0.793590
1	0.1	0.1	3	1	1	0.778710

Table 6. Rapid convergence of the Spectral Weighted Residual Method.

N	δ	ε	α	β	γ	H	λ_C
5	1	0.1	0.1	1	1	1	0.961880
10	1	0.1	0.1	1	1	1	0.861512
15	1	0.1	0.1	1	1	1	0.861553
20	1	0.1	0.1	1	1	1	0.861553
25	1	0.1	0.1	1	1	1	0.861553
30	1	0.1	0.1	1	1	1	0.861553

Figure 2 relates the influence of the viscosity variation parameter on the velocity profile. The graphical result shows that flow velocity rises with increasing value of the viscosity variation parameter, showing that viscous fluid becomes lighter with thermal effect. In other words, the heat transfer to the fluid from an exothermic chemical reaction has a thinning effect on the fluid viscosity. In Figure 3, the effect of the viscosity variation parameter on the temperature distribution within the flow channel is presented. The result shows that the viscosity parameter increases the maximum temperature. This effect is directly connected with the irreversible energy conversion from kinetic to heat. The effect of nanoparticles on the flow and thermal distribution is displayed n Figures 4 and 5. These nanoparticles could present as drugs for a specific treatment, polymer additives to enhance the fluid rheological properties of some lubricants, etc. As these nano-sized particles are added, internal friction is created, and the velocity of the fluid decreases, as shown in the inverse of the couple stress parameter. Evidently, as the fluid thickens, inter-molecular interaction declines. This is also associated with a decline in fluid temperature as observed in the plot of temperature distribution with couple stress inverse.

Figures 6 and 7 demonstrate the impact of modified Hartmann number (H) on both velocity and temperature distributions, respectively. It is noticed in Figure 6 that an improvement in the velocity profile occurred as the modified Hartmann number (H) increased. This is because the external magnetic field strengthens as the values of the modified Hartmann number (H) become more significant, consequently improving the flow motion. This is a well-behaved solution since $+H^2e^{\eta y}$ represents a positive definite function as used in Equation (12) for all values of $Ha,\ \eta$. Moreover, increasing the same parameter (H) leads to an upsurge in temperature distribution. This effect can be seen in the momentum transfer through viscous dissipation, which acts as a heat source in the flow domain, as seen in Figure 7.

The influence of the velocity and temperature field against the Frank–Kameneskii parameter (λ) is displayed in Figures 8 and 9, respectively. The emerged parameter results from the Arrhenius kinetics that releases energy from heat into the fluid stream. As depicted in Figure 8, an increment in the Frank–Kameneskii parameter is noticed to enhance flow velocity. This is physically correct due to the heat of the reaction that produces a melting effect on the viscous fluid. The thinning effect guarantees enhanced flow down the inclined channel. The exothermic nature of the Arrhenius kinetics is seen to increase the temperature distribution as the Frank–Kameneskii parameter (λ) value increases, as shown in Figure 9. The diagram for the bifurcation slice that is typical of all laminar flames in combustion problems is illustrated in Figure 10. This shows the effect of the maximum temperature against the Frank–Kameneskii parameter (λ). In the region I and II where $\lambda_c < 0.861553$ two values of λ exist for the maximum temperature. Only one solution exists at the point when $\lambda_c = 0.861553$ while no solution is possible beyond $\lambda_c > 0.861553$ where heat generation is far greater than heat dissipation, this corresponds to the hazardous blow-up point where spontaneous ignition occurs. Accurate determination of critical values is vital in ensuring the safety of lives and properties

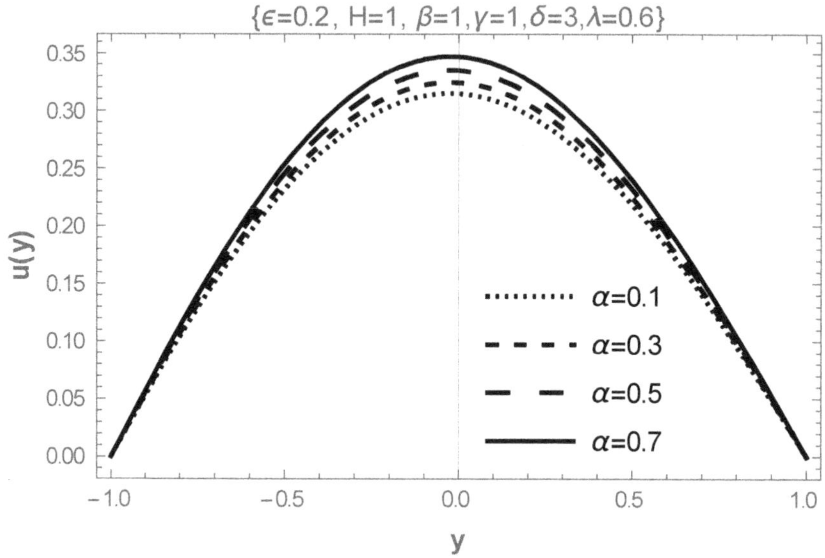

Figure 2. Velocity profile with viscosity variation parameter (α).

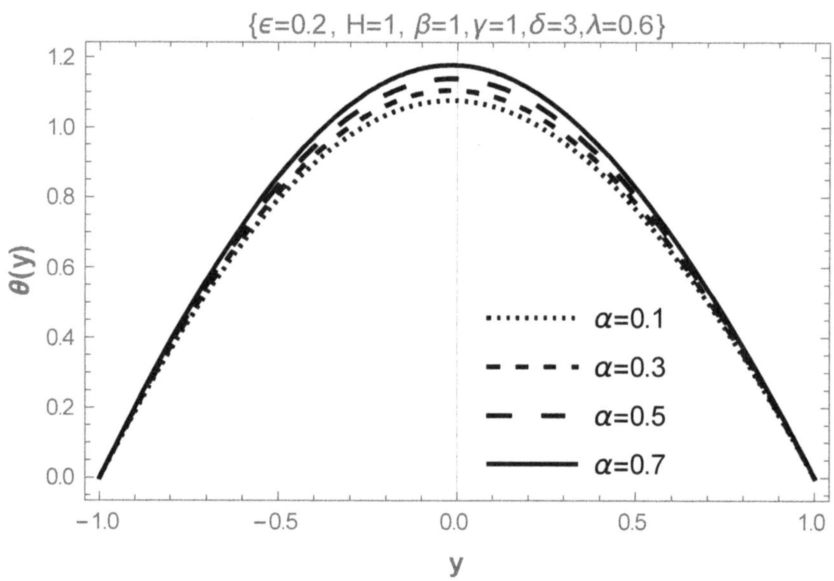

Figure 3. Thermal distribution with viscosity variation parameter (α).

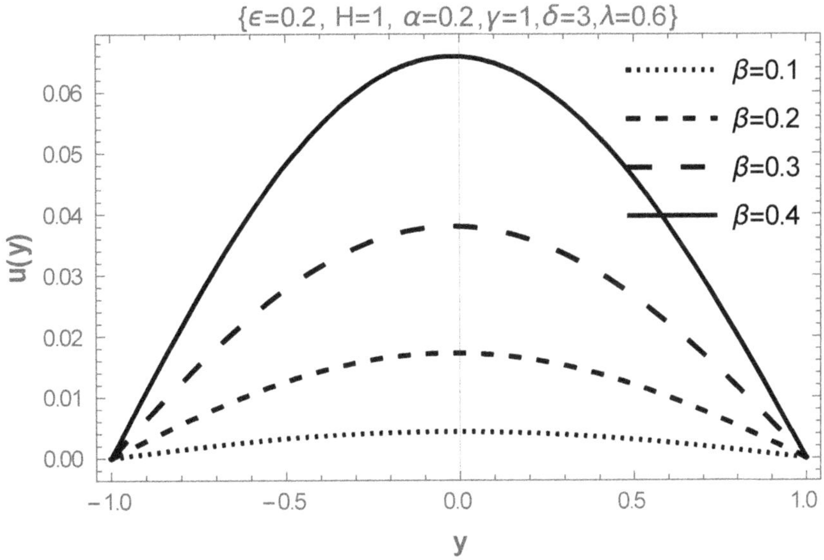

Figure 4. Velocity profile with couple stress inverse parameter (β).

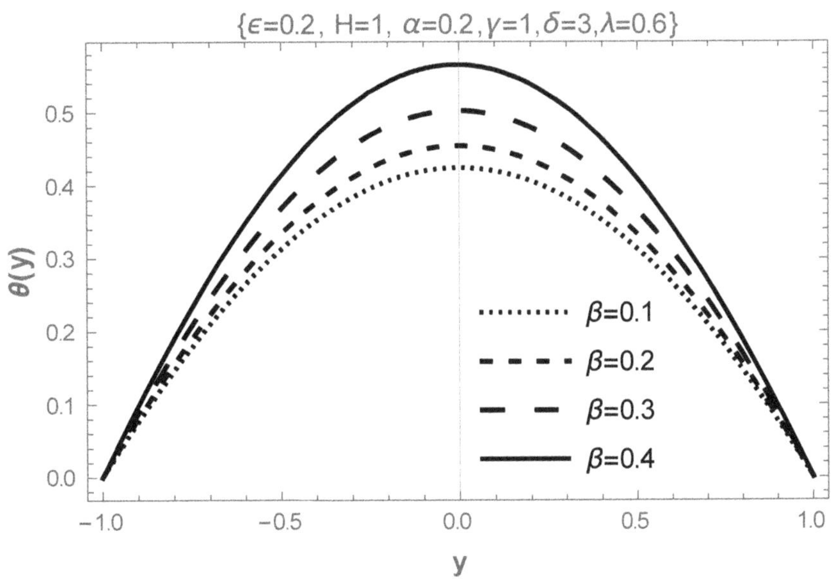

Figure 5. Temperature distribution with couple stress inverse parameter (β.

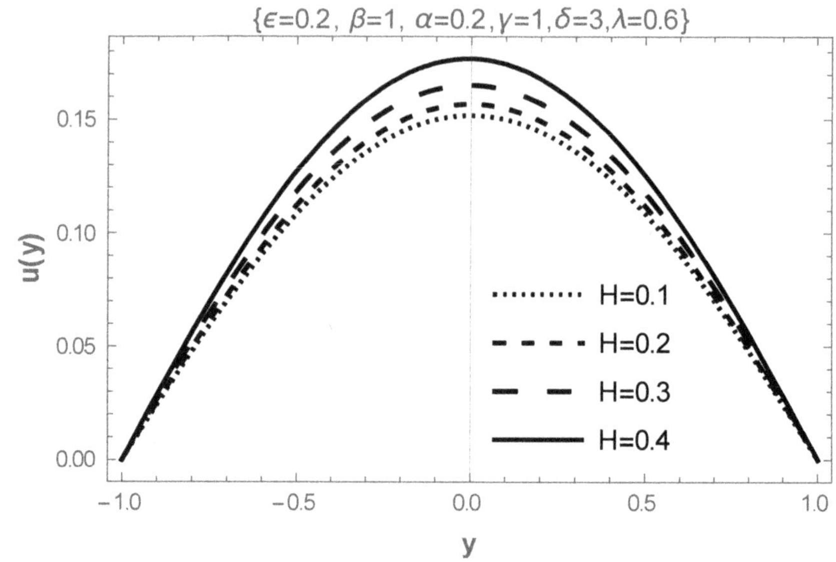

Figure 6. Velocity profile with modified Hartman number (*H*).

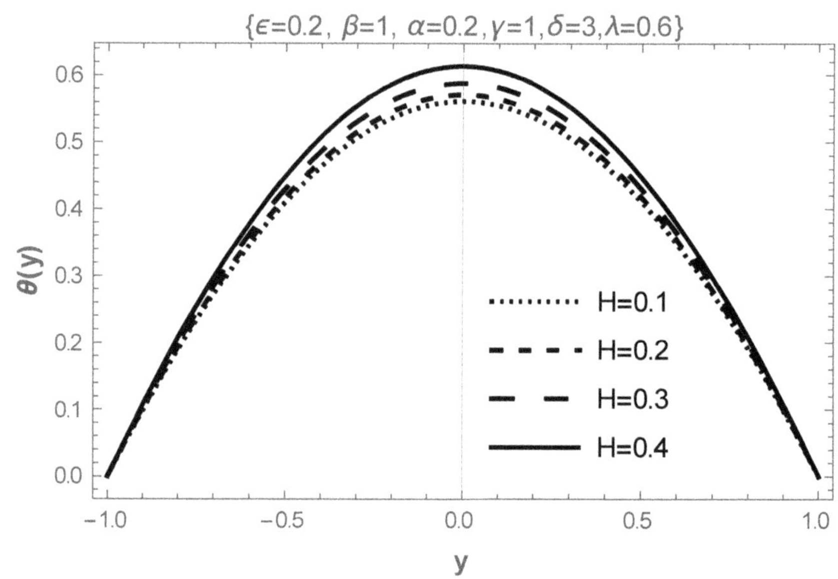

Figure 7. Temperature distribution with modified Hartman number (*H*).

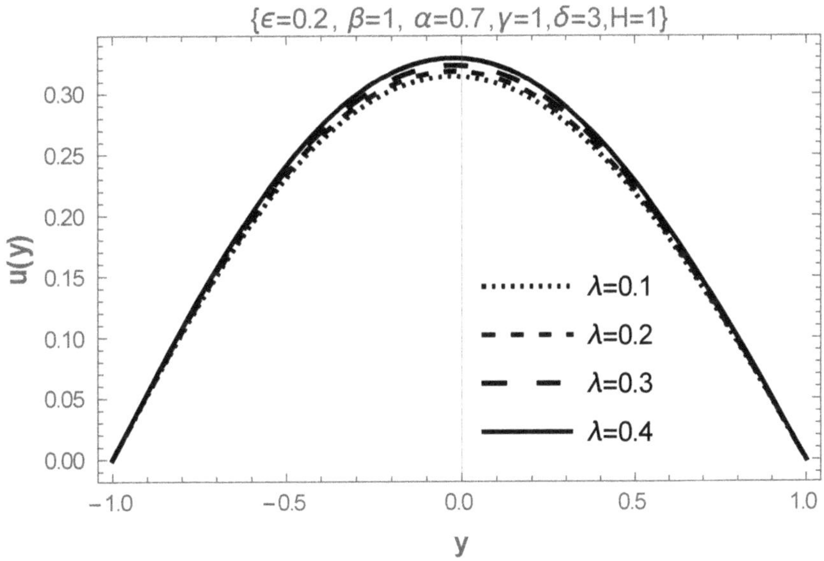

Figure 8. Velocity profile with Frank–Kameneskii parameter (λ).

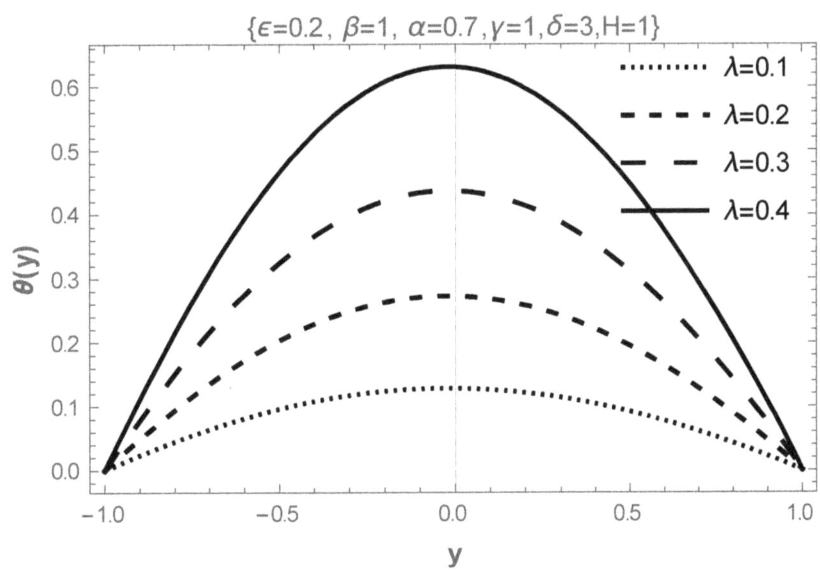

Figure 9. Temperature distribution with Frank–Kameneskii parameter (λ).

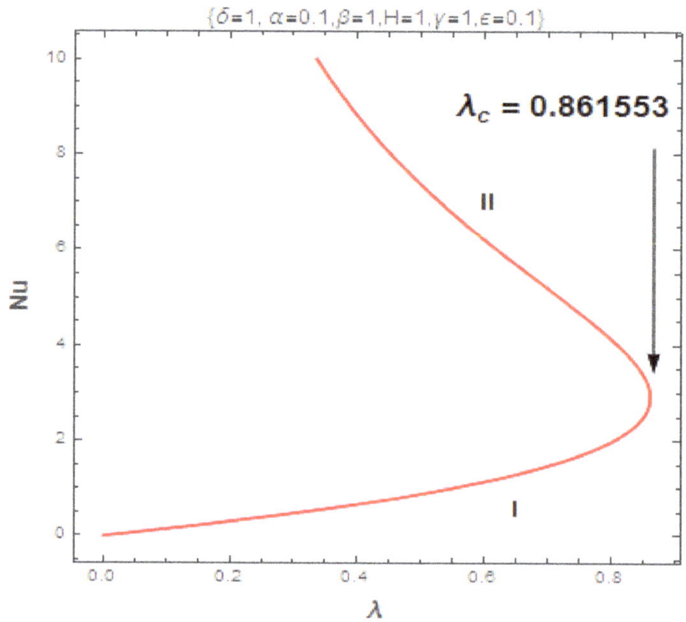

Figure 10. Bifurcation slice.

5. Conclusions

A mathematical model of a magnetized reactive variable viscous couple stress fluid flow down an inclined Riga plate has been investigated. Solutions are obtained numerically using the spectral collocation technique. Runge–Kutta–Fehlberg (RKF) integration technique is additionally employed in the model to validate the spectral collocation method. Tables and graphical results are available and analyzed. The significant outcomes of this current investigation are as follows:

- The fluid motion strengthened with an increase in the modified Hartman number and the couple stress parameter.
- As the fluid viscosity varies, the fluid motion and the heat balance enhance along the inclined channel.
- Large values of the Frank–Kameskii parameter significantly boost the thermal field. Also, the flow becomes thermal stable with an activation energy parameter increase.
- The velocity gradient at the lower wall increases for larger values of modified Hartman and couple stress parameters while slightly dropping for increasing values of the viscous heating parameter.
- At the lower wall, the heat transfer rate dramatically improves as the values of modified Hartman, viscous heating, and couple stress parameters enhance.

Author Contributions: Conceptualization, S.O.A.; Data curation, T.A.Y.; Formal analysis, S.O.A. and R.S.L.; Funding acquisition, R.S.L.; Investigation, T.A.Y.; Methodology, S.O.A.; Project administration, R.S.L.; Resources, T.A.Y.; Software, S.O.A.; Writing original draft, S.O.A. All authors have read and agreed to the published version of the manuscript.

Funding: This research received no external funding.

Conflicts of Interest: The authors declare no conflict of interest.

Nomenclature

M	Inclination angle of channel
M_0	magnetization of the magnets
J_0	applied current density in the electrodes
g	acceleration due to gravity
H^2	modified Hartman number
Q	heat of reaction
A	rate constant
C_0	initial concentration of the reactant species
E	activation energy
R	universal gas constant
T	fluid temperature [K]
K	thermal conductivity $[Wm^{-1} K^{-1}]$
l	magnets and electrodes width
u', u	dimensinal and dimensionless velocity $[ms^{-1}]$
h	channel width [m]
x, y	Cartesian coordinates [m]
p	fluid pressure
T_0	geometry wall temperature
Greek	
μ'	dynamic viscosity [Pas]
μ_0	viscosity constant
σ	electrical conductivity
ρ	density of the fluid $[kgm^{-3}]$
α	temperature-dependent viscous parameter
α'	temperature-dependent viscous parameter
β	couple stress parameter
θ	dimensionless temperature
γ	dimensionless parameter
λ	Frank–Kameneskii parameter
ε	activation energy parameter,
δ	viscous heating parameter
η	coefficient of couple stress

References

1. Odenbach, S. Recent progress in magnetic fluid research. *J. Phys. Condens. Matter* **2004**, *16*, R1135–R1150. [CrossRef]
2. Kozissnik, B.; Bohórquez, A.; Dobson, J.; Rinaldi, C. Magnetic fluid hyperthermia: Advances, challenges, and opportunity. *Int. J. Hyperth.* **2013**, *29*, 706–714. [CrossRef] [PubMed]
3. Rinaldi, C.; Chaves, A.; Elborai, S.; He, X.; Zahn, M. Magnetic fluid rheology and flows. *Curr. Opin. Colloid Interface Sci.* **2005**, *10*, 141–157. [CrossRef]
4. Abbas, Z.; Hasnain, J.; Sajid, M. Hydromagnetic Mixed Convective Two-Phase Flow of Couple Stress and Viscous Fluids in an Inclined Channel. *Z. Nat. A* **2014**, *69*, 553–561. [CrossRef]
5. Stokes, V.K. Couple stresses in fluids. *Phys. Fluids* **1966**, *9*, 1709–1715. [CrossRef]
6. Hayat, T.; Aziz, A.; Muhammad, T.; Ahmad, B. Influence of Magnetic Field in Three-Dimensional Flow of Couple Stress Nanofluid over a Nonlinearly Stretching Surface with Convective Condition. *PLoS ONE* **2015**, *10*, e0145332. [CrossRef]
7. Makinde, O.D.; Eegunjobi, A.S. Entropy Generation in a Couple Stress Fluid Flow Through a Vertical Channel Filled with Saturated Porous Media. *Entropy* **2013**, *15*, 4589–4606. [CrossRef]
8. Adesanya, S.O.; Makinde, O.D. Effects of couple stresses on entropy generation rate in a porous channel with convective heating. *Comput. Appl. Math.* **2015**, *34*, 293–307. [CrossRef]
9. Tripathi, D.; Yadav, A.; Beg, O.A. Electro-osmotic flow of couple stress fluids in a micro-channel propagated by peri-stalsis. *Eur. Phys. J. Plus* **2013**, *132*, 173. [CrossRef]
10. Sui, J.; Zheng, L.; Zhang, X.; Chen, G. Mixed convection heat transfer in power law fluids over a moving conveyor along an inclined plate. *Int. J. Heat Mass Transf.* **2015**, *85*, 1023–1033. [CrossRef]
11. RamReddy, C.; Naveen, P.; Srinivasacharya, D. Nonlinear Convective Flow of Non-Newtonian Fluid over an Inclined Plate with Convective Surface Condition: A Darcy–Forchheimer Model. *Int. J. Appl. Comput. Math.* **2018**, *4*, 51. [CrossRef]
12. Yusuf, T.; Kumar, R.N.; Prasannakumara, B.; Adesanya, S. Irreversibility analysis in micropolar fluid film along an incline porous substrate with slip effects. *Int. Commun. Heat Mass Transf.* **2021**, *126*, 105357. [CrossRef]

13. Ahmad, F.; Nazeer, M.; Ali, W.; Saleem, A.; Sarwar, H.; Suleman, S.; Abdelmalek, Z. Analytical Study on Couple Stress Fluid in an inclined Channel. *Sci. Iran.* **2021**, *28*, 2164–2175. [CrossRef]
14. Gailitis, A.; Lielausis, O. On a possiblilty to reduce the hydrodynamical resistance of a plate in an electrolyte. *Appl. Magnetohydrodyn.* **1961**, *12*, 143–146.
15. Zaib, A.; Haq, R.U.; Chamkha, A.J.; Rashidi, M.M. Impact of partial slip on mixed convective flow towards a Riga plate comprising micropolar TiO_2-kerosene/water nanoparticles. *Int. J. Numer. Methods Heat Fluid Flow* **2019**, *29*, 1647–1662. [CrossRef]
16. Nayak, M.; Shaw, S.; Makinde, O.D.; Chamkha, A.J. Investigation of Partial Slip and Viscous Dissipation Effects on the Radiative Tangent Hyperbolic Nanofluid Flow Past a Vertical Permeable Riga Plate with Internal Heating: Bungiorno Model. *J. Nanofluids* **2019**, *8*, 51–62. [CrossRef]
17. Nadeem, S.; Malik, M.Y.; Abbas, N. Heat transfer of three-dimensional micropolar fluid on a Riga plate. *Can. J. Phys.* **2019**, *98*, 32–38. [CrossRef]
18. Islam, R.; Nasrin, S. Micropolar Fluid Flow Along with an Inclined Riga Plate Through a Porous Medium. *Int. J. Heat Technol.* **2021**, *39*, 1123–1133. [CrossRef]
19. Abbas, N.; Nadeem, S.; Malik, M. Theoretical study of micropolar hybrid nanofluid over Riga channel with slip conditions. *Phys. A Stat. Mech. Its Appl.* **2020**, *551*, 124083. [CrossRef]
20. Naseem, A.; Shafiq, A.; Zhao, L.; Farooq, M.U. Analytical investigation of third grade nanofluidic flow over a Riga plate using Cattaneo-Christov model. *Results Phys.* **2018**, *9*, 961–969. [CrossRef]
21. Khashi'Ie, N.S.; Arifin, N.; Pop, I. Mixed Convective Stagnation Point Flow towards a Vertical Riga Plate in Hybrid Cu-Al_2O_3/Water Nanofluid. *Mathematics* **2020**, *8*, 912. [CrossRef]
22. Chinyoka, T.; Goqo, S.; Olajuwon, B. Computational analysis of gravity driven flow of a variable viscosity viscoelastic fluid down an inclined plane. *Comput. Fluids* **2013**, *84*, 315–326. [CrossRef]
23. Nadeem, S.; Ahmed, Z.; Saleem, S. Carbon nanotubes effects in magneto nanofluid flow over a curved stretching surface with variable viscosity. *Microsyst. Technol.* **2019**, *25*, 2881–2888. [CrossRef]
24. Megahed, A.M.; Reddy, M.G. Numerical treatment for MHD viscoelastic fluid flow with variable fluid properties and viscous dissipation. *Indian J. Phys.* **2021**, *95*, 673–679. [CrossRef]
25. Mai-Duy, N.; Tanner, R. A spectral collocation method based on integrated Chebyshev polynomials for two-dimensional biharmonic boundary-value problems. *J. Comput. Appl. Math.* **2007**, *201*, 30–47. [CrossRef]
26. Khater, A.H.; Temsah, R.S. Numerical solutions of some nonlinear evolution equations by Chebyshev spectral collocation methods. *Int. J. Comput. Math.* **2007**, *84*, 305–316. [CrossRef]
27. Khater, A.; Temsah, R.; Hassan, M. A Chebyshev spectral collocation method for solving Burgers'-type equations. *J. Comput. Appl. Math.* **2008**, *222*, 333–350. [CrossRef]
28. Mai-Duy, N. An effective spectral collocation method for the direct solution of high-order ODEs. *Commun. Numer. Methods Eng.* **2005**, *22*, 627–642. [CrossRef]
29. Dehghan, M.; Fakhar-Izadi, F. The spectral collocation method with three different bases for solving a nonlinear partial differential equation arising in modeling of nonlinear waves. *Math. Comput. Model.* **2011**, *53*, 1865–1877. [CrossRef]

mathematics

MDPI

Article

Finite Element Analysis of Generalized Thermoelastic Interaction for Semiconductor Materials under Varying Thermal Conductivity

Aatef Hobiny [1] and Ibrahim Abbas [1,2,*]

1 Mathematics Department, Faculty of Science, King Abdulaziz University, Jeddah 21589, Saudi Arabia
2 Mathematics Department, Faculty of Science, Sohag University, Sohag 82524, Egypt
* Correspondence: ibrabbas7@science.sohag.edu.eg

Abstract: In this work, we consider the problem of a semiconductor half-space formed of varying thermal conductivity materials with and without Kirchhoff's transforms. Specifically, we deal with one thermal relaxation time within the context of generalized photothermoelastic theory. It is expected that the thermal conductivity of the material will vary with temperature. The finite element method is used to numerically solve this problem. The Laplace transform and the eigenvalues method are used to determine analytical solutions to the linear problem. Various hypotheses are investigated, both with and without the use of Kirchhoff's transformations, to consider the influence of thermal conductivity change. To verify the accuracy of the proposed approach, we provide a comparison of numerical and analytical results by ignoring the new parameters and investigating the behaviors of physical quantities for numerical outcomes.

Keywords: finite element method; variable thermal conductivity; semiconductor materials; thermal relaxation time

MSC: 35Q81

Citation: Hobiny, A.; Abbas, I. Finite Element Analysis of Generalized Thermoelastic Interaction for Semiconductor Materials under Varying Thermal Conductivity. *Mathematics* **2022**, *10*, 4676. https://doi.org/10.3390/math10244676

Academic Editor: Ramoshweu Solomon Lebelo

Received: 16 November 2022
Accepted: 6 December 2022
Published: 9 December 2022

1. Introduction

The versatility of semiconducting materials in modern technology makes them a vital topic for study in the sciences. Most of these studies concentrate on investigating various forms of renewable energy. Semiconductors are a model for the utilization of renewable energy when they are exposed to sunshine. Neither the internal structures of these materials when subjected to external fields and activated by a laser beam, nor the link between thermal conductivity and temperature were considered in most previous experiments. Deformations during the microinertia of the microelement contribute to the body's temperature rise, in addition to external and internal thermal causes. In many of these uses [1], the effects of sunlight or laser beams on the surface of semiconductors materials are examined without considering the media's internal structures. Different materials, especially temperature-dependent semiconductor devices, require different thermal load analyses. The micromechanical structures of thermal, elastic and plasma field in Green and Naghdi theory has been previously analyzed by Todorovic et al. [2–4]. The theory developed by Lord and Shulman [5] uses a single relaxation time to compute the motion generated by a finitely fast thermal field. Marin et al. [6] studied the extensions of the domain of influences theory for generalized thermoelastic of anisotropic materials with voids. Ezzat and El-Bary [7] studied the effects of fractional derivatives and magnetic field in thermoelastic material under phase-lag GN models. Abbas [8] applied eigenvalue approaches to the fractional order model of thermo-diffusion problems for an unbounded elastic medium with spherical cavities. Abbas et al. [9] studied the propagations of waves in a generalized thermoelastic plane by an eigenvalue approach. Alharbi et al. [10] studied the influences of initial stresses and varying thermal conductivity on fiber-reinforced magneto-thermoelastic media under microtemperatures. Shuangquan and Tianhu [11] presented a study on the transient responses of porous mediums with strain and thermal relaxations. Abouelregal and Tiwari [12] studied the effects of memory-dependent heat conduction on thermoelastic vibrations of a nano-sized rotating beam with varying thermal properties under axial load. Several authors [13–20] have proposed solutions to various problems by using the thermoelasticity theory. Ailawalia and Kumar [21] investigated how photothermal interactions manifest in semiconductor media due to ramp-type heating. Abbas et al. [22] looked at photothermal interactions in semiconductors using the DPL model. Heat transfer in convective fins of varying thermal conductivity and heating generation was studied by Ghasemi et al. [23]. Energy pile displacement under thermal and mechanical loading was the subject of a numerical study by Yang et al. [24]. Using photo-thermoelastic excitations, Lotfy et al. [25] discussed the Thomson and electromagnetic effects of laser pulses on semiconductor materials. The effects of ramp-type heating on photo-thermo-elastic waves in a semi-conductor have been investigated by Hobiny et al. [26]. Mohamed et al. [27] looked at the absorption illuminations of semi-infinite thermoelastic materials with a rotator in two dimensions using a modified GL model.

Most thermoelastic studies consider thermal conductivity independent of temperature, which is only the case for some situations. At higher temperatures, as common in pipes conveying hot flow, missiles, nuclear reactors, etc., material properties may not remain constant. In the view of variable thermal conductivity with Kirchhoff's transforms, Youssef et al. [28] looked at the temperature dependence of the thermal conductivity and elastic modulus of a material in an unbounded medium containing spherical cavities. In [29], Sherief and Hamza model a thermoelastic hollow cylinder with variable thermal conductivity. Khoukhi et al. [30] evaluated the impact of varying thermal conductivity inside wall-encased insulations. Zenkour and Abbas [31] used a finite element model to analyze the nonlinear thermal transient stress of a temperature-dependent hollow cylinder. Mahdy et al. [32] studied the influences of variable thermal conductivity on wave propagations for ramp-type heating semiconductors in magneto-rotator hydrostatic stress media during photo-excited micro temperature processes. Abbas et al. [33] investigated the photothermal interaction in semiconductor media with cylindrical holes and varying thermal conductivity. In addition, the authors [34–41] applied Kirchhoff's transforms to solve nonlinear problems as linear problems.

The purpose of this study is to investigate how variations in thermal conductivity influence the transmission of waves through semiconductors. The nonlinear issue was solved using the finite element approach (without the use of Kirchhoff's transform). The linear problem (with Kirchhoff's transform) was solved using the Laplace transform and the eigenvalues technique. All physical quantities have numerical outcomes that are graphically shown. The accuracy of the suggested technique is confirmed by comparing the numerical solution to previously obtained analytic solutions by others while neglecting the new parameters, and by exploring the behavior of the solutions.

2. Mathematical Model

For homogeneous and isotropic semiconductor materials, the basic formulations are as follows, assuming the absence of any external heat source and body force [42–44]:

$$\mu u_{i,jj} + (\lambda + \mu) u_{j,ij} - \gamma_n N_{,i} - \gamma_t T_{,i} = \rho \frac{\partial^2 u_i}{\partial t^2}. \tag{1}$$

$$\left(K T_{,j}\right)_{,j} + \frac{E_g}{\tau} N = \left(1 + \tau_o \frac{\partial}{\partial t}\right)\left(\rho c_e \frac{\partial T}{\partial t} + \gamma_t T_o \frac{\partial u_{j,j}}{\partial t}\right), \tag{2}$$

$$D_e N_{,jj} - \frac{N}{\tau} + \frac{k}{\tau} T = \frac{\partial N}{\partial t}. \tag{3}$$

$$\sigma_{ij} = \mu\left(u_{i,j} + u_{j,i}\right) + (\lambda u_{k,k} - \gamma_n N - \gamma_t T)\delta_{ij}, \tag{4}$$

where ρ is the density of material, $N = n - n_o$, n_o refer to the carrier concentration at equilibrium, $i,j,k = 1,2,3$, c_e points to the heat specific at constant strain, λ, μ are the Lame's constants, $\gamma_n = (3\lambda + 2\mu)d_n$, d_n is the coefficient of electronic deformation, $\gamma_t = (3\lambda + 2\mu)\alpha_t$, α_t refers to the coefficients of linear thermal expansion, T_o is the reference temperature, t refers to the time, D_e refers to the carrier diffusion coefficient, u_i refers to the components of displacement, σ_{ij} are the stresses components, $T = T^* - T_o$, T^* is the increment of temperature, $k = \frac{\partial n_o}{\partial T}$ is the coupling parameter of thermal activation [42], and τ is the photo-generated carrier lifetime. Considered to be temperature-dependent, K denotes thermal conductivity. We shall analyze the linear expression for the thermal conductivity as in [45]:

$$K(T) = K_o(1 + K_s T), \tag{5}$$

where K_o is the thermal conductivity when $T = T_o$ and $K_s \leq 0$ points to the non-positive parameter. Considering the case of infinite isotropic semiconductor mediums whose states may be represented as functions of the spatial variable x and the time variable t leads to the formulation of the Equations (1)–(4) by [29]:

$$(\lambda + 2\mu)\frac{\partial^2 u}{\partial x^2} - \gamma_t \frac{\partial T}{\partial x} - \gamma_n \frac{\partial N}{\partial x} = \rho \frac{\partial^2 u}{\partial t^2}, \tag{6}$$

$$K_o(1 + K_s T)\frac{\partial^2 T}{\partial x^2} + K_o K_s \left(\frac{\partial T}{\partial x}\right)^2 + \frac{E_g}{\tau} N = \left(1 + \tau_o \frac{\partial}{\partial t}\right)\left(\rho c_e \frac{\partial T}{\partial t} + \gamma_t T_o \frac{\partial^2 u}{\partial t \partial x}\right), \tag{7}$$

$$\frac{\partial N}{\partial t} = D_e \frac{\partial^2 N}{\partial x^2} + \frac{k}{\tau} T - \frac{N}{\tau}, \tag{8}$$

$$\sigma_{xx} = (\lambda + 2\mu)\frac{\partial u}{\partial x} - \gamma_t T - \gamma_n N. \tag{9}$$

3. Application

The starting condition should be homogenous, and the boundary conditions at $x = 0$ are given by

$$u(0,t) = 0, \tag{10}$$

$$T(0,t) = T_1 H(t), \tag{11}$$

$$D_e \left.\frac{\partial N(x,t)}{\partial x}\right|_{x=0} = s_1 N(0,t), \tag{12}$$

where T_1 is the constant temperature, s_1 is the surface recombination velocity and $H(t)$ is the Heaviside unit function. The non-dimensional parameters may be conveniently stated as:

$$(x', u') = \eta c(x,u),\ K_s' = T_o K_s, N' = \frac{N}{n_o}, T' = \frac{T}{T_o}, \sigma_{xx}' = \frac{\sigma_{xx}}{\lambda + 2\mu}, (t', \tau', \tau_o') = \eta c^2(t, \tau, \tau_o), \tag{13}$$

where $\eta = \frac{\rho c_e}{K}$ and $c = \sqrt{\frac{\lambda + 2\mu}{\rho}}$.

Now, if we ignore the dashes, we may write out the governing equations in a non-dimensional form using parameters (13).

$$\frac{\partial^2 u}{\partial x^2} - a_1 \frac{\partial N}{\partial x} - a_2 \frac{\partial T}{\partial x} = \frac{\partial^2 u}{\partial t^2}, \tag{14}$$

$$\frac{\partial^2 N}{\partial x^2} - \frac{a_3}{\tau} N + \frac{\alpha}{\tau} T = a_3 \frac{\partial N}{\partial t}, \tag{15}$$

$$(1+K_sT)\frac{\partial^2 T}{\partial x^2}+K_s\left(\frac{\partial T}{\partial x}\right)^2+\frac{a_4}{\tau}N=\left(1+\tau_o\frac{\partial}{\partial t}\right)\left(\frac{\partial T}{\partial t}+a_5\frac{\partial^2 u}{\partial t\partial x}\right), \tag{16}$$

$$\sigma_{xx}=\frac{\partial u}{\partial x}-a_1N-a_2T, \tag{17}$$

$$u(0,t)=0,\ T(0,t)=T_1H(t),\ \left.\frac{\partial N(x,t)}{\partial x}\right|_{x=0}=a_6N(0,t), \tag{18}$$

where $a_1=\frac{\gamma_n n_o}{\lambda+2\mu}$, $a_2=\frac{\gamma_t T_o}{\lambda+2\mu}$, $a_3=\frac{1}{\eta D_e}$, $a_4=\frac{E_g n_o}{T_o\rho c_e}$, $a_5=\frac{\gamma_t}{\rho c_e}$, $a_6=\frac{s_1}{D_e\eta c}$ and $\alpha=\frac{T_o k}{n_o c^2 D_e\eta^2}$.

4. Numerical Solution (Finite Element Method)

In this section, we establish the fundamental forms of the equations, which are nonlinear partial differential equations. The potential solutions to this problem are investigated using the finite element technique (FEM). Similar to [46,47], this strategy employs the standard weak formulation techniques. The weak formulations of the essential equations are fixed in a non-dimensional setting. The sets of independent weight functions are presented, which include the carrier density N, temperature T, and displacement u. Integrating across the spatial domain involves multiplying the basic equations by the independent weight functions, as dictated by the problem's boundary conditions. So, we may express the carrier density, temperature, and displacement values at each node as follows:

$$T=\sum_{j=1}^{m}M_jT_j(t),\ N=\sum_{j=1}^{m}M_jN_j(t),\ u=\sum_{j=1}^{m}M_ju_j(t), \tag{19}$$

where m denotes the node's number of elements, and M refers to the shape functions, where the shape functions and weight function are identical to Galerkin's standard methods. Therefore,

$$\delta T=\sum_{j=1}^{m}M_j\delta T_j,\ \delta N=\sum_{j=1}^{m}M_j\delta N_j,\ \delta u=\sum_{j=1}^{m}M_j\delta u_j. \tag{20}$$

The implicit techniques should be used to derive the time derivatives of the unknown variables in the following phase. The weak formulations for FEM that correspond to (14)–(16) are now as follows:

$$\int_0^L\frac{\partial\delta u}{\partial x}\left(\frac{\partial u}{\partial x}-a_1N-a_2T\right)dx+\int_0^L\delta u\left(\frac{\partial^2 u}{\partial t^2}\right)dx=\delta u\left(\frac{\partial u}{\partial x}-a_1N-a_2T\right)_0^L, \tag{21}$$

$$\int_0^L\frac{\partial\delta N}{\partial x}\left(\frac{\partial N}{\partial x}\right)dx+\int_0^L\delta u\left(\frac{a_3}{\tau}N-\frac{\alpha}{\tau}T+a_3\frac{\partial N}{\partial t}\right)dx=\delta u\left(\frac{\partial N}{\partial x}\right)_0^L. \tag{22}$$

$$\int_0^L\frac{\partial\delta T}{\partial x}(1+K_sT)\frac{\partial T}{\partial x}dx+\int_0^L\delta T\left(-\frac{a_4}{\tau}N+\left(1+\tau_o\frac{\partial}{\partial t}\right)\left(\frac{\partial T}{\partial t}+a_5\frac{\partial^2 u}{\partial t\partial x}\right)\right)dx=\delta T\left((1+K_sT)\frac{\partial T}{\partial x}\right)_0^L, \tag{23}$$

5. Linear Cases (with Kirchhoff's Transforms)

In order to change the basic equations from a nonlinear form into a linear one, we use Kirchhoff's transforms mapping [45] to varying thermal conductivity, which is shown in the equation. This allows us to convert the essential forms (5)

$$\theta=\frac{1}{K_o}\int_0^T K(T)dT, \tag{24}$$

where the recently added function represents the conduction of heat. We may get [45] by substituting from Equation (24) in (5), then integrating.

$$\theta = T + \frac{1}{2}K_s T^2. \tag{25}$$

The following may be concluded from Equations (24) and (25):

$$K_o\theta_{,i} = K(T)T_{,i},\ K_o\theta_{,ii} = (K(T)T_{,i})_{,i},\ K_o\frac{\partial\theta}{\partial t} = K(T)\frac{\partial T}{\partial t}. \tag{26}$$

Consequently, the governing Equations (14)–(18), may be stated in the linear form:

$$\frac{\partial^2 u}{\partial x^2} - a_1\frac{\partial N}{\partial x} - a_2\frac{\partial\theta}{\partial x} = \frac{\partial^2 u}{\partial t^2}, \tag{27}$$

$$\frac{\partial^2 N}{\partial x^2} - \frac{a_3}{\tau}N + \frac{\alpha}{\tau}\theta = a_3\frac{\partial N}{\partial t}, \tag{28}$$

$$\frac{\partial^2\theta}{\partial x^2} + \frac{a_4}{\tau}N = (1+\tau_o\frac{\partial}{\partial t})(\frac{\partial\theta}{\partial t} + a_5\frac{\partial^2 u}{\partial t\partial x}), \tag{29}$$

$$\sigma_{xx} = \frac{\partial u}{\partial x} - a_1 N - \frac{a_2}{K_s}\left(-1+\sqrt{1+2K_s\theta}\right), \tag{30}$$

$$u(0,t) = 0,\ \theta(0,t) = T_1 H(t) + \frac{1}{2}K_s(T_1 H(t))^2,\ \left.\frac{\partial N(x,t)}{\partial x}\right|_{x=0} = a_6 N(0,t), \tag{31}$$

6. Analytical Solution

For the $g(x,t)$ function, Laplace transforms were written as

$$\overline{g}(x,s) = L[g(x,t)] = \int_0^\infty g(x,t)e^{-st}dt, \tag{32}$$

where s is the Laplace transformation parameter. Thus, the essential equations may be rewritten in the following ways:

$$\frac{d^2\overline{u}}{dx^2} = s^2\overline{u} + a_1\frac{d\overline{N}}{dx} + x_2\frac{d\overline{\theta}}{dx}, \tag{33}$$

$$\frac{d^2\overline{N}}{dx^2} = a_3\left(s+\frac{1}{\tau}\right)\overline{N} - \frac{\beta}{\tau}\overline{\theta}, \tag{34}$$

$$\frac{d^2\overline{\theta}}{dx^2} = s(1+\tau_o s)\theta - \frac{a_4}{\tau}\overline{N} + a_5 s(1+\tau_o s)\frac{d\overline{u}}{dx}, \tag{35}$$

$$\overline{\sigma}_{xx} = \frac{d\overline{u}}{dx} - a_1\overline{N} - \frac{a_2}{K_s}\left(-1+\sqrt{1+2K_s\overline{\theta}}\right), \tag{36}$$

$$\overline{u}(0,t) = 0,\ \left.\frac{d\overline{N}(x,t)}{dx}\right|_{x=0} = x_6\overline{N}(0,t),\ \overline{\theta}(0,t) = \frac{T_1}{s}\left(1+\frac{1}{2s}T_1 K_s\right), \tag{37}$$

Now, using the eigenvalue approach provided in [48–50], we will get the solutions of the coupled differential system (33)–(35) with the boundary conditions (37). We may get the matrices and vectors from Equations (33)–(35) as

$$\frac{dV}{dx} = AV, \tag{38}$$

where $V = \begin{bmatrix} \overline{u} & \overline{N} & \overline{\phi} & \frac{d\overline{u}}{dx} & \frac{d\overline{N}}{dx} & \frac{d\overline{\phi}}{dx} \end{bmatrix}^T$ and $A = \begin{bmatrix} a_{ij} \end{bmatrix}_{6\times 6}$ with $a_{ij} = 0$, expect $a_{14} = 1$, $a_{25} = 1$, $a_{36} = 1$, $a_{41} = s^2$, $a_{45} = a_1$, $a_{46} = a_2$, $a_{52} = a_3\left(s + \frac{1}{\tau}\right)$, $a_{53} = -\frac{\beta}{\tau}$, $a_{62} = -\frac{a_4}{\tau}$, $a_{63} = s(1 + \tau_o s)$, $a_{64} = s(1 + \tau_o s)a_5$.

The equations that characterize matrix A are given

$$\omega^6 - b_1\omega^4 + b_2\omega^2 + b_3 = 0, \tag{39}$$

where $b_1 = a_{52} + a_{63} + a_{41} + a_{46}a_{64}$, $b_3 = a_{41}a_{62}a_{53} - a_{63}a_{41}a_{52}$, $b_2 = -a_{64}a_{45}a_{53} + a_{41}a_{63} + a_{64}a_{46}a_{52} + a_{52}a_{41} + a_{63}a_{52} - a_{53}a_{62}$. The six roots of Equation (40) are the six eigenvalues of matrix A, which are written as $\pm\omega_1$, $\pm\omega_2$ and $\pm\omega_3$. Thus, the eigenvectors Y are computed as: $Y_1 = (a_{52} - \omega^2)a_{46}\omega - a_{45}a_{53}\omega$, $Y_2 = a_{53}(a_{41} - \omega^2)$, $Y_3 = (a_{52} - \omega^2)(\omega^2 - a_{41})$, $Y_4 = \omega Y_1$, $Y_5 = \omega Y_2$, $Y_6 = \omega Y_3$.

The solutions of Equation (40) have the following form:

$$\mathrm{V}(x, s) = \sum_{i=1}^{3} \left(A_i Y_i e^{-\omega_i x} + A_{i+1} Y_{i+1} e^{\omega_i x}\right) \tag{40}$$

The rising exponential nature of the variable x has been removed to infinity due to the regularity constraint of the solution. Hence, the general solutions (40) may be shown as

$$\mathrm{V}(x, s) = \sum_{i=1}^{3} A_i Y_i e^{-\omega_i x} \tag{41}$$

where A_1, A_2 and A_3 are constants which can be calculated through the use of the problem's boundary conditions. To get the final solutions of displacement, temperature, carrier density, and stresses distributions, the Fourier series approximation [51] may be employed as a numerical inversion approach.

7. Discussion of Numerical Results

The results are theoretically investigated using the physical constants and physical characteristics of silicon as an elastic semiconductor material. Calculations and explanations of results from numerical simulations are made possible using constants derived from silicon (Si). The constants of Si are [52]:

$$\lambda = 3.64 \times 10^{10} \frac{N}{m^2}, E_g = 1.11\ eV,\ \mu = 5.46 \times \frac{10^{10} N}{m^2}, T_o = 300\ \mathrm{k}, T_s = 1,$$

$$c_e = 695\ J/kgk, d_n = -9 \times 10^{-31} m^3,\ \alpha_t = 3 \times 10^{-6} k^{-1}, \tau = 5 \times 10^{-5} s,$$

$$s_f = 2\frac{\mathrm{m}}{\mathrm{s}},\ \rho = 2330\ \mathrm{kg}/m^3, n_o = 10^{20} m^{-3}, D_e = 2.5 \times 10^{-3} m^3/s.$$

Using these values, we can do numerical simulations of the physical variables over the distance x to evaluate the effect of varying thermal conductivity within the context of the coupled photothermal theory with one relaxation time (see Figures 1–16). The field distributions, such as carrier density distributions (plasma waves), stress distributions (mechanical wave distributions), displacement distributions (strain wave distributions), and thermal wave distributions (thermal temperature distribution), are used in the numerical calculations. For the time $t = 0.8$, a numerical calculation is performed. Figure 1 predicts the increment of temperature along the distance x. For the generalized photo-thermal theory, it has been observed that the temperature begins at its maximum value, according to the applied boundary condition, then decreases with increasing x. It then decreases gradually as the distance x increases until it approaches zero beyond a wavefront.

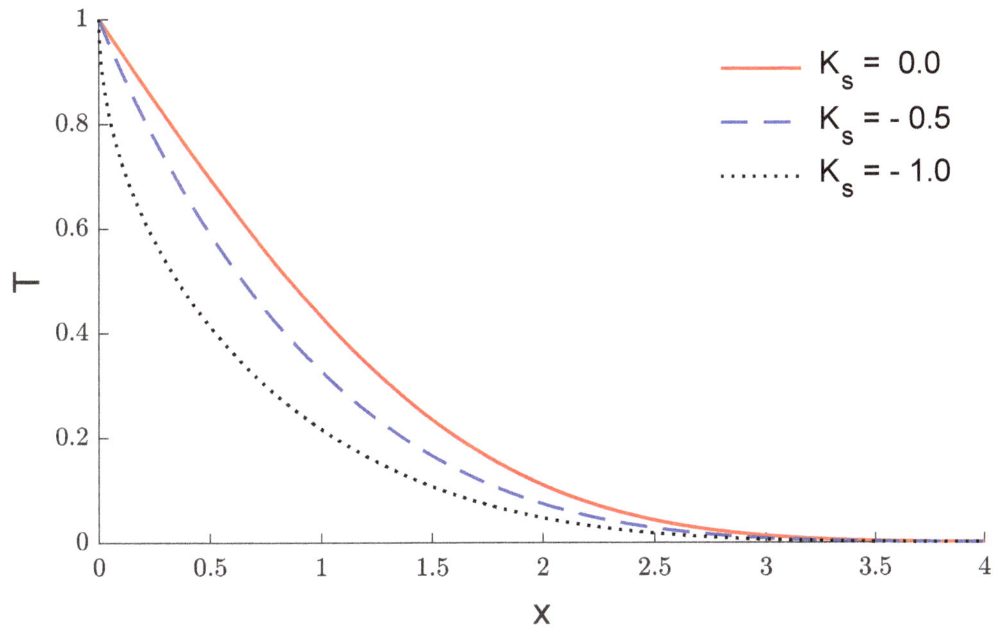

Figure 1. The variation of temperature under various values of K_s.

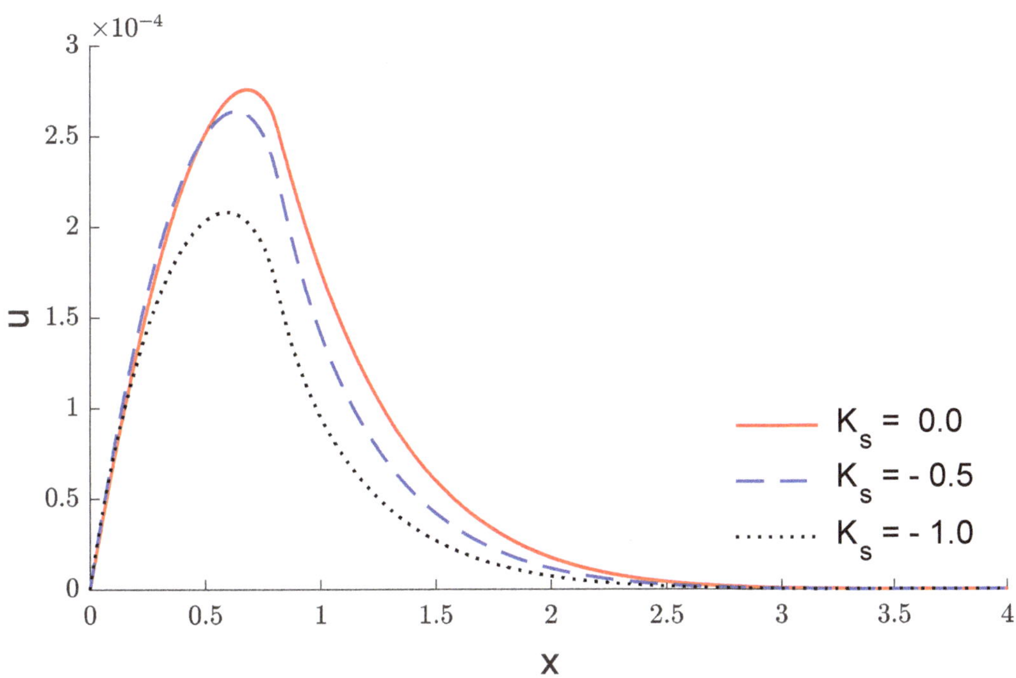

Figure 2. The displacement variations under various values of K_s.

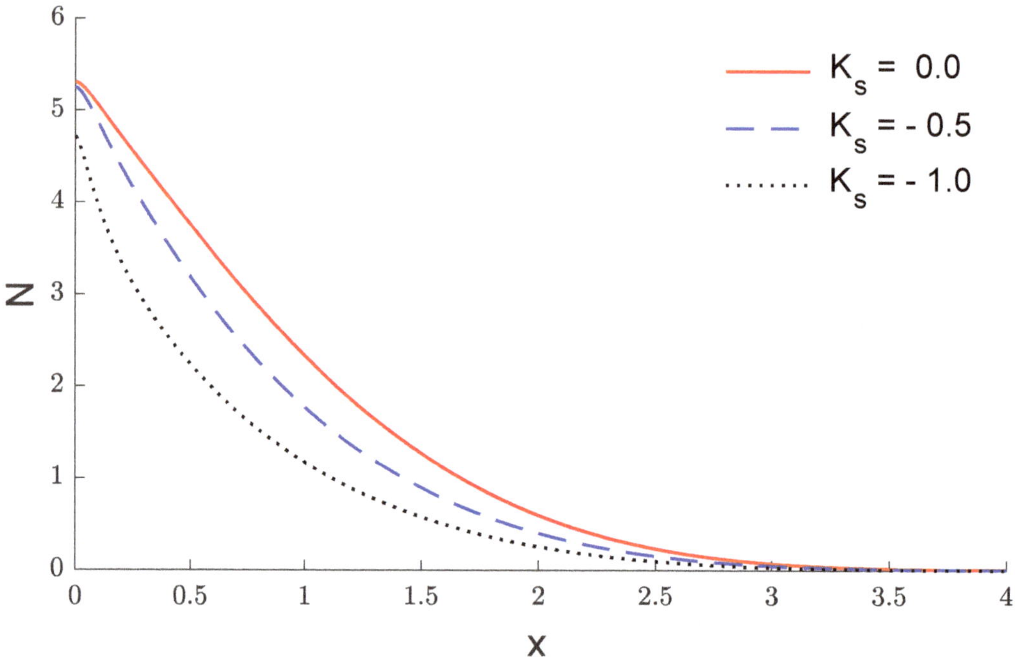

Figure 3. The carrier density variations under various values of K_s.

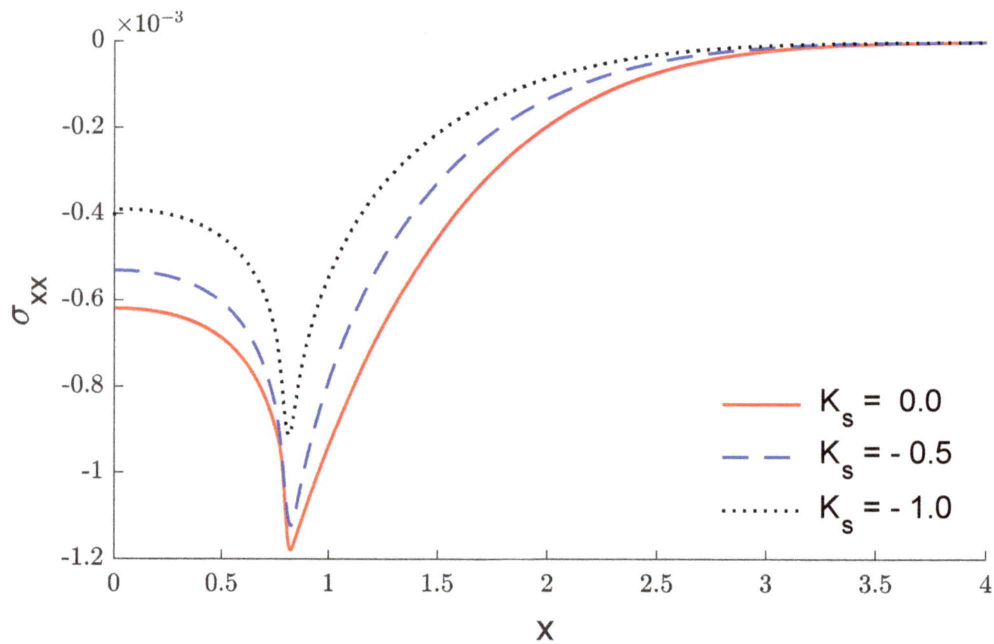

Figure 4. The stress variations under various values of K_s.

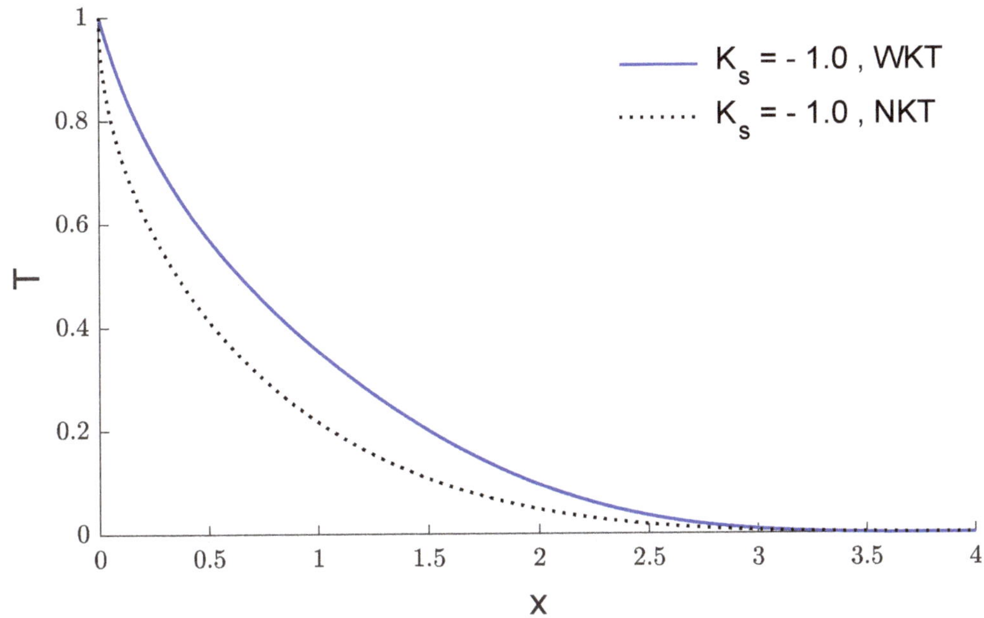

Figure 5. The variations of temperature with and without Kirchhoff transforms when $K_s = -1$.

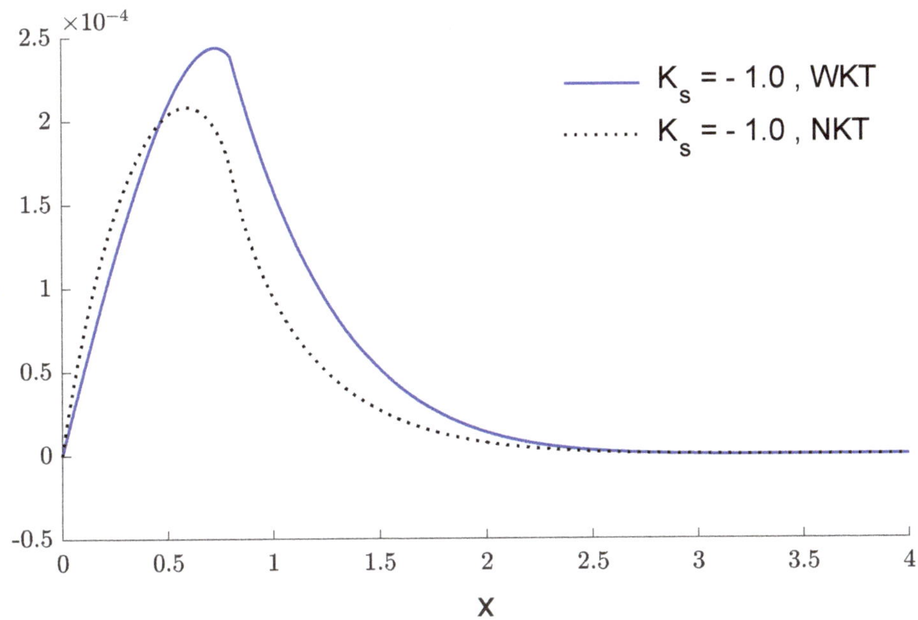

Figure 6. The variations of displacement with and without Kirchhoff transforms when $K_s = -1$.

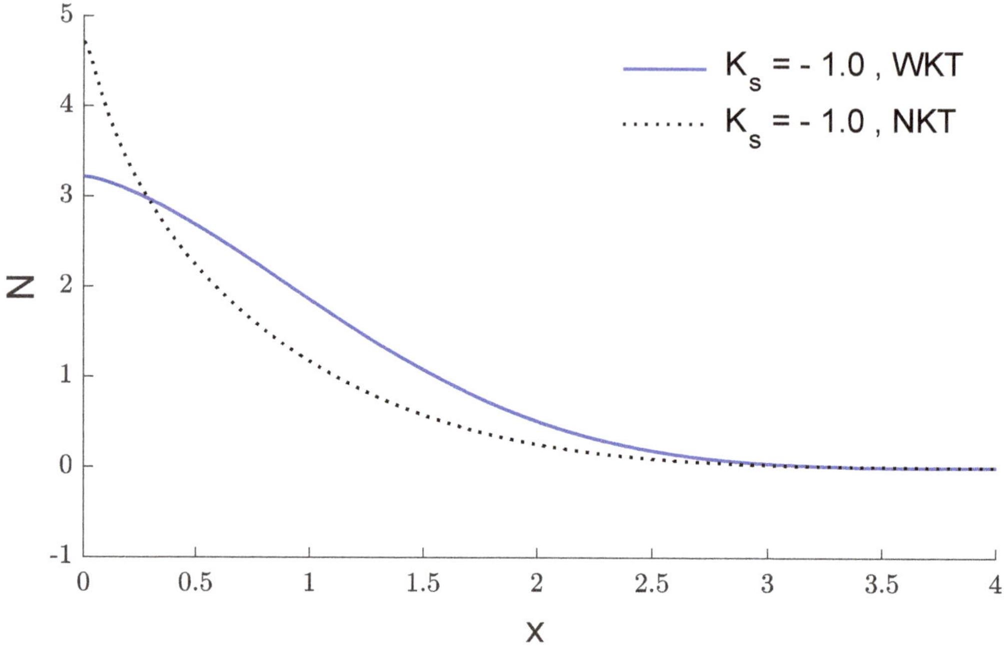

Figure 7. The variations of carrier density with and without Kirchhoff transforms when $K_s = -1$.

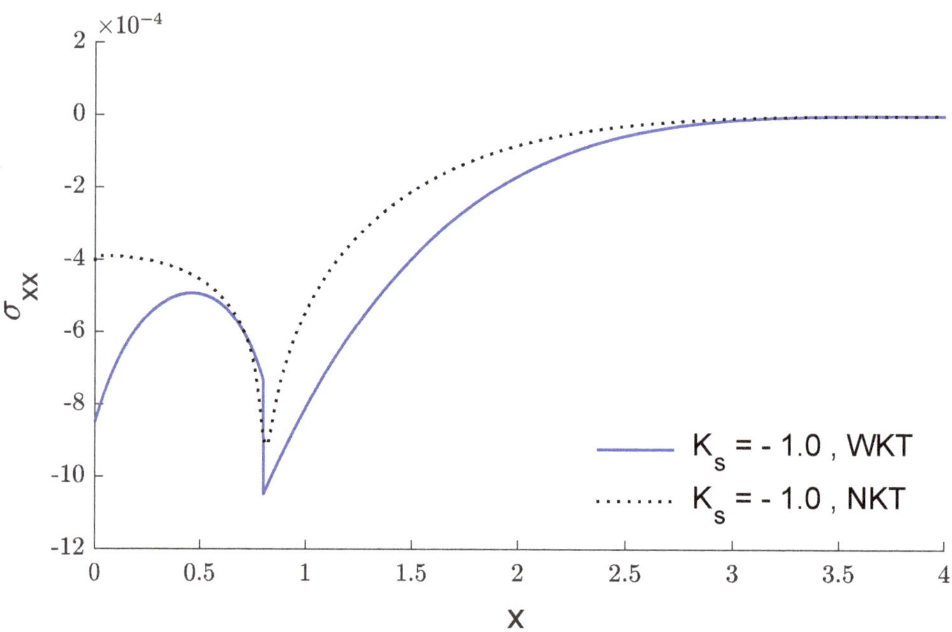

Figure 8. The variations of stress with and without Kirchhoff transforms when $K_s = -1$.

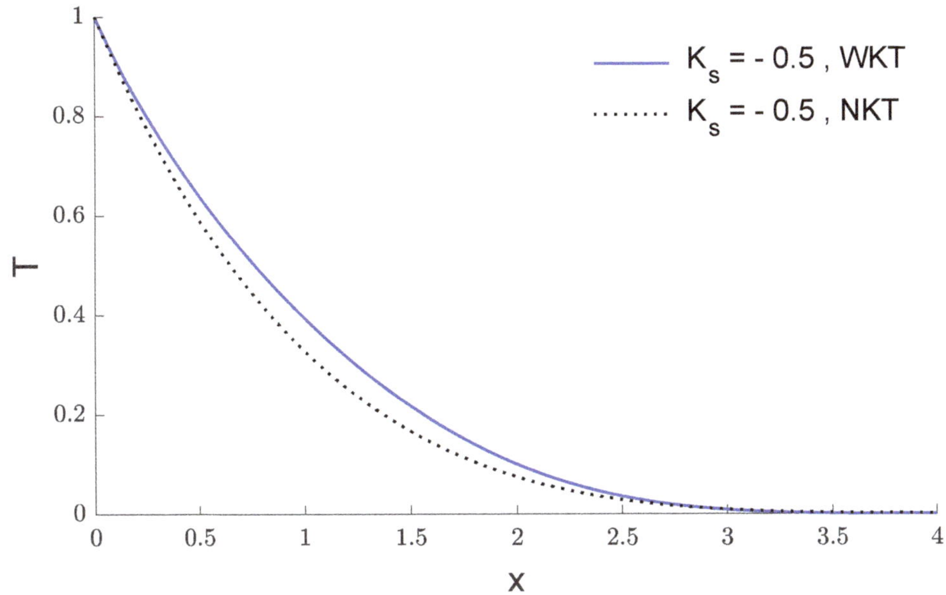

Figure 9. The variations of temperature with and without Kirchhoff transforms when $K_s = -0.5$.

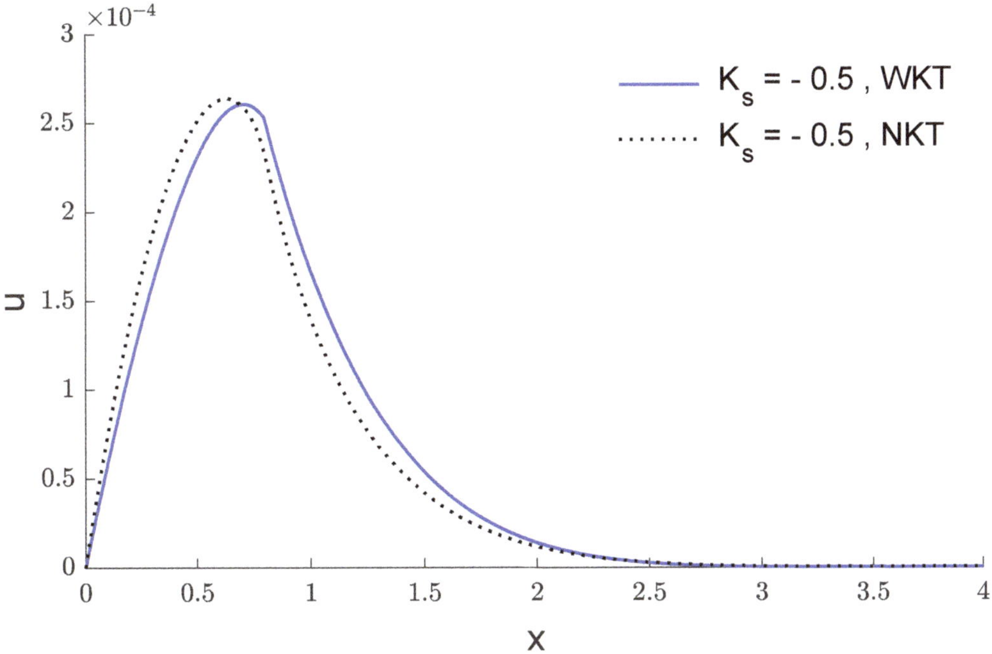

Figure 10. The variations of displacement with and without Kirchhoff transforms when $K_s = -0.5$.

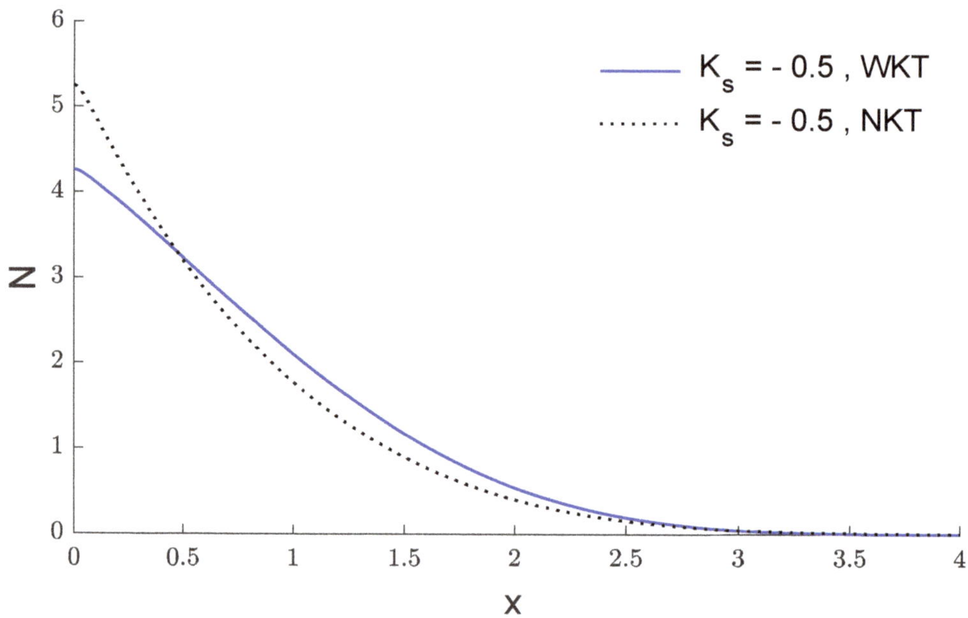

Figure 11. The variations of carrier density with and without Kirchhoff transforms when $K_s = -0.5$.

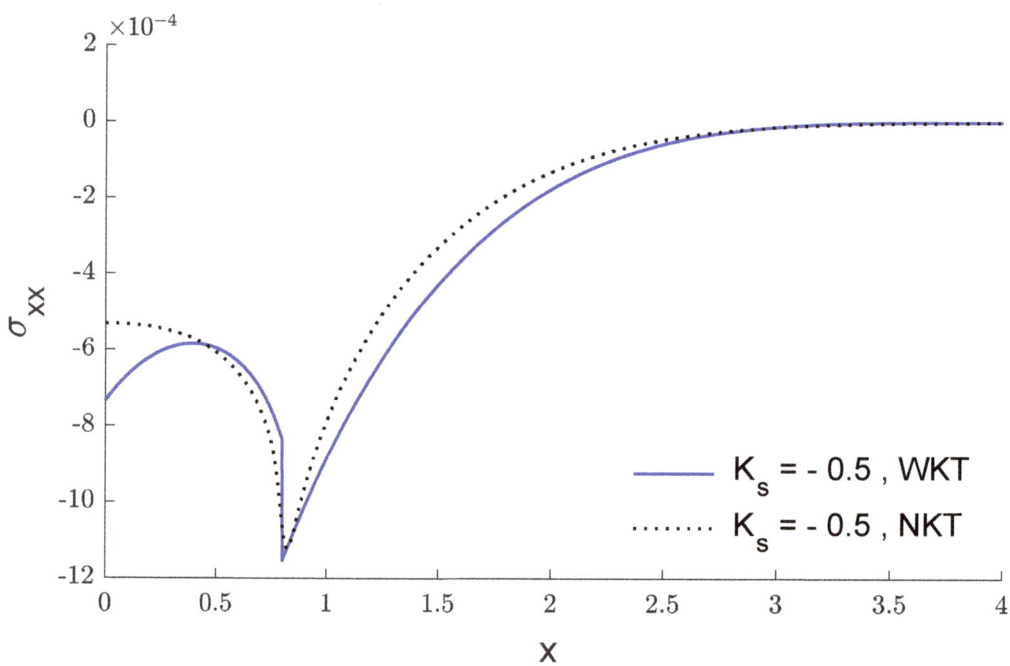

Figure 12. The stress variations with and without Kirchhoff transforms when $K_s = -0.5$.

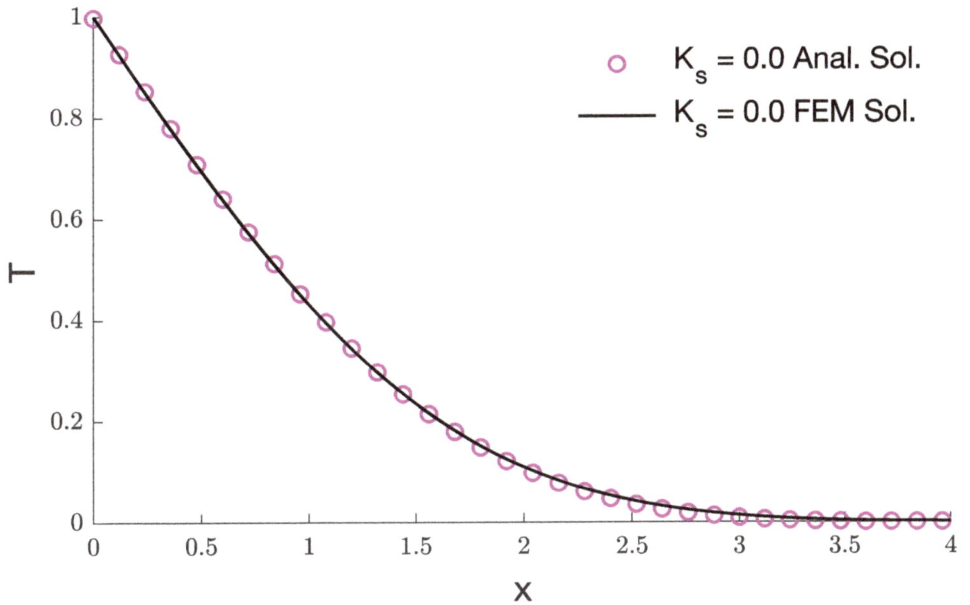

Figure 13. The study on temperature comparison.

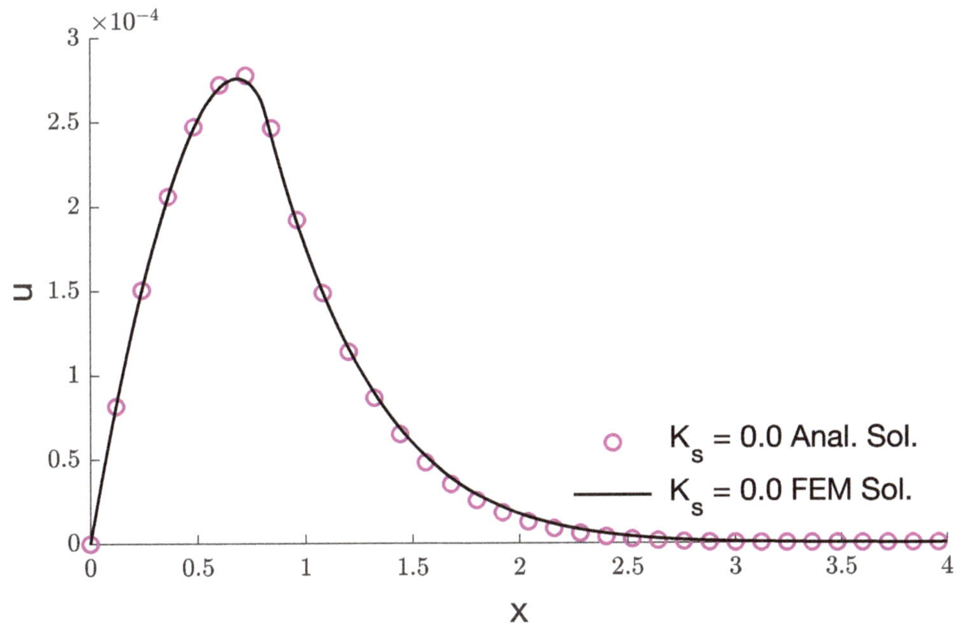

Figure 14. The study on displacement comparison.

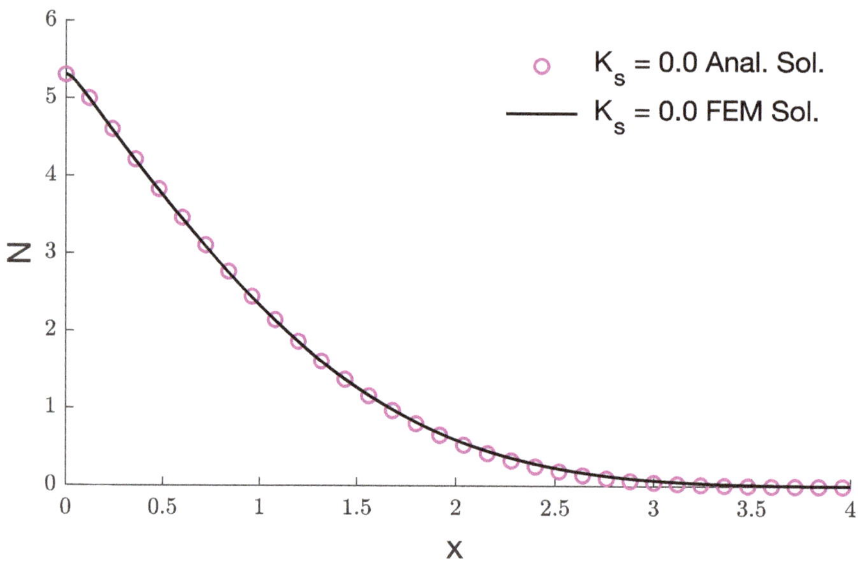

Figure 15. The study on carrier density comparison.

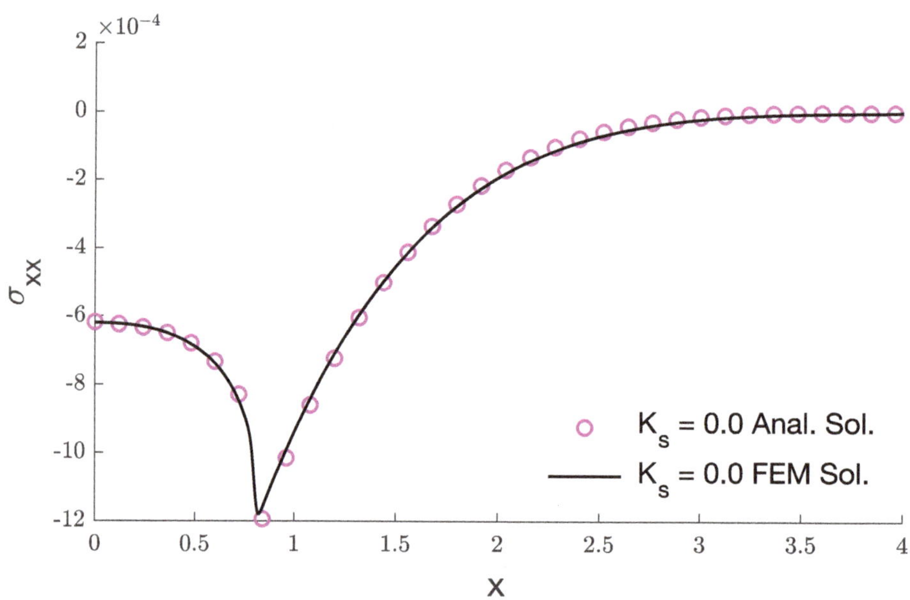

Figure 16. The study on stress comparison.

Figure 2 shows how the displacement changes with respect to the distance x. It was observed that the displacement of the zero values, according to the applied boundary condition, then rises with rising x, reaches a peak at a certain point relatively near to the surface, and then gradually falls to zero. Different carrier densities are shown as a function of x distance in Figure 3. At $x = 0$, where the surface is located, the carrier density is at its highest. As x increases, the carrier density slowly decreases until it is close to zero.

Figure 4 shows how the stress changes as x gets farther away. It is clear that the stress reaches some negative values, then slowly goes up until it reaches a peak of negative values, and then slowly goes back down to zero. A comparison of the Kirchhoff transforms (WKT) and non-Kirchhoff transforms (NKT) results are shown in Figures 5–16. When $K_s = -1$, temperature, displacement, carrier density, and stress are all shown to vary along x. The usage of the Kirchhoff transforms (WKT) is shown by the solid line, whereas the absence of the transforms in the nonlinear situation is denoted by the dotted line (NKT). Figure 4 shows that the curves coincide at the surface since the temperature boundary condition is $T_1 = 1$. After that, the difference ratio grows with distance until x = 0.74, before falling to zero at x = 2.55. Figures 5–16 show the difference between using the Kirchhoff transforms (WKT) and not using the Kirchhoff transforms (NKT).

Figures 5–8 display how the temperature, displacement, carrier density, and stress change with respect to the distance x when $K_s = -1$. The solid line depicts the case when Kirchhoff transforms (WKT) are used, while the dotted line show the case when Kirchhoff transforms are not used (NKT). As shown in Figure 5, the curves are the same at the surface because the temperature boundary condition is $T_1 = 1$. After that, the difference ratio goes up as the distance goes up until $x = 1$, and then it goes down until it reaches zero at $x = 3.15$. Figure 6 depicts the displacement variation with and without Kirchhoff transforms. The curves coincide at the surface under the displacement boundary condition $(u = 0)$, where the difference ratio grows with distance until $x = 1$, and then reduces to zero at $x = 2.5$.

Figure 7 shows the carrier density variation with Kirchhoff transforms (WKT) and without Kirchhoff transforms (NKT), in which the curves have the ratio of a maximum difference on the surface $x = 0$. Figure 8 shows the variations of stress with Kirchhoff transforms (WKT) and without Kirchhoff transforms (NKT), where on the surface $x = 0$, the curves have the greatest difference ratio. When $K_s = -0.5$, the changes in temperature, displacement, carrier density, and stress along x are shown in Figures 9–12. It was discovered that, when comparing results obtained with and without Kirchhoff transforms (WKT), the differences are striking (NKT).

When $K_s = 0$, the analytical solutions (Laplace transforms and eigenvalue method with Kirchhoff transforms) are shown to be superior to the numerical solutions (finite element method without Kirchhoff transforms) in Figures 13–16. The analytical data strongly agree with the numerical results of temperature change, displacement variation, carrier density variation, and stress variation over x. As expected, the variable thermal conductivity parameter has significant effects on the speed of the wave propagation of all studied fields.

8. Conclusions

In this paper, the mathematical implications of changing thermal conductivity in semiconductor media with and without Kirchhoff's transformations are investigated. Without using Kirchhoff's transformations, the finite element approach yields numerical solutions for nonlinear equations. For nonlinear equations using Kirchhoff's transformations, the eigenvalue approach is used to provide an analytical solution. It was determined that the variable thermal conductivity has a considerable effect on the deformation behaviors of various physical field components. The numerical results and findings reported in this study should be helpful for scholars working on the advancement of solid mechanics as well as those in scientific and technical domains. Numerous thermodynamics problems may be solved by using the approaches presented in this article. The theoretical conclusions presented here can be of interest for experimental scientists and researchers working on this topic.

Author Contributions: Formal analysis, I.A.; Funding acquisition, A.H.; Investigation, A.H.; Methodology, I.A.; Visualization, I.A. All authors have read and agreed to the published version of the manuscript.

Funding: This research work was funded by Institutional Fund Projects under grant no. (IFPIP: 7-130-1443). The authors gratefully acknowledge technical and financial support provided by the Ministry of Education and King Abdulaziz University, DSR, Jeddah, Saudi Arabia.

Data Availability Statement: Not applicable.

Conflicts of Interest: The authors declare no conflict of interest.

References

1. He, C.F.; Lu, G.G.; Shan, X.N.; Sun, Y.F.; Li, T.; Qin, L.; Yan, C.L.; Ning, Y.Q.; Wang, L.J. Theoretical analysis of 980 nm high power Vertical External-cavity Surface-emitting Semiconductor Laser (VECSEL). In Proceedings of the ICO20: Optical Devices and Instruments, Changchun, China, 21–26 August 2005; pp. 204–213.
2. Todorović, D.M. Plasma, thermal, and elastic waves in semiconductors. *Rev. Sci. Instrum.* **2003**, *74*, 582–585. [CrossRef]
3. Todorović, D.M. Photothermal and electronic elastic effects in microelectromechanical structures. *Rev. Sci. Instrum.* **2003**, *74*, 578–581. [CrossRef]
4. Song, Y.; Cretin, B.; Todorovic, D.M.; Vairac, P. Study of photothermal vibrations of semiconductor cantilevers near the resonant frequency. *J. Phys. D Appl. Phys.* **2008**, *41*, 15516. [CrossRef]
5. Lord, H.W.; Shulman, Y. A generalized dynamical theory of thermoelasticity. *J. Mech. Phys. Solids* **1967**, *15*, 299–309. [CrossRef]
6. Marin, M.; Othman, M.I.A.; Abbas, I.A. An Extension of the Domain of Influence Theorem for Generalized Thermoelasticity of Anisotropic Material with Voids. *J. Comput. Theor. Nanosci.* **2015**, *12*, 1594–1598. [CrossRef]
7. Ezzat, M.A.; El-Bary, A.A. Fractional magneto-thermoelastic materials with phase-lag Green-Naghdi theories. *Steel Compos. Struct.* **2017**, *24*, 297–307. [CrossRef]
8. Abbas, I.A. Eigenvalue approach on fractional order theory of thermoelastic diffusion problem for an infinite elastic medium with a spherical cavity. *Appl. Math. Model.* **2015**, *39*, 6196–6206. [CrossRef]
9. Abbas, I.A.; Abdalla, A.E.N.N.; Alzahrani, F.S.; Spagnuolo, M. Wave propagation in a generalized thermoelastic plate using eigenvalue approach. *J. Therm. Stress.* **2016**, *39*, 1367–1377. [CrossRef]
10. Alharbi, A.M.; Said, S.M.; Abd-Elaziz, E.M.; Othman, M.I.A. Influence of Initial Stress and Variable Thermal Conductivity on a Fiber-Reinforced Magneto-Thermoelastic Solid with Micro-Temperatures by Multi-Phase-Lags Model. *Int. J. Struct. Stab. Dyn.* **2022**, *22*, 2250007. [CrossRef]
11. Li, Y.; He, T. Investigation of a half-space heated by laser pulses based on the generalized thermoelastic theory with variable thermal material properties. *Waves Random Complex Media* **2022**, *32*, 120–136. [CrossRef]
12. Abouelregal, A.E.; Tiwari, R. The thermoelastic vibration of nano-sized rotating beams with variable thermal properties under axial load via memory-dependent heat conduction. *Meccanica* **2022**, *57*, 2001–2025. [CrossRef]
13. Li, C.L.; Tian, X.G.; He, T.H. Analytical study of transient thermo-mechanical responses in a fractional order generalized thermoelastic diffusion spherical shell with variable thermal conductivity and diffusivity. *Waves Random Complex Media* **2021**, *31*, 1083–1106. [CrossRef]
14. Othman, M.I.A.; Zidan, M.E.M.; Mohamed, I.E.A. Dual-phase-lag model on thermo-microstretch elastic solid Under the effect of initial stress and temperature-dependent. *Steel Compos. Struct.* **2021**, *38*, 355–363. [CrossRef]
15. Zhang, L.; Bhatti, M.M.; Michaelides, E.E.; Marin, M.; Ellahi, R. Hybrid nanofluid flow towards an elastic surface with tantalum and nickel nanoparticles, under the influence of an induced magnetic field. *Eur. Phys. J. Spec. Top.* **2021**, *231*, 521–533. [CrossRef]
16. Scutaru, M.L.; Vlase, S.; Marin, M.; Modrea, A. New analytical method based on dynamic response of planar mechanical elastic systems. *Bound. Value Probl.* **2020**, *2020*, 104. [CrossRef]
17. Abouelregal, A.E.; Marin, M. The response of nanobeams with temperature-dependent properties using state-space method via modified couple stress theory. *Symmetry* **2020**, *12*, 1276. [CrossRef]
18. Hobiny, A.; Abbas, I.A. A study on the thermoelastic interaction in two-dimension orthotropic materials under the fractional derivative model. *Alex. Eng. J.* **2022**, *in press*. [CrossRef]
19. Alzahrani, F.; Hobiny, A.; Abbas, I.; Marin, M. An eigenvalues approach for a two-dimensional porous medium based upon weak, normal and strong thermal conductivities. *Symmetry* **2020**, *12*, 848. [CrossRef]
20. Abouelregal, A.E.; Marin, M. The Size-Dependent Thermoelastic Vibrations of Nanobeams Subjected to Harmonic Excitation and Rectified Sine Wave Heating. *Mathematics* **2020**, *8*, 1128. [CrossRef]
21. Ailawalia, P.; Kumar, A. Ramp Type Heating in a Semiconductor Medium under Photothermal Theory. *Silicon* **2019**, *12*, 347–356. [CrossRef]
22. Abbas, I.A.; Alzahrani, F.S.; Elaiw, A. A DPL model of photothermal interaction in a semiconductor material. *Waves Random Complex Media* **2018**, *29*, 328–343. [CrossRef]
23. Ghasemi, S.E.; Hatami, M.; Ganji, D.D. Thermal analysis of convective fin with temperature-dependent thermal conductivity and heat generation. *Case Stud. Therm. Eng.* **2014**, *4*, 1–8. [CrossRef]
24. Yang, W.; Zhang, L.; Zhang, H.; Wang, F.; Li, X. Numerical investigations of the effects of different factors on the displacement of energy pile under the thermo-mechanical loads. *Case Stud. Therm. Eng.* **2020**, *21*, 100711. [CrossRef]
25. Lotfy, K.; Hassan, W.; El-Bary, A.A.; Kadry, M.A. Response of electromagnetic and Thomson effect of semiconductor medium due to laser pulses and thermal memories during photothermal excitation. *Results Phys.* **2020**, *16*, 102877. [CrossRef]

26. Hobiny, A.D.; Alzahrani, F.S.; Abbas, I.A. A study on photo-thermo-elastic wave in a semi-conductor material caused by ramp-type heating. *Alex. Eng. J.* **2021**, *60*, 2033–2040. [CrossRef]
27. Mohamed, M.S.; Lotfy, K.; El-Bary, A.; Mahdy, A.M.S. Absorption illumination of a 2D rotator semi-infinite thermoelastic medium using a modified Green and Lindsay model. *Case Stud. Therm. Eng.* **2021**, *26*, 101165. [CrossRef]
28. Youssef, H.M.; Abbas, I.A. Thermal shock problem of generalized thermoelasticity for an infinitely long annular cylinder with variable thermal conductivity. *Comput. Methods Sci. Technol.* **2007**, *13*, 95–100. [CrossRef]
29. Sherief, H.H.; Hamza, F.A. Modeling of variable thermal conductivity in a generalized thermoelastic infinitely long hollow cylinder. *Meccanica* **2015**, *51*, 551–558. [CrossRef]
30. Khoukhi, M.; Abdelbaqi, S.; Hassan, A. Transient temperature change within a wall embedded insulation with variable thermal conductivity. *Case Stud. Therm. Eng.* **2020**, *20*, 100645. [CrossRef]
31. Zenkour, A.M.; Abbas, I.A. Nonlinear Transient Thermal Stress Analysis of Temperature-Dependent Hollow Cylinders Using a Finite Element Model. *Int. J. Struct. Stab. Dyn.* **2014**, *14*, 1450025. [CrossRef]
32. Mahdy, A.M.S.; Lotfy, K.; El-Bary, A.; Atef, H.M.; Allan, M. Influence of variable thermal conductivity on wave propagation for a ramp-type heating semiconductor magneto-rotator hydrostatic stresses medium during photo-excited microtemperature processes. *Waves Random Complex Media* **2021**, 1–23. [CrossRef]
33. Abbas, I.; Hobiny, A.; Marin, M. Photo-thermal interactions in a semi-conductor material with cylindrical cavities and variable thermal conductivity. *J. Taibah Univ. Sci.* **2020**, *14*, 1369–1376. [CrossRef]
34. Said, S.M.; Othman, M.I.A. The effect of gravity and hydrostatic initial stress with variable thermal conductivity on a magneto-fiber-reinforced. *Struct. Eng. Mech.* **2020**, *74*, 425–434. [CrossRef]
35. Ezzat, M.A.; El-Bary, A.A. Effects of variable thermal conductivity and fractional order of heat transfer on a perfect conducting infinitely long hollow cylinder. *Int. J. Therm. Sci.* **2016**, *108*, 62–69. [CrossRef]
36. Othman, M.I.A.; Abouelregal, A.E. Magnetothermoelstic analysis for an infinite solid cylinder with variable thermal conductivity due to harmonically varying heat. *Microsyst. Technol.* **2017**, *23*, 5635–5644. [CrossRef]
37. Lotfy, K.; Tantawi, R.S.; Anwer, N. Response of Semiconductor Medium of Variable Thermal Conductivity Due to Laser Pulses with Two-Temperature through Photothermal Process. *Silicon* **2019**, *11*, 2719–2730. [CrossRef]
38. Zenkour, A.M.; Abouelregal, A.E. Thermoelastic Interactions in an Infinite Orthotropic Continuum of a Variable Thermal Conductivity with a Cylindrical Hole. *Iran. J. Sci. Technol. Trans. Mech. Eng.* **2017**, *43*, 281–290. [CrossRef]
39. Mahdy, A.M.S.; Lotfy, K.; El-Bary, A.A.; Roshdy, E.M.; Abd El-Raouf, M.M. Variable thermal conductivity during photo-thermoelasticy theory of semiconductor medium induced by laser pulses with hyperbolic two-temperature theory. *Waves Random Complex Media* **2021**, 1–23. [CrossRef]
40. Khamis, A.K.; Lotfy, K.; El-Bary, A.A. Effect of variable thermal conductivity of semiconductor elastic medium during photothermal excitation subjected to thermal ramp type. *Waves Random Complex Media* **2022**, *32*, 78–90. [CrossRef]
41. Khamis, A.K.; El-Bary, A.A.; Lotfy, K. Electromagnetic Hall current and variable thermal conductivity influence for microtemperature photothermal excitation process of semiconductor material. *Waves Random Complex Media* **2022**, *32*, 406–423. [CrossRef]
42. Mandelis, A.; Nestoros, M.; Christofides, C. Thermoelectronic-wave coupling in laser photothermal theory of semiconductors at elevated temperatures. *Opt. Eng.* **1997**, *36*, 459–468. [CrossRef]
43. Song, Y.Q.; Bai, J.T.; Ren, Z.Y. Study on the reflection of photothermal waves in a semiconducting medium under generalized thermoelastic theory. *Acta Mech.* **2012**, *223*, 1545–1557. [CrossRef]
44. Youssef, H.M.; El-Bary, A.A. Theory of hyperbolic two-temperature generalized thermoelasticity. *Mater. Phys. Mech.* **2018**, *40*, 158–171. [CrossRef]
45. Youssef, H. State-space approach on generalized thermoelasticity for an infinite material with a spherical cavity and variable thermal conductivity subjected to ramp-type heating. *Can. Appl. Math. Q.* **2005**, *13*, 4.
46. Abbas, I.A. Generalized magneto-thermoelasticity in a nonhomogeneous isotropic hollow cylinder using the finite element method. *Arch. Appl. Mech.* **2009**, *79*, 41–50. [CrossRef]
47. Abbas, I.A. A two-dimensional problem for a fibre-reinforced anisotropic thermoelastic half-space with energy dissipation. *Sadhana* **2011**, *36*, 411–423. [CrossRef]
48. Das, N.C.; Lahiri, A.; Giri, R.R. Eigenvalue approach to generalized thermoelasticity. *Indian J. Pure Appl. Math.* **1997**, *28*, 1573–1594.
49. Abbas, I.A. A dual phase lag model on thermoelastic interaction in an infinite fiber-reinforced anisotropic medium with a circular hole. *Mech. Based Des. Struct. Mach.* **2015**, *43*, 501–513. [CrossRef]
50. Lahiri, A.; Das, B.; Sarkar, S. Eigenvalue approach to thermoelastic interactions in an unbounded body with a spherical cavity. In Proceedings of the World Congress on Engineering, London, UK, 30 June–2 July 2010; pp. 1881–1886.
51. Crump, K.S. Numerical Inversion of Laplace Transforms Using a Fourier Series Approximation. *J. ACM* **1976**, *23*, 89–96. [CrossRef]
52. Song, Y.Q.; Todorovic, D.M.; Cretin, B.; Vairac, P.; Xu, J.; Bai, J.T. Bending of Semiconducting Cantilevers Under Photothermal Excitation. *Int. J. Thermophys.* **2014**, *35*, 305–319. [CrossRef]

mathematics

MDPI

Article

Reduce-Order Modeling and Higher Order Numerical Solutions for Unsteady Flow and Heat Transfer in Boundary Layer with Internal Heating

Muhammad Bilal [1], Muhammad Safdar [1], Safia Taj [2], Amad Zafar [3], Muhammad Umair Ali [4,*] and Seung Won Lee [5,*]

1 School of Mechanical and Manufacturing Engineering (SMME), National University of Sciences and Technology (NUST), H-12, Islamabad 44000, Pakistan
2 College of Electrical and Mechanical Engineering (CEME), National University of Sciences and Technology (NUST), H-12, Islamabad 44000, Pakistan
3 Department of Intelligent Mechatronics Engineering, Sejong University, Seoul 05006, Republic of Korea
4 Department of Unmanned Vehicle Engineering, Sejong University, Seoul 05006, Republic of Korea
5 Department of Precision Medicine, Sungkyunkwan University School of Medicine, Suwon 16419, Republic of Korea
* Correspondence: umair@sejong.ac.kr (M.U.A.); lsw2920@gmail.com (S.W.L.)

Abstract: We obtain similarity transformations to reduce a system of partial differential equations representing the unsteady fluid flow and heat transfer in a boundary layer with heat generation/absorption using Lie symmetry algebra. There exist seven Lie symmetries for this system of differential equations having three independent and three dependent variables. We use these Lie symmetries for the reduced-order modeling of the flow equations by constructing invariants corresponding to linear combinations of these Lie point symmetries. This procedure reduces one independent variable of the concerned fluid flow model when applied once. Double reductions are achieved by employing invariants twice that lead to ordinary differential equations with one independent and two dependent variables. Similarity transformations are constructed using these two sets of derived invariants corresponding to linear combinations of the Lie point symmetries. These similarity transformations have not been obtained earlier for this flow model. Moreover, the corresponding reduced systems of ordinary differential equations are different from those which exist in the literature for fluid flow and heat transfer that we have been dealing with. We obtain multiple similarity transformations which lead us to new classes of systems of ordinary differential equations. Accurate numerical solutions of these systems are obtained using the combination of an adaptive fourth-order Runge–Kutta method and shooting procedure. Effects of variation of unsteadiness parameter, Prandtl number and heat generation/absorption on fluid velocity, skin friction, surface temperature and heat flux are studied and presented with the help of tables and figures.

Keywords: boundary layer unsteady flow; reduce-order modeling; Lie symmetry; Runge–Kutta; shooting method; heat and mass transfer

MSC: 76M60; 58J70; 35A30; 34B15

Citation: Bilal, M.; Safdar, M.; Taj, S.; Zafar, A.; Ali, M.U.; Lee, S.W. Reduce-Order Modeling and Higher Order Numerical Solutions for Unsteady Flow and Heat Transfer in Boundary Layer with Internal Heating. *Mathematics* **2022**, *10*, 4640. https://doi.org/10.3390/math10244640

Academic Editor: Ramoshweu Solomon Lebelo

Received: 17 October 2022
Accepted: 4 December 2022
Published: 7 December 2022

1. Introduction

With the advancements in industrial manufacturing processes, the accurate prediction of flow and heat transfer is of prime importance. Many industrial and food processes involve heat transfer in thin film flow, e.g., polymer coatings, metal sheets extractions, wire coatings, heat exchangers, reactor fluidization, surface paint processing, etc. In most of these processes, the surface finish, thickness and quality of coatings depend on the fluid flow and heat transfer in the boundary layer/thin films. Moreover, in numerous applications, fluid flow and heat transfer in the boundary layer involve heat generation/absorption effects, such as electric kettles, air conditioning, aerodynamic heating, solar water heaters, cooling in electronic devices using heat sinks, and temperature control techniques in the stacks of batteries. In many of these applications, the heat transfer rate and fluid velocity changes with time, thereby making the behavior of the problem unsteady.

The unsteady boundary layer flows involving heat transfer are modeled using Navier–Stokes equations. Normally, their solution could only be possible using numerical integration [1] due to the non-existence of the exact solutions in most of the cases for such nonlinear equations. However, if we restrict the motion of fluid to a specific group of coordinates and time dependence, we can derive similarity transformations to map flow equations into their simpler and analytically solvable forms. By using similarity transformations, the system of partial differential equations (PDEs) representing the unsteady fluid and heat transfer in the boundary layer flow is mapped into a system of coupled nonlinear ordinary differential equations (ODEs). Such reductions fall into the category of reduced-order modeling. Reduced-order modeling is a mathematical procedure that reduces the computational complexity of the concerned systems. The similarity transformations provide reductions of the dependent and independent variables of the fluid flow model, which brings down the computational complexity of these models. Numerous exact, analytic or/and approximate solution techniques for these systems of ODEs are available as compared to system PDEs for fluid flow and heat transfer in thin films.

In [2], the idea of [3] is implemented to model the unsteady fluid flow in thin film and [4] incorporated the heat transfer effects in it. Many researchers have studied these flow and heat transfer models [5–15] using a few similarity transformations. These studies have been conducted under multiple physical conditions, e.g., unsteady heat transfer in non-Newtonian fluid using power-law, fluid flow in an unsteady sheet by integrating thermocapillarity effects, fluid flow in an unsteady sheet by incorporating thermocapillarity effects with variable fluid properties, MHD flow with heat transfer in an unsteady stretching sheet with a non-uniform heat source, heat transfer with viscous dissipation on an unsteady stretching sheet, fluid flow and heat transfer on an unsteady surface with thermocapillarity and radiation effects, fluid flow and heat transfer on an unsteady stretching surface in the presence of radiation and with variable fluid properties, heat transfer in nanofluid thin film on an unsteady stretching sheet, etc. The reductions in the above cited works through similarity transformations are valid for a specific time interval. An extensive numerical and analytic treatment of flow in thin films has been completed to present optimum flow and heat transfer to acquire the desired refinements of many industrial products depending on such flows.

The Lie symmetry method is a mathematical technique by which one obtains similarity reductions for differential equations (DEs) if there exist Lie point symmetries [16–18] for these DEs. Previously, researchers used Lie point symmetries to derive similarity transformations for differential equations [19–28], e.g., for modified 1D shallow-water equations, the spatial motion on a rotating plane of incompressible fluid on shallow water, free convective nanofluid flow with heat generation/absorption on a chemically reacting sheet in porous medium, the Green Naghdi model hyperbolic and shallow water equations, a Schwartzian $(2+1)$-dimensional wave equation with a variable coefficient for shallow water, rotating shallow water equations, 2D shallow water equations in Lagrangian coordinates with a constant Coriolis parameter, a family of $1+1$ 5th-order PDEs, unsteady boundary layer flow on a vertical sheet with free convection and shallow water equations with Coriolis force, etc. However, in most of these studies, either single reduction is made, or double reduction is completed using general boundary conditions.

In this study, we derive Lie symmetries for heat transfer and fluid flow in an unsteady stretching sheet in the presence of heat generation/absorption. These Lie point symmetries may reduce the dependent and/or independent variables of flow equations considered through functions that remain invariant under Lie symmetry generators that are called invariants. In these flow equations, we have three independent and three dependent variables subject to specific boundary conditions. Using invariants associated with linear combinations of derived symmetries that is again a Lie symmetry, we reduce the PDEs of the flow and heat transfer into systems of equations with two independent and three dependent variables. Repeating the same procedure on these first reductions of PDEs, i.e., we obtain Lie symmetries of these systems, and using invariants corresponding to the obtained symmetries, we provide another reduction which finally leads to systems of ODEs. By combining invariants employed in these two reductions, we construct similarity transformations. These similarity transformations can map the flow equations straight to ODEs; for a detailed procedure, the reader is referred to [29]. A similar study is conducted by [30], using a similarity transformation of the form employed by, e.g., [2,4]. These are different from those presented here. Hence, the systems of ODEs deduced by using them and the ranges in which these similarity transformations are applicable are also different from those imposed in [30]. Moreover, we present the velocity and heat profiles in those ranges of the parameters involved, which either have not been considered earlier or solutions have not been approximated there.

In Section 2, the mathematical formulation of the flow, construction of Lie symmetries, invariants, similarity transformations and reductions to ODEs of the flow equations are presented. Section 3 discusses the numerical solution procedure. In Section 4, we have presented the results, which are followed by the conclusions.

2. Mathematical Formulation, Lie Symmetries, Similarity Transformations and Reductions of Flow Equations

An incompressible, viscous, laminar and unsteady fluid coming out of origin of the coordinate system on a thin horizontal surface along with the heat transfer is considered here, as shown in Figure 1. In addition to that, it is also assumed that the temperature variations are small, and thus, the viscosity of fluid remains constant. The pressure and gravitational effects are also negligible. Initially, the temperature and velocity are taken as arbitrary functions of x-coordinate and time t. It is further assumed that the flow is free from any kind of surface waves, and streamwise diffusion is negligible. Under the stated assumptions, the governing 2D boundary layer equations with uniform heat generation/absorption are written as

$$\begin{aligned} \frac{\partial u}{\partial x}+\frac{\partial v}{\partial y} &= 0, \\ \frac{\partial u}{\partial t}+u\frac{\partial u}{\partial x}+v\frac{\partial u}{\partial y} &= \nu\left(\frac{\partial^2 u}{\partial y^2}\right), \\ \rho C_p\left(\frac{\partial T}{\partial t}+u\frac{\partial T}{\partial x}+v\frac{\partial T}{\partial y}\right) &= \kappa\left(\frac{\partial^2 T}{\partial y^2}+H\right), \end{aligned} \tag{1}$$

subject to

$$\begin{aligned} \text{at } y &= 0: \ u = U,\ v=0,\ T = T_s, \\ \text{at } y &= h(t): \ v = \frac{dh}{dt},\ \frac{\partial u}{\partial y} = \frac{\partial T}{\partial y} = 0. \end{aligned} \tag{2}$$

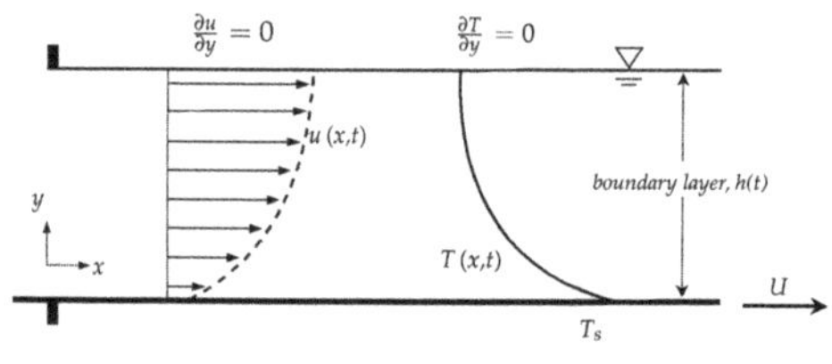

Figure 1. Schematic of flow in boundary layer.

In (1) and (2) x, y are the coordinates parallel and normal to the stretching surface and u, v are the velocities in these directions, respectively. T is the temperature, t is the time, ρ is the density, ν is the kinematic viscosity, C_p is the specific heat at constant pressure, κ is the thermal diffusivity and H is the heat generation/absorption per unit volume and is defined as

$$H = \left(\frac{\tilde{U}(T - T_0)}{x\nu}\right)G^*, \tag{3}$$

where $\tilde{U}$ is considered as velocity in the x-direction due to the flow in [30], which implies

$$\tilde{U} = U = \frac{bx}{(1-at)}. \tag{4}$$

In (3), T_0 is temperature at the origin and G^* is the temperature-dependent heat generation/absorption parameter [30]. After simplifying, H is observed to be a function of temperature, t^{-1} and G^*. For heat addition, $G^* > 0$, and for heat absorption, $G^* < 0$. By restricting the motion in its own horizontal plane and imposing specific time dependence, the surface velocity U and temperature T_s in [30] are written as (4) and

$$T_s = T_0 - \left(\frac{dx^{r_1}}{\nu}\right)T_{ref}(1-at)^{-r_2}, \tag{5}$$

respectively, where T_{ref} is a reference temperature, r_1 and r_2 are positive constants and d is a positive proportionality constant with dimension length^{2-r}time^{-1} [9]. A similarity transformation that is compatible with (4) and (5) is

$$\begin{aligned} \eta &= \frac{y}{\beta}\sqrt{\frac{b}{\nu(1-at)}}, \quad u = \frac{bx}{(1-at)}f'(\eta), \; v = -\sqrt{\frac{b\nu}{(1-at)}}\beta f(\eta), \\ T &= T_0 - \left(\frac{dx^{r_1}}{\nu}\right)T_{ref}(1-at)^{-r_2}\vartheta(\eta), \end{aligned} \tag{6}$$

where a and b are positive constants and have the dimensions of t^{-1}. β is the dimensionless boundary layer/film thickness, f is the stream function, ϑ is the dimensionless temperature and η is the similarity variable. Aziz et al. [30] employed these transformations on (1) and obtained the system of ODEs as

$$\begin{aligned} f''' + \lambda\left(ff'' - S(f' + \frac{\eta}{2}f'') - f'^2\right) &= 0, \\ \frac{\vartheta''}{Pr} + \lambda\left(f\vartheta' - r_1 f'\vartheta - S(\frac{\eta}{2}\vartheta' + r_2 S\vartheta) + \frac{1}{Pr}G^*\vartheta\right) &= 0. \end{aligned} \tag{7}$$

Likewise, considering (4) and (5) in (2) and transforming them via (6), we obtain

$$\begin{aligned} at\ \eta &= 0: \; f = 0, \, f' = 1, \, \vartheta = 1, \\ at\ \eta &= 1: \; f'' = 0, \, \vartheta' = 0, \, f = \frac{1}{2}S, \end{aligned} \tag{8}$$

where prime denotes the derivative with respect to η, $S = a/b$ denotes the dimensionless unsteadiness parameter, $Pr = \frac{\rho\nu C_p}{\kappa}$ is the Prandtl number and $\lambda = \beta^2$ is the dimensionless film thickness. It is important to note that the above similarity transformations are valid only for $t < a^{-1}$. In the subsequent sections, we are performing reduced-order modeling using Lie symmetry algebra.

2.1. Lie Symmetries and Invariants

To derive similarity transformations for the system of PDEs (1), we first obtain Lie point symmetry generators for this system that is a vector field as

$$\mathbf{X} = \xi_i \frac{\partial}{\partial \psi_i} + \phi_i \frac{\partial}{\partial \zeta_i}, \tag{9}$$

where $i = 1,2,3$, ξ and ϕ are infinitesimal coordinates and they are functions of independent x, y, t and dependent variables u, v, T, respectively. The system (1) and boundary conditions (2) contain both first and second-order partial derivatives. For this, we require first and second extensions of (9) to operate on them. MAPLE contains an algebraic procedure to derive Lie point symmetries for DEs; here, we use it to obtain Lie point symmetries of system (1). In (3), we consider $\tilde{U} = \frac{bx}{at}$, that is different from $U(t,x)$, given in conditions (2) which provides

$$H = \frac{b}{avt} G^*(T - T_0). \tag{10}$$

The stretchable sheet velocity $U(t,x)$, is obtained in the subsequent section by applying the Lie symmetry generators. Hence, it is not expected to be similar for all symmetries that are derived using (10) in system (1). System (1) admits an infinite dimensional Lie point symmetry algebra that is spanned by the following symmetry generators

$$\begin{aligned}
\mathbf{X}_1^\infty &= \frac{\partial}{\partial x} + f_1(t,x)\frac{\partial}{\partial y} + (f_{1,t} + u f_{1,x})\frac{\partial}{\partial v}, \\
\mathbf{X}_2^\infty &= t\frac{\partial}{\partial x} + f_2(t,x)\frac{\partial}{\partial y} + \frac{\partial}{\partial u} + (f_{2,t} + u f_{2,x})\frac{\partial}{\partial v}, \\
\mathbf{X}_3^\infty &= x\frac{\partial}{\partial x} + f_3(t,x)\frac{\partial}{\partial y} + u\frac{\partial}{\partial u} + (f_{3,t} + u f_{3,x})\frac{\partial}{\partial v}, \\
\mathbf{X}_4^\infty &= f_4(t,x)\frac{\partial}{\partial y} + (T - T_0)\frac{\partial}{\partial T} + (f_{4,t} + u f_{4,x})\frac{\partial}{\partial v}, \\
\mathbf{X}_5^\infty &= t\frac{\partial}{\partial t} + (\frac{y}{2} + f_5(t,x))\frac{\partial}{\partial y} - u\frac{\partial}{\partial u} + (f_{5,t} + u f_{5,x} - \frac{v}{2})\frac{\partial}{\partial v}, \\
\mathbf{X}_6^\infty &= f_6(t,x)\frac{\partial}{\partial y} + t^{\frac{b\kappa G^*}{aC_p\rho v}}\frac{\partial}{\partial T} + (f_{6,t} + u f_{6,x})\frac{\partial}{\partial v}, \\
\mathbf{X}_7^\infty &= \frac{\partial}{\partial t} + f_7(t,x)\frac{\partial}{\partial y} + \frac{(T - T_0)b\kappa G^*}{aC_p\rho v t}\frac{\partial}{\partial T} + (f_{7,t} + u f_{7,x})\frac{\partial}{\partial v}, \\
\mathbf{X}_8^\infty &= f_8(t,x)\frac{\partial}{\partial y} + (f_{8,t} + u f_{8,x})\frac{\partial}{\partial v}.
\end{aligned} \tag{11}$$

By considering $f_i(t,x) = 0$, for $i = 1,2,\ldots,8$, we obtain a finite dimensional algebra. The reason to use finite dimensional algebra is to extract scaling transformations to perform the reduced-order modeling. Scaling transformations are the most suitable mappings that are employed for reduction of the independent variables of the flow models. Table 1 presents the finite dimensional symmetry algebra and corresponding invariants for system (1). These symmetry generators $\mathbf{X}_1, \ldots, \mathbf{X}_7$ and their linear combinations leave system (1) and associated conditions invariant. The boundary conditions (2) also remain invariant under these generators. Both U and T at $y = 0$ are functions of x-coordinate and time t. There, invariant forms under $\mathbf{X}_1, \ldots, \mathbf{X}_7$ are determined by applying these generators on them and evaluating the resulting expressions on these conditions. However, when a single symmetry is used in this procedure, both U and T become either functions of x or t. In this work, we want to keep them functions of both x and t. We achieved it through linear combinations of the symmetry generators $\mathbf{X}_1, \ldots, \mathbf{X}_7$ by adding two at a time, which also leave boundary conditions (2) invariant. We obtain seven such combinations in which these conditions remain functions of both space and time, as shown in Table 2.

Table 1. Lie point symmetry generators and invariants.

Symmetry	Invariants-Conserved Quantities
$\mathbf{X}_1 = \frac{\partial}{\partial x}$	$x,\ y,\ T,\ u,\ v$
$\mathbf{X}_2 = t\frac{\partial}{\partial x} + \frac{\partial}{\partial u}$	$t,\ y,\ T,\ \frac{u}{x},\ v$
$\mathbf{X}_3 = x\frac{\partial}{\partial x} + u\frac{\partial}{\partial u}$	$t,\ y,\ T,\ \frac{u}{x},\ v$
$\mathbf{X}_4 = (T - T_0)\frac{\partial}{\partial T}$	$t,\ x,\ y,\ u,\ v$
$\mathbf{X}_5 = t\frac{\partial}{\partial t} + \frac{y}{2}\frac{\partial}{\partial y} - u\frac{\partial}{\partial u} - \frac{v}{2}\frac{\partial}{\partial v}$	$x,\ \frac{y}{t},\ T,\ ut,\ v$
$\mathbf{X}_6 = \left(t^{\frac{b\kappa G^*}{aC_p\rho v}}\right)\frac{\partial}{\partial T}$	$t,\ x,\ y,\ u,\ v$
$\mathbf{X}_7 = \frac{\partial}{\partial t} + \frac{b(T-T_0)\kappa G*}{aC_p\rho vt}\frac{\partial}{\partial T}$	$x,\ y,\ \frac{T-T_0}{t},\ u,\ v$

Table 2. First invariants for similarity transformations.

Case	Symmetry and Invariants	Corresponding Boundary Conditions
1	$\mathbf{X}_3 + \mathbf{X}_4$ $t, y, \frac{u}{x}, v, \frac{T-T_0}{x}$	$at\ y=0:\ \ v=0,\ \ u=x\bar{U}(t),\ \ T=T_0+x\bar{T}_s(t)$ $at\ y=h(t):\ \ v=\frac{\partial h}{\partial t},\ \ \frac{\partial u}{\partial y}=\frac{\partial T}{\partial y}=0$
2	$\mathbf{X}_2 + \mathbf{X}_5$ $x-t, \frac{y}{\sqrt{t}}, ut-t, \sqrt{t}v, T$	$at\ y=0:\ \ v=0,\ \ u=1+\frac{\bar{U}(x-t)}{t},\ \ T=\bar{T}_s(x-t)$ $at\ y=a_3\sqrt{t}:\ \ v=\frac{a_3}{2\sqrt{t}},\ \ \frac{\partial u}{\partial y}=\frac{\partial T}{\partial y}=0$
3	$\mathbf{X}_3 + \mathbf{X}_5$ $\frac{x}{t}, \frac{y}{\sqrt{t}}, u, \sqrt{t}v, T$	$at\ y=0:\ \ v=0,\ \ u=\bar{U}(\frac{x}{t}),\ \ T=\bar{T}_s(\frac{x}{t})$ $at\ y=a_3\sqrt{t}:\ \ v=\frac{a_3}{2\sqrt{t}},\ \ \frac{\partial u}{\partial y}=\frac{\partial T}{\partial y}=0$
4	$\mathbf{X}_3 + \mathbf{X}_6$ $t, y, \frac{u}{x}, v, T - ln(x)\left(t^{\frac{b\kappa G^*}{aC_p\rho v}}\right)$	$at\ y=0:\ \ v=0,\ \ u=x\bar{U}(t),\ \ T=\bar{T}_s(t)+\ln(x)t^{\frac{b\kappa G^*}{aC_p\rho v}}$ $at\ y=h(t):\ \ v=\frac{\partial h}{\partial t},\ \ \frac{\partial u}{\partial y}=\frac{\partial T}{\partial y}=0$
5	$\mathbf{X}_4 + \mathbf{X}_5$ $x, \frac{y}{\sqrt{t}}, tu, \sqrt{t}v, \frac{T-T_0}{t}$	$at\ y=0:\ \ v=0,\ \ u=\frac{\bar{U}(x)}{t},\ \ T=T_0+t\bar{T}_s(x)$ $at\ y=a_3\sqrt{t}:\ \ v=\frac{a_3}{2\sqrt{t}},\ \ \frac{\partial u}{\partial y}=\frac{\partial T}{\partial y}=0$
6	$\mathbf{X}_5 + \mathbf{X}_6$ $x, \frac{y}{\sqrt{t}}, tu, \sqrt{t}v, T - \frac{aC_p\rho v}{b\kappa G^*}t^{\frac{b\kappa G^*}{aC_p\rho v}}$	$at\ y=0:\ \ v=0,\ \ u=\frac{\bar{U}(x)}{t},\ \ T=\frac{aC_p\rho v}{b\kappa G^*}t^{\frac{b\kappa G^*}{aC_p\rho v}}+\bar{T}_s(x)$ $at\ y=a_3\sqrt{t}:\ \ v=\frac{a_3}{2\sqrt{t}},\ \ \frac{\partial u}{\partial y}=\frac{\partial T}{\partial y}=0$
7	$\mathbf{X}_5 + \mathbf{X}_7$ $x, \frac{y}{\sqrt{1+t}}, u(t+1), \sqrt{1+t}v,$ $(1+t)^{\frac{b\kappa G^*}{aC_p\rho v}}(T-T_0)t^{-\frac{b\kappa G^*}{aC_p\rho v}}$	$at\ y=0:\ \ v=0,\ \ u=\frac{\bar{U}(x)}{t+1},\ \ T=T_0+(1+t)^{-\frac{b\kappa G^*}{aC_p\rho v}}t^{\frac{b\kappa G^*}{aC_p\rho v}}\bar{T}_s(x)$ $at\ y=a_3\sqrt{1+t}:\ \ v=\frac{a_3}{2\sqrt{1+t}},\ \ \frac{\partial u}{\partial y}=\frac{\partial T}{\partial y}=0$

2.2. Double Reductions and Construction of Similarity Transformations

Similarity transformations are derived through double reductions of differential equations (DEs) using Lie point symmetry generators. Consider Case 1 in Table 2, which is $\mathbf{X}_3 + \mathbf{X}_4$. Except for $u = U(t, x)$ and $T = T_s(t, x)$, it leaves all other boundary conditions (2) invariant. The conditions $u = U(t, x)$ and $T = T_s(t, x)$ when inserted in the invariance criterion read as

$$\begin{aligned} [\mathbf{X}_3 + \mathbf{X}_4](u - U(t,x))|_{u=U(t,x)} &= 0, \\ [\mathbf{X}_3 + \mathbf{X}_4](T - T_s(t,x))|_{T=T_s(t,x)} &= 0. \end{aligned} \tag{12}$$

Applying these symmetry generators and expanding the resulting expressions at $u = U(t, x)$ and $T = T_s(t, x)$ provides the following linear PDEs

$$\begin{aligned} x\frac{\partial U(x,t)}{\partial x} - u &= 0, \\ x\frac{\partial T_s(x,t)}{\partial x} - T + T_0 &= 0. \end{aligned} \tag{13}$$

Solving these equations, we obtain

$$u = x\bar{U}(t), \qquad \text{and} \qquad T = T_0 + x\bar{T}_s(t). \tag{14}$$

Now, for the derivation of 0th-order differential invariants, we apply $\mathbf{X}_{34} = \mathbf{X}_3 + \mathbf{X}_4$, in the following invariance criterion

$$\mathbf{X}_{34}\, J(t, x, y, u, v, T) = 0, \tag{15}$$

which leads to the following PDE

$$x\frac{\partial J}{\partial x} + u\frac{\partial J}{\partial u} + (T - T_0)\frac{\partial J}{\partial T} = 0. \tag{16}$$

Solving it using MAPLE, we obtain five invariants $\{t, y, \frac{u}{x}, v, \frac{T-T_0}{x}\}$ with two independent t, y and three dependent variables $\frac{u}{x}, v, \frac{T-T_0}{x}$. We obtain first components of the claimed similarity transformations by renaming these invariants as follows

$$c_1 = t,\ c_2 = y,\ P = \frac{u}{x},\ Q = v,\ R = \frac{T - T_0}{x}. \tag{17}$$

This maps the system of PDEs (1) and conditions (2) to

$$\begin{aligned} P + \frac{\partial Q}{\partial c_2} &= 0, \\ \frac{\partial P}{\partial c_1} + P^2 + Q\frac{\partial P}{\partial c_2} &= \nu\left(\frac{\partial^2 P}{\partial {c_2}^2}\right), \\ \rho C_p\left(\frac{\partial R}{\partial c_1} + PR + Q\frac{\partial R}{\partial c_2}\right) &= \kappa\left(\frac{\partial^2 R}{\partial {c_2}^2} + H\right), \end{aligned} \tag{18}$$

and

$$\begin{aligned} \text{at } c_2 &= 0 : P = F(c_1),\ Q = 0,\ R = G(c_1), \\ \text{at } c_2 &= h : Q = \frac{dh}{dc_1},\ \frac{\partial P}{\partial c_2} = \frac{\partial R}{\partial c_2} = 0. \end{aligned} \tag{19}$$

For the second reduction, symmetry generators for system (18) are obtained that admits a *three*-dimensional Lie symmetry algebra

$$\mathbf{Y}_1 = R\frac{\partial}{\partial R}, \ \mathbf{Y}_2 = c_1\frac{\partial}{\partial c_1} + \frac{c_2}{2}\frac{\partial}{\partial c_2} - P\frac{\partial}{\partial P} - \frac{Q}{2}\frac{\partial}{\partial Q},$$

$$\mathbf{Y}_3 = \frac{\partial}{\partial c_1} + \left(\frac{bR\kappa G*}{aC_p c_1 \rho \nu}\right)\frac{\partial}{\partial R}. \tag{20}$$

The combination $\mathbf{Y}_1 + \mathbf{Y}_2$ further converts the boundary conditions (19) to

$$\begin{aligned} &\text{at } c_2 = 0: \ P = \frac{a_1}{c_1}, \ Q = 0, \ R = a_2 c_1, \\ &\text{at } c_2 = a_3\sqrt{c_1}: \ Q = \frac{a_3}{2}, \ \frac{\partial P}{\partial c_2} = \frac{\partial R}{\partial c_2} = 0. \end{aligned} \tag{21}$$

The invariants obtained using these symmetries $\mathbf{Y}_1 + \mathbf{Y}_2$ are $\{\frac{c_2}{\sqrt{c_1}}, Pc_1, Q\sqrt{c_1}, \frac{R}{c_1}\}$, where now $\frac{c_2}{\sqrt{c_1}}$ is the new independent variable and $Pc_1, Q\sqrt{c_1}, \frac{R}{c_1}$ are the new dependent variables, which are given the following notations

$$\chi = \frac{c_2}{\sqrt{c_1}}, \ Pc_1 = L, \ Q\sqrt{c_1} = M, \ \frac{R}{c_1} = N. \tag{22}$$

Using (22), the second reductions are performed that transform the system (18) and associated conditions (21) to

$$\begin{aligned} L + M' &= 0, \\ L^2 - L - \frac{\chi}{2}L' + ML' &= \nu L'', \\ LN + N - \frac{\chi}{2}N' + MN' &= \frac{\kappa}{\rho C_p}N'' + \frac{b\kappa G^* N}{a\rho C_p \nu}, \end{aligned} \tag{23}$$

and

$$\begin{aligned} &\text{at } \chi = 0: \ L = a_1, \ M = 0, \ N = a_2, \\ &\text{at } \chi = a_3: \ M = \frac{a_3}{2}, \ L' = N' = 0, \end{aligned} \tag{24}$$

where prime denotes differentiation with respect to χ. In system (7), we have Pr, S and β. We introduce these variables in the similarity transformations constructed here by combining (17), (22) and

$$\chi = \eta\beta\sqrt{\frac{a\nu}{b}}, \ L = -\frac{b}{a}f'(\eta), \ M = \beta\sqrt{\frac{b\nu}{a}}f(\eta), \ N = \vartheta(\eta), \tag{25}$$

which leads to claimed similarity transformations

$$u = -\frac{bx}{at}f'(\eta), \ v = \beta\sqrt{\frac{b\nu}{at}}f(\eta), \ T = T_0 + xt\vartheta(\eta), \ \eta = \frac{1}{\beta}\sqrt{\frac{b}{at\nu}}y. \tag{26}$$

The set of similarity transformations (26) maps the system of PDEs (1) and boundary conditions (2) into system of ODEs as

$$\begin{aligned} f''' + \lambda\left(Sf' - ff'' + \frac{S\eta}{2}f'' + f'^2\right) &= 0, \\ \frac{1}{Pr}\vartheta'' + \lambda\left(-f\vartheta' + \frac{S\eta}{2}\vartheta' - S\vartheta + f'\vartheta + \frac{1}{Pr}\vartheta G^*\right) &= 0, \end{aligned} \tag{27}$$

and

$$\begin{aligned} at\ \eta &= 0: \ f = 0,\ f' = 1,\ \vartheta = 1, \\ at\ \eta &= 1: \ f = \frac{1}{2}S,\ f'' = 0,\ \vartheta' = 0. \end{aligned} \tag{28}$$

Using linear combinations of symmetries $\mathbf{X}_1 - \mathbf{X}_7$, we perform double reductions and obtain a set of similarity transformations against each combination that reduces the system of PDEs (1) into systems of ODEs given in Table 3. The reason to consider only these linear combinations in Table 3 is the form of $U(t,x)$ and $T_s(t,x)$ they provide; i.e., both of them are functions of the x-coordinate and time-t in all cases. Once invariance of the conditions (2) under the admitted Lie point symmetries of the system (1) has been established, then it implies that any linear combination of these symmetries also leaves the associated conditions invariant.

Table 3. Similarity transformations and systems of ODEs.

Case	Symmetry Generator and Similarity Transformation	System of ODEs
1	$\mathbf{X}_3 + \mathbf{X}_4$ $v = \beta\sqrt{\frac{b\nu}{at}}f(\eta),\ u = -\frac{bx}{at}f'(\eta)$ $\eta = \frac{1}{\beta}\sqrt{\frac{b}{at\nu}}y,\ T = T_0 + xt\vartheta(\eta)$	$f''' + \lambda\left(Sf' - ff'' + f'^2 + \frac{S\eta}{2}f''\right) = 0$ $\frac{1}{\Pr}\vartheta'' + \lambda\left(-f\vartheta' + \frac{S\eta}{2}\vartheta' - S\vartheta + f'\vartheta + \frac{1}{\Pr}\vartheta G^*\right) = 0$
2	$\mathbf{X}_2 + \mathbf{X}_5$ $v = \beta\sqrt{\frac{b\nu}{at}}f(\eta),\ u = 1 - \frac{b(x-t)}{at}f'(\eta)$ $\eta = \frac{1}{\beta}\sqrt{\frac{b}{at\nu}}y,\ T = T_0 + (x-t)\vartheta(\eta)$	$f''' + \lambda\left(Sf' - ff'' + f'^2 + \frac{S\eta}{2}f''\right) = 0$ $\frac{1}{\Pr}\vartheta'' + \lambda\left(-f\vartheta' + \frac{S\eta}{2}\vartheta' + f'\vartheta + \frac{1}{\Pr}\vartheta G^*\right) = 0$
3	$\mathbf{X}_3 + \mathbf{X}_5$ $v = \beta\sqrt{\frac{b\nu}{at}}f(\eta),\ u = -\frac{bx}{at}f'(\eta)$ $\eta = \frac{1}{\beta}\sqrt{\frac{b}{at\nu}}y,\ T = T_0 + \frac{x}{t}\vartheta(\eta)$	$f''' + \lambda\left(Sf' - ff'' + f'^2 + \frac{S\eta}{2}f''\right) = 0$ $\frac{1}{\Pr}\vartheta'' + \lambda\left(-f\vartheta' + \frac{S\eta}{2}\vartheta' + S\vartheta + f'\vartheta + \frac{1}{\Pr}\vartheta G^*\right) = 0$
4	$\mathbf{X}_3 + \mathbf{X}_6$ $v = \beta\sqrt{\frac{b\nu}{at}}f(\eta),\ u = -\frac{bx}{at}f'(\eta)$ $\eta = \frac{1}{\beta}\sqrt{\frac{b}{at\nu}}y,\ T = T_0 + (\vartheta(\eta) + ln(x))t^{\frac{b\kappa G^*}{aC_p\rho\nu}}$	$f''' + \lambda\left(Sf' - ff'' + f'^2 + \frac{S\eta}{2}f''\right) = 0$ $\frac{1}{\Pr}\vartheta'' + \lambda\left(-f\vartheta' + \frac{S\eta}{2}\vartheta' + f'\right) = 0$
5	$\mathbf{X}_4 + \mathbf{X}_5$ $v = \beta\sqrt{\frac{b\nu}{at}}f(\eta),\ u = -\frac{bx}{at}f'(\eta)$ $\eta = \frac{1}{\beta}\sqrt{\frac{b}{at\nu}}y,\ T = T_0 + xt\vartheta(\eta)$	$f''' + \lambda\left(Sf' - ff'' + f'^2 + \frac{S\eta}{2}f''\right) = 0$ $\frac{1}{\Pr}\vartheta'' + \lambda\left(-f\vartheta' + \frac{S\eta}{2}\vartheta' - S\vartheta + f'\vartheta + \frac{1}{\Pr}\vartheta G^*\right) = 0$
6	$\mathbf{X}_5 + \mathbf{X}_6$ $v = \beta\sqrt{\frac{b\nu}{at}}f(\eta),\ u = -\frac{bx}{at}f'(\eta)$ $\eta = \frac{1}{\beta}\sqrt{\frac{b}{at\nu}}y,\ T = T_0 + x\vartheta(\eta) + \frac{aC_p\rho\nu}{b\kappa G^*}t^{\frac{b\kappa G^*}{aC_p\rho\nu}}$	$f''' + \lambda\left(Sf' - ff'' + f'^2 + \frac{S\eta}{2}f''\right) = 0$ $\frac{1}{\Pr}\vartheta'' + \lambda\left(-f\vartheta' + \frac{S\eta}{2}\vartheta' + f'\vartheta + \frac{1}{\Pr}\vartheta G^*\right) = 0$
7	$\mathbf{X}_5 + \mathbf{X}_7$ $v = \beta\sqrt{\frac{b\nu}{a(1+t)}}f(\eta),\ u = -\frac{bx}{a(1+t)}f'(\eta)$ $\eta = \frac{1}{\beta}\sqrt{\frac{b}{a\nu(1+t)}}y,\ T = T_0 + x\vartheta(\eta)(1+t)^{-\frac{b\kappa G^*}{aC_p\rho\nu}}t^{\frac{b\kappa G^*}{aC_p\rho\nu}}$	$f''' + \lambda\left(Sf' - ff'' + f'^2 + \frac{S\eta}{2}f''\right) = 0$ $\frac{1}{\Pr}\vartheta'' + \lambda\left(-f\vartheta' + \frac{S\eta}{2}\vartheta' + f'\vartheta + \frac{1}{\Pr}\vartheta G^*\right) = 0$

The skin friction C_f and heat flux q_s are important physical parameters, and they are written as

$$C_f = \left(\frac{2\tau_s}{\rho u^2}\right), \tag{29}$$

$$q_s = -\kappa\left(\frac{\partial T}{\partial y}\right)_{y=0}, \tag{30}$$

where τ_s is the shear stress

$$\tau_s = \mu\left(\frac{\partial u}{\partial y}\right)_{y=0}, \tag{31}$$

Using the similarity transformations (26), i.e., for Case 1, we have

$$C_f = \frac{2}{\beta\sqrt{Re_x}} f''(0), \tag{32}$$

$$q_s = -\frac{x\kappa}{\beta\mu}\sqrt{\frac{b(1+x)\rho^3}{a}}\vartheta'(0), \tag{33}$$

where Re_x is the local Reynolds number.

3. Numerical Solutions

The solution of a nonlinear coupled system of ODEs, e.g., (7) subject to (8) is obtained by using the combination of most efficient Runge–Kutta Fehlberg numerical integration technique and shooting method. It is a fourth-order O(h^4) accurate scheme with the fifth-order O(h^5) error estimation. This method is known as RKF45. This method automatically varies the step size at specified locations based on the approximation accuracy required. This adaptive grid sizing helps to reduce the computational cost [31]. In system (7), if dimensionless film thickness λ is known, the solution can be approximated by using only the first five conditions from (8). We write system (7) in the form of system of five 1st coupled ODEs by considering

$$f = y_1,\ f' = y_2,\ f'' = y_3,\ f''' = y_3',\ \vartheta = y_4,\ \vartheta' = y_5,\ \vartheta'' = y_5'. \tag{34}$$

By substituting the above assumptions (34) in system (7) and boundary conditions (8), we obtain

$$y_1' = f',\ y_2' = f'',\ y_3' = \lambda\left(-y_1y_3 + \frac{S\eta}{2}y_3 + {y_2}^2 + Sy_2\right),\ y_4' = \vartheta',$$
$$y_5' = \lambda Pr\left(-y_1y_5 + 2y_2y_4 + \frac{S\eta}{2}y_5 + y_4\left(\frac{3}{2}S - \frac{1}{Pr}B^*\right)\right), \tag{35}$$

and

$$y_1(0) = 0,\ y_2(0) = 1,\ y_3(0) = b_1,\ y_4(0) = 0,\ y_5(0) = b_2. \tag{36}$$

Three simultaneous shooting techniques are applied to transform the boundary value problem into an initial value problem. The transformed initial conditions b_1 and b_2 are found iteratively by using Newton's method until the error is 10^{-12}. In the RKF45 integration procedure, the numerical integration in performed until the error is less than 10^{-10}.

As the film/boundary layer thickness β is unknown, so the value of λ is approximated iteratively until the last condition of (8); that is, $f(1) = \frac{S}{2}$ is satisfied within a range of less than 10^{-9}. The film thickness λ varies with the unsteadiness parameter S, so at different values of S, first, the film thickness λ is approximated before analyzing the effects of Prandtl number Pr and heat generation/absorption G^*. The results obtained for system (7) with conditions (8) using this procedure are compared with the 10th order Homotopy Analysis Method (analytical method) employed by Wang [32]. In Table 4, the effects of unsteadiness on film thickness λ and skin friction $f''(0)$ are compared with the analytical results. In Table 5, the effects of change in Prandtl number Pr on surface temperature $\vartheta(1)$ and heat flux $-\vartheta'(0)$ are compared. It is clear from Tables 4 and 5 that the numerical approach used in the present study is in good agreement with the analytical method.

Table 4. Validation of numerical results.

S	Present Study		Wang [32]	
	β	$-f''(0)$	β	$-f''(0)$
1.2	1.1277809	1.4426253	1.127780	1.442631
1.3	0.9642181	1.2183196	0.964219	1.218322
1.4	0.8210322	1.0127802	0.821032	1.012784
1.5	0.6931444	0.8218421	0.693144	0.821842
1.6	0.5761730	0.6423970	0.576173	0.642397

Table 5. Validation of numerical Results at $S = 1.2$ and $G^* = 0$.

Pr	Present Study		Wang [32]	
	$\vartheta(1)$	$-\vartheta'(0)$	$\vartheta(1)$	$-\vartheta'(0)$
0.01	0.9823314	0.0377342	0.982331	0.037734
0.1	0.8462218	0.3439312	0.843622	0.343931
1.0	0.2867165	1.9995915	0.286717	1.999590
2.0	0.1281219	2.9759051	0.128124	2.975450
3.0	0.0676448	3.7013202	0.067658	3.698830

4. Results

From Table 3, it is evident that despite having the unique symmetry generators and invariants (except for Case 1 and 5 for which the corresponding systems of ODEs are also similar), the transformed systems of ODEs for Case 2, Case 6 and Case 7 are the same. The system of ODEs in Case 4 is not containing any heat generation/absorption parameter G^* and thus is not considered further for a solution. It is important to note that the Lie similarity transformations in Table 3 are valid at any time interval, i.e., for $t > 0$. Moreover, the ranges of S for which we are providing the variations in film thickness, velocity of the flow and temperature have not been revealed in [30].

4.1. *Effect of Unsteadiness on Film Thickness and Fluid Velocity*

The first equation is not coupled with the second one and is the same for all cases of Table 3. It controls the dimensionless boundary layer thickness, fluid velocity and skin friction. The variable in these equations is the dimensional unsteadiness parameter S. Table 6 shows the effect of variation of unsteadiness S on film thickness λ, surface velocity $f'(1)$ and skin friction $f''(0)$. Figure 2 shows the variation of velocity distribution $f'(\eta)$ in the boundary layer with unsteadiness parameter S. The film thickness β is observed to decrease with the increase in unsteadiness S in the flow. The surface velocity $f'(1)$ and so the skin friction $f''(0)$ are observed to increase with the increase in unsteadiness S.

Table 6. Variation of velocity $f'(1)$ and dimensionless film thickness β with unsteadiness parameter S.

S	β	$f'(1)$	$f''(0)$
4.0	0.46222258	2.54527934	2.64859856
6.0	0.42256741	4.11506337	5.12669108
8.0	0.38196786	5.68906454	7.57845493
10.0	0.34939989	7.26442893	10.0221543

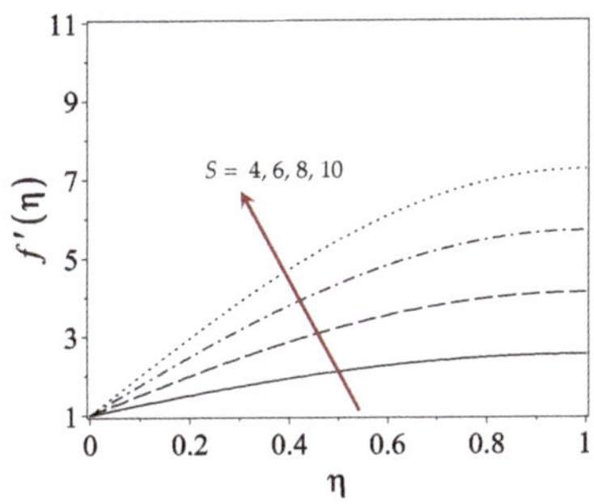

Figure 2. Velocity distribution $f'(\eta)$ variation with unsteadiness parameter S.

4.2. *Effect of Unsteadiness on Temperature*

As the unsteadiness in the fluid increases, the flow velocity $f'(\eta)$ increases. This also increases the heat flux $-\vartheta'(0)$, and thus, a drop in the surface temperature $\vartheta(1)$ is observed for Case 1 and Case 5. For Cases 2, 3, 6, and 7, the surface temperature $\vartheta(1)$ increases with the increase in unsteadiness S in the fluid. Figure 3 and Table 7 show the effects of unsteadiness S on surface temperature $\vartheta(1)$ and heat flux $-\vartheta'(0)$ for all the cases.

Table 7. Variation of temperature distribution with unsteadiness parameter at $Pr = 1$ and $G^* = 1$.

S	Case 1 and 5		Case 2, 6 and 7		Case 3	
	$\vartheta(1)$	$-\vartheta'(0)$	$\vartheta(1)$	$-\vartheta'(0)$	$\vartheta(1)$	$-\vartheta'(0)$
4.0	0.9254323	0.2063995	1.5018941	0.8804902	3.2395043	3.7582109
6.0	0.8800376	0.3375851	1.6365029	1.0733105	5.6566690	7.5566639
8.0	0.8559376	0.4088891	1.6905412	1.1414694	8.0118998	11.224076
10	0.8412676	0.4528668	1.7180535	1.1725929	10.128307	14.510949

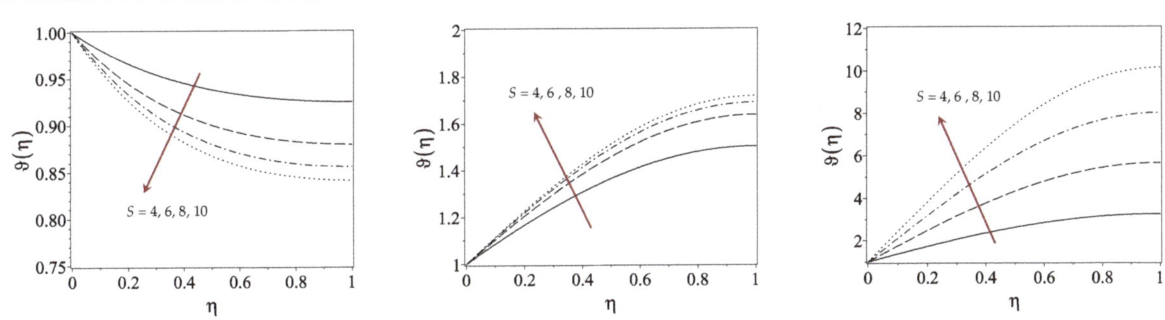

(**a**) Case 1 and 5 (**b**) Case 2, 6 and 7 (**c**) Case 3

Figure 3. Variation of temperature distribution $\vartheta(\eta)$ with the unsteadiness parameter S at $Pr = 1$ and $G^* = 1$.

4.3. Effect of Prandtl Number on Temperature

As the Prandtl number increases, the ratio of momentum diffusivity to thermal diffusivity increases. This increases the heat flux $-\vartheta'(0)$, and thus, a temperature $\vartheta(1)$ drop is observed for Case 1 and Case 5. For Cases 2, 3, 6, and 7, the surface temperature $\vartheta(1)$ increases with the increase in Prandtl number Pr in the fluid. Figure 4 and Table 8 show the effect of Prandtl number Pr on the temperature distribution $\vartheta(\eta)$, surface temperature $\vartheta(1)$, and heat flux $-\vartheta'(0)$.

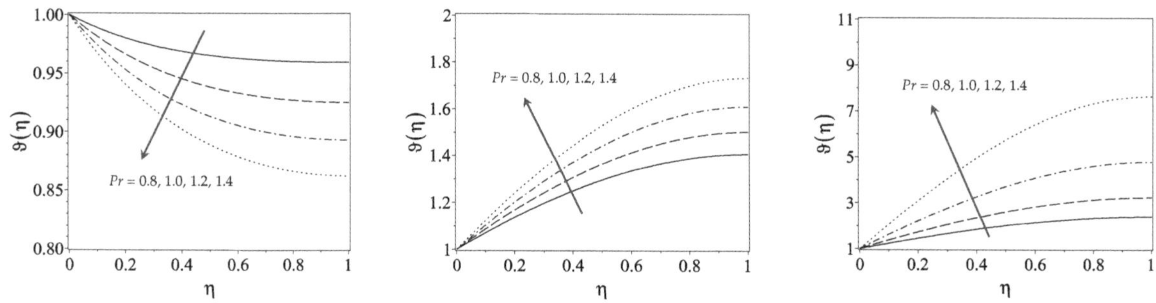

(**a**) Cases 1 and 5 (**b**) Cases 2, 6 and 7 (**c**) Case 3

Figure 4. Variation of temperature distribution $\vartheta(\eta)$ with Prandtl number Pr at $S = 4$ and $G^* = 1$.

Table 8. Variation of temperature distribution $\vartheta(\eta)$ with Prandtl number at $S = 4$ and $G^* = 1$.

Pr	Case 1 and 5		Case 2, 6 and 7		Case 3	
	$\vartheta(1)$	$-\vartheta'(0)$	$\vartheta(1)$	$-\vartheta'(0)$	$\vartheta(1)$	$-\vartheta'(0)$
0.8	0.9595775	0.1261086	1.4071290	0.7190225	2.4052188	2.3063102
1.0	0.9254323	0.2063995	1.5018941	0.8804902	3.2395043	3.7582109
1.2	0.8930815	0.2836747	1.6093876	1.0579964	4.7864442	6.2626166
1.4	0.8623943	0.3581463	1.7320289	1.2589990	7.6471113	10.783625

4.4. Effect of Heat Generation/Absorption on Temperature

When fluid generates heat, i.e., $G^* > 0$, the surface temperature $\vartheta(1)$ increases and when it absorbs heat, i.e., $G^* < 0$, the surface temperature is $\vartheta(1)$. This corresponds to the increase and decrease in heat flux $-\vartheta'(0)$, respectively. Similar effects are observed here for all cases as shown in Figure 5 and Table 9.

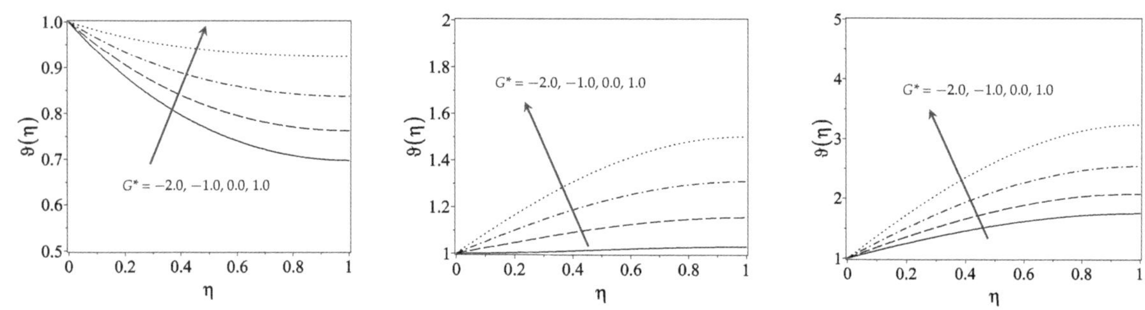

(**a**) Cases 1 and 5 (**b**) Cases 2, 6 and 7 (**c**) Case 3

Figure 5. Variation of temperature distribution $\vartheta(\eta)$ with heat generation/absorption parameter G^* at $S = 4$ and $Pr = 1$.

Table 9. Variation of temperature distribution $\vartheta(\eta)$ with heat generation/absorption parameter G^* at $S = 4$ and $Pr = 1$.

G^*	Case 1 and 5		Case 2, 6 and 7		Case 3	
	$\vartheta(1)$	$-\vartheta'(0)$	$\vartheta(1)$	$-\vartheta'(0)$	$\vartheta(1)$	$-\vartheta'(0)$
−2.0	0.6980416	0.6846427	1.0296499	0.0020397	1.7502717	1.3163411
−1.0	0.7626680	0.5501907	1.1551679	0.2443599	2.0816556	1.8831544
0.0	0.8376079	0.3886397	1.3090560	0.5316521	2.5452882	2.6576106
1.0	0.9254323	0.2063995	1.5018941	0.8804902	3.2395043	3.7582109

5. Conclusions

Reduced-order modeling has been performed by systematically deriving the similarity transformations using Lie symmetry algebra to map the system of PDEs representing heat transfer in unsteady flow with heat generation/absorption to ODEs. Similarity transformations are deduced through invariants corresponding to each linear combination of the Lie symmetries (considering two at a time) of the flow equations. These similarity transformations are used to perform double reductions to map the said system of PDEs into the system of ODES. We present only those cases here in which specific boundary conditions remain functions of both space and time variables. Seven such cases are obtained. Case 4 has not been pursued here because it has no heat generation/absorption parameter in the corresponding system of ODEs.

In all cases, the film thickness $\lambda = \beta^2$ decreases, and the flow velocity $f'(\eta)$ increases with an increase in unsteadiness parameter S. For Cases 1 and 5, the surface temperature $\vartheta(1)$ decreases with an increase in unsteadiness S, Prandtl number Pr and heat absorption $G^* < 0$. While for Cases 2, 3, 6 and 7, the surface temperature $\vartheta(1)$ increases with increase in unsteadiness S, Prandtl number Pr and heat generation $G^* > 0$. The Lie symmetry method provides more than one type of similarity transformation and correspondingly reduces the system of ODEs, which enables a comprehensive study of the flow and heat transfer through approximate solutions of the systems of ODEs corresponding to concerned flow equations.

In this study, we show that there exists more than one type of similarity transformation which provides three different systems of ODEs when employed on PDEs describing the unsteady fluid flow and heat transfer in a boundary layer with heat generation/absorption. To the best of our knowledge, the similarity transformations and corresponding reductions of the flow model are different from those which exist in the literature. We have considered linear combinations of two symmetries by assigning a unique positive value, i.e., 1 to each coefficient in these linear combinations. Keeping arbitrary constant coefficients in these linear combinations may lead to more general forms of the similarity transformations and corresponding systems of ODEs. The inclusion of arbitrary constant coefficients in the linear combinations may yield similarity transformations and corresponding systems of ODEs with these constant coefficients. With the involvement of the arbitrary constants in resulting systems of ODEs, the convergence of the analytic solutions can be controlled, i.e., the flow and heat transfer rates can be altered with a variation in the arbitrary constants. Moreover, by constructing optimal systems of Lie sub-algebras, the classes of ODEs derived in this work can be retrieved along with maybe a few more. The inequivalence of these classes of systems of ODEs can also be established. Although the construction of optimal systems has not been included in the scope of the present study, it may lead to more general results.

Author Contributions: Conceptualization M.S. and S.T.; data curation, M.B.; formal analysis, M.B., M.S. and S.T.; software, M.B. and M.S.; validation, M.B. and M.S.; visualization M.B. and M.S.; writing—original draft preparation, M.B. and M.S.; writing—review and editing, M.B., M.S., S.T., A.Z., M.U.A. and S.W.L.; funding acquisition S.W.L. All authors have read and agreed to the published version of the manuscript.

Funding: This work was supported by the National Research Foundation of Korea (NRF) grant funded by the Korea government (MSIT) (NRF2021R1I1A2059735).

Institutional Review Board Statement: Not applicable.

Informed Consent Statement: Not applicable.

Data Availability Statement: The data that support the findings of this study are available from the corresponding author upon reasonable request.

Conflicts of Interest: The authors have no conflict to disclose.

References

1. Ferziger, J.H.; Peric, M.; Leonard, A. *Computational Methods for Fluid Dynamics*; California Institute of Technology: Pasadena, CA, USA, 1997.
2. Wang, C.Y. Liquid Film on an Unsteady Stretching Surface. *Q. Appl. Math.* **1990**, *48*, 601–610. [CrossRef]
3. Crane, L.J. Flow past a stretching plate. *Z. Angew. Math. Und Phys. ZAMP* **1970**, *21*, 645–647. [CrossRef]
4. Andersson, H.I.; Aarseth, J.B.; Dandapat, B.S. Heat transfer in a liquid film on an unsteady stretching surface. *Int. J. Heat Mass Transf.* **2000**, *43*, 69–74. [CrossRef]
5. Chen, C.H. Heat transfer in a power-law fluid film over a unsteady stretching sheet. *Heat Mass Transf.* **2003**, *39*, 791–796. [CrossRef]
6. Dandapat, B.S.; Santra, B.; Andersson, H.I. Thermocapillarity in a liquid film on an unsteady stretching surface. *Int. J. Heat Mass Transf.* **2003**, *46*, 3009–3015. [CrossRef]
7. Chen, C.H. Effect of viscous dissipation on heat transfer in a non-Newtonian liquid film over an unsteady stretching sheet. *J. Non-Newton. Fluid Mech.* **2006**, *135*, 128–135. [CrossRef]
8. Dandapat, B.S.; Santra, B.; Vajravelu, K. The effects of variable fluid properties and thermocapillarity on the flow of a thin film on an unsteady stretching sheet. *Int. J. Heat Mass Transf.* **2007**, *50*, 991–996. [CrossRef]
9. Liu, I.C.; Andersson, H.I. Heat transfer in a liquid film on an unsteady stretching sheet. *Int. J. Therm. Sci.* **2008**, *47*, 766–772. [CrossRef]
10. Abel, M.S.; Tawade, J.; Nandeppanavar, M.M. Effect of non-uniform heat source on MHD heat transfer in a liquid film over an unsteady stretching sheet. *Int. J. Non-Linear Mech.* **2009**, *44*, 990–998. [CrossRef]
11. Noor, N.F.M.; Hashim, I. Thermocapillarity and magnetic field effects in a thin liquid film on an unsteady stretching surface. *Int. J. Heat Mass Transf.* **2010**, *53*, 2044–2051. [CrossRef]
12. Aziz, R.C.; Hashim, I. Liquid film on unsteady stretching sheet with general surface temperature and viscous dissipation. *Chin. Phys. Lett.* **2010**, *27*, 110202. [CrossRef]
13. Aziz, R.C.; Hashim, I.; Abbasbandy, S. Effects of thermocapillarity and thermal radiation on flow and heat transfer in a thin liquid film on an unsteady stretching sheet. *Math. Probl. Eng.* **2012**, *2012*, 127320. [CrossRef]
14. Liu, I.C.; Megahed, A.M. Numerical study for the flow and heat transfer in a thin liquid film over an unsteady stretching sheet with variable fluid properties in the presence of thermal radiation. *J. Mech.* **2012**, *28*, 291–297. [CrossRef]
15. Aziz, R.C.; Hashim, I.; Abbasbandy, S. Flow and heat transfer in a nanofluid thin film over an unsteady stretching sheet. *Sains Malays.* **2018**, *47*, 1599–1605. [CrossRef]
16. Ibragimov, N.H. *Elementary Lie Group Analysis and Ordinary Differential Equations*; Wiley: New York, NY, USA, 1999.
17. Hydon, P.E. *Symmetry Methods for Differential Equations: A Beginner's Guide*; Cambridge University Press: Cambridge, UK, 2000.
18. Olver, P.J. *Applications of Lie Groups to Differential Equations*; Springer: New York, NY, USA, 2000.
19. Szatmari, S.; Bihlo, A. Symmetry analysis of a system of modified shallow-water equations. *Commun. Nonlinear Sci. Numer. Simul.* **2014**, *19*, 530–537. [CrossRef]
20. Chesnokov, A.A. Symmetries of shallow water equations on a rotating plane. *J. Appl. Ind. Math.* **2010**, *4*, 24–34. [CrossRef]
21. Rashidi, M.M.; Momoniat, E.; Ferdows, M.; Basiriparsa, A. Lie group solution for free convective flow of a nanofluid past a chemically reacting horizontal plate in a porous media. *Math. Probl. Eng.* **2014**, *2014*, 239082. [CrossRef]
22. Siriwat, P.; Kaewmanee, C.; Meleshko, S.V. Symmetries of the hyperbolic shallow water equations and the Green Naghdi model in Lagrangian coordinates. *Int. J. Non-Linear Mech.* **2016**, *86*, 185–195. [CrossRef]
23. Xin, X.; Zhang, L.; Xia, Y.; Liu, H. Nonlocal symmetries and exact solutions of the (2+1)-dimensional generalized variable coefficient shallow water wave equation. *Appl. Math. Lett.* **2019**, *94*, 112–119. [CrossRef]
24. Paliathanasis, A. Lie symmetries and similarity solutions for rotating shallow water. *Z. Naturforschung-Sect. A J. Phys. Sci.* **2019**, *74*, 869–877.
25. Meleshko, S.V. Complete group classification of the two-Dimensional shallow water equations with constant coriolis parameter in Lagrangian coordinates. *Commun. Nonlinear Sci. Numer. Simul.* **2020**, *89*, 105293. [CrossRef]
26. Paliathanasis, A. Shallow-water equations with complete Coriolis force: Group properties and similarity solutions. *Math. Methods Appl. Sci.* **2021**, *44*, 6037–6047. [CrossRef]
27. Abd-El-malek, M.B.; Badran, N.A.; Amin, A.M.; Hanafy, A.M. Lie symmetry group for unsteady free convection boundary-layer flow over a vertical surface. *Symmetry* **2021**, *13*, 175. [CrossRef]

28. Paliathanasis, A. Lie symmetries and similarity solutions for a family of 1+1 fifth-order partial differential equations. *Quaest. Math.* **2021**, *45*, 1099–1114.
29. Safdar, M.; Ijaz Khan, M.; Taj, S.; Malik, M.Y.; Shi, Q.H. Construction of similarity transformations and analytic solutions for a liquid film on an unsteady stretching sheet using lie point symmetries. *Chaos Solitons Fractals* **2021**, *150*, 111115–111121. [CrossRef]
30. Aziz, R.C.; Hashim, I.; Alomari, A.K. Thin film flow and heat transfer on an unsteady stretching sheet with internal heating. *Meccanica* **2011**, *46*, 349–357. [CrossRef]
31. John, H.; Mathews, K.D.F. *Numerical Methods Using MATLAB*; Pearson Prentice Hall: Hoboken, NJ, USA, 2004.
32. Wang, C. Analytic solutions for a liquid film on an unsteady stretching surface. *Heat Mass Transf.- Und Stoffuebertragung* **2006**, *42*, 759–766. [CrossRef]

mathematics

MDPI

Article

Numerical Investigation of MWCNT and SWCNT Fluid Flow along with the Activation Energy Effects over Quartic Auto Catalytic Endothermic and Exothermic Chemical Reactions

Yasir Mehmood [1], Ramsha Shafqat [1,*], Ioannis E. Sarris [2,*], Muhammad Bilal [3], Tanveer Sajid [4] and Tasneem Akhtar [1]

1 Department of Mathematics and Statistics, The University of Lahore, Sargodha 40100, Pakistan
2 Department of Mechanical Engineering, University of West Attica, 250 Thivon & P. Ralli Str., 12244 Egaleo, Greece
3 Department of Mathematics, The University of Chenab, Gujrat 50700, Pakistan
4 Department of Mathematics, Capital University of Science and Technology, Islamabad 44000, Pakistan
* Correspondence: ramshawarriach@gmail.com (R.S.); sarris@uniwa.gr (I.E.S.)

Abstract: A mathematical model is created to analyze the impact of Thompson and Troian slip boundaries over a contracting/expanding surface sustaining nanofluid-containing carbon nanotubes along a stagnation point flow. Both multi-wall (MWCNTs) and single-wall (SWCNTs) carbon nanotubes are taken into consideration, with water serving as the base liquid. The flow is obtained due to the stretching or contracting of the surface. The thermal radiation, activation energy, buoyancy impacts, and chemical processes called quartic autocatalysis are additionally added to the original mathematical model. The MATLAB-constructed bvp4c function involving the three-stage Lobatto IIIa formula for the numerical results of dimensionless velocity, concentration, and temperature profiles are used. By contrasting it against a published paper in this limited instance, it is determined whether the suggested mathematical model is legitimate. In this sense, a remarkable consensus is achieved. Graphical representations are used to depict the behavior of many non-dimensional flow variables, such as the slip velocity parameter, the inertia coefficient, the porosity parameter, and the solid volume fraction. Surface drag force computations are reported to examine the effects at the permeable stretching surface. It has been shown that increasing the slip velocity factor increases the fluid streaming velocity while decreasing the surface drag force. If the endothermic/exothermic coefficient increases, the local thermal transfer efficiency falls. For nanofluids, the changing viscosity factor increases axial velocity while decreasing temperature distribution. Additionally, the solid volumetric fraction improves the temperature distributions by lowering the concentration profile and speed.

Keywords: stagnation point; magnetohydrodynamics; endothermic and exothermic reaction; heat generation/absorption; activation energy; carbon nano tubes

MSC: 76W05; 76D05; 7604

Citation: Mehmood, Y.; Shafqat, R.; Sarris, I.E.; Bilal, M.; Sajid, T.; Akhtar, T. Numerical Investigation of MWCNT and SWCNT Fluid Flow along with the Activation Energy Effects over Quartic Auto Catalytic Endothermic and Exothermic Chemical Reactions. *Mathematics* **2022**, *10*, 4636. https://doi.org/10.3390/math10244636

Academic Editor: Theodore E. Simos

Received: 12 October 2022
Accepted: 29 November 2022
Published: 7 December 2022

1. Introduction

When industries and scientists were seeking out greater and better thermal characteristics in fluids used on a regular basis for diverse tasks, Mesuda et al. [1] suggested the introduction of nano-sized particles in ordinary fluids in 1993. Later, the term nanofluid was officially defined by Choi and Eastman [2], and these fluids became a prime focus of researchers. Nanoparticles are typically 1–100 nm in size, but this might vary significantly depending on their sizes and shapes. Water, ethylene glycol, and oil are commonly used base fluids. A nanofluid is a combination of nanosized materials and a base liquid. These nanoparticles are suspended in a base fluid that is colloidal in nature and has weaker thermal conductivity. The major goal of nanosized particles is intended to improve the thermal conductivity of fluids, as well as increase heat transmission. Because of their structure, nanosized particles have unique physical and chemical features and contribute to the development of thermophysical systems. Nanofluids have a wide range of applications, such as in nano-drug delivery, pharmaceutical operations, heating/cooling appliances, fuel cells, and microelectronics. Nanofluids are utilized as coolants in the thermal exchange systems of automobiles and nuclear reactors. However, they are also useful due to their regulated optical features.

Carbon nanotubes are cylinder-type shapes that are formed by rolling or folding a graphite sheet. They have unique mechanical, thermophysical, and chemical properties. Because of their cylindrical form, huge surface area, and small size, carbon nanotubes offer advantages over other macro/nanoparticles. Carbon nanotubes are categorized as SWCNTs or MWCNTs based on their number of graphene layers. The effects of the nanofluid in the physiological examination of cilia were highlighted by Sadaf and Nadeem [3]. Sivasankaran et al. [4] investigated the heat production of nanofluids in the cavity. Ahmed et al. [5] investigated the flow of multi-walled and single-walled carbon nanotubes (MWCNT and SWCNT) across a circular stretchable semi-infinite zone containing water as the base fluid. Hosseinzadeh et al. [6] concentrated on the MWCNTs and SWCNTs combined in ethylene glycol flowing between the two rotating discs with extensible qualities in it; the effects of MHD and thermal radiations were considered, and findings revealed that the fluid system's instability and the volume fraction of the nanosized particles decreases as the radiation increases. Ramzan et al. [7] investigated a physical system of gyrotactic microorganisms and CNTs submerged in water that was flowing on the top of a vertical cone immersed in a permeable medium by using the bvp4c MATLAB software and discovered that increasing the suction parameter reduces the nanofluid stream velocity. They also considered chemical reactions, thermal radiation, and species stratification. Khan et al. [8] examined the flow at the stagnation point of carbon nanotubes moving across an extended surface in the applied magnetic field, as well as thermal radiation, homogeneous and heterogeneous reactions, and heat absorption/generation. By applying the shooting method, numerical consequences revealed that MWCNTs had a greater induced magnetic field than SWCNTs. Ramazan et al. [9] studied CNTs and gyrotactic microorganisms in a fluid moving through a vertical cone enclosed by porous media. Joule heating, thermal radiation, MHD, and the homogeneous and heterogeneous reactions were all thought to be important inside the fluid system. According to the results obtained by using bvp4c MATLAB software, the flow of the fluid decreases with increasing magnetic force. Khan et al. [10] used the homotopy analysis method (HAM) to study a radiant bioconvective MHD nanofluid flow across an elongated oscillating plane and discovered that the temperature of the nanofluid increases as the buoyancy ratio increases. Sergii et al. [11] presented an analytical theory for the electrostatic interactions between two spherical dielectric particles with arbitrary charge distributions expanded in multipolar terms submerged in a polarizable ionic solvent and with arbitrary radii and dielectric constants. Yu [12] presented a new formalism regarding reciprocity and arbitrarily accommodated many dielectric spheres of different dielectric constants and sizes while being rigorous at the Debye–Hückel level. Some further prominent articles highlighting the use of nanofluid are as follows: refs. [13–17]. A chemical reaction is the interaction of two or more chemicals, which leads to the composition of one or more new chemical substances. A chemical reaction is the breaking of old bonds in order to form a new bond chemically. This is also known as the chemical change caused by the interaction of two or more chemical substances. The temperature of the system rises or decreases when energy is transferred to or from the environment during a chemical process. Exothermic reactions are chemical processes for which energy is discharged into the atmosphere (i.e., outside of the system). Energy is typically conveyed as heat energy, which causes the atmosphere to heat up. The burning process is an illustrative example of an exothermic reaction. Endothermic reactions are chemical reactions that extract energy from their environment. Normally, the energy is shifted as heat, raising the temperature of the reaction mixture. Chemically reactive models, such as biological systems and combustion, are represented by homogeneous/heterogeneous reactions. The surfaces of the catalysts experience heterogeneous reactions, but the fluid itself experiences homogenous reactions. In practice, homogeneous and heterogeneous reactions can be observed in a number of different fields, including air pollution, food processing, ignition, and biological processes. The species of chemical reactions together with the activation energy phenomenon plays a major role in various engineering fields. Mass transport and chemical processes can be seen in the consumption as well as production of reactant species. Regarding the existence of thermal dissipation, Maleque [18] was interested in the study of the impact of endothermic/exothermic chemical processes having Arrhenius activation energy on magnetohydrodynamic-free convective and the mass transfer flow. Recently, Bejawada et al. [19] investigated a magnetohydrodynamic Casson fluid flow with chemical reaction properties along with a porous Forchheimer medium over a non-linear sheet. Suleman et al. [20] used the shooting technique to show how the concentration of nanoparticles in a silver-water nanofluid mixture decayed due to upsurging homogeneous and heterogeneous reactions influenced by viscous dissipation, MHD, thermal radiation that was not linear, and Joule heating moving through a non-linear extending cylinder. Imtiaz et al. [21] explored the streaming of the two-dimensional magnetohydrodynamic viscous fluid flow passing over a stretched sheet. The solution was approximated with the use of a quasi-linearization method and the implicit finite difference approach, taking into account the significant impact of homogeneous and heterogeneous reactions, thermal radiation, and Joule heating. The study discovered that viscous fluids had lower fluid speed and concentration than viscoelastic fluids, and both homogeneous and heterogeneous reactions have a negative effect on fluid viscosity. Suleman et al. [22] used the shooting approach to analyze silver-water nanofluids with MHD, nonlinear heat radiation, and homogeneous and heterogeneous reactions through a nonlinear extended cylinder. Despite a greater radiation impact, the results showed better thermal conditions. Doh et al. [23] further investigated homogeneous and heterogeneous reactions with silver water nanofluids on a revolving permeable disc with changing disc thickness.

During fluid flow research, it is frequently thought that small-scale slips can happen at a fluid-solid interface as a result of uncertainty at the intense stress levels in methodologies such as polymer extraction. Fluid motion at a geometry surface is affected by such fluid slip impacts. Khan et al. [24] investigated the viscous hydromagnetic fluid flow across the permeable rotatable disk with non-linear thermal radiation and partial slip by considering the shooting method. The findings revealed a clear decline in surface friction as slip estimation increased. Using the Crank–Nicolson approach, Hamid et al. [25] investigated the natural convection of the Prandtl fluid, which was flowing at a point of stagnation across an infinite elongated plate. The geometry of the model is presented on the Figure 1. When studying slip at the sheet surface and MHD, it was discovered that when a slip impact was combined with a low magnetic field, the velocity increased dramatically. Reddy et al. [26] investigated magnetohydrodynamic Eyring–Powell fluid flow regarding nonlinear radiation, solutal slippage, temperature, velocity, and chemical processes using a Range–Kutta 4th order scheme. Kiyasatfar [27] investigated the convective slip flow of a non-Newtonian fluid among parallel plates with circular microchannels using the power-law model. The results show that for both geometries, lowered fluid stream speed and increased heat exchange rates and molecule stability occurred in rising slip situations.

According to the aforementioned studies, fewer expeditions are explored for the study of the Thompson and Troian slip forms. In the existence of suction/injection, for nanofluids based on the Yamada and Ota model consisting of two nanoparticles, i.e., single and multi-wall carbon nanotubes suspended in a base fluid (water), the flow velocity, temperature, and concentration of nanofluid in the presence of the Cattaneo–Christov heat flux model are taken into account. Stagnation point flow is taken over the infinitely expanding sheet. Chemical reactions, nonlinear heat generation, and activation energy are the main key points that are focused on the current problem. The modeled equations are solved by the bvp4c technique. Graphs are drawn for the different parameters to better understand their impact on velocity and temperature.

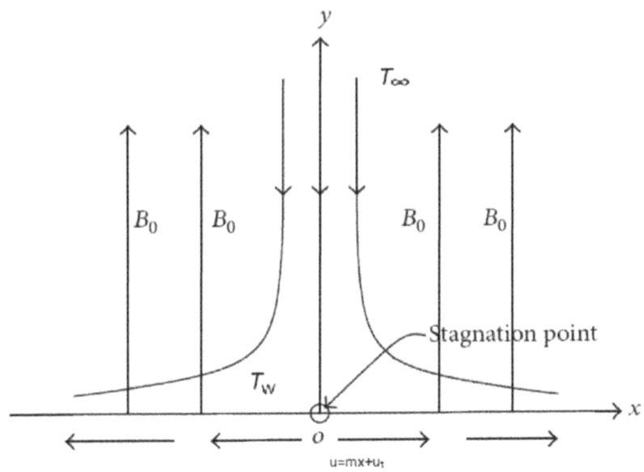

Figure 1. Geometry of the model [25].

2. Mathematical Model

Here, we consider the two-dimensional Newtonian nanofluid flowing over the sheet which is linearly stretched along the x-axis. A stagnation point flow is considered along with a magnetic field which is applied normally to the flow. Heat generation, chemical reaction, activation energy, and convection are also prominent impacts that are taken into account. Importantly Thompson and Troian slip mechanisms are employed on the surface. Single and multi-wall carbon nanotubes are taken as nanoparticles which are mixed in water. The fluid flow is compressible, and the Boussinesq approximation is valid for the case of the present problem. Free convection behavior is taken due to buoyancy forces. The momentum equation comprises the buoyancy phenomenon, stagnation point, and free convection phenomenon. Heat transfer is determined in a more precise manner with the utilization of the Cattaneo–Christov heat flux expression instead of classical Fourier law expression in the energy equation. Sometimes, extra energy is required to proceed with a chemical reaction due to the slow collision of fluid molecules. This is why an activation energy expression is utilized in energy as well as concentration equations. The governing modeled PDEs are derived from the law of conservation of mass, Newton's second law of motion, the second law of thermodynamics, and Fick's second law of diffusion. After accounting for the boundary layer estimation, the system can be represented as follows [6–8,18].

$$\frac{\partial v}{\partial y} + \frac{\partial u}{\partial x} = 0, \tag{1}$$

$$v\frac{\partial u}{\partial y} + u\frac{\partial u}{\partial x} = \nu_{nf}\frac{\partial^2 u}{\partial y^2} + U_\infty\frac{dU_\infty}{dx} + g\frac{(1-\phi)\beta_f\rho_f + \phi\beta_s\rho_s}{\rho_{nf}}(T - T_\infty) - \frac{\sigma_{nf}B^2(x)}{\rho_{nf}}(u - U_\infty), \tag{2}$$

$$\begin{aligned} &v\frac{\partial T}{\partial y} + u\frac{\partial T}{\partial x} + \tau\left\{\begin{array}{c} u\dfrac{\partial u}{\partial x}\dfrac{\partial T}{\partial x} + v\dfrac{\partial v}{\partial y}\dfrac{\partial T}{\partial y} + u^2\dfrac{\partial^2 T}{\partial x^2} + v^2\dfrac{\partial^2 T}{\partial y^2} \\ +2uv\dfrac{\partial^2 T}{\partial x\partial y} + v\dfrac{\partial u}{\partial y}\dfrac{\partial T}{\partial x} + u\dfrac{\partial v}{\partial x}\dfrac{\partial T}{\partial y} \end{array}\right\} = \alpha_{nf}\frac{\partial^2 T}{\partial y^2} \\ &+ \frac{Q_0}{(\rho c_p)_{nf}}(T - T_\infty) + \beta K_r{}^2\left(\frac{T}{T_\infty}\right)^n exp\left[-\frac{E_a}{KT}\right](C - C_\infty), \end{aligned} \tag{3}$$

$$v\frac{\partial C}{\partial y} + u\frac{\partial C}{\partial x} = D_m\frac{\partial^2 C}{\partial y^2} - {K_r}^2\left(\frac{T}{T_\infty}\right)^m exp\left[-\frac{E_a}{KT}\right](C - C_\infty), \tag{4}$$

where T represents the fluid temperature, C represents the concentration, and (v, u) reresents the velocity components in the (y, x). The parameter D_m stands for the mass diffusion coefficient, K_r which is the rate of limiting factor for the chemical process, $(\beta = \pm 1)$ is the endothermic/exothermic factor, and $\left(\frac{T}{T_\infty}\right)^n exp\left[-\frac{E_a}{KT}\right]$ represents the Arrhenius expression, where n represents a unit-less rate constant $(-1 < n < 1)$. The thermophysical characteristics of the nanofluid are presented in Table 1. The corresponding boundary conditions are as follows:

$$\begin{aligned} &u|_{y=0} = u_t + u_w(x) = mx + \gamma^*(1 - \xi^*\frac{\partial u}{\partial y})^{-1/2}\frac{\partial u}{\partial y},\ v|_{y=0} = 0,\ C|_{y=\infty} \to C_\infty,\ C|_{y=0} = C_w, \\ &u|_{y=\infty} \to U_\infty(x),\ T|_{y=\infty} \to T_\infty,\ T = T_w(x) = (T_0 x + T_\infty)|_{y=0}, \end{aligned} \tag{5}$$

Table 1. Thermophysical characteristics of nanofluid and base.

Physical Attributes	Base Fluid (H_2O)	MWCNT	SWCNT
$C_p(J/kgK)$	4179	796	425
$\rho(kg/m^3)$	997	1600	2600
$K(W/mK)$	0.613	3000	6600
$\sigma(\Omega m)^{-1}$	5.5×10^{-6}	10^7	10^6

The thermophysical properties are represented as defined from the Yamada and Ota model:

$$\begin{aligned} &\mu_{nf} = \frac{\mu_f}{(1-\phi)^{2.5}},\ \alpha_{nf} = \frac{k_{nf}}{\rho_{nf}(C_p)_{nf}},\ \nu_{nf} = \frac{\mu_{nf}}{\rho_{nf}}, \\ &\rho_{nf} = (1-\phi)\rho_f + \phi\rho_{CNT},\ \frac{k_{nf}}{k_f} = \frac{2\phi\frac{k_{CNT}}{k_{CNT}-k_f}\ln\frac{k_{CNT}+k_f}{2k_f} + (1-\phi)}{2\phi\frac{k_f}{k_{CNT}-k_f}\ln\frac{k_{CNT}+k_f}{2k_f} + (1-\phi)}, \\ &\frac{\sigma_{nf}}{\sigma_f} = 1 + \frac{3\left(\frac{\sigma_s}{\sigma_f} - 1\right)\phi}{\left(\frac{\sigma_s}{\sigma_f} + 2\right) - \left(\frac{\sigma_s}{\sigma_f} - 1\right)\phi}. \end{aligned} \tag{6}$$

Introducing dimensionless variables, we obtain:

$$\eta = y\sqrt{\frac{c}{\nu_f}},\ v = -\sqrt{\nu_f c} f(\eta),\ u = U_\infty f'(\eta),\ \theta(\eta) = \frac{T - T_\infty}{T_w - T_\infty},\ \phi(\eta) = \frac{C - C_\infty}{C_w - C_\infty} \tag{7}$$

After simplification, we convert the system of PDEs into the following ODEs:

$$\frac{1}{(1-\phi+\phi\frac{\rho_{CNT}}{\rho_f})(1-\phi)^{2.5}}f''' + f''f - f'^2 + 1 + \frac{\phi\frac{\rho_{CNT}\beta_{CNT}}{\rho_f\beta_f} - \phi + 1}{(1-\phi+\phi\frac{\rho_{CNT}}{\rho_f})}\lambda\theta + \frac{\left(1 + \frac{3\left(\frac{\sigma_s}{\sigma_f}-1\right)\phi}{\left(\frac{\sigma_s}{\sigma_f}+2\right)-\left(\frac{\sigma_s}{\sigma_f}-1\right)\phi}\right)M}{(1-\phi+\phi\frac{\rho_{CNT}}{\rho_f})}(1-f') = 0, \tag{8}$$

$$\begin{aligned} &\frac{k_{nf}}{k_f}\theta'' + Pr\left(1 - \phi + \phi\frac{(\rho C_p)_{CNT}}{(\rho C_p)_f}\right)[(f\theta' + D_c\theta - f'\theta) - \gamma(f^2\theta'' + ff'\theta') \\ &+ \beta\sigma\lambda_1(1+\gamma_T\theta)^n exp\left[-\frac{E}{1+\gamma_T\theta}\right]\phi] = 0, \end{aligned} \tag{9}$$

$$\phi_c'' + S_c f \phi_c' - \sigma S_c (1+\gamma_T\theta)^m exp\left[\frac{-E}{1+\gamma_T\theta}\right]\phi_c = 0. \tag{10}$$

The transformed boundary conditions of (5) are:

$$\begin{gathered} f(0)=0,\ f'(0)=\epsilon+\gamma_1[1-\xi f''(0)]^{-1/2} f''(0),\ \phi_c(0)=1,\ \theta(0)=1, \\ f'(\eta)=1,\ \theta(\eta)=0,\ \phi_c(\eta)=0\ at\ \eta\to\infty. \end{gathered} \tag{11}$$

The different non-dimensional parameters seen in Equations (8)–(11) are explained as:

$$\begin{aligned} M &= \frac{\sigma_{nf} B^2(x)}{c\rho_f},\ Pr=\frac{v_f}{\alpha_f},\ D_c=\frac{Q_0}{c(\rho c_p)_f},\ \gamma=\tau c, \\ \gamma_T &= \frac{T_w - T_\infty}{T_\infty},\ \epsilon=\frac{m}{c},\ \xi=c\sqrt{\frac{c}{v_f}}b^*,\ E=\frac{E_a}{KT_\infty}, \\ \sigma &= \frac{K_r{}^2}{c},\ S_c=\frac{v_f}{D_A},\ \lambda_1=\beta\frac{C_w-C_\infty}{T_w-T_\infty},\ \gamma_1=a\sqrt{\frac{c}{v_f}}. \end{aligned} \tag{12}$$

The skin friction coefficient C_f and local Nusselt number Nu_x are explained as:

$$C_f = \frac{\tau_w}{\rho_f U^2},\ \tau_w = (\frac{\partial u}{\partial y}\mu_{nf})_{y=0}, \tag{13}$$

$$Nu_x = \frac{x q_w}{k_f(T_w - T_0)},\ q_w = (-\frac{\partial T}{\partial y}k_{nf})_{y=0}, \tag{14}$$

The dimensionless form of the surface drag and the heat transfer rate is given below:

$$R_e^{1/2} C_f = \left(\frac{1}{(1-\phi)^{2.5}}\right) f''(0),\ R_e^{-1/2} Nu_x = \left(-\frac{k_{nf}}{k_f}\right)\theta'(0)\ and\ R_e = \frac{cx^2}{v_f}. \tag{15}$$

3. Numerical Solution

With the help of the bvp4c technique, which is a built-in function in MATLAB, the mathematical Equations (8)–(10) subject to condition (11) are solved numerically after setting $\eta = \eta_{max}$, where η_{max} is different for different combinations of the physical parameters. Alternatively, the $[0, \infty)$ domain is restricted to $[0, \eta_{max}]$. For the purpose of the solution, we first convert the system of equations into first-order ODEs. The following variables are used for the purposes of conversion.

$$y_7 = \phi',\ y_6=\phi,\ y_5=\theta',\ y_4=\theta,\ y_3=f'',\ y_2=f',\ y_1=f. \tag{16}$$

Hence, the system of equations are transformed as:

$$\left.\begin{array}{ll}
y_2 = y_1', & 0 = y_1(0) \\
y_3 = y_2', \qquad \epsilon + \gamma_1[1- & \xi y_3(0)]^{-1/2} y_3(0) = y_2(0) \\
(1-\phi)^{2.5}(1-\phi+\phi\frac{\rho_{CNT}}{\rho_f})[y_2^2 - y_1 y_3 - 1- & \\
\frac{\phi\frac{\rho_{CNT}\beta_{CNT}}{\rho_f\beta_f}+1-\phi}{(1-\phi+\phi\frac{\rho_{CNT}}{\rho_f})}\lambda y_4 - \frac{\left(1+\frac{3\left(\frac{\sigma_s}{\sigma_f}-1\right)\phi}{\left(\frac{\sigma_s}{\sigma_f}+2\right)-\left(\frac{\sigma_s}{\sigma_f}-1\right)\phi}\right)M}{(1-\phi+\phi\frac{\rho_{CNT}}{\rho_f})}(1-y_2)] = y_3', & 1 = y_2(\eta) \\
y_5 = y_4', & 1 = y_4(0), \\
\frac{-k_f}{k_{nf}}\left(\frac{Pr(1-\phi+\phi\frac{(\rho C_p)_{CNT}}{(\rho C_p)_f})}{1-\gamma y_1^2 Pr(1-\phi+\phi\frac{(\rho C_p)_{CNT}}{(\rho C_p)_f})}\right)[y_1 y_5 + D_c y_4 - y_2 y_4 & \\
-\gamma y_1 y_2 y_5 + \beta\sigma\lambda_1(1+\gamma_T y_4)^n exp\left[\frac{-E}{1+\gamma_T y_4}\right] y_6] = y_5', & 0 = y_4(\eta) \\
y(7) = y_6', & 1 = y_6(0)
\end{array}\right\} \quad (17)$$

$$S_c\sigma(1+\gamma_T y_4)^m exp\left[\frac{-E}{1+\gamma_T y_4}\right] y_6 - S_c y_1 y_7 = y_7'. \quad 0 = y_6(\eta)\}$$

Every numerical solution is obtained by putting $\epsilon = 10^{-5}$, where ϵ is defined as the tolerance.

4. Step-by-Step Graphical Detail of the Problem

4.1. Problem Formulation

The governing modeled PDEs are highly nonlinear in nature and derived from the law of conservation, Newton's second law of motion, the second law of thermodynamics, and Fick's second law of diffusion. Yamada and Ota's nanofluid has been employed in the case of SWCNT and MWCNT in order to check the behavior of nanoparticles on fluid flow. The momentum of fluid flow is scrutinized with the inclusion of MHD, stagnation point, and buoyancy effects. A heat transfer analysis has been carried out with the inclusion of heat generation, Cattaneo–Christov and activation energy effects, whereas mass transfer analysis is studied with the utilization of the activation energy phenomenon.

4.2. Modeling

The governing modeled PDEs are highly nonlinear in nature and embedded with various physical effects. The PDEs are renovated into ODEs with the utilization of similarity transformations in order to dimensionalize the PDEs. It is easy to understand the behavior of fluid in the case of dimensionless parameters such as the Prandtl number Pr, Schmidt number Sc, Nusselt number Nu, etc.

4.3. Numerical Process

The dimensionless system of equations can be handled numerically with the utilization of the Lobatto IIIA scheme incorporated with the MATLAB built-in bvp4c scheme. During this procedure, the nonlinear modeled PDEs with the inclusion of various effects in momentum, energy, and concentration equations are transformed into ODEs with the help of similarity variables. In the second step, these dimensionless ODEs are stepped down into first-order ODEs for the Lobatto IIIA scheme. The tolerance level for the case of the present problem was 10^{-6}, with an interval of computation of [0,4] instead of [0,∞]. All the numerical results have been obtained by considering η = 4. The detailed procedure of the proposed numerical scheme is presented in the table mentioned below.

4.4. Numerical Results

The behavior of the obtained numerical outcomes was scrutinized in terms of its impact on velocity, temperature, and concentration fields. The physical quantities of interest, such as heat transfer, Nusselt number, and Sherwood number, were computed as a result of magnification in various dimensionless numbers obtained during the numerical simulation of the problem. The impact of the dimensionless parameters on the velocity, temperature, and concentration fields is portrayed in terms of figures and tables.

4.5. Analysis

The accuracy and convergence of the proposed numerical scheme have been checked with the comparison of obtained numerical results with the existing literature. In order to obtain the convergence criterion, the tolerance level for the case of the present problem was 10^{-6}, and the domain for the case of the numerical solution was taken to be $\eta = 4$ instead of $\eta = \infty$. The convergence criterion was achieved if the value of the obtained outcome was less than the tolerance level.

5. Results and Discussions

After employing the bvp4c model, we obtained the required results in the form of graphs and tables that highlight the impact of various parameters on the velocity, temperature, concentration, skin friction, and Nusselt number. The values of the dimensionless parameters used in this study are in the range of $0.1 \leq K \leq 1$; for the magnetic number, the range is $0.1 \leq M \leq 2$, while the range for the shear rate is $0.1 \leq \xi \leq 0.8$. The volume fraction of nanoparticles is in the range of $0.01 \leq \phi \leq 0.06$, the Schmidt number is in the range of $0.5 \leq S_c \leq 1.5$, the Prandtl number has a value in $3 \leq Pr \leq 9$, the velocity ratio parameter is in the range of $0.1 \leq \epsilon \leq 0.5$, the velocity slip lies in $0.1 \leq \gamma_1 \leq 0.5$, the chemical reaction effect is in the range of $0.1 \leq \beta \leq 0.5$, the activation energy lies in $0.1 \leq E \leq 1$, the power law index is in the range of $0.1 \leq m \leq 0.5$, and the fitted rate constant is in the range of $0.1 \leq n \leq 0.5$.

Figures 2 and 3 show the impact of the solid volume fraction of CNTs on the thermal situation and axial velocity. For higher ϕ, the fluid stream speed decreases, while temperature increases because of the direct relationship between the concentration of nanoparticles and their thermal conductivity. Higher values of the parameter ϕ have a favorable influence on a system having higher thermal conductivity, resulting in an improved temperature profile. Figure 4 shows the influence of a velocity ratio parameter ϵ on the fluid speed. The velocity profile improves as a result of the direct impact of ϵ based on the stream flow rate. Figure 5 shows the effect of a slip velocity parameter γ_1 on $f'(\eta)$, indicating the positive influence of γ_1 on the stream velocity of the fluid. As the slip effects on a wall become more significant, there is a small amount of friction and hence small resistance to fluid motion. Figure 6 shows an increasing behavior in the velocity field against increasing magnetic parameters. Usually, the magnetic field acts as an opposing agent and resists the fluid motion due to the presence of external Lorentz forces, but here, due to the employed slip mechanism, the behavior reverses, and an inclination in fluid velocity is observed.

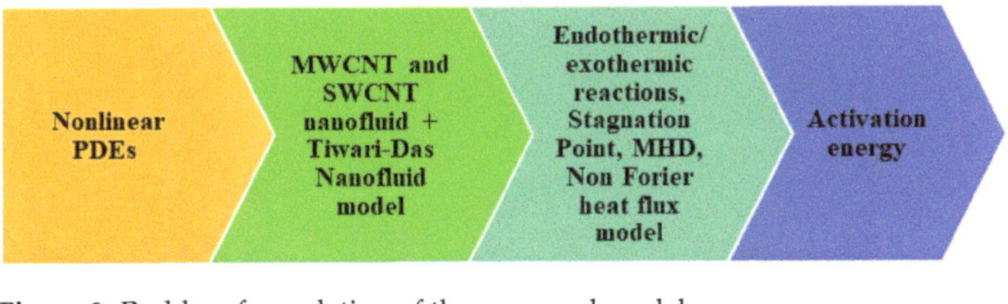

Figure 2. Problem formulation of the proposed model.

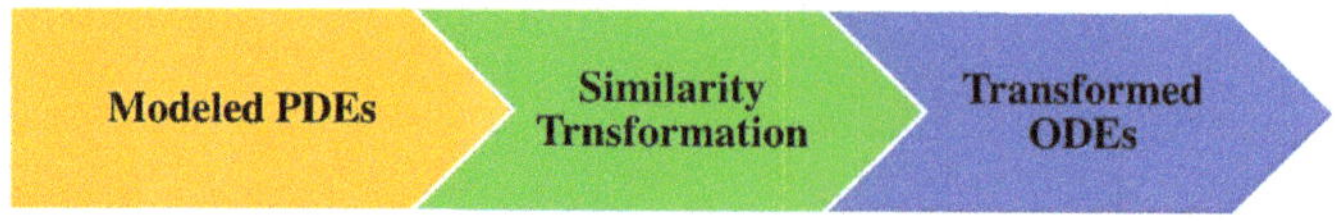

Figure 3. Modelling of the proposed modeled PDEs and their conversion into ODEs.

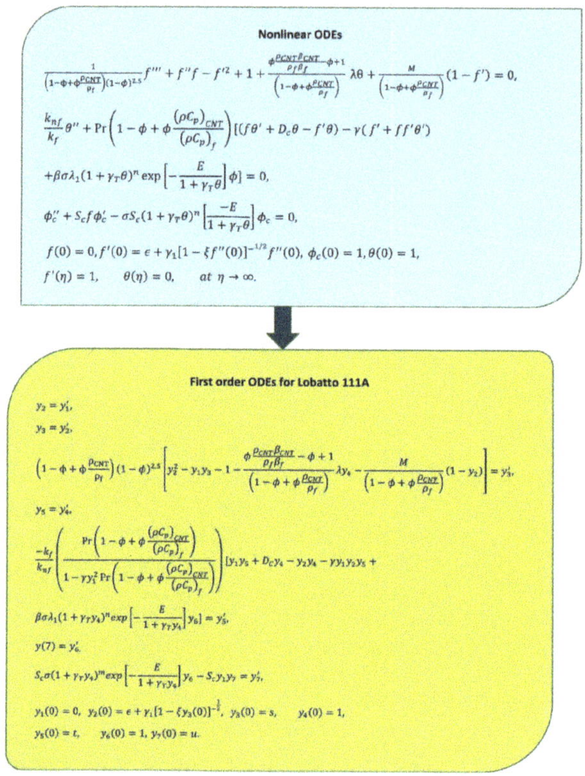

Figure 4. Conversion of PDEs into first-order ODEs and Lobatto111a scheme.

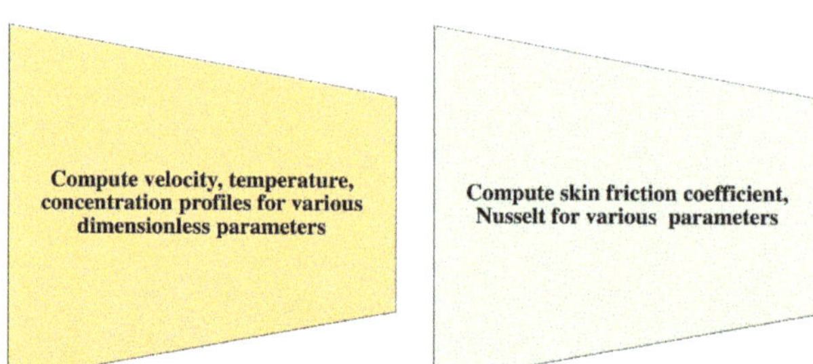

Figure 5. Impact of obtained numerical results on velocity, temperature and concentration fields.

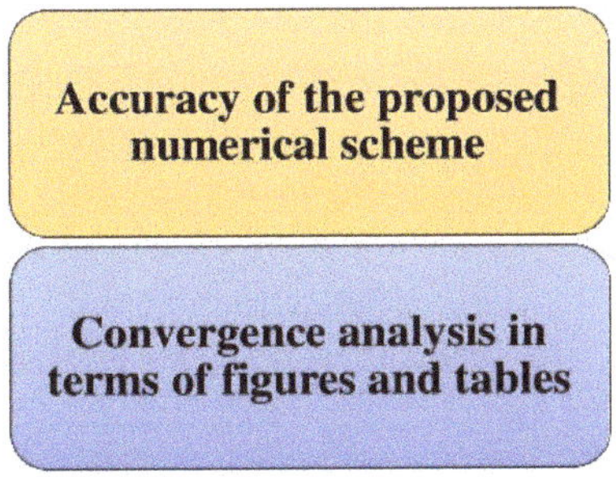

Figure 6. Analysis of obtained results.

Figure 7 shows the thermal behavior caused by heat generation through D_c. The heat transfer surrounded by the adjoining fluid layers and surface improves as D_c increases. Higher parameters ultimately generate more heat internally, which causes an increase in the temperature of the nanofluid. Figure 8 shows the effect of the Prandtl number P_r on the temperature profile. The temperature is found to be decreased for expanding P_r. The Prandtl number is the quotient of momentum to thermal diffusivity, and it is used to measure the heat transfer within the solid surface and moving liquid. As the Prandtl number rises, the thermal diffusivity becomes weaker as the fluid temperature decreases. The thermal relaxation parameter γ has an effect on temperature distribution, as shown in Figure 9. For greater values of γ, the temperature, as well as the thickness of the boundary layer, is found to decrease. Figure 10 depicts the effect of the shear rate ξ on a velocity. A high shear rate indicates a lower viscosity, which increases the fluid velocity. Because of the inverse relationship between the mass diffusivity and Schmidt number (S_c), Figure 11 shows an increasing trend in the nanoparticle concentration for increasing values of S_c.

As seen in Figure 12, the value of β, which represents the strength of the chemical reaction, increases as the concentration field $\phi_c(\eta)$ decreases. The chemical reaction reduces the movement of the mass of the fluid. Physically, this is accurate since a chemical reaction is called an exothermic reaction when energy is released into the environment, and an endothermic reaction is when energy is taken from the environment. Figure 13 shows how the concentration profile improves as the activation energy parameter increases. The species B, which contains nanoparticles embedded in it, is amplified by the chemical reaction factor (λ_1). Therefore, the manufacturing of nanoparticles increases. This explains why the system $\phi_c(\eta)$ is increased, as shown in Figure 14. The concentration profile decreases as the rate constant increases, as shown in Figure 15. This is owing to the fact that as σ increases, so does the destructive intensity of chemical reactions. As can be seen in Figure 16, the concentration profile declines as m is magnified.

Figures 17–19 show the velocity and thermal behavior caused by exothermic/ endothermic parameters through β. Exothermic reactions are when energy is released due to the interference of two chemical species, while in endothermic reactions, energy is absorbed. If the ratio of energy absorption versus energy release is the same, then the state of this chemical reaction is called isothermic because, in it, the overall energy is balanced. Heat transfer surrounded by the adjoining fluid layers and surface improves as β increases, leading to a velocity profile $f'(\eta)$ that rises. The effect of a velocity ratio parameter ϵ on the skin friction coefficient $C_f R_e^{1/2}$ is seen in Figure 20. A decreasing behavior in skin friction when increasing the values of ϵ is noted, and a reverse pattern is observed for the augmented values in λ. The value of ϵ increases and the free stream velocity overpowers the extending velocity, generating an enlarged motion about the stagnation point, lowering the drag force on a surface and causing $C_f R_e^{1/2}$ to drop.

The effect of the thermal relaxing time γ and velocity ratio parameter ϵ on the Nusselt number $\theta'(0)$ is shown in Figure 21. $\theta'(0)$ becomes higher for escalating γ. With increasing γ, the time it would take to transport heat between neighboring particles increases, resulting in a decrement in heat transfer characteristics. On the other hand, an opposing relationship is noted for ϵ, which improves the rate of heat transfer characteristics by increasing the fluid speed. The idea behind the plotting of the Figures 22–26 is inspired from the following references [13,15,16].

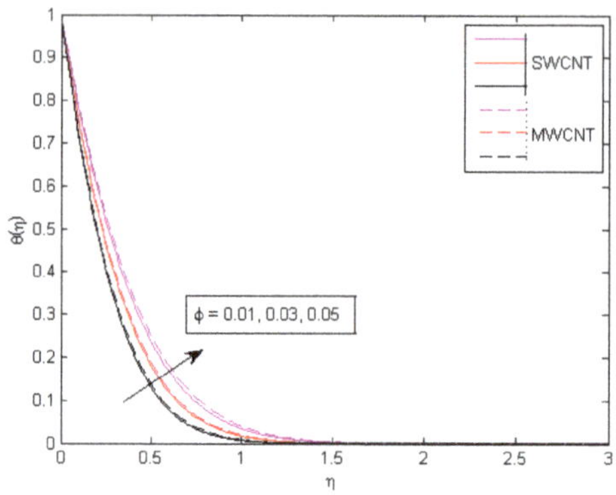

Figure 7. Influence of ϕ on temperature pattern.

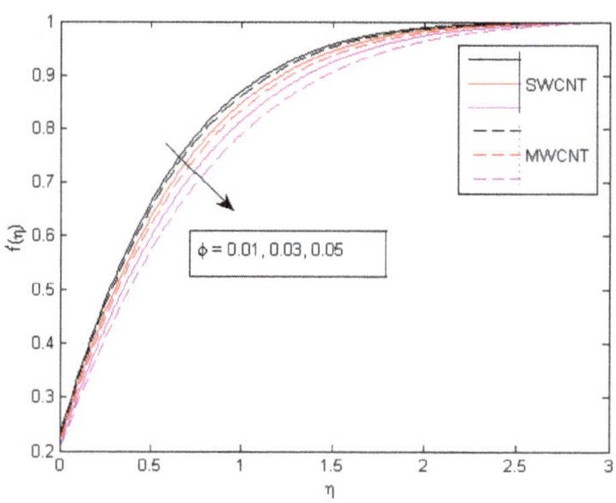

Figure 8. Influence of ϕ on velocity behavior.

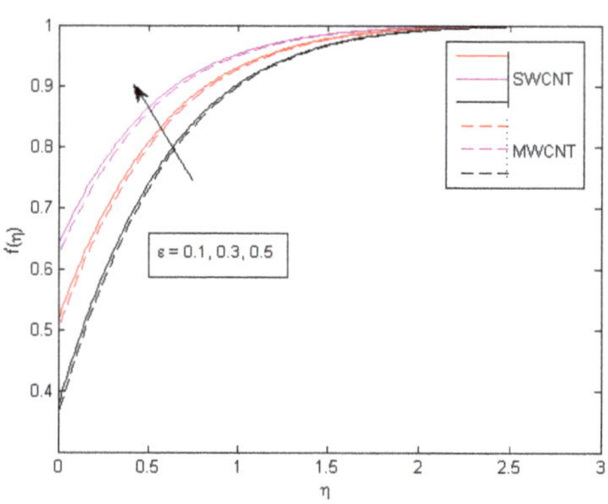

Figure 9. Influence of ϵ on velocity profile.

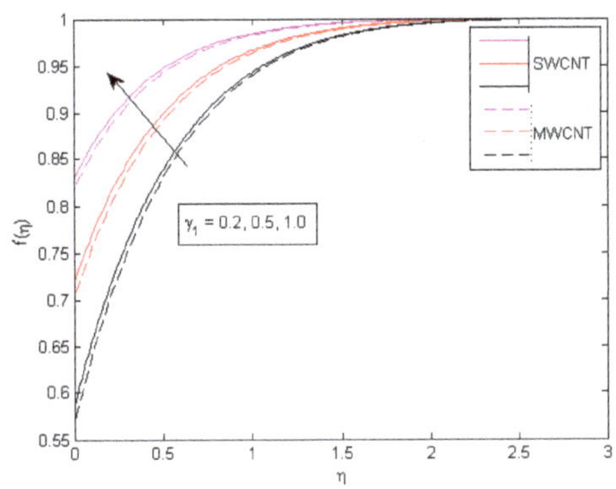

Figure 10. Impact of γ_1 on $f'(\eta)$.

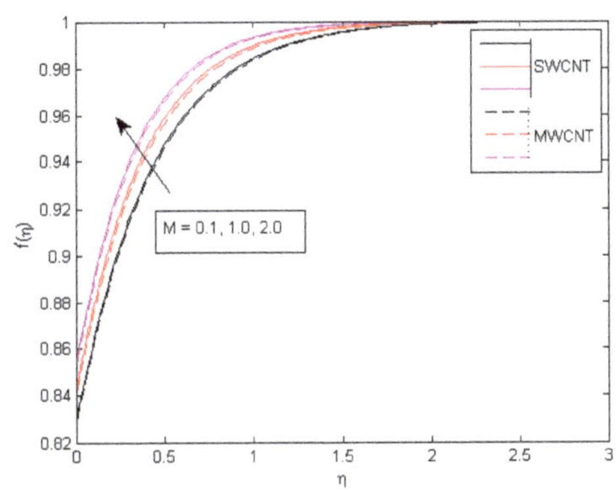

Figure 11. Consequences of M on velocity pattern.

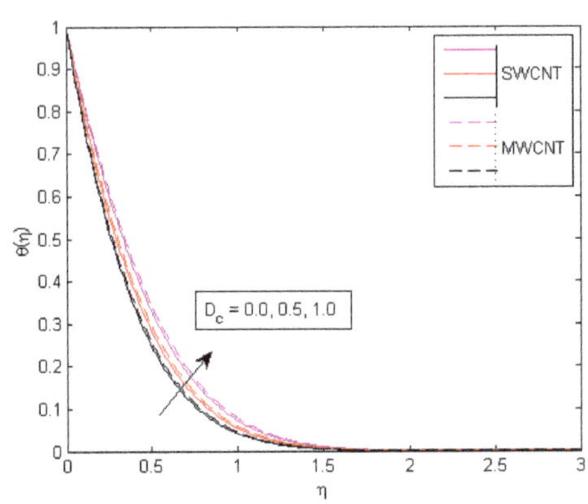

Figure 12. Consequences of D_c on temperature pattern.

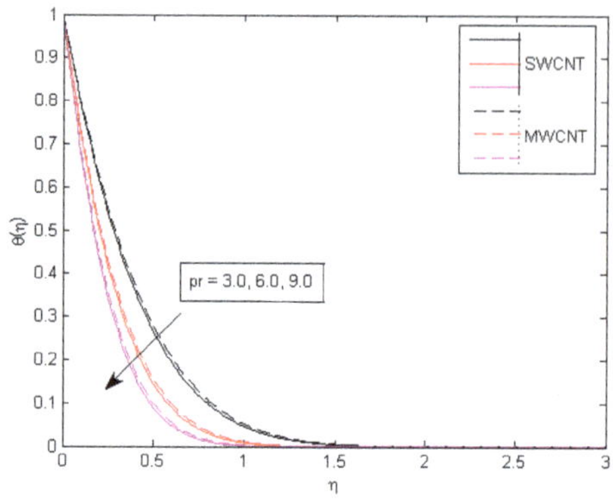

Figure 13. Consequences of Pr on temperature behavior.

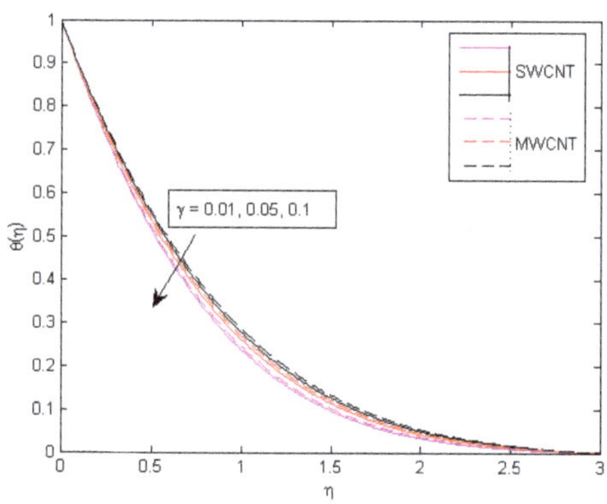

Figure 14. Consequences of γ on temperature pattern.

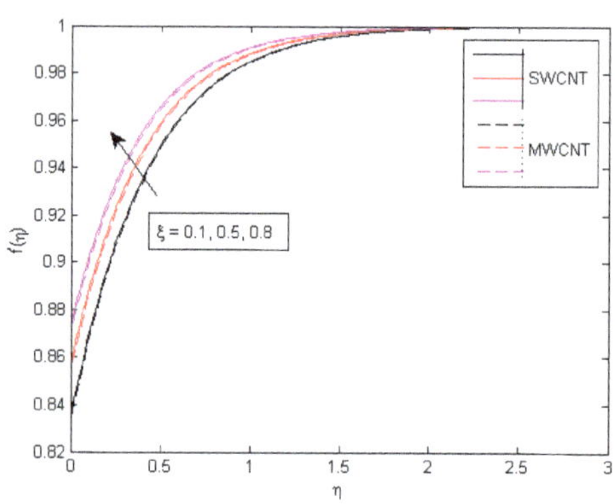

Figure 15. Consequences of ξ on velocity profile.

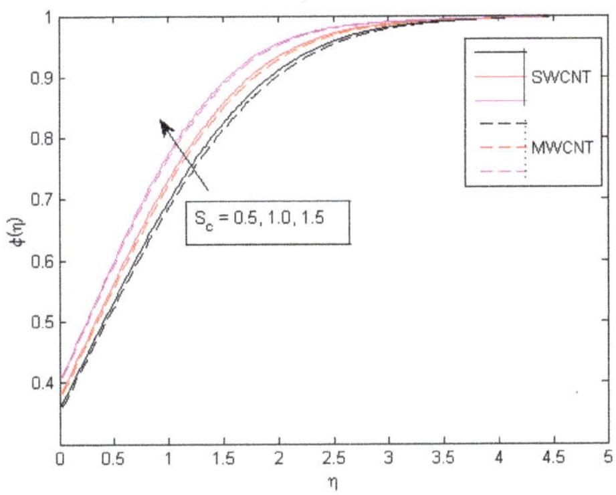

Figure 16. Consequences of S_c on concentration pattern.

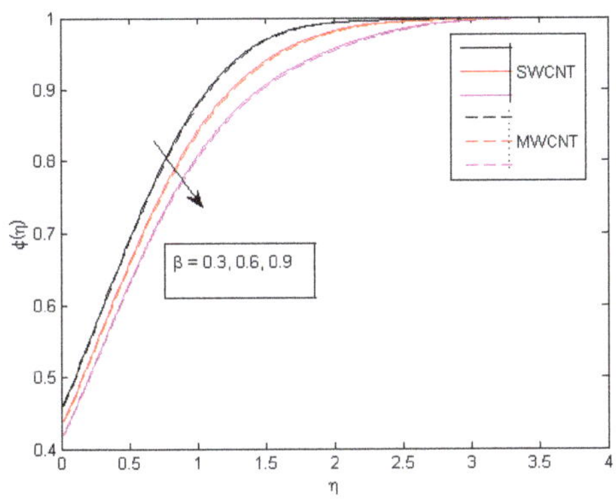

Figure 17. Consequences of β on concentration behavior.

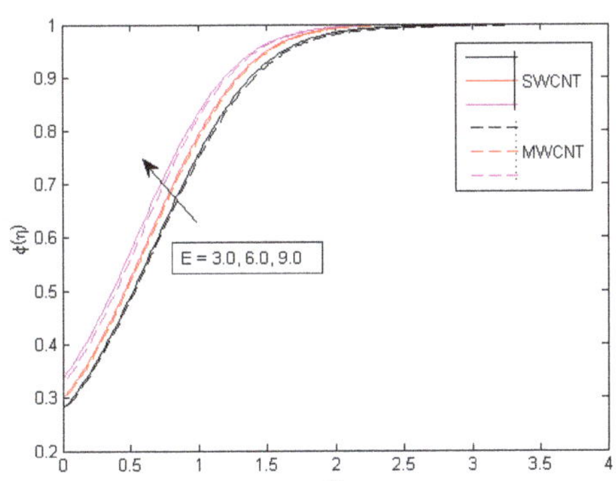

Figure 18. Consequences of E on concentration behavior.

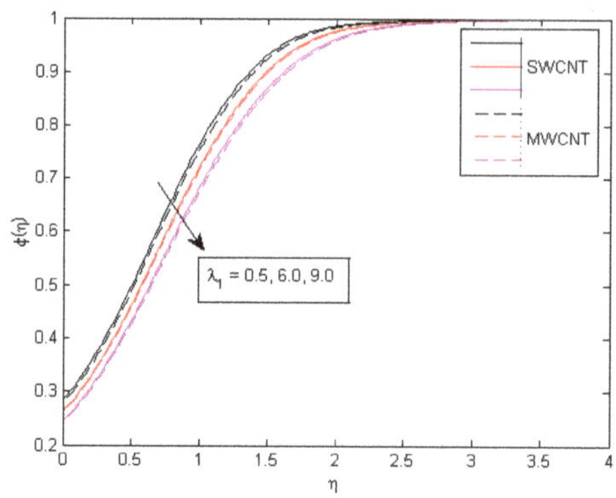

Figure 19. Consequences of λ_1 on concentration behavior.

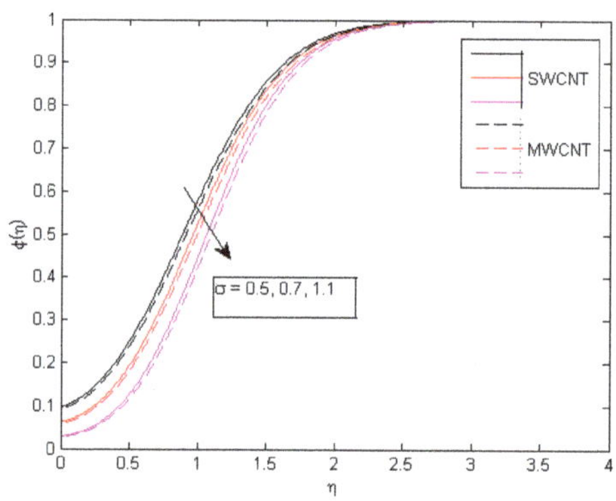

Figure 20. Consequences of σ on concentration behavior.

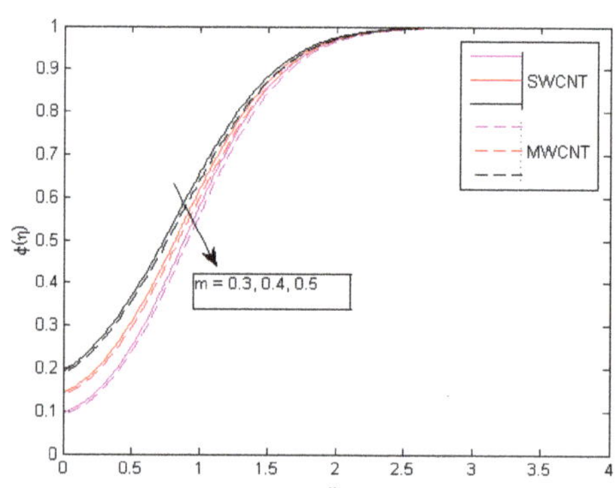

Figure 21. Consequences of m on concentration behavior.

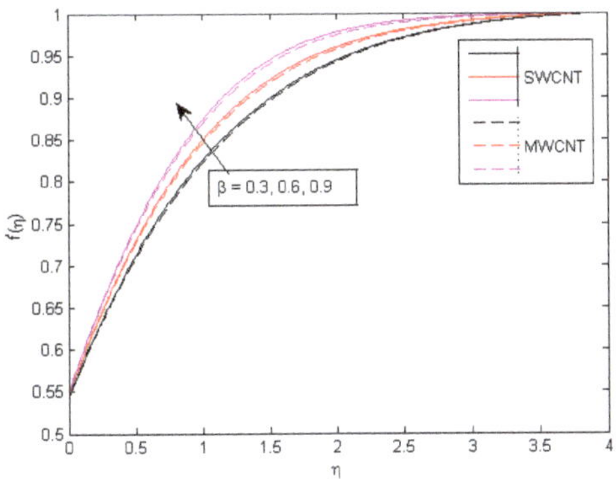

Figure 22. Consequences of β on velocity portfolio.

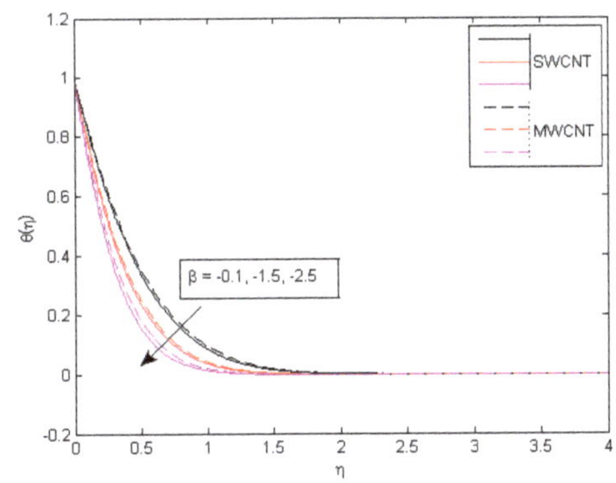

Figure 23. Consequences of β on temperature portfolio.

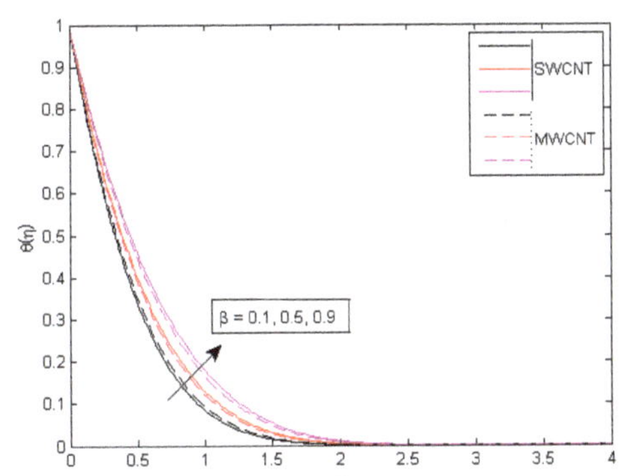

Figure 24. Consequences of β on temperature portfolio.

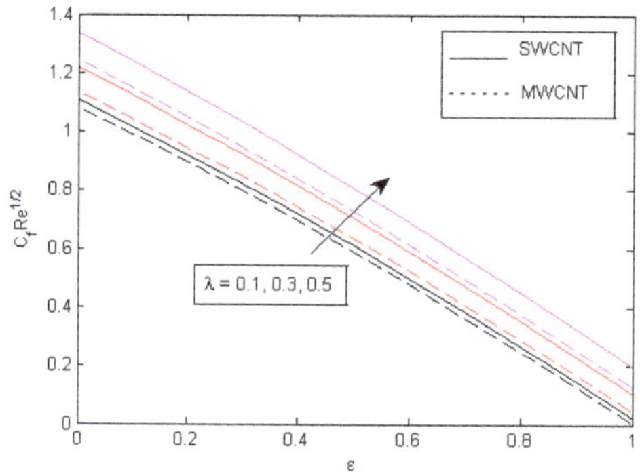

Figure 25. Consequences of λ on skin friction.

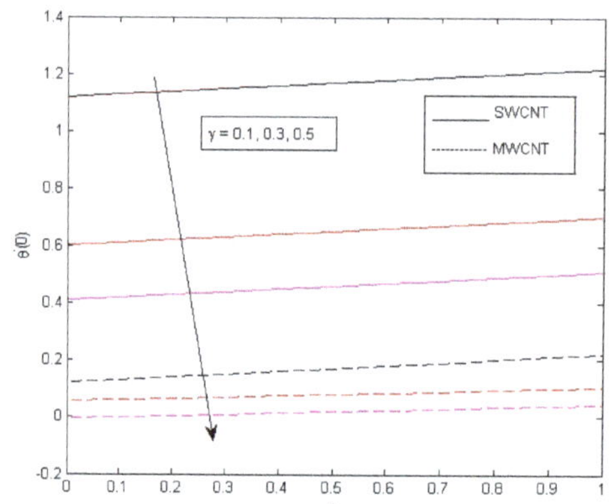

Figure 26. Consequences of γ on temperature gradient.

The findings from bvp4c MATLAB software and the earlier work of Ramzan et al. [7] for increasing ϕ values exhibit great consistency, as shown in Table 2. From comparison analysis, it is quite clear that the obtained results are quite reliable.

Table 2. The work of Ramzan et al. [7]'s limited case is compared with statistical data on surface drag force as well as local Nusselt number versus Prandtl number.

	Ramzan et al. [7] $f''(0)$		**Ramzan et al.** [7] $-\theta'(0)$		**Present** $f''(0)$	**Results**	**Present** $-\theta'(0)$	**Results**
ϕ	SWCNT	MWCNT	SWCNT	MWCNT	SWCNT	MWCNT	SWCNT	MWCNT
0.01	0.338910	0.337270	1.105710	1.079040	0.338995	0.337276	1.105710	1.079043
0.1	0.408120	0.390070	4.806290	4.277160	0.408107	0.390084	4.806290	4.277160
0.2	0.504530	0.464660	12.30352	10.56796	0.504522	0.464669	12.30358	10.56796

Table 3 displays a comparison of numerically achieved outcomes with Othman et al. [28] and Wang [29] for diverse values of ϵ by keeping other parameters $M = \lambda = \gamma_1 = \xi = 0$. From the comparison analysis, it is quite evident the proposed numerical is quite trustworthy, and the obtained outcomes are quite accurate.

Table 3. Comparison of current numerical outcomes with Othman et al. [28] and Wang [29].

$R_e^{1/2}C_f$				
Parameters		**Comparison Analysis**		
ϕ	ϵ	**Othman et al.** [28]	**Wang** [29]	**Current**
0	2	−1.887306668	−1.88731	−1.88795
0	1	0	0	0
0	0.5	0.71329495	0.7133	0.7136
0	0	1.232587647	1.232588	1.232600
0	−0.5	1.495669739	1.49567	1.49590
0	−1	1.328816861	1.32882	1.32900

The effects of the volume fraction of the nano-size particle ϕ, non-dimensional velocity profile parameter ϵ, slip factor γ_1, Schmidt number S_c, exothermic/endothermic parameter β, activation energy E, dimensionless chemical reaction rate constant λ_1 and σ, unitless rate constant m and n, as well as λ, are statistically shown in Table 4, which refers to $C_f Re^{1/2}$. The pattern shows increasing drag force corresponding to λ, E, S_c and ϕ, but $C_f Re^{1/2}$ decreases as the influence of m, n, γ_1, σ, and ϵ increases.

Table 4. Rheological numerics of $-(1+\beta)f''(0)$ and $-\theta'(0)$.

										$R_e^{1/2}C_f$	
ϵ	γ_1	ϕ_2	λ	S_c	β	E	m	σ	n	**SWCNT**	**MWCNT**
0.1	0.1	0.01	0.1	0.5	0.5	1	0.5	0.1	0.1	1.118505	1.113907
0.3										0.909755	0.906056
0.5										0.678612	0.675895
0.2	0.1									1.178579	1.175287
	0.2									1.037153	1.034589
	0.3									0.924627	0.922584
	0.5	0.01								0.719589	0.718489
		0.03								0.753620	0.750208
		0.05								0.790372	0.784373
		0.01	0.2							0.709207	0.707394
			0.3							0.721306	0.719536
			0.4							0.733302	0.731576
			0.1	0.1						0.796844	0.793677
				0.3						0.908853	0.905215
				0.5						1.015271	1.011183
				0.5	0.1					1.045206	1.044045
					0.5					0.933344	0.921626
					0.5					0.815149	0.805263
					0.9	0.1				0.631555	0.618866
						0.5				0.755968	0.743733
						1				0.875441	0.863706
						1	0.2			0.835110	0.825900
							0.3			0.745517	0.737508
							0.4			0.655203	0.648064
							0.1	0.1		0.559715	0.552366
								0.5		0.465787	0.459695
								0.9		0.364019	0.365172
								0.1	0.2	0.745179	0.737164
									0.3	0.654127	0.647323
									0.4	0.555421	0.5558216

6. Conclusions

The bvp4c MATLAB software was used to examine the buoyant flow of a nanofluid containing carbon nanotubes, including homogeneous and heterogeneous reactions as well as heat absorption/generation. The fluid is in a stagnation point flow past a porous shrinking/expanding plane, and the Thompson and Torian slip situations have also been taken into consideration at the boundary. Through naturally occurring factors, the fluid streaming speed, thermodynamic conditions, CNT nano-size particle density, heat transfer rates, and surface drag were investigated. The main observations are summarised as follows:

- Larger magnetic parameters, slip parameters, and velocity ratio factors all cause fluid flow to speed up, but the solid volume fraction causes it to slow down.
- As with the measurements of the heat generation and solid volume ratio, the system is observed to gradually cool down.
- When increasing the slip parameters and velocity ratio, fluid tends to flow smoothly, whereas for the solid volume fractions, surface roughness increased.
- The concentration profile decreases for the larger values of activation energy and exothermic/endothermic parameters.
- The process of heat transmission inside the system was influenced in opposing ways by the velocity ratio parameter as well as the thermal expansion parameter.

Author Contributions: Conceptualization, Y.M.; Data curation, T.A.; Formal analysis, T.S.; Funding acquisition, I.E.S.; Investigation, R.S.; Resources, R.S.; Software, Y.M.; Supervision, I.E.S.; Validation, R.S.; Visualization, T.S.; Writing-review and editing, M.B.; Methodology, M.B.; Writing original draft, T.A. All aithors have read and agreed to the published version of the manuscript.

Funding: This work is not funded by the government or any private agency.

Data Availability Statement: No new data have been created for this study.

Acknowledgments: The authors would like to acknowledge The University of Lahore for the provision of the research plate form to complete this research work.

Conflicts of Interest: The authors declare that they have no known competing financial interests or personal relationships that could have appeared to influence the work reported in this paper.

Nomenclature

Symbols	
u, v	Velocity component along the x and y directions
Q_0	Volumetric rate of a heat source
Pr	Prandtl number
$U_\infty(x)$	Free-stream velocity of the fluid
S_c	Schmidt number
C_f	Surface drag force
N_u	Local heat transfer
f'	Dimensionless stream velocity
E	Activation energy
D_c	Dimensionless heat generation parameter
m, n	Unitless rate constants
Greek Symbols	
ρ_{nf}	Density of nanofluid
ρ_f	Density of fluid
γ^*	Navier slip length density
ϵ	Velocity ratio parameter
μ_{nf}	Dynamic viscosity shear stress
μ_f	Dynamic viscosity shear stress
τ_w	Dynamic viscosity shear stress
α_{nf}	Thermal diffusivity of nanofluid

τ	Ratio of specific heats
ξ^*	Reciprocal of some critical shear rate
ξ	Critical shear rate
$(\rho C_p)_{nf}$	Heat capacity of nanofluid
β_f, β_s	Coefficient of thermal expansion
$(\rho C_p)_f$	Heat capacity of fluid
γ_1	Non-dimensional slip velocity parameter
σ_{nf}	Electric conductivity of fluid
σ_f	Electric conductivity of fluid
σ_f	Electric conductivity of nanofluid
β_{CNT}	Coefficient of thermal expansion of carbon nanotubes
ϕ	Nanofluid volume fraction
γ	Dimensionless thermal relaxation time
β	Exothermic/endothermic parameter
σ, λ_1	Dimensionless chemical reaction rate
k_{nf}	Thermal conductivity of nanofluid
k_f	Thermal conductivity of fluid
ν_{nf}	Kinematic viscosity of nanofluid
k_{CNT}	Thermal conductivity of carbon nanotubes
ρ_{CNT}	Density of carbon nanotubes
$(\rho C_p)_{CNT}$	Heat capacity of carbon nanotubes
$B^2(x)$	Magnetic field strength

References

1. Masuda, H.; Ebata, A.; Teramae, K. Alteration of thermal conductivity and viscosity of liquid by dispersing ultra-fine particles. Dispersion of Al_2O_3, SiO_2 and TiO_2 ultra-fine particles. *Sci. Inf. Database* **1993**. [CrossRef]
2. Choi, S.U.S.; Eastman, J.A. *Enhancing Thermal Conductivity of Fluids with Nanoparticles;* Argonne National Lab: Argonne, IL, USA, 1995.
3. Sadaf, H.; Nadeem, S. Influences of slip and Cu-blood nanofluid in a physiological study of cilia. *Comput. Methods Programs Biomed.* **2016**, *131*, 169–180. [CrossRef] [PubMed]
4. Sivasankaran, S.; Alsabery, A.; Hashim, I. Internal heat generation effect on transient natural convection in a nanofluid-saturated local thermal non-equilibrium porous inclined cavity. *Phys. A Stat. Mech. Its Appl.* **2018**, *509*, 275–293. [CrossRef]
5. Ahmed, N.; Khan, U.; Mohyud-Din, S.T. Modified heat transfer flow model for SWCNTs-H_2O and MWCNTs-H_2O over a curved stretchable semi-infinite region with thermal jump and velocity slip: A numerical simulation. *Phys. A Stat. Mech. Its Appl.* **2020**, *545*, 123431. [CrossRef]
6. Hosseinzadeh, K.; Asadi, A.; Mogharrebi, A.R.; Khalesi, J.; Mousavisani, S.M.; Ganji, D.D. Entropy generation analysis of $(CH_2OH)_2$ containing CNTs nanofluid flow under effect of MHD and thermal radiation. *Case Stud. Therm. Eng.* **2019**, *14*, 100482. [CrossRef]
7. Ramzan, M.; Mohammad, M.; Howari, F.; Chung, J.D. Entropy analysis of carbon nanotubes based nanofluid flow past a vertical cone with thermal radiation. *Entropy* **2019**, *21*, 642. [CrossRef]
8. Khan, M.I.; Hayat, T.; Shah, F.; Haq, F. Physical aspects of CNTs and induced magnetic flux in stagnation point flow with quartic chemical reaction. *Int. J. Heat Mass Transf.* **2019**, *135*, 561–568. [CrossRef]
9. Ramzan, M.; Mohammad, M.; Howari, F. Magnetized suspended carbon nanotubes based nanofluid flow with bio-convection and entropy generation past a vertical cone. *Sci. Rep.* **2019**, *9*, 1–15. [CrossRef]
10. Khan, S.U.; Rauf, A.; Shehzad, S.A.; Abbas, Z.; Javed, T. Study of bioconvection flow in Oldroyd-B nanofluid with motile organisms and effective Prandtl approach. *Phys. A Stat. Mech. Its Appl.* **2019**, *527*, 121179. [CrossRef]
11. Siryk, S.V.; Bendandi, A.; Diaspro, A.; Rocchia, W. Charged dielectric spheres interacting in electrolytic solution: A linearized Poisson–Boltzmann equation model. *J. Chem. Phys.* **2021**, *155*, 114114. [CrossRef]
12. Yu, Y.K. Electrostatics of charged dielectric spheres with application to biological systems. III. Rigorous ionic screening at the Debye-Hückel level. *Phys. Rev. E* **2020**, *102*, 1–5. [CrossRef]
13. Bilal, M.; Mazhar, S.Z.; Ramzan, M.; Mehmood, Y. Time-dependent hydromagnetic stagnation point flow of a Maxwell nanofluid with melting heat effect and amended Fourier and Fick's laws. *Heat Transf.* **2021**, *50*, 4417–4434. [CrossRef]
14. Bilal, M.; Arshad, H.; Ramzan, M.; Shah, Z.; Kumam, P. Unsteady hybrid-nanofluid flow comprising ferrousoxide and CNTs through porous horizontal channel with dilating/squeezing walls. *Sci. Rep.* **2021**, *11*, 12637. [CrossRef] [PubMed]
15. Bilal, M.; Ramzan, M.; Mehmood, Y.; Alaoui, M.K.; Chinram, R. An entropy optimization study of non-Darcian magnetohydrodynamic Williamson nanofluid with nonlinear thermal radiation over a stratified sheet. *Proc. IMechE Part E J. Process. Mech. Eng.* **2021**, *235*, 1883–1894. [CrossRef]

16. Bilal, M.; Ramzan, M.; Mehmood, Y.; Sajid, T.; Shah, S.; Malik, M.Y. A novel approach for EMHD Williamson nanofluid over nonlinear sheet with double stratification and Ohmic dissipation. *Proc. IMechE Part E J. Process. Mech. Eng.* **2021**, 1–16. [CrossRef]
17. Bilal, M.; Ramzan, M.; Siddique, I.; Anum, A. A numerical simulation of electrically conducting micro-channel nanofluid flow with thermal slip effects. *Waves Random Complex Media* **2022**, 1–25. [CrossRef]
18. Maleque, K. Effects of exothermic/endothermic chemical reactions with Arrhenius activation energy on MHD free convection and mass transfer flow in presence of thermal radiation. *J. Thermodyn.* **2013**, *2013*, 692516. [CrossRef]
19. Bejawada, S.G.; Reddy, Y.D.; Jamshed, W.; Nisar, K.S.; Alharbi, A.N.; Chouikh, R. Radiation effect on MHD Casson fluid flow over an inclined non-linear surface with chemical reaction in a Forchheimer porous medium. *Alex. Eng. J.* **2022**, *61*, 8207–8220. [CrossRef]
20. Suleman, M.; Ramzan, M.; Ahmad, S.; Lu, D.C. Numerical simulation for homogeneous–heterogeneous reactions and Newtonian heating in the silver-water nanofluid flow past a nonlinear stretched cylinder. *Phys. Scr.* **2019**, *94*, 085702. [CrossRef]
21. Imtiaz, M.; Mabood, F.; Hayat, T.; Alsaedi, A. Homogeneous-heterogeneous reactions in MHD radiative flow of second grade fluid due to a curved stretching surface. *Int. J. Heat Mass Transf.* **2019**, *145*, 118781. [CrossRef]
22. Suleman, M.; Ramzan, M.; Ahmad, S.; Lu, D.C.; Muhammad, T.; Chung, J.D. A numerical simulation of silver-water nanofluid flow with impacts of newtonian heating and homogeneous–heterogeneous reactions past a nonlinear stretched cylinder. *Symmetry* **2019**, *11*, 295. [CrossRef]
23. Doh, D.H.; Muthtamilselvan, M.; Swathene, B.; Ramya, E. Homogeneous and heterogeneous reactions in a nanofluid flow due to a rotating disk of variable thickness using HAM. *Math. Comput. Simul.* **2020**, *168*, 90–110. [CrossRef]
24. Khan, M.I.; Hayat, T.; Khan, M.I.; Waqas, M.; Alsaedi, A. Numerical simulation of hydromagnetic mixed convective radiative slip flow with variable fluid properties: a mathematical model for entropy generation. *J. Phys. Chem. Solids* **2019**, *125*, 153–164. [CrossRef]
25. Hamid, M.; Zubair, T.; Usman, M.; Khan, Z.H.; Wang, W. Natural convection effects on heat and mass transfer of slip flow of time-dependent Prandtl fluid. *J. Comput. Des. Eng.* **2019**, *6*, 584–592. [CrossRef]
26. Reddy, S.R.R.; Reddy, P.B.A.; Bhattacharyya, K. Effect of nonlinear thermal radiation on 3D magneto slip flow of Eyring-Powell nanofluid flow over a slendering sheet with binary chemical reaction and Arrhenius activation energy. *Adv. Powder Technol.* **2019**, *30*, 3203–3213. [CrossRef]
27. Kiyasatfar, M. Convective heat transfer and entropy generation analysis of non-Newtonian power-law fluid flows in parallel-plate and circular microchannels under slip boundary conditions. *Int. J. Therm. Sci.* **2018**, *128*, 15–27. [CrossRef]
28. Othman, M.N.; Jedia, A.; B., A.; A., N. MHD Stagnation Point on Nanofluid Flow and Heat Transfer of Carbon Nanotube over a Shrinking Surface with Heat Sink Effect. *Molecules* **2021**, *26*, 7441. [CrossRef]
29. Wang, C. Stagnation flow towards a shrinking sheet. *Int. J. -Non-Linear Mech.* **2008**, *43*, 377–382. [CrossRef]

mathematics MDPI

Article

Statistical Descriptions of Inhomogeneous Anisotropic Turbulence

J. J. H. Brouwers

Romico Hold A.V.V., 6226 GV Maastricht, The Netherlands; j.j.h.brouwers@gmail.com

Abstract: Descriptions are given of the Langevin and diffusion equation of passively marked fluid particles in turbulent flow with spatially varying and anisotropic statistical properties. The descriptions consist of the first two terms of an expansion in powers of C_0^{-1}, where C_0 is an autonomous Lagrangian-based Kolmogorov constant: $C_0 \approx 7$. Solutions involve the application of methods of stochastic analysis while complying with the basic laws of physics. The Lagrangian-based descriptions are converted into Eulerian-based fixed-point expressions through asymptotic matching. This leads to novel descriptions for the mean values of the fluctuating convective terms of the conservation laws of continua. They can be directly implemented in CFD codes for calculating fluid flows in engineering and environmental analysis. The solutions are verified in detail through comparison with direct numerical simulations of turbulent channel flows at large Reynolds numbers.

Keywords: statistical turbulence; Langevin and diffusion equation; nonlinear convection statistics

MSC: 37M10

Citation: Brouwers, J.J.H. Statistical Descriptions of Inhomogeneous Anisotropic Turbulence. *Mathematics* **2022**, *10*, 4619. https://doi.org/10.3390/math10234619

Academic Editor: Ramoshweu Solomon Lebelo

Received: 17 October 2022
Accepted: 30 November 2022
Published: 6 December 2022

Publisher's Note: MDPI stays neutral with regard to jurisdictional claims in published maps and institutional affiliations.

1. Introduction

Fluid flow that exhibits turbulence is more of a rule than an exception. It occurs when the Reynolds number of the flow Re is sufficiently large. The number is specified as

$$Re = ULv^{-1}, \tag{1}$$

where U [m/s] is the fluid velocity, ν [m^2/s] is the kinematic viscosity of the fluid and L [m] is the spatial dimension of the flow configuration, e.g., the diameter of a tube, length of an air foil, or height above the earth's surface. Values for ν in the cases of water and air are typically 10^{-6} and 10^{-5} m^2/s, with the corresponding velocities 0.1 and 1 m/s. A configuration where $L = 0.1$ m results in a value of Re of 10^5. This exceeds, by far, the critical value of approximately 10^3, where turbulence starts to occur.

Turbulence can be considered as a statistical process. General descriptions of the statistical parameters have yet to be found. What is known are partial results, such as the solutions for the log layer by Von Karman and the theory of the small viscous scales by Kolmogorov: e.g., Monin and Yaglom [1]. However, a general description for the statistical parameters of the large scale is missing. The key problem is the description of the statistics of the fluctuations of the convective accelerations in the governing conservation equations.

The averaged representations of the conservation equations lack well-founded statistical descriptions of the non-linear convective terms. Instead, semi-empirical versions are used, which are adapted and calibrated from case to case. These are a common feature of the methods used in fluid mechanics, including computational fluid mechanics (CFD), which are widely used in engineering and environmental analysis (Bernard and Wallace [2], Hanjalic and Launder [3]). The presented analysis does not resort to empirical construction. Instead, statistical descriptions are derived by applying the methods of stochastic analysis (Stratonovich [4], Van Kampen [5]) and obeying the basic laws of physics.

The first part of the analysis is a presentation and update of previous work [6–9] concerning the Langevin and diffusion equations for the motion of passively marked fluid particles. In the Langevin equation, the autonomous universal Lagrangian-based

Kolmogorov constant C_0 appears. Its reciprocal value is about 0.14. Its smallness forms the basis for approximation. Solutions are given in descending powers of C_0^{-1} where the leading and next to leading terms are retained.

All these solutions comply with the known laws of physics. The Lagrangian-based descriptions are subsequently converted into Eulerian-based fixed-point expressions by matching using C_0^{-1} as the small parameter. They can directly be employed in the averaged equations of fluid mechanics. The outcome of the C_0^{-1} expansion is tested through comparison with the results of Hoyas et al. [10,11] and Kuerten et al. [12] for direct numerical simulations (DNS) of turbulent channel flow at a high Reynolds number.

2. Langevin Equation Including Kolmogorov Similarity

Turbulent flow occurs for large values of Reynolds numbers, *Re*, a situation that is frequently encountered in practice. For $Re \gg 1$, the time over which fluid particle accelerations decorrelate compares to the decorrelation times of particle velocity as $Re^{-1/2}$ to 1, e.g., [1]. This forms the basis for assuming that the velocity process can be represented by a Markov process, where accelerations are modelled as delta correlated. The corresponding Langevin equation reads as

$$\frac{dv'_i}{dt} = a_i(\mathbf{v}',\mathbf{y}) + b_{ij}(\mathbf{v}',\mathbf{y})w_j(t), \tag{2}$$

where the time-dependent position of the moving fluid particle is described by

$$\frac{dy_i}{dt} = u_i^0(\mathbf{y}(t)) + v'_i, \tag{3}$$

and $i, j = 1, 2, 3$. In the above equations:

$t =$ time.
$v'_t =$ a statistical representation of the fluctuating fluid particle velocity at time t.
$y_i(t) =$ a statistical representation of the particle position at time t.
$a_i(\mathbf{v}',\mathbf{y}) =$ a damping function.
$b_{ij}(\mathbf{v}',\mathbf{y}) =$ the amplitude of white noise.
$w_j(t) =$white noise of unit intensity.
$u_i^0(\mathbf{y}(t)) =$ the velocity based on the mean Eulerian velocity evaluated at the particle position $\mathbf{y}(t)$.

Fluid velocities at a fixed point in a fixed frame of reference using the Eulerian description are indicated by **u**, while velocities of fluid particles that move with the flow using the Lagrangian description, are indicated by **v**. The coordinate **x** is used to denote a fixed position in the non-moving fixed coordinate system, while $\mathbf{y}(t)$ is the position of a moving particle. The turbulent flow field is considered to be stationary in a fixed frame of reference. Statistical averages of Eulerian flow variables can be calculated by time averaging, which is indicated by angled brackets or superscript ^{0}. The white-noise amplitude can be specified by implementing the Lagrangian version of Kolmogorov's similarity theory of 1941, also referred to as K-41 theory: [13] and [1] Section 21.3. This yields

$$b_{ij}(\mathbf{v}',\mathbf{y})w_j(t) = \{C_0\epsilon(\mathbf{y})\}^{\frac{1}{2}}w_i(t), \tag{4}$$

where C_0 is a universal Lagrangian-based Kolmogorov constant, and $\epsilon = \epsilon(\mathbf{y})$ is the mean energy dissipation rate averaged at a fixed position **x** and evaluated at particle position $\mathbf{y}(t)$ when applied in Equation (4):

$$\epsilon = \frac{1}{2}\nu\left\langle\left(\frac{\partial u'_i}{\partial x_j} + \frac{\partial u'_j}{\partial x_i}\right)^2\right\rangle, \tag{5}$$

where u_i' is a fluctuating component of Eulerian velocity at fixed position $\mathbf{x}$.

The observation that second-order correlations of fluid particle accelerations tend to those of a delta-correlated process, when $Re \gg 1$, is, in itself, not sufficient to justify the Langevin model [9]. The description of the forcing term by Gaussian white noise leads to applying ordinary non-intermittent Kolmogorov (K-41) theory. The effects of intermittency, apparent in corrections in higher-order structural functions, are not accounted for in the Langevin model [9]. For that purpose, one can adopt a fractal model based on Kolmogorov's refined similarity theory: [1] Section 25.2.

However, the statistical averages of particle displacement that determine turbulent dispersion change little under such an approach: [1] and Borgas [14]. The effect of intermittency is apparent in small viscous scales, which govern the acceleration process, rather than in large energetic scales, which govern the velocity process of turbulence. In many applications, a Langevin model resting on K-41 theory can be considered to be a sound approach for describing the mean dispersion on distances of large-scale turbulence.

Individual values of displacement $\mathbf{y}(t)$ and velocity $\mathbf{v}(t)$ obtained from Equations (2) and (3) do not represent the actual values of fluctuating displacements and velocities of fluid particles as they occur in turbulent flow. Instead, they are dummy variables that enable the specification of statistical averages of actual fluid flow. This is achieved by generating many realizations using $\mathbf{w}(t)$ as a random generator and averaging the results. In the case of passive marking of fluid particles all starting at position $\mathbf{y} = \mathbf{x}_0$ at $t = 0$, the fluctuating velocities $\mathbf{v}'$ should, for every realization, be selected randomly in accordance with the distribution of the Eulerian fluctuating velocity at position x_0:

$$t = 0; \quad \mathbf{y} = \mathbf{x}_0; \quad \mathbf{v}' = \mathbf{u}' \tag{6}$$

During a simulation, the coefficients in Equations (2) and (3) vary in magnitude with the particle position in accordance with their value at $\mathbf{y} = \mathbf{x}$. A probabilistic description of particle displacement and its velocity is obtained after performing many simulations and averaging the result at every moment in time. This enables evaluating the average spatial distribution of particles with time. This type of Lagrangian averaging is denoted by an overbar: In the case of the simulated variable $f_n(t)$, it can be written as

$$\overline{f(t)} = \lim_{N\to\infty} \sum_{n=1}^{N} f_n(t), \tag{7}$$

where $f_n(t)$ is the value of f at time t in the case of simulation n.

As alternative to time simulation using the Langevin equation the same statistical distributions of fluid particle velocity and position can be obtained from the Fokker–Planck equation associated with Equations (2) and (3). It is given by

$$\frac{\partial p}{\partial t} + u_i^0 \frac{\partial p}{\partial y_i} + v_i' \frac{\partial p}{\partial y_i} = -\frac{\partial}{\partial v_i'}(a_i(\mathbf{v}',\mathbf{y})p) + \frac{1}{2} C_0 \epsilon \frac{\partial^2 p}{\partial v_i' \partial v_i'} \tag{8}$$

where $p = p(\mathbf{v}, \mathbf{y}, t)$ is the joint probability density function of velocity and position at time t.

3. Specification of Damping Function by C_0^{-1}-Expansion

Thus far, I have not specified the damping term $a_i(\mathbf{v}', \mathbf{y})$ in the Langevin equation. The specification of the damping term in a form that is generally applicable has long been an issue [6–9,15]. A method was proposed in which Kolmogorov constant C_0 is used as the basis for an expansion. Solutions are described in terms of an expansion [7–9] in consecutive powers of C_0^{-1}. The expansion is not related to a dimensionless combination of parameters, which can attain a vanishingly small or large value. Such a combination does not exist. Instead, C_0 is used as a scaling parameter, facilitated by its autonomous position in statistical turbulence at a large Reynolds number [9].

The scaling parameter enters by the white-noise term and results in specific powers of C_0 in each of the terms on the basis of the required balances between them. The accuracy of the expansion depends on the truncation of subsequent terms. According to the measurements and data from numerical simulations, C_0 has a value of about 7: Sawford [16] and Section 9. The accuracy of the resulting expressions is discussed in Sections 8 and 9.

Realistic solutions in the limit of $C_0^{-1} \to 0$ are obtained from the Langevin equation when all terms scale in the same manner with C_0. For this to happen, the damping term must scale as C_0, and the time of correlation, which is the statistically relevant time as C_0^{-1}; thereby, noting that the white noise term $w(t)$ scales as $C_0^{-1/2}$. The displacement due to fluctuations during correlation scales is C_0^{-1}. This initial scaling allows for a number of approximations [9]. To the leading order in C_0^{-1}, the displacement of a particle is small, and values of fixed-point statistical quantities used in the parameters of the Langevin equation can be represented by their values at the marking point $\mathbf{x} = \mathbf{x}_0$.

We can thus discuss a homogeneous statistical process in the initial stages after marking [6–9]. During that short time, the dissipation of energy by viscous action is small. The change in the Hamiltonian by viscous dissipation $(d/dt)H \approx \epsilon(\mathbf{x}_0)$ is small and proportional to C_0^{-1}. The statistical process is initially one that can be described by Einstein's fluctuation theory, e.g., Reichl [17]. In the leading order formulation in powers of C_0^{-1}, the damping term is linear in velocity, satisfies Onsager symmetry, and its magnitude is determined by the fluctuation–dissipation theorem [8,9]. As a result,

$$a_i' = -\frac{1}{2} C_0 \lambda_{ij} \epsilon v_j', \tag{9}$$

where λ_{ij} is the inverse of the covariance tensor of the Eulerian velocity field

$$\lambda_{ij} = \sigma_{ij}^{-1} = \left\langle u_i' u_j' \right\rangle^{-1} \tag{10}$$

4. Higher-Order Formulation of the Langevin Equation

Until now, attention has been focused on the leading-order term in the expansion with respect to C_0^{-1}. The resulting descriptions involve a truncation error of $\mathcal{O}(C_0^{-1})$. Such an error will become smaller, the larger C_0 is. However, in turbulence, the value of C_0 is limited to about 7. This corresponds to $C_0^{-1} = 0.14$ and implies that the truncation error can become large. Deriving expressions for higher-order terms is, thus, desired [7]. For that purpose, one can resort to the well-mixed principle of Thomson [15]. Given an initial distribution, particles will, in the course of time, mix up with the fluid and attain the distribution of fluid velocity. This equilibrium distribution satisfies the Eulerian interpretation of Equation (8), which is given by [7]

$$u_i^0 \frac{\partial p_E}{\partial x_i} = \frac{1}{2} \epsilon^0 C_0 \frac{\partial^2 p_E}{\partial u_i' \partial u_i'} - \frac{\partial}{\partial u_i'} (a_i p_E) - u_i' \frac{\partial p_E}{\partial x_i} \tag{11}$$

where $\boldsymbol{a_i} = \boldsymbol{a_i}(\mathbf{u}',\mathbf{x})$ and $\boldsymbol{p_E} = \boldsymbol{p_E}(\mathbf{u}')$ is the distribution of the fluctuating component of the fixed-point Eulerian fluid velocity $\mathbf{u}'$. There is no time derivative in Equation (11) when considering stationary turbulence: Statistical averages at a fixed point do not vary with time. Note further that $\mathbf{x}$ is not a statistical variable but a fixed position. Statistical averages can be obtained from time averaging at fixed point $\mathbf{x}$. Derivatives of p_E with respect to $\mathbf{x}$ attain values whenever the statistical parameters of p_E (covariances, etc.) vary in space (inhomogeneous turbulence).

The Eulerian distribution p_E is equivalent to the non-equilibrium steady state distribution in statistical mechanics. Equation (11) represents the general form of the fluctuation–dissipation theorem that is appropriate for turbulence. Given $p_E(u_i')$, Equation (11) can be used to derive expressions for the damping function a_i'. Noting the leading-order formulation with respect to C_0, cf. Equation (9), we have

$$a_i = -\frac{1}{2} C_0 \lambda_{ij} \epsilon \mathbf{u}_j' + a_i', \tag{12}$$

where $a_i' = a_i'(\mathbf{u}',\mathbf{x})$ is to be determined. The Eulerian velocity distribution can be taken as Gaussian to the leading order,

$$p_E = p_G + C_o^{-1} f_e p_G, \tag{13}$$

where $p_G = p_G(\mathbf{u}')$ is the zero-mean Gaussian while $f_c = f_c(\mathbf{u}')$ is the correction on the Gaussian behavior. Values of the zero, first-, and second-order moments are fully captured by the Gaussian part of the description,

$$\begin{aligned}\int_{-\infty}^{+\infty} f_c(\mathbf{u}')p_G(\mathbf{u}')d\mathbf{u}' &= \int_{-\infty}^{+\infty} u_i' f_c(\mathbf{u}')p_G(\mathbf{u}')d\mathbf{u}' \\ &= \int_{-\infty}^{+\infty} u_m' u_n' f_c(\mathbf{u}')p_G(\mathbf{u}')d\mathbf{u}' = 0,\end{aligned} \tag{14}$$

values of cumulants higher than second order are determined by $f_c(\mathbf{u}')$. Substituting the description for p_E and Equation (12) into Equation (11), one obtains, for a_i', the equation

$$\begin{aligned}&\frac{\partial a_i'}{\partial u_i'} - \lambda_{ij} u_j' a_i' \\ &= \frac{1}{2}\left[\lambda^{-1}(u_i^0 + u_i')\frac{\partial \lambda}{\partial x_i} - (u_i^0 + u_i')\frac{\partial \lambda_{mn}}{\partial x_i} u_m' u_n'\right] \\ &+ \frac{1}{2}\epsilon\left(\frac{\partial^2 f_c}{\partial u_i' \partial u_i'} - \lambda_{ij} u_j' \frac{\partial f_c}{\partial u_i'}\right),\end{aligned} \tag{15}$$

where there are dropped terms of relative magnitude $\mathcal{O}(C_0^{-1})$ in the contributions due to non-Gaussianity, i.e., the second term on the right-hand side of Equation (15). Equation (15) is exact, i.e., it does not involve any approximation or truncation with regard to C_0 in the case of Gaussian Eulerian velocities ($f_c = 0$). The solution of Equation (15) is [7]

$$a_i' = \frac{1}{2}\lambda_{jm} u_k^0 \frac{\partial \sigma_{mi}}{\partial x_k} u_j' + \frac{1}{2}\lambda_{jn}\frac{\partial \sigma_{ij}}{\partial x_m}(u_m' u_n' + \sigma_{mn}) + g_i, \tag{16}$$

where $g_i = g_i(\mathbf{u}',\mathbf{x})$,

$$g_i = \frac{1}{2}\epsilon \frac{\partial f_c}{\partial u_i'} + a_i'^H, \tag{17}$$

where $a_i'^H = a_i'^H(\mathbf{u}',\mathbf{x})$ is the solution of the homogeneous problem

$$(\partial/\partial u_i')(a_i'^H p_G) = 0 \quad \text{or} \quad (\partial/\partial u_i')a_i'^H = \lambda_{ij} u_j' a_i'^H. \tag{18}$$

A variety of solutions exists for $a_i'^H$, linear and nonlinear in $\mathbf{u}'$; however, each of them contains a degree of indeterminacy apparent in unspecified constants. When confining the damping function to linear representations in u_i, the solution of Equation (18) is

$$a_i'^H = b_k \sigma_{ij} \epsilon_{kij} u_j' \tag{19}$$

where ϵ_{kij} is the alternating unit tensor. Solution (19) constitutes an antisymmetric extension to the symmetric damping tensor derived in the previous section as described by the first term of solution (12). In this solution b_k are three dimensionless constants whose values are unknown. It is a reflection of the nonuniqueness problem: Except for isotropic turbulence, it is impossible to fully specify the damping function on the basis of a specified fixed-point Eulerian velocity distribution. Yet, there is a practical way out of the nonuniqueness problem [7–9].

It appears that $a_i'^H$ yields only contributions of relative magnitude $\mathcal{O}(C_0^{-2})$ compared to the previously determined leading terms in the statistical distributions of particle displacement. This conclusion is arrived at when deriving the diffusion equation from the Langevin equation: see Section 5. This reveals the contributions of relative magnitude $\mathcal{O}(C_0^{-2})$ in diffusivity and convection only. The same result is obtained for the other term in solution (17), which describes the effect of non-Gaussianity. In general, the contribution of g_i in solution (16) can be disregarded in any description, which allows for a relative error of $\mathcal{O}(C_0^{-2})$ in the diffusion limit. Setting $g_i = 0$, we arrive at a Langevin model, which has, as a damping function,

$$a_i = -\frac{1}{2}C_0\lambda_{ij}\epsilon v'_j + \frac{1}{2}\lambda_{jm}u_k^0\frac{\partial\sigma_{mi}}{\partial x_k}v'_j + \frac{1}{2}\lambda_{jn}\frac{\partial\sigma_{ij}}{\partial x_m}(v'_m v'_n + \sigma_{mn}). \tag{20}$$

While the first term in this solution corresponds to the result of the Hamiltonian base case, the second and third terms represent the correction due to inhomogeneity in an otherwise locally homogeneous statistical field. The corrections can be related to the change of energy, which was disregarded in the leading-order formulation where underlying particle mechanics can be considered Hamiltonian [7,8]. The second term describes the change of energy due to changes of covariances in the direction of the mean flow. Accelerating or decaying the mean flow results in non-zero values of the second term. The third term describes the effects of the spatial gradient of the fluid velocity covariance. This can be associated with shearing due to external forcing.

Solution (20) corresponds to a previous result of Thomson [15]. It was one of several proposals made for the damping functions, which all satisfy the well-mixed criterion and which correspond to an entirely Gaussian Eulerian velocity distribution. This is a reflection of indeterminacy because of the nonuniqueness. The present analysis provides an answer. It reveals descriptions for statistical displacement obtained from Equation (20), which are unique up to an error of $\mathcal{O}(C_0^{-2})$.

5. The Diffusion Limit

The diffusion limit concerns the description of random particle displacements on a time scale that is much larger than the correlation time of the fluctuating velocity. As indicated by a balance between the acceleration term and damping term in the Langevin equation, the correlation time can be expressed as

$$\tau_c = C_0^{-1}\tau_E, \quad \tau_E = |u'|^2\epsilon^{-1} \tag{21}$$

where $|u'|$ is the magnitude of velocity flucutations and τ_E is the characteristic time of large scales or of the eddy turn-over time. The description of the time scale $t \gg \tau_c$ is known as coarse graining, e.g., [1] vol.I, Section 10.3. The magnitude of the fluctuating fluid particle displacement during correlation can be represented by

$$l_c = \tau_c|u'| = C_0^{-1}|u'|^3\epsilon^{-1} \tag{22}$$

where $|u'|^3\epsilon^{-1}$ represents the size of the eddies, which is also the distance over which the statistical parameters vary in magnitude.

The Langevin model is centered around the fluctuating particle velocity relative to the mean Eulerian velocity: cf. Equations (2) and (3). In line with this representation, the displacement of a fluid particle by the sum of a component due to the mean flow and a component representing the zero-mean random displacement are described (see also [18] and [5] Section XVI.5):

$$y_i(t) = y'_{i0} + y'_i(t) \tag{23}$$

where y'_{i0} is the particle track according to the Eulerian mean velocity:

$$\frac{dy'_{i0}}{dt} = u_i^0(y'_0) \quad , \quad y'_{i0} = \mathbf{x}_0 \quad \text{at} \quad t = 0 \tag{24}$$

For general inhomogeneous turbulent flow, the Eulerian-based coefficients in the Langevin model vary in magnitude with the space coordinates. This makes the coefficients time-dependent in the Langrangian-based description of the Langevin model. Representing displacement by Equation (23), the time dependency occurs in two ways [8]: (i) through spatial variations when following the particle according to the mean velocity $\mathbf{y}_0'$ and (ii) through dependency on random displacement $\mathbf{y}'$

$$\begin{aligned} \lambda_{ij} &= \lambda_{ij}(\mathbf{y}_0' + \mathbf{y}'), \quad \epsilon = \epsilon(\mathbf{y}_0' + \mathbf{y}'), \quad \sigma_{ij} = \sigma_{ij}(\mathbf{y}_0' + \mathbf{y}'), \\ u_i^0 &= u_i^0(\mathbf{y}_0' + \mathbf{y}'), \quad a_i'^H = a_i'^H(\mathbf{v}', \mathbf{y}_0' + \mathbf{y}') \end{aligned} \tag{25}$$

In the next analysis, I shall disregard the dependency on $\mathbf{y}'$. Furthermore, I disregard the non-linear third term in the damping function as well as $a_i'^H$. Requiring the Gaussian behavior in the leading order formulation and mixing for next-to-leading order, all these terms yield contributions of relative magnitude $\mathcal{O}(C_0^{-2})$ in the diffusion model: see Appendix of [8]. The Langevin model, which specifies diffusion to the leading order and next-to-leading order now follows from Equations (2) and (3) as

$$\begin{aligned} \frac{dv_j'}{dt} =& \left(-\frac{1}{2} C_0 \lambda_{ij}(\mathbf{y}_0') \epsilon(\mathbf{y}_0') + \frac{1}{2} \lambda_{jm}(\mathbf{y}_0') u_k^0(\mathbf{y}_0') \frac{\partial \sigma_{mi}(\mathbf{y}_0')}{\partial y_{k0}'} \right) v_j' \\ &+ \left(C_0 \epsilon(\mathbf{y}_0') \right)^{1/2} w_i(t) \end{aligned} \tag{26}$$

From Equations (3), (23) and (24), we obtain

$$v_i' = \frac{dy_i'}{dt} \tag{27}$$

Fluctuating Equations (26) and (27) can be transformed into a Fokker–Planck equation for the joint probability of $\mathbf{v}'$ and $\mathbf{y}'$. The solution is a multi-dimensional Gaussian distribution with time-dependent parameters: [5] Section VIII.6. The zero-mean probability density distribution for the fluid particle position in the fixed coordinate system $\mathbf{x}'$, which moves with the mean Eulerian velocity $\mathbf{u}^0$ is specified by the diffusion equation

$$\frac{\partial p(\mathbf{x}', t')}{\partial t'} = \frac{\partial}{\partial x_i'} \left(\overline{y_k' v_i'} \frac{\partial p(\mathbf{x}', t')}{\partial x_k'} \right) \tag{28}$$

subject to a suitably chosen initial distribution at $t' = 0$, i.e., the delta pulse $\delta(\mathbf{x}')$ in the case of passive marking of particles at $t' = 0$ and $\mathbf{x}' = \mathbf{0}$. Note that the time derivative in the above Eulerian description applies to the coordinate system, which moves with the mean velocity according to Equations (23) and (24) ($\partial/\partial t' = \partial/\partial t + u_i^0 \partial/\partial x_i$). To evaluate the diffusion coefficient $\overline{y_k' v_i'}$, note that

$$\frac{d}{dt} \overline{y_k' v_i'} = \overline{v_k' v_i'} + \overline{y_k' \frac{dv_i'}{dt}} = \sigma_{ki}(\mathbf{y}_0') + \overline{y_k' \frac{dv_i'}{dt}} \tag{29}$$

where the latter term can be calculated by multiplying Equation (26) with x_k' and averaging

$$\overline{y_k' \frac{dv_i'}{dt}} = -\frac{1}{2} C_0 \lambda_{ij}(\mathbf{y}_0') \epsilon(\mathbf{y}_0') \overline{y_k' v_j'} + \frac{1}{2} \lambda_{jm}(\mathbf{y}_0') u_n^0(\mathbf{y}_0') \frac{\partial \sigma_{mi}(\mathbf{y}_0')}{\partial y_{n0}'} \overline{y_k' v_j'} \tag{30}$$

There is no contribution of the last term of Langevin Equation (26) because $w_i(t)$ is only correlated with $v_i'(t)$. Substituting Equation (30) into the r.h.s. of Equation (29) results in the following first-order differential equation for the diffusion coefficient

$$\begin{aligned} &\frac{d}{dt} \left(\overline{y_k' v_i'} \right) + \frac{1}{2} C_0 \lambda_{ij}(\mathbf{y}_0') \epsilon(\mathbf{y}_0') \overline{y_k' v_i'} = \\ &\qquad \sigma_{ki}(\mathbf{y}_0') + \frac{1}{2} \lambda_{jm}(\mathbf{y}_0') u_n^0(\mathbf{y}_0') \frac{\partial \sigma_{mi}(\mathbf{y}_0')}{\partial y_{n0}'} \overline{y_k' v_j'} \end{aligned} \tag{31}$$

subject to the initial condition $\overline{y_k' v_i'} = 0$ at $t' = 0$. The equation describes the transient of the diffusion coefficient towards its value valid in the diffusion limit when $t \gg \tau_c$. This limit value can be time-dependent on the time-scale $t \gg \tau_c$ and can be obtained by iteration using C_0^{-1} as the small parameter [8]. The leading order follows from a balance between the second term on the l.h.s. and the first term on the r.h.s. Substituting this solution into the neglected other terms and noting that, according to our definitions, $\overline{y_k' v_i'} = D_{ki}$ when $t \gg \tau_c$, we obtain, in terms of the Eulerian coordinates of the non-moving frame [8]

$$D_{kn} = 2C_0^{-1}\epsilon^{-1}\sigma_{ki}\sigma_{in} + 2C_0^{-2}\epsilon^{-2}\sigma_{in}\sigma_{km}u_l^0\frac{\partial\sigma_{mi}}{\partial x_l} - 4C_0^{-2}\epsilon^{-1}\sigma_{in}u_l^0\frac{\partial}{\partial x_l}\left(\epsilon^{-1}\sigma_{km}\sigma_{mi}\right) \tag{32}$$

The leading order term in the diffusion tensor is symmetric; however, the terms that are next to the leading order are not. However, the non-symmetric part of the tensor is found to make contributions of $\mathcal{O}(C_0^{-2})$ in the convection of fluid particles and admixture only: [8]. The non-symmetric part makes no contribution to the next-to-leading order terms in the diffusion coefficient.

In the above derivation, I considered the limit $t \gg \tau_c$ by which velocities de-correlated from their initial value at $t = 0$. At the same time, one can take $t \ll |u'|^2\epsilon^{-1}$, which is the time scale of the large eddies and the time scale of inhomogeneous behavior. Under this condition, the values of parameters can be represented by their values at the initial marking: $\mathbf{y}_0^t = \mathbf{x}_0$. As one can repeat the derivation for any other point of marking, one can replace $\mathbf{x}_0$ by $\mathbf{x}$: i.e., $u_i^0 = u_i^0(\mathbf{x}), \sigma_{ij} = \sigma_{ij}(\mathbf{x})$ and $\epsilon = \epsilon(\mathbf{x})$ in (32).

The diffusion equation in a non-moving Eulerian frame now follows from Equations (28) and (32) as

$$\frac{\partial p}{\partial t} + u_i^0\frac{\partial p}{\partial x_i} = \frac{\partial}{\partial x_k}\left(D_{kn}\frac{\partial p}{\partial x_n}\right) \tag{33}$$

where $p = p(\mathbf{x}, t)$ is the probability density of a marked fluid particle at position $\mathbf{x}$ and time t. The probability distribution applies equally to parameters whose values are linearly connected to the value of the particle position: i.e., concentrations of passive or almost passive admixtures, such as aerosols or the temperature in incompressible or almost incompressible fluids; see also Section 7. To determine the distributions from (33), the mean values $\mathbf{u}_i^0$, co-variances σ_{ij} and mean dissipation rates ϵ need to be known. These can be obtained using techniques of Computational Fluid Dynamics.

6. Statistical Descriptions of Momentum Flux

Momentum flux plays a central role in the conservation equations of fluid mechanics. An issue is the specification of the Reynolds stresses, i.e., the mean value of the fluctuating components of the momentum flux tensor. The conservation equations are typically formulated with respect to a fixed coordinate system, viz. the Eulerian formulation. The aim of the present analysis is to derive expressions for the mean value of the fluctuating components that fit in the Eulerian frame. First, the Lagrangian-based momentum flux tensor $v_i(t)v_j(t)$ is considered where $v_i(t)$ are the velocities of moving fluid particles that all pass at $t = 0$ through the surface at x_{j0} (alternatively, one can choose the velocity $v_j(t)$ and the surface x_{i0} but with ultimately the same Eulerian result due to the symmetry of the diffusion tensor).

Statistical averages are determined at close distance from x_{j0} using Lagrangian-based expressions for $v_i(t)$, which were derived in the previous sections. Taking the diffusion limit of the Lagrangian-based solutions and letting the distance from x_{j0} approach zero on the coarse scale of the diffusion approximation, a connection can be made with the Eulerian-based value of the tensor: $< u_i(x_0, t)u_j(x_0, t) >$. This enables the completion of the description of the averaged representation of the conservation equations of momentum for fluid mechanics.

The displacement of a marked fluid particle that is at position $\mathbf{x}_0$ at time $t = 0$ follows from (3) as

$$y_i(t) = \int_0^t u_i^0(\mathbf{y})dt + y_i'(t) + x_{i0}, \tag{34}$$

where $u^0(y)$ is value of the Eulerian mean velocity at particle position $\mathbf{y} = \mathbf{y}(t)$ and

$$y_i'(t) = \int_0^t v_i'(t)dt. \tag{35}$$

To describe the position of the particle at times close to $t = 0$, expand the r.h.s. of (35) as:

$$y_i(t) = x_{i0} + u_{i0}^0 t + y_i'(t) + \frac{\partial u_{i0}^0}{\partial x_{n0}}\left(\frac{1}{2}u_{n0}^0 t^2 + \int_0^t y_n'(t)dt\right), \tag{36}$$

where u_{i0}^0 is the Eulerian mean velocity at $\mathbf{x}_0$. Neglected terms on the r.h.s. of (36) are larger than the quadratic in t and $\mathbf{y}'$. It can be shown that these terms only contribute to $\mathcal{O}(C_0^{-3})$ in the diffusion approximation. In accordance with the above expansion, the particle velocity is described by

$$v_i(t) = \frac{dy_i(t)}{dt} = u_{i0}^0 + v_i'(t) + \frac{\partial u_{i0}^0}{\partial x_{n0}}\left(u_{n0}^0 t + y_n'(t)\right). \tag{37}$$

The objective is to describe the average momentum of particles that approach the surface at x_{j0} with velocity $v_i'(t)$. The particles are situated in an area that is small compared to the size of the large eddies so that Eulerian statistical averages can be treated as homogeneous in space. Furthermore, the area considered is large compared to the area where the particle velocities are correlated. For these conditions to be satisfied, $C_0^{-1} \ll t/\tau_E \ll 1$, which is the condition for the diffusion limit to apply. This involves a limit process whereby time approaches zero but on the time scale of coarse graining of the diffusion limit: $t \gg \tau_c, t \to 0$, where $\tau_c = C_0^{-1}\tau_E$ is the correlation time of the particle velocities.

The momentum for small negative times is given by

$$\begin{aligned} v_i(-t)v_j(-t) = {} & u_{i0}^0 u_{j0}^0 + u_{j0}^0 \frac{\partial u_{i0}^0}{\partial x_{n0}}\left(-tu_{n0}^0 + y_n'(-t)\right) + u_{j0}^0 v_i'(-t) \\ & + u_{i0}^0 \frac{\partial u_{j0}^0}{\partial x_{k0}}\left(-tu_{k0}^0 + y_k'(-t)\right) \\ & + \frac{\partial u_{i0}^0}{\partial x_{n0}}\frac{\partial u_{j0}^0}{\partial x_{k0}}\left(-tu_{n0}^0 + y_n'(-t)\right)\left(-tu_{k0}^0 + y_k'(-t)\right) \\ & + v_i'(-t)\frac{\partial u_{j0}^0}{\partial x_{k0}}\left(-tu_{k0}^0 + y_k'(-t)\right) + v_j'(-t)u_{i0}^0 \\ & + v_j'(-t)\frac{\partial u_{i0}^0}{\partial x_{n0}}\left(-tu_{n0}^0 + y_n'(-t)\right) + v_i'(-t)v_j'(-t). \end{aligned} \tag{38}$$

The average value of the momentum of all particles passing the surface x_{j0} is

$$\begin{aligned} \overline{v_i(-t)v_j(-t)}\Big|_{-t\gg\tau_c, t\to 0} = {} & u_{i0}^0 u_{j0}^0 + \overline{v_i'(-t)v_j'(-t)}\Big|_{-t\gg\tau_c, t\to 0} \\ & + \overline{v_i'(-t)y_k'(-t)}\Big|_{-t\gg\tau_c, t\to 0}\frac{\partial u_{j0}^0}{\partial x_{k0}} \\ & + \overline{v_j'(-t)y_n'(-t)}\Big|_{-t\gg\tau_c, t\to 0}\frac{\partial u_{i0}^0}{\partial x_{n0}}, \end{aligned} \tag{39}$$

with the property that $\overline{v_i'}(t) = 0$ and $\overline{y_i'}(t) = 0$ at $t = 0$. Fluid particles will cross the plane x_{j0} at different positions. However, the particles under consideration are in an area whose size is limited. The spatial variations of the mean Eulerian velocities are small and can be disregarded within the order of approximation of the developed perturbation scheme. In (39), they are taken to be equal to the value at the point of the crossing x_0 of (38).

When applying Langevin Equation (2) to negative values of t, the damping term has to change sign in order to yield the required decay with $t \to -\infty$. Hence, $v'_j(-t) = v'_j(t)$, $y'_j(-t) = -y'_j(t)$, and

$$\overline{v'_i(-t)y'_j(-t)} = -\overline{v'_i(t)y'_j(t)}. \tag{40}$$

The correlation $\overline{v'_i(t)y'_j(t)}$ can be determined in accordance with (31) and (32), where the energy dissipation rate and the co-variances can be taken to be equal to their values at x_0 under the limit process of the diffusion limit. The result equals the expression for the diffusion coefficient of (32).

$$\overline{y'_j(t)v'_i(t)}\Big|_{-t\gg\tau_c,t\to 0} = D_{ij}(x_0). \tag{41}$$

When shear and mean flow gradients are absent, an isotropic state exists, a feature that is seen in grid turbulence. In this case,

$$\overline{v'_i(-t)v'_j(-t)} = \frac{2}{3}k_0\delta_{ij}, \tag{42}$$

where k_0 is the kinetic energy of the isotropic state. Invoking (40)–(42) in (39), we have

$$\overline{v_i(-t)v_j(-t)}\Big|_{-t\gg\tau_c,-t\to 0} = u^0_{i0}u^0_{j0} + \frac{2}{3}k_0\delta_{ij} - D_{ik}\frac{\partial u^0_{j0}}{\partial x_{k0}} - D_{jk}\frac{\partial u^0_{i0}}{\partial x_{k0}} \tag{43}$$

Result (43) applies in an area where the diffusion limit holds. The area is of volume l^3 where $C_0^{-1}L \gg l \gg L$ and where $C_0^{-1}L$ is the length of velocity correlations and L is the size of the large eddies or flow configuration. The presented descriptions are valid in the limit of $C_0^{-a} \to 0$, $0 < a < 1$. The smallness of C_0^{-1} is limited: $C_0^{-1} \approx 1/7$. Yet, comparison with a range of results of measurements and direct numerical simulations shows fairly good agreement (Section 9). The reason is that terms of order C_0^{-2} are incorporated into the expansion, and the correlations decay exponentially with time C_0t/τ_E where τ_E is the eddy turnover time (cf. Equation (21)).

Reducing the volume of the area l^3 to zero, it becomes identical to a point in the Eulerian description of the flow field. We can, thus, take

$$\overline{v_i(-t)v_j(-t)}\Big|_{-t\gg\tau_c,-t\to 0} = < u_i(\mathbf{x}_0,t)u_j(\mathbf{x}_0,t) > . \tag{44}$$

Noting that

$$< u_i(\mathbf{x}_0,t)u_j(\mathbf{x}_0,t) >= u^0_{i0}u^0_{j0} + < u'_i(\mathbf{x}_0,t)u'_j(\mathbf{x}_0,t) >, \tag{45}$$

where

$$< u'_i(\mathbf{x}_0,t)u'_j(\mathbf{x}_0,t) >= \sigma_{ij}(\mathbf{x}_0) \tag{46}$$

is covariance or Reynolds stress, we have from (43)–(46)

$$\sigma_{ij}(\mathbf{x}_0) = \frac{2}{3}k_0 - D_{ik}\frac{\partial u^0_{j0}}{\partial x_{k0}} - D_{jk}\frac{\partial u^0_{i0}}{\partial x_{k0}}. \tag{47}$$

The mean value of the fluctuating kinetic energy k is given by

$$k = \frac{1}{2} < u'^2_1(\mathbf{x}_0,t) + u'^2_2(\mathbf{x}_0,t) + u'^2_3(\mathbf{x}_0,t) >= \frac{1}{2}(\sigma_{11}+\sigma_{22}+\sigma_{33}) = \frac{1}{2}\sigma_{nn}, \tag{48}$$

where repeated indices n imply summation. Substituting (47) into (48), one obtains

$$k = k_0 - D_{nk}\frac{\partial u^0_{n0}}{\partial x_{k0}}, \tag{49}$$

which can be used to eliminate k_0 from (47) with the result

$$\sigma_{ij}(\mathbf{x}_0) = \frac{2}{3}\left(k + D_{nk}\frac{\partial u_{n0}^0}{\partial x_{k0}}\right)\delta_{ij} - D_{ik}\frac{\partial u_{j0}^0}{\partial x_{k0}} - D_{jk}\frac{\partial u_{i0}^0}{\partial x_{k0}} \tag{50}$$

Similar to the analysis in the previous section, one can repeat the above procedures for any other point $\mathbf{x}_0$ and extend the results (45) and (50) to all positions $\mathbf{x}$ by replacing $\mathbf{x}_0$ by $\mathbf{x}$. The resulting statistical descriptions account for inhomogeneity of the turbulence field.

Equation (50) allows all six co-variances to be determined for given values of k, ϵ, D_{ij} and u_i^0. Having implemented Equations (32), (45), (46) and (50), the values of the mean velocities u_i^0 can be derived from the averaged versions of the equations of conservation of momentum. To obtain a closed set of equations, two equations determining k and ϵ have to be included. In this respect, it is noted that relation (48) is implied by (50) and does not represent an extra relation for k. The extra equations are provided by the two equations of the $k-\epsilon$ model that describes these variables [2,3]. The closed system of coupled equations, thus, obtained is a straightforward extension of the equations of the widely used $k-\epsilon$ model. The model may be termed the anisotropic $k-\epsilon$ model, and this enables the mean values of the statistical parameters of an anisotropic inhomogeneous turbulent flow to be calculated.

7. Statistical Descriptions of Scalar Flux

Examples of scalar flux are the dispersion of substances immersed in fluids and of temperature distributions in incompressible and almost incompressible fluids. Turbulence is known to have a significant effect on these phenomena. Similar to the analysis of the previous section, consider an area that is small to the area of inhomogeneity but large compared to the area where particle velocities are correlated. The Lagrangian scalar flux is described by $v_i(t)\phi(t)$, where $v_i(t)$ are the velocities of marked fluid particles that all pass at $t=0$ through a surface at x_{j0}, and $\phi(t)$ is the value of the scalar quantity at the position of each moving particle. When considering the velocities of particles at a short distance, the time from the surface of passing (37) can be employed. For the value of the scalar quantity at the position of the particle, we have

$$\phi(t) = \theta_0^0 + (y_n' + u_{n0}^0 t)\frac{\partial \theta_0^0}{\partial x_{n0}} + \phi'(t). \tag{51}$$

The first and second term on the right-hand side represent dispersion of the scalar quantity due to fluid particle displacement whereby the scalar does not vary in magnitude while moving with the fluid particle. The third term is autonomous random changes of the value of the scalar quantity while moving with the fluid particle. For the first term, take the Eulerian-based mean value at $\mathbf{x}_0$. Similar to the analysis of the previous section, particles pass through different positions at the surface x_{j0}. However, all these positions are at a limited distance from each other in accordance with the coarse graining of the diffusion limit.

On this scale, spatial variations in value of the first term of the expansion can be disregarded. They can be taken to be equal to the Eulerian mean value at the single point $\mathbf{x}_0$. Furthermore, the first term is allowed to vary deterministically with time t^*, where t^* is the time in the Eulerian fixed frame of reference: $\theta_0^0 = \theta_0^0(x_0, t^*)$. Lagrangian averaging can take place by adding the simulation results of the Langevin equations at a fixed value of t^* and subsequently repeating for every other value of t^*. The variation with t^* is considered to be slow compared to the rapid variation of the random fluctuations of τ_c.

Multiplying the right-hand sides of Equations (37) and (51), replacing t by $-t$, applying Lagrangian averaging and letting $-t \to 0$ after applying the diffusion limit, one obtains (similar to the procedure of the previous section)

$$\overline{\phi(-t)v_i(-t)}\Big|_{-t\gg\tau_c,-t\to 0} = \theta_0^0 u_{i0}^0 + \overline{y_n'(-t)v_i'(-t)}\Big|_{-t\gg\tau_c,-t\to 0}\frac{\partial\theta_0^0}{\partial x_{n0}} + \overline{\phi'(-t)y_n'(-t)}\Big|_{-t\gg\tau_c,-t\to 0}\frac{\partial u_{i0}^0}{\partial x_{n0}} + \overline{\phi'(-t)v_i'(-t)}\Big|_{-t\gg\tau_c,-t\to 0}. \tag{52}$$

The third and fourth terms on the right-hand side are contributions due to autonomous fluctuation of the scalar when moving with the fluid particles. To determine the values of these terms, fluctuation equations of $\phi(t)$ need to be known. No attempts will be made to derive such equations. We assume that $\phi'(-t)$ is a conserved quantity whose value does not change while moving with the fluid particle: $\phi'(-t) = 0$. Noting that

$$\overline{y_n'(-t)v_i'(-t)} = -\overline{y_n'(t)v_i'(t)} \tag{53}$$

and implementing (41) then yields

$$\overline{\phi(-t)v_i(-t)}\Big|_{-t\gg\tau_c,-t\to 0} = \theta_0^0 u_{i0}^0 - D_{in}\frac{\partial\theta_0^0}{\partial x_{n0}}, \tag{54}$$

where $\overline{\phi(-t)v_i(-t)}$ equals the Eulerian-based value at $\mathbf{x}_0$. As the relation holds for every position x, we have

$$<\theta u_i> = \theta^0 u_i^0 - D_{in}\frac{\partial\theta^0}{\partial x_n}, \tag{55}$$

where θ is a conserved quantity that satisfies the Eulerian-based conservation equation

$$\frac{\partial\theta}{\partial t^*} + \frac{\partial}{\partial x_i}(\theta u_i) = 0. \tag{56}$$

Applying equation ensemble averaging to the above, substituting Equation (55) and replacing t^* by t, yields

$$\frac{\partial\theta^0}{\partial t} + u_i^0\frac{\partial\theta^0}{\partial x_i} = \frac{\partial}{\partial x_i}\left(D_{in}\frac{\partial\theta^0}{\partial x_n}\right), \tag{57}$$

where I employed the averaged version of continuity: $(\partial/\partial x_i)u_i^0 = 0$. The above result equals the equation for fluid particle distribution given by Equation (33). This is consistent with the feature that the distribution of the particle must be equal to the distribution of a conserved quantity whose value does not change with value of $\mathbf{x}$ following the path of a fluid particle.

8. Decaying Grid Turbulence

Decaying grid turbulence has been studied many times during the previous century, and many results are available. Turbulence is generated by a uniform mean flow that passes through a grid of squarely spaced bars. The grid is perpendicular to the incoming mean flow. At some distance behind the grid, a homogeneous field of isotropic turbulence develops and decays in the downstream direction with the mean flow. For grid turbulence, exact results for the Langevin and diffusion equations are known. In the present section, I recapitulate these results and compare them with the present results based on the two-term C_0^{-1} expansion.

For convenience in presentation, turbulence is described in a frame that moves with the uniform mean velocity. I thus describe the equivalent situation where the grid moves with constant speed from right to left through a fluid that is initially at rest. When the grid has passed, a uniform field of isotropic zero-mean Gaussian fluctuations exists, which decays in time. The strength is the same in all three coordinate directions. Therefore, analysis is restricted to fluctuations in one direction only. Corresponding variables are indicated by a subscript of 1. Regarding the independence of fluctuations in three directions, a problem of nonuniqueness, as discussed in Section 5, does not exist. The appropriate Langevin equation in one-dimensional form can be written as

$$\frac{dv'_1}{dt} = -\frac{1}{2}\left(\frac{\epsilon C_0}{\sigma_1} - \frac{1}{\sigma_1}\frac{\partial \sigma_1}{\partial t}\right)v'_1 + (\epsilon C_0)^{1/2}w(t), \tag{58}$$

where σ_1 is the Eulerian mean square of the fluctuations

$$\sigma_1 =< u_1'^2 > . \tag{59}$$

Expressions for σ_1 and ϵ can be derived from the Von Karman-Howarth equation for conservation of the mean kinetic energy of fluctuations. For large Reynolds numbers, these expressions are: [19]

$$\epsilon = \epsilon_0(t/t_0)^{-2}; \quad \sigma_1 = \sigma_{10}(t/t_0)^{-1}; \quad \epsilon_0 = \frac{3}{2}\sigma_{10}t_0^{-1}, \tag{60}$$

where t_0 is the reference time, i.e., a moment in time where the grid has passed the observer at a fixed position. The value of σ_{10} depends on the dimensioning of the grid and can be established by measurements at time t_0. Implementing (60) into (58), we have

$$\frac{dv'_1}{dt} = -\frac{\epsilon}{2\sigma_1}(C_0 + 2/3)v'_i + (C_0\epsilon)^{1/2}w(t). \tag{61}$$

From this result, one can derive (analogous to the derivation in Section 5) the diffusion equation

$$\frac{\partial p}{\partial t} = D_1\frac{\partial^2 p}{\partial x^2}, \tag{62}$$

where the diffusion coefficient is given by

$$D_1 = \frac{2\sigma_{10}^2}{\epsilon_0 C_0(1 + (2/3)C_0^{-1})}. \tag{63}$$

Note that the diffusion coefficient does not decrease in the stream-wise direction. A decay in the strength of fluctuations is compensated for by an increase in the correlation time.

Results (61)–(63) were obtained without using the C_0^{-1} expansion. The Langevin equation according to a two term C_0^{-1} expansion is given by Equations (2) and (20). Introducing the features of grid turbulence results in an equation that is the same as Equation (61). The diffusion coefficient according to the two term C_0^{-1} expansion is given by Equation (32). This reduces, in the case of grid turbulence, to $D_1 = 2\sigma_{10}^2(1 - (2/3)C_0^{-1})/(\epsilon_0 C_0)$. Expanding the exact result given by Equation (63) in powers of C_0^{-1} yields $D_1 = 2\sigma_{10}^2(1 - (2/3)C_0^{-1} + (4/9)C_0^{-2} + ...)/(\epsilon_0 C_0)$.

The first two terms in the diffusion coefficient of the exact result thus agree with the two-term C_0^{-1} expansion. The third term amounts to a relative contribution of 0.9% when $C_0 = 7$. In conclusion, the two-term expansion complies with the corresponding expansion of the exact result, and the error of truncating the third term is small. The latter conclusion is, however, of limited value as the grid turbulence is isotropic and only slightly inhomogeneous in the stream-wise direction. In practice, turbulence is mostly anisotropic and appreciably inhomogeneous. The next section analyses such a case.

9. Turbulent Channel Flow

Turbulence is a well-known feature of flows in pipes and channels and in boundary layers along walls, including the boundary layers along the earth's surface. A representative case for such flows is a developed turbulent flow in a channel of two parallel flat plates. The statistical values are constant in the direction of the mean flow between the plates and in the direction that is parallel to the plates and perpendicular to the mean flow but changes significantly in magnitude in the direction normal to the plates. The fluctuations are strongly anisotropic.

9.1. Exact Results

Some exact results can be derived from the averaged Navier–Stokes (N-S) Equations: ([1] vol I, p. 268). The averaged equations are given by

$$\frac{\partial}{\partial x_i} < u_i u_j >= -\frac{1}{\rho}\frac{\partial < p >}{\partial x_j} + \nu \frac{\partial^2 < u_j >}{\partial x_i \partial x_i}, \tag{64}$$

where p is the pressure relative to the pressure of the fluid at rest, and ρ is the density. In the case of a developed turbulent channel flow, the mean values involving fluctuating velocities and pressure gradients vary only with the wall normal coordinate x_2. The averaged N-S equations then reduce to

$$-\frac{1}{\rho}\frac{\partial p^0}{\partial x_1} = \frac{\partial}{\partial x_2}\sigma_{12} \tag{65}$$

$$-\frac{1}{\rho}\frac{\partial p^0}{\partial x_2} = \frac{\partial}{\partial x_2}(\sigma_{22}), \tag{66}$$

where x_1 is the coordinate of the mean flow direction, $\sigma_{ij} =< u_i' u_j' >$ are the co-variances of fluctuating velocities, and $p^0 =< p >$ is the mean pressure. The contribution of the viscous stress represented by the last term in Equation (64) was disregarded in the above equations. The effect is limited to thin viscous layers near the wall. Their effect on the flow outside these thin layers can be accounted for by the boundary condition imposed on the shear stress σ_{12} at the wall. From Equations (65) and (66), one obtains the solutions

$$\frac{1}{\rho}p^0 = -u_\tau^2 x_1/H - \sigma_{22} \tag{67}$$

$$\sigma_{12} = -u_\tau^2(1 - x_2/H), \tag{68}$$

where u_τ is the shear velocity and $2H$ is the distance between the parallel plates. The shear velocity u_τ can be related to the pressure drop in the channel by solving the flow in the boundary layer at the wall. The relationship is also known from measurements: e.g., [1]. The value of u_τ is representative for the magnitude of the fluctuations.

9.2. Results from the C_0^{-1}-Expansion

The exact results of Section 9.1 can be extended by supplementing the expressions for the turbulent momentum diffusion of Equation (50). For the channel flow, these become

$$\sigma_{22} = \frac{2}{3}k + \frac{2}{3}D_{12}\frac{du_1^0}{dx_2} \tag{69}$$

$$\sigma_{33} = \sigma_{22} \tag{70}$$

$$\sigma_{11} = \frac{2}{3}k - \frac{4}{3}D_{12}\frac{du_1^0}{dx_2} \tag{71}$$

$$\sigma_{12} = -D_{22}\frac{du_1^0}{dx_2} \tag{72}$$

where $u_1^0 = u_1^0(x_2)$ is the mean flow in the channel and where the diffusion coefficients are given by

$$D_{12} = \frac{2}{\epsilon C_0}\sigma_{12}(\sigma_{11} + \sigma_{22}) \tag{73}$$

$$D_{22} = \frac{2}{\epsilon C_0}\left(\sigma_{12}^2 + \sigma_{22}^2\right) \tag{74}$$

Equations (67)–(74) constitute eight relations for 10 variables: $p^0, \sigma_{11}, \sigma_{22}, \sigma_{33}, \sigma_{12}, k, D_{12}, D_{22}, \epsilon$ and u_1^0. A closed system of equations requires two extra equations. These are provided by the conservation equations for kinetic energy k and dissipation rate ϵ known from CFD models [2,3]. Our aim is not to study a complete and closed system of equations but to analyze all those components that describe turbulent transport in such equations. For this purpose, one can calculate the values of the left-hand sides and right-hand sides of the developed relations using the data of direct numerical simulations and compare them with each other. This provides a direct test of the outcome of the C_0^{-1} expansion. An article in which the complete set of equations is formulated and analyzed is in preparation.

9.3. Comparison with the DNS Results

Super computers have created the possibility to simulate turbulent fluid flows through direct numerical simulations (DNS) of the equations that govern fluid flow, i.e., the Navier–Stokes equations. Initially, attention was focused on grid turbulence at modest values of Reynolds numbers. The calculation power has increased with time. This allows handling flows at larger Reynolds numbers and with more complex configurations of channel flow. Hoyas et al. [11] recently published results for channel flow at a friction Reynolds number Re of 10^4. This corresponds to a bulk flow Reynolds number of about 3×10^5. DNS is the most reliable technique to study turbulence, and its outcome can be considered as exact. The results of Hoyas et al. provide an excellent opportunity to verify the present results.

9.3.1. Statistical Values of Fluctuations

Making u_1^0 dimensionless by u_τ, σ_{ij} and k by u_τ^2, x_2 by H and P and ϵ by u_τ^3/H, and dropping the subscript 2 from x_2, one can derive, from Equations (67)–(74), the relations

$$\sigma_{22} = \sigma_{33} = (1-x)(\gamma^{-1}-1)^{1/2} \tag{75}$$

$$\sigma_{11} = \sigma_{22}(1+2\gamma)/(1-2\gamma) \tag{76}$$

where

$$\gamma = \frac{2}{C_0} P/\epsilon \tag{77}$$

and P is the production of energy defined as

$$P = (1-x)\frac{d}{dx}u_1^0 \tag{78}$$

From Equations (75)–(77), it can be verified that, at $x = 1$: $\sigma_{11} = \sigma_{22} = \sigma_{33} = \frac{2}{3}k_0$. This is consistent with the solution for a zero mean flow gradient. At $x = 0$, the solutions for the log law apply, according to which, $P/\epsilon = 1$. From Equations (75)–(77), one then finds

$$\sigma_{22} = \sigma_{33} = (\frac{C_0}{2} - 1)^{1/2} \quad \text{at } x = 0 \tag{79}$$

$$\sigma_{11} = \sigma_{22}(C_0+4)/(C_0-4) \quad \text{at } x = 0 \tag{80}$$

These reveal anisotropy whose magnitude depends on the magnitude of C_0.

Figure 1 shows the values of the root mean square of fluctuations $\sigma_{11}, \sigma_{22}, \sigma_{33}$ according to Equations (75)–(77) versus x for P/ϵ taken from DNS and $C_0 = 7$. The values are compared with the corresponding DNS values of these parameters. Close to the wall at $x = 0$, the effect of the viscous layer is seen. Its thickness is about $100/Re_\tau$, which amounts to 1% of the height of the channel. The results of the C_0^{-1} expansion only apply outside this area. Here, it is seen that strong anisotropy in the longitudinal direction is predicted.

A difference between fluctuations in normal and the span-wise direction as forecast by DNS is not revealed. Differences between the longitudinal fluctuations and normal fluctuations near the axis $x = 1$ are not revealed either. Near $x = 0$, differences between rms values in the normal and span-wise direction are at maximum. The differences between longitudinal and transverse fluctuations are at a maximum at $x = 1$. Otherwise, the differences between the DNS and C_0^{-1} expansion are rather limited—keeping in mind the limited smallness of the perturbation parameter C_0^{-1}.

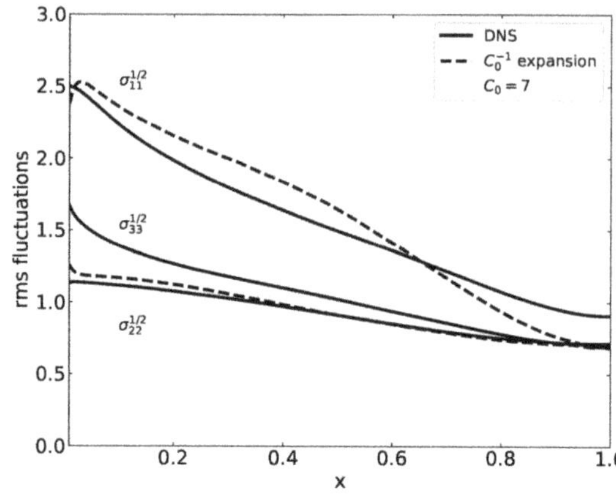

Figure 1. The root mean square values of the velocity fluctuations versus the dimensionless distance from the wall. The root mean square values obtained from DNS are represented by full lines. The root mean square values of the C_0^{-1} expansion are represented by broken lines. They result from Equations (75) and (76) in which the right-hand sides were evaluated using the DNS values. Differences between full and broken lines can be ascribed to truncation of the C_0^{-1} expansion.

9.3.2. Statistical Values of Turbulent Fluxes

An issue in turbulence theory is the statistical description of the non-linear fluctuating convective terms in the equations of conservation of momentum and energy. The issue is known as the closure problem. The present analysis provided an answer by the expressions for turbulent flux and turbulent diffusion coefficients. Figures 2 and 3 present the results obtained for these terms and compare them with the DNS results.

Turbulent fluxes according to the C_0^{-1} expansion are present in the descriptions of the diffusion terms of Equations (69)–(71). The accuracy of these descriptions has been been tested by the DNS results. The results are shown in Figure 2.

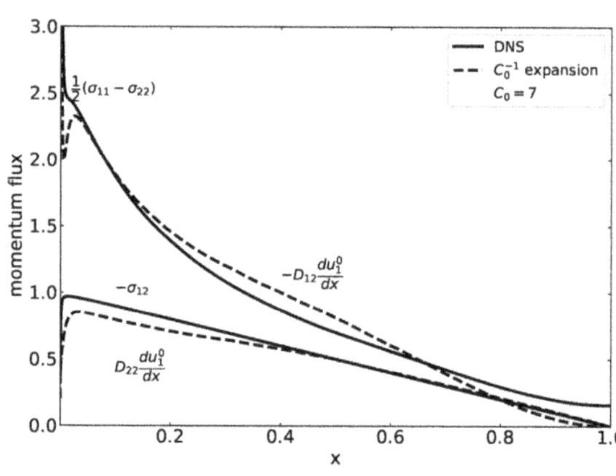

Figure 2. The mean momentum fluxes versus the dimensionless distance from wall. The values of σ_{11}, σ_{22} and σ_{12} are obtained using DNS and are represented by full lines. The values of $D_{12}(d/dx)u_1^0$ and $D_{22}(d/dx)u_1^0$ result from the C_0^{-1} expansion and are represented by broken lines. They follow from Equations (69), (71) and (72) in which the right-hand sides were evaluated using the DNS values.

9.3.3. Diffusion Coefficients

Coefficients of diffusion in the wall normal direction are compared in Figure 3. Diffusion of both momentum using the data of Hoyas et al. [11] and of the conserved scalar temperature using the DNS data of Kuerten et al. [12] are analyzed.

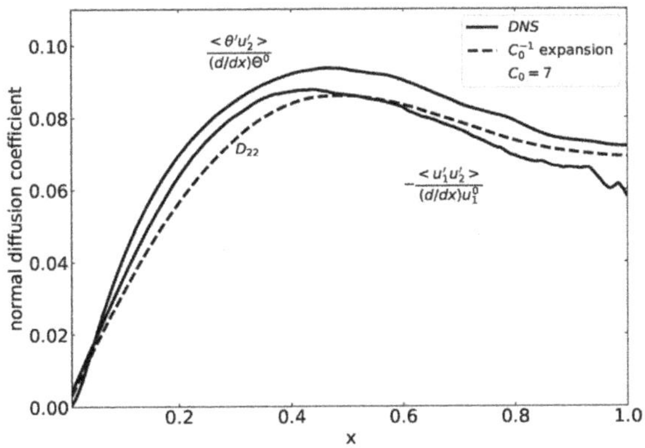

Figure 3. Coefficients of diffusion in the wall normal direction for momentum transport $< u_1' u_2' >$ and heat transport $< \theta' u_2' >$ versus dimensionless distance from wall. The values of $< u_1' u_2' > /(d/dx)u_1^0$ and $< \theta' u_2' > /(d/dx)\theta^0$ are obtained using DNS and are presented by full lines. The values of D_{22} result from the C_0^{-1} expansion. They follow from Equation (74) in which the right-hand side was evaluated using the DNS values. They are represented by a broken line.

9.3.4. Kolmogorov Constant

In general, it is found that a value of 7 for C_0 gives a good fit to the DNS. This value is somewhat higher than the value of 6.2 mentioned previously considering the DNS of turbulent channel flow at *Re* of 0.2×10^4 [8]. A value of C_0 of 7 at high Reynolds number has been claimed by Sawford referring to DNS of grid turbulence [16].

10. Conclusions

The presented statistical descriptions fit within the asymptotic structure of turbulence at a large Reynolds number with Kolmogorov theory. The descriptions apply to large scales, which determine the main flow outside small viscous boundary layers at adjoining walls of the configuration considered. The given representations of the velocity and position of marked fluid particles are Lagrangian-based and concern Langevin and diffusion equations. In these equations, the universal Kolmogorov constant C_0 appears with a value of about 7. This is used as an autonomous parameter in developing solutions by the first two terms of perturbation expansions in powers of C_0^{-1}. The leading solution complies with the conditions of Hamiltonian dynamics, Gaussian behavior and Onsager symmetry as $C_0^{-1} \to 0$. The second term of the solution satisfies mixing with the Eulerian-based statistical distribution of the flow field.

The Lagrangian-based descriptions were connected to Eulerian statistics through asymptotic matching. When considering a small but sufficiently large area around a fixed point in space where the diffusion limit applies, shrinking this area to a point accomplishes matching the Eulerian description at the corresponding point. The matching involves the limit process $C_0^{-a} \to 0$, $0 < a < 1$ where $0 < a$ is required for obtaining the diffusion limit, and $a < 1$ to ensure that the considered area is much smaller than the area of inhomogeneous behavior of the main flow.

The two-term descriptions meet the requirements that follow from the laws of physics and the methods of stochastic analysis. The presented solutions reveal the functional relationships between the statistical averages of various fluctuating quantities, such as turbulent diffusivity. They do not rely on semi-empirical hypotheses and fitted constants. Limiting factors include inaccuracies due to truncation of the higher order terms. For slowly decaying grid turbulence, these are small. However, in the case of strong inhomogenity, the matching of Lagrangian and Eulerian results appears to require small values of C_0^{-a}, $0 < a < 1$. Yet, comparison with the DNS of turbulent channel flow at high Reynolds number reveals deviations of limited magnitude despite large inhomogeneity and anisotropy.

An underlying reason is likely the inclusion of next to leading terms and exponential decay of velocity correlations by $C_0 t / r_E$, where τ_E is the eddy turnover time or characteristic time of large-scale turbulence: (21).

The results for the main Eulerian statistical parameters in the case of channel flow are shown in Figures 1–3. They reveal fairly good agreement between the predictions of the C_0^{-1} model when compared with those of DNS. This conclusion applies to turbulent channel flow, which is a case of turbulence that is significantly anisotropic and inhomogeneous. As the C_0^{-1} model has a general basis, this entails the prospect of yielding reliable results for other cases of anisotropic inhomogeneous turbulent flow.

Funding: This research received no external funding.

Informed Consent Statement: Not applicable.

Data Availability Statement: Not applicable.

Acknowledgments: B.G.J. Ruis is acknowledged for processing the DNS data of [10,11] and preparing the manuscript.

Conflicts of Interest: The author declares no conflict of interest.

References

1. Monin, A.S.; Yaglom, A.M. *Statistical Fluid Mechanics*; Dover: New York, NY, USA, 2007; Volumes 1–2.
2. Bernard, P.S.; Wallace, J.K. *Turbulent Flow: Analysis, Measurement and Prediction*; Wiley: New Jersey, NJ, USA, 2002.
3. Hanjalić, K.; Launder, B. *Modelling Turbulence in Engineering and the Environment: Second-Moment Routes to Closure*; Cambridge University Press: Cambridge, UK, 2011. [CrossRef]
4. Stratonovich, R.L. *Topics in the Theory of Random Noise*; Gordon and Breach: New York, NY, USA, 1967; Volume 1.
5. van Kampen, N.G. *Stochastic Processes in Physics and Chemistry*, 3rd ed.; Elsevier: New York, NY, USA, 2007.
6. Brouwers, J.J.H. Langevin and diffusion equation of turbulent fluid flow. *Phys. Fluids* **2010**, *22*, 085102. [CrossRef]
7. Brouwers, J.J.H. Statistical description of turbulent dispersion. *Phys. Rev. E Stat. Nonlinear Soft Matter Phys.* **2012**, *86*, 066309. [CrossRef] [PubMed]
8. Brouwers, J.J.H. Statistical Models of Large Scale Turbulent Flow. *Flow Turbul. Combust.* **2016**, *97*, 369–399. [CrossRef]
9. Brouwers, J.J.H. Statistical Model of Turbulent Dispersion Recapitulated. *Fluids* **2021**, *6*, 190. [CrossRef]
10. Hoyas, S.; Jiménez, J. Scaling of the velocity fluctuations in turbulent channels up to Reτ=2003. *Phys. Fluids* **2006**, *18*, 011702. [CrossRef]
11. Hoyas, S.; Oberlack, M.; Alcántara-Ávila, F.; Kraheberger, S.V.; Laux, J. Wall turbulence at high friction Reynolds numbers. *Phys. Rev. Fluids* **2022**, *7*, 014602. [CrossRef]
12. Kuerten, J.; Brouwers, J.J.H. Lagrangian statistics of turbulent channel flow at Reτ = 950 calculated with direct numerical simulation and Langevin models. *Phys. Fluids* **2013**, *25*, 105108. [CrossRef]
13. Kolmogorov, A. The Local Structure of Turbulence in Incompressible Viscous Fluid for Very Large Reynolds' Numbers. *Akad. Nauk. Sssr Dokl.* **1941**, *30*, 301–305.
14. Borgas, M.S. The Multifractal Lagrangian Nature of Turbulence. *Philos. Trans. Phys. Sci. Eng.* **1993**, *342*, 379–411.
15. Thomson, D.J. Criteria for the selection of stochastic models of particle trajectories in turbulent flows. *J. Fluid Mech.* **1987**, *180*, 529–556. [CrossRef]
16. Sawford, B.L. Reynolds number effects in Lagrangian stochastic models of turbulent dispersion. *Phys. Fluids A Fluid Dyn.* **1991**, *3*, 1577–1586. [CrossRef]
17. Reichl, L.E. *A Modern Course in Statistical Physics*; Wiley-VCH: Hoboken, NJ, USA, 2004.
18. Brouwers, J.J.H. On diffusion theory in turbulence. *J. Eng. Math.* **2002**, *44*, 277–295. [CrossRef]
19. George, W.K. The decay of homogeneous isotropic turbulence. *Phys. Fluids A* **1992**, *4*, 1492–1509. [CrossRef]

mathematics MDPI

Article

Nonlinear Mixed Convection in a Reactive Third-Grade Fluid Flow with Convective Wall Cooling and Variable Properties

Samuel Olumide Adesanya [1,*], Tunde Abdulkadir Yusuf [2] and Ramoshweu Solomon Lebelo [3]

1 Department of Mathematics and Statistics, Faculty of Natural Sciences, Redeemer's University, Ede 232101, Nigeria
2 Department of Mathematics, Faculty of Science, Adeleke University, Ede 232104, Nigeria
3 Education Department, Vaal University of Technology, Private Bag X021, Vanderbijlpark 1911, South Africa
* Correspondence: adesanyas@run.edu.ng

Abstract: Energy management and heat control whenever a reactive viscous fluid is the working medium has been one of the greatest challenges encountered by many in the field of chemical and industrial engineering. A mathematical approach to thedetermination of critical points beyond which the working environment becomes hazardous is presented in the present investigation together with the entropy generation analysis that guarantees the efficient management of expensive energy resources. In this regard, the nonlinear mixed convective flow behavior of a combustible third-grade fluid through a vertical channel with wall cooling by convection is investigated. The mathematical formulation captures the nonlinearities arising from second-order Boussinesq approximation and exponential dependence of internal heat generation, viscosity, and thermal conductivity on temperature. The resulting nonlinear boundary value problems were solved based on the spectral Chebyshev collocation method (SCCM) and validated with the shooting-Runge–Kutta method (RK4). The nonlinear effects on the flow velocity, temperature distribution, entropy generation, and Bejan heat irreversibility ratio are significant. Further analyses include the thermal stability of the fluid. Findings from the study revealed that flow, temperature, and entropy generation are enhanced byincreasing values of the Grashof number, the quadratic component of buoyancy, and the Frank-Kameneskii parameter, but are reducedbyincreasing the third-grade material parameter. Moreover, it was shown that increasing values of the third-grade parameter encourages the thermal stability of the flow, while increasing values of the linear and nonlinear buoyancy parameter destabilizes the flow. The present result is applicable to thick combustible polymers with increased molecular weight.

Keywords: thermal stability; entropy generation; nonlinear buoyancy; variable properties

MSC: 37M20

Citation: Adesanya, S.O.; Yusuf, T.A.; Lebelo, R.S. Nonlinear Mixed Convection in a Reactive Third-Grade Fluid Flow with Convective Wall Cooling and Variable Properties. *Mathematics* **2022**, *10*, 4276. https://doi.org/10.3390/math10224276

Academic Editor: Vasily Novozhilov

Received: 6 October 2022
Accepted: 9 November 2022
Published: 15 November 2022

1. Introduction

A renewed interest in energy utilization and management has led to a significant increase in studies centered on a broad range of reactive non-Newtonian fluids in the last few decades. A good number of researchers gave mathematical explanations from a fluid dynamics viewpoint on how best to maximize exergy when working with reactive fluids. For instance, Salawu et al. [1] reported the transient flow of a reactive fourth-order fluid through a horizontal channel

Permeated by a magnetic field, Cuiet al.in [2] investigated the two-dimensional reacting flow of hydromagnetic nano-Oldroyd-B fluid driven by mixed convection. In [3], Salawu and his cohorts considered the slip flow of combustible MHD couple stress fluids with variable properties. Sadiq and Hayat [4] presented a model for optimizing entropy generation in Reiner–Rivlin flow over a stretched rotating disc. Okoya [5] reported a numerical investigation of a time-independent variable viscous third-grade fluid flow through a cylinder. Adesanya et al. [6] analyzed the flow of third-grade fluid between two solid boundaries under Arrhenius kinetics. The study in [7] focused on the reactive flow under the Erying–Powell constitutive model. More studies on reactive flows of third-grade fluid can be found in [8–14] and the literature cited therein.

The quantity of energy absorbed by the flow is defined by the degree of friction caused by viscosity, which plays a crucial role in the creation of fluid measures for flow measurements. The changing viscosity in Couette and Poiseuille flows of a tangential hyperbolic fluid via an inclined channel with porous media was explored by Zehra et al. [15]. The fluid's viscosity has been assumed in this research to be inversely proportional to a linear function of pressure. It should go without saying that the viscosity and thermal conductivity of any fluid may be alteredas the temperature rises. This phenomenon therefore plays a crucial role in the rate of heat transport near the surface, much as other thermophysical features. It is acknowledged that certain characteristics, particularly fluid viscosity, may vary with temperature. This change in viscosity and thermal conductivity must be taken into consideration in order to correctly forecast the flow and heat transfer rates. Jeffrey fluid flow, heat transfer, and mass transfer via a permeable wave-like channel with variable viscosity and thermal conductivity were studied by Manjunatha et al. in their study [16]. Qasim et al.'s [17] investigation on the impact of dissipation of a fluid flow through a tiny needle takes these fluid parameters into account. Recently, Saraswathy et al. [18] investigated these phenomena in order to use a numerical approach to investigate the solution for an asymmetric flow and heat transfer of a rotating fluid with the influence of Arrhenius energy.

Additionally, while the first law of thermodynamics has been used by certain writers, its efficacy has been shown to be inferior to that of the second law. Thermal efficiency decreases and system entropy increases as a result of energy losses in engineering and thermal equipment. It is well known that the development of entropy during any thermal process determines the degree of irreversibility. The optimization procedures have recently paid a lot of attention to this examination. Many thermal systems may include irreversible processes. The second law of thermodynamics has been extensively employed in the literature to enhance these types of irreversibility. In this light, Khan et al. [19] analyzed the formation of entropy on an Erying–Powell liquid flowing via a permeable channel using a homotopy analysis technique. Singh et al. [20] investigated the entropy generation impact in a micropolar fluid flowing through an aligned tube. Here, it is claimed that the fluid's viscosity and thermal conductivity both change with temperature. The conclusion of their analysis shows that temperature-dependent thermal and viscous conductivity factors enhance the entropy production profile. Similar studies on entropy production rates and irreversibility in convection-based heat transport may be found in references [21–23].

In modeling the flow driving mechanism, the nonlinear temperature-density relationship has been established over the years to give a better approximation than the linearized form of heat transfer in a wide range of buoyancy-driven flows due to the considerable difference in the fluid temperature and the ambient. Given these numerous applications, Adesanyaet al. [24] focused on the entropy generation, flow, and nonlinear heat transfer to a couple stress fluid flow through aporous medium. In the Xiaet al. [25] report, a mathematical explanation for the nonlinear bio-convective flow of hydromagnetic fluid containing microorganisms with Hall effect is discussed. Ibrahim and Gadisa [26] presented a finite element residual approximation for Oldroyd-B flow with internal heat generation. Patilet al. [27] constructed the numerical approximation to nonlinear convective flow problems with Brownian and thermophoresis. Waqaset al. [28] reported the steady developing flow of Williamson fluid with variable diffusion and thermal conductivity. A semi-analytical approach is employed by Yusuf et al. [29] to examine the magnetohydrodynamic (MHD) nonlinear convective flow of a reactive non-Newtonian fluid model with convective heating. Other vital studies on nonlinear convective flows include refs. [30–32] and the cited references.

Despite the huge body of literature surrounding heat irreversibility in a combustible non-Newtonian fluid flow, the nonlineardependence of some fluid parameters on temperatureremains poorly understood with respect to thermal stability for the determination of safe working conditions and energy management or efficiency. The proposed nonlinear coupled model will ensure that the thermal system is not underdetermined when used to predict buoyancy-induced flow and thermal behavior. The nonlinearity of the model suggests the non-existence of a close-form solution; therefore, the present work focuses on a fast-converging numerical approach totackling the coupled nonlinear convective flow and heat transfer problem. To the best of our knowledge, the problem presented here has not been undertaken byany researcher, and that explains the novelty of the work.

Including the nonlinear buoyancy approximation, nonlinear internal heat, and exponential dependence of fluid viscosity and thermal conductivity is expected to produce a better approximation of reactive flow and heat transfer problems, especially during a number of engineering processes, such as advanced powder synthesis by self-propagating high-temperature, heat exchangers, loop/fluidized bed reactors, treatment of reactive hydrocarbons, combustion chambers and many more engineering applications. In the next section, the mathematical analysis of the problem is presented.

2. Mathematical Formulation

In the model formulation, we consider a fully-developed flow of a reactive third-grade fluid flow through a vertical channel with Newtonian cooling at the walls. The buoyancy effect is felt upstream in the vertical x-axis while the y-axis is perpendicular to it, as shown in Figure 1. All the fluid properties like viscosity, thermal conductivity, and internal heat generation are assumed to vary with temperature in an exponential manner, except density which has quadratic dependence.

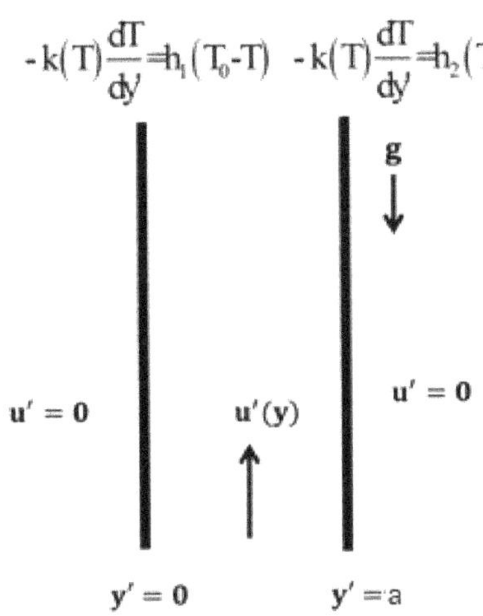

Figure 1. Flow geometry.

Under the flow assumptions, the vector form for the momentum equation becomes:

$$\rho \frac{DV}{Dt} = \rho g + div\tau \tag{1}$$

where the operator, $\frac{D(.)}{Dt} = \left(\frac{\partial}{\partial t} + V.\nabla\right)(.)$ is the material derivate acting on the velocity vector V, and ρ, g-density and gravitational force, respectively. The last term τ is the Cauchy stress tensor defined as [14,33]

$$\tau = -PI + \mu S_1 + \alpha_1 S_2 + \alpha_2 S_1^2 + \beta_1 S_3 + \beta_2(S_1S_2 + S_2S_1) + \beta_3\left(trS_1^2\right)S_1 \tag{2}$$

In (2), α_n, $(n = 1,2,3), \beta_n$, $(n = 1,2,3)$ are material effects, (P, I, μ) are the pressure, tensor identity and fluid viscosity, respectively while S_n, $(n = 1,2,3)$ represents the Rilvin–Ericksen tensors and are defined as

$$\begin{aligned} S_1 &= \nabla V + (\nabla V)^T \\ S_n &= \frac{DS_{n-1}}{Dt} + S_{n-1}\nabla V + (\nabla V)^T S_{n-1}, \quad n > 1 \end{aligned} \tag{3}$$

The Clausius–Duhem inequality condition $\mu \geq 0,\ \alpha_1 \geq 0\,, |\alpha_1 + \alpha_2| \leq \sqrt{24\mu\beta_3}$ ensures the thermodynamic process is consistent when $\beta_1 = \beta_2 = 0$ implies that $\beta_3 \geq 0$.

Therefore, (2) reduces to

$$\tau = -PI + \left(\mu + \beta_3\left(trS_1^2\right)\right)S_1 + \alpha_1 S_2 + \alpha_2 S_1^2 \tag{4}$$

From (4), it is easy to note that the effective viscosity, which is assumed to be a nonlinear function of temperature, becomes

$$\mu_{eff} = \mu(T) + \beta_3\left(trS_1^2\right) \tag{5}$$

As a consequence of (5), the momentum equation with the nonlinear temperature-dependent fluid density becomes

$$0 = -\frac{dP}{dx} + \frac{d}{dy'}\left(\mu(T)\frac{du'}{dy'} + 2\beta_3\left(\frac{du'}{dy'}\right)^3\right) + \rho g\phi_0(T - T_0) + \rho g\phi_1(T - T_0)^2 \tag{6}$$

While the balanced energy equation with variability in thermal conductivity becomes

$$0 = \frac{d}{dy'}\left(k(T)\frac{dT}{dy'}\right) + QC_0A\left(\frac{hT}{vl}\right)^m e^{-\frac{E}{RT}} + \left(\mu(T) + 2\beta_3\left(\frac{du'}{dy'}\right)^2\right)\left(\frac{du'}{dy'}\right)^2 \tag{7}$$

The fact that the heat transfer to the viscous non-Newtonian fluid is irreversible suggests that the entropy generation profile is a nonlinear combination of heat transfer and fluid friction, that is,

$$E_G = \frac{k(T)}{T_0^2}\left(\frac{dT'}{dy'}\right)^2 + \frac{1}{T_0}\left(\frac{du'}{dy'}\right)^2\left(\mu(T) + 2\beta_3\left(\frac{du'}{dy'}\right)^2\right) \tag{8}$$

Togetherwith the essential and temperature dependent natural boundary conditions

$$\begin{aligned} u' &= 0, \quad -k(T)\tfrac{dT}{dy'} = h_1(T_0 - T) \qquad y' = 0 \\ u' &= 0, \quad -k(T)\tfrac{dT}{dy'} = h_2(T - T_0) \qquad y' = a \end{aligned} \tag{9}$$

Of engineering interest is the wall skin friction and Nusselt number for heat transfer at the wall

$$C_f = \mu_0 e^{-\bar{\alpha}(T-T_0)}\frac{du'}{dy'} + 2\beta_3\left(\frac{du'}{dy'}\right)^3\Bigg|_{y'=0}, \quad q_w = k_0 e^{-\bar{\eta}(T-T_0)}\frac{dT}{dy'}\Bigg|_{y'=0} \tag{10}$$

The Reynold's exponential dependence of viscosity and thermal conductivity on temperature is assumed to take the form

$$\mu(T) = \mu_0 e^{-\bar{\alpha}(T-T_0)},\ k(T) = k_0 e^{-\bar{\eta}(T-T_0)} \tag{11}$$

With the introduction of the following variables and parameters

$$\left.\begin{array}{l} y=\frac{y'}{a}, \quad u=\frac{u'}{UG}, \quad \theta=\frac{E(T-T_0)}{RT_0^2}, \alpha=\frac{\overline{\alpha}RT_0^2}{E}, \gamma=\frac{\beta_3 U^2G^2}{\mu_0 a^2}, \delta=\left(\frac{\upsilon\iota}{\hbar T_0}\right)^m \frac{\mu_0 U^2 G^2}{QAC_0a^2}e^{\frac{E}{RT_0}} \\ G=-\frac{a^2}{\mu_0 UG}\frac{dP}{dx}=1, Gr=\frac{\rho g\phi_0 a^2RT_0^2}{\mu_0 EUG}, \sigma=\frac{\phi_1 RT_0^2}{\phi_0 E}, \lambda=\left(\frac{\hbar T_0}{\upsilon\iota}\right)^m \frac{QEAa^2C_0}{k_0RT_0^2}e^{-\frac{E}{RT_0}}, Bi_{1,2}=\frac{h_{1,2}a}{k_0} \end{array}\right\} \tag{12}$$

We obtainthe following dimensionless, nonlinear, coupled, system of boundary value problems

$$\left.\begin{array}{l} 0=1+\frac{d}{dy}\left(e^{-\alpha\theta}\frac{du}{dy}\right)+6\gamma\left(\frac{du}{dy}\right)^2\frac{d^2u}{dy^2}+Gr(\theta+\sigma\theta^2), \ u(0)=0=u(1); \\ 0=\frac{d}{dy}\left(e^{-\eta\theta}\frac{d\theta}{dy}\right)+\lambda\left((1+\varepsilon\theta)^m e^{\frac{\theta}{1+\varepsilon\theta}}+\delta\left(\frac{du}{dy}\right)^2\left(e^{-\alpha\theta}+2\gamma\left(\frac{du}{dy}\right)^2\right)\right), \\ \qquad \left(e^{-\eta\theta}\frac{d\theta}{dy}-Bi_1\theta\right)\Big|_{y=0}=0=\left(e^{-\eta\theta}\frac{d\theta}{dy}+Bi_2\theta\right)\Big|_{y=1}. \end{array}\right\} \tag{13}$$

Please note that the special case of the coupled Equation (13) is the non-reactive Newtonian flow when $\alpha=\gamma=Gr=\eta=0$, that was reported numerically in [33] by using shooting-Runge–Kutta scheme within Maple environment. In its dimensionless form, the expression for the positive definite entropy generation becomes

$$N_S = \underbrace{e^{-\eta\theta}\left(\frac{d\theta}{dy}\right)^2}_{N_h} + \underbrace{\frac{\lambda\delta}{\varepsilon}\left(\left(\frac{du}{dy}\right)^2\left(e^{-\alpha\theta}+2\gamma\left(\frac{du}{dy}\right)^2\right)\right)}_{N_f} \tag{14}$$

Component-wise, the Bejan irreversibility ratio becomes

$$Be=\frac{N_h}{N_s}=\frac{N_h}{N_h+N_f} \tag{15}$$

While the expressions for the dimensionless skin friction and Nusselt number becomes

$$S_f=\frac{\rho a^2 C_f}{\mu_0^2}=e^{-\alpha\theta}\frac{du}{dy}+2\gamma\left(\frac{du}{dy}\right)^3\Bigg|_{y=0}, \quad Nu=\frac{aEq_w}{k_0RT_0^2}=e^{-\eta\theta}\frac{d\theta}{dy}\Bigg|_{y=0} \tag{16}$$

3. Solution by Weighted Residual Method

A collocation method based on Chebyshev polynomials is considered here to obtain numerical solutions of a differential equationin $[0, L]$. This technique approximatesa solution to a differential equation

$$f(\phi(x))+g(x)=0, \text{ in domain, } D=[0,L] \tag{17}$$

The approximate solution needs to satisfy the boundary conditions and it is given as

$$\psi(x)=\sum_{i=0}^{N}k_i\dddot{T}_i\left(\tfrac{2x}{L}-1\right) \tag{18}$$

where $\phi(x)$ is the unknown dependent function, $f(x)$ is the source term, k_i are undetermined coefficients, $\dddot{T}_i\left(\frac{2x}{L}-1\right)$ are shifted Chebyshev base functions from $[-1,1]$ to $[0, L]$. To acquire the values of k_i, Equation (18) is inserted into Equation (17) to yield residual error $R_\psi(x,k_i)$. $R_\psi(x,k_i)$ is forced to be closer to zero using the collocation method as follows;

$$\text{for } \delta(x-x_j)=\begin{cases}1, & x=x_j\\ 0, & \text{otherwise,}\end{cases}$$

$$\int_0^1 R_\psi(x,k_i)\delta(x-x_j)dx = R_\psi(x,k_i) = 0, \quad \text{for } j = 1,\, 2, \ldots N-1 \tag{19}$$

where $x_j = \frac{1}{2}\left(1-\cos\left(\frac{j\pi}{N}\right)\right)$ are shifted Gauss–Lobatto points. Therefore, Equation (14) as well as the equations generated from the boundary conditions constitute a system of algebraic equations that must be solved in order to obtainthe constant coeficient values, k_i. In order to obtain the solutions of Equation (13), solutions are assumed for $u(y)$ and $\theta(y)$, respectively, in the form:

$$u(y) = \sum_{i=0}^{N} a_i \dddot{T}_i(2y-1), \qquad \theta(y) = \sum_{i=0}^{N} b_i \dddot{T}_i(2y-1) \tag{20}$$

where a_i and b_i are constants to be determined, and $\dddot{T}_i(2y-1)$ are the shifted Chebyshev base functions. To obtain the values of a_i and b_i, to satisfy the boundary conditions, Equation (20) is substituted into the boundary conditions in Equation (13) and thus have

$$\left.\sum_{i=0}^{N} \dddot{T} a_i(2y-1)\right|_{y=0} = 0 = \left.\sum_{i=0}^{N} \dddot{T} a_i(2y-1)\right|_{y=1}, \tag{21}$$

$$\begin{aligned} &\left. e^{-\eta\left(\sum_{i=0}^{N} \dddot{T} b_i(2y-1)\right)} \sum_{i=0}^{N} \frac{d\dddot{T}_i}{dy} b_i(2y-1) - Bi_1 \sum_{i=0}^{N} \dddot{T} b_i(2y-1)\right|_{y=0} = 0 \\ &= \left. e^{-\eta\left(\sum_{i=0}^{N} \dddot{T} b_i(2y-1)\right)} \sum_{i=0}^{N} \frac{d\dddot{T}_i}{dy} b_i(2y-1) - Bi_2 \sum_{i=0}^{N} \dddot{T} b_i(2y-1)\right|_{y=1} \end{aligned} \tag{22}$$

Equation (15) is also substituted into Equations (9) and (10), resulting in $res_u(y, a_i)$ and $res_\theta(y, a_i, b_i)$, respectively, as residues. At this juncture, the residues are forced to be zero by applying collocation method as follows

$$\begin{aligned} \int_0^1 res_u \delta(y-y_j)dy &= res_u(y_j, a_i) = 0, \text{ for } j = 1,2,\ldots N-1 \\ \int_0^1 res_\theta \delta(y-y_j)dy &= res_\theta(y_j, a_i, b_i) = 0, \quad \text{for } j = 1,2,\ldots N-1 \end{aligned} \tag{23}$$

where $y_j = \frac{1}{2}\left(1-\cos\left(\frac{j\pi}{N}\right)\right)$. Thus, Equations (16)–(18) form $2N + 2$ system of algebraic equations containing $2N + 2$ undetermined coefficients (a_i, b_i). This procdeure is coded in a mathematicalsymbolic software to solve the generated equations using the Newton technique.

4. Solution by Shooting-Runge–Kutta Method

To achieve a numerical solution by the Runge–Kutta scheme, we first apply the shooting technique to the coupled problem by converting the second-order differential equations to a set of first-order ordinary Lipshitz continuous differential equations.

$$z_1 = u,\ z_2 = u',\ z_3 = \theta,\ z_4 = \theta' \tag{24}$$

With (8), the coupled BVP becomes a coupled system of first order differential equations

$$\begin{pmatrix} z_1 \\ z_2 \\ z_3 \\ z_4 \end{pmatrix}' = \begin{pmatrix} z_2 \\ \frac{\alpha e^{-\alpha z_3} - Gr\left(z_3 + \sigma z_3^2\right) - 1}{e^{-\alpha z_3} + 6\gamma z_2^2} \\ z_4 \\ e^{\eta z_3}\left(\eta e^{-\eta z_3} z_4^2 - \lambda\left\{(1+\varepsilon z_3)^m e^{\frac{z_3}{1+\varepsilon z_3}} + \delta z_2^2\left(e^{-\alpha z_3} + 2\gamma e^{-\alpha z_3}\right)\right\}\right) \end{pmatrix} \tag{25}$$

with the set of initial conditions

$$\begin{pmatrix} z_1(0) \\ z_2(0) \\ z_3(0) \\ z_4(0) \end{pmatrix} = \begin{pmatrix} 0 \\ D_1 \\ \frac{z_4(0)e^{-\eta z_3}}{Bi_1} \\ D_2 \end{pmatrix} \tag{26}$$

Since the derivatives are not specified at the initial point, $D_{1,2}$ are therefore given initial guess values that ensure the boundary conditions are okay at the wall $y = 1$. It is easy to see that the sufficient condition that confirms the existence of a unique solution to the system of Equations (24) and (25)is satisfied as long as the Lipchitz constant $\frac{\partial f_i}{\partial x_j} \leq K$ is bounded. The first-order equations are then carefully coded and solved by Runge–Kutta using the Mathematica package.

5. Results and Discussion

In this section, a physical interpretation of the numerical results is presented and discussed as regards the flow and thermal behavior. Tables 1 and 2 represent the validation of the point collocation weighted residual method with the shooting-Runge–Kutta procedure for the dimensionless momentum and energy equations. A good agreement is observed between the two computations. This confirmed the uniqueness of the numerical solutions. Table 3 shows the convergence of the of the spectral collocation solution. The convergence of the solution to the highly nonlinear boundary value problem is seen to improve with an increasing number of terms N used in the approximation. Table 4 represents the computational result of the thermal stability analysis of the fluid. As observed from the plot, an increase in the nonlinear component of the Grashof number is seen to destabilize the flow thermal structure due increasing buoyancy force over viscous forces in the flow field. Additionally, an increase in the viscosity variation parameter is also observed to have a destabilizing effect on the fluid flow. This is true, since viscosity is expected to drop and fluid experiences shear thinning, which aids heat generation by kinetic theory of matter. Moreover, a rise in the third-grade parameter is seen to enhance thermal stability due to the thickening effect of the fluid. The thermal conductivity variation parameter increase is seen to discourage the thermal stability of the fluid flow. A similar behavior is seen with increasing values of the Grashof number due to increasing buoyancy forces. Tables 5 and 6 revealed the accuracy of the solution when compared with already-published results in the special case when the fluid is Newtonian, non-reactive, with constant physical properties.

Table 1. Validation of numerical results for velocity $\delta = 0.5, \eta = 0.1, \alpha = 0.3, \alpha = 0.1, N = 30$, $\gamma = 0.3, Gr = 1, Bi_1 = Bi_2 = 10, \lambda = 0.4, \varepsilon = 0.1, m = 0.5$.

y	$u(y)_{\mathbf{SCCM}}$	$u(y)_{\mathbf{RK4}}$	$\lvert u(y)_{\mathbf{CWRM}} - u(y)_{\mathbf{RK4}}\rvert$
0	$-6.96195549259\times10^{-18}$	0.000000000000000	$6.961955492590054 \times 10^{-18}$
0.1	0.04313941885340904	0.04313941993774605	$1.084337016010739 \times 10^{-9}$
0.2	0.07813634077297339	0.07813635190993772	$1.11369643229775 \times 10^{-8}$
0.3	0.10406977160813229	0.10406978921035206	$1.760221977897824 \times 10^{-9}$
0.4	0.12006990360382683	0.12006992567673559	$2.207290876465872 \times 10^{-8}$
0.5	0.12548527366302917	0.12548528888798596	$1.522495679529001 \times 10^{-9}$
0.6	0.12006990360382683	0.12006990533241987	$1.728593046479432 \times 10^{-9}$
0.7	0.1040697716081323	0.10406977022133693	$1.386795372981808 \times 10^{-9}$
0.8	0.07813634077297338	0.07813634228509685	$1.512123468105919 \times 10^{-9}$
0.9	0.043139418853409044	0.04313941955035403	$6.96944987832459 \times 10^{-10}$
1.0	$-7.69196285588 \times 10^{-18}$	$-9.12736796 \times 10^{-10}$	$9.12736788740352 \times 10^{-10}$

Table 2. Validation of numerical results for temperature.

y	$\theta(y)_{\mathbf{SCCM}}$	$\theta(y)_{\mathbf{RK4}}$	$\lvert\theta(y)_{\mathbf{SCCM}} - \theta(y)_{\mathbf{RK4}}\rvert$
0	0.022151849024484603	0.022151849374097372	$3.496127692903528 \times 10^{-10}$
0.1	0.04208489447368267	0.04208489042284373	$4.050838935121259 \times 10^{-9}$
0.2	0.05751544434774708	0.057515440006678895	$4.341068185476082 \times 10^{-9}$
0.3	0.06849268440471615	0.06849268031195316	$4.092762989627019 \times 10^{-9}$
0.4	0.07505930588225566	0.07505930197957113	$3.902684536649659 \times 10^{-9}$
0.5	0.07724467496042003	0.07724467142322092	$3.537199116943057 \times 10^{-9}$
0.6	0.07505930588225566	0.07505930282648429	$3.055771372051374 \times 10^{-9}$
0.7	0.06849268440471615	0.06849268167736071	$2.727355438714163 \times 10^{-9}$
0.8	0.05751544434774707	0.05751544180643439	$2.541312688064678 \times 10^{-9}$
0.9	0.04208489447368267	0.04208489220244641	$2.271236268502896 \times 10^{-9}$
1.0	0.022151849024484603	0.02215184702121924	$2.003265362621187 \times 10^{-9}$

Table 3. Fast convergence of critical values by weighted residual method when $Gr = 3, \delta = 0.2$, $Bi_1 = 10 = Bi_2, m = 0.5, \varepsilon = 0.1$.

N	α	σ	γ	η	λ_c
5	0.1	0.1	0.3	0.1	3.369629994329486
10	0.1	0.1	0.3	0.1	3.329979429762467
15	0.1	0.1	0.3	0.1	3.3377915382719980
20	0.1	0.1	0.3	0.1	3.3375451276368135
25	0.1	0.1	0.3	0.1	3.3376391452777585
30	0.1	0.1	0.3	0.1	3.3376334297519272
35	0.1	0.1	0.3	0.1	3.3376356319779727

Table 4. Thermal stability with values when $\delta = 0.2, Bi_1 = 10 = Bi_2, m = 0.5, \varepsilon = 0.1$.

σ	α	γ	η	Gr	λ_c
0.1	0.1	0.3	0.1	3	3.337791538271998
0.3	0.1	0.3	0.1	3	3.294547099649542
0.5	0.1	0.3	0.1	3	3.251342585235665
0.1	0.3	0.3	0.1	3	3.327113416799457
0.1	0.5	0.3	0.1	3	3.3180031889548034
0.1	0.1	0.5	0.1	3	3.354693697420452
0.1	0.1	0.8	0.1	3	3.370237078964153
0.1	0.1	0.3	0.01	3	3.5150936154732557
0.1	0.1	0.3	0.001	3	3.5341579072708558
0.1	0.1	0.3	0.1	5	3.1911369048073452
0.1	0.1	0.3	0.1	7	3.0387490555917243

Table 5. Validation with previous result when $Bi_1 = Bi_2 = 10, Gr = 0 = \gamma = \eta, \alpha = 0.1$.

y	$u_1(y)$ **Makinde & Aziz [34]**	$u_2(y)$ **Present Result**	$\lvert u_1(y) - u_2(y) \rvert$
0	0.0000000000000000	$1.9996146847 \times 10^{-18}$	$1.999614684734152 \times 10^{-18}$
0.1	0.04500260897270728	0.04500261769036607	$8.717658789292315 \times 10^{-9}$
0.2	0.08000542199231503	0.08000543151384462	$9.521529592548816 \times 10^{-9}$
0.3	0.10500767322725557	0.10500768307611696	$9.848861393102482 \times 10^{-9}$
0.4	0.12000907000446212	0.12000908068447211	$1.068000998749596 \times 10^{-8}$
0.5	0.12500954265746725	0.12500954933196154	$6.674494290592747 \times 10^{-9}$
0.6	0.12000907650344211	0.12000908068447211	$4.181029994443364 \times 10^{-9}$
0.7	0.10500768080412677	0.10500768307611698	$2.271990207081131 \times 10^{-9}$
0.8	0.08000543108796186	0.08000543151384461	$4.258827457359615 \times 10^{-10}$
0.9	0.045002617803451495	0.04500261769036607	$1.130854229702826 \times 10^{-10}$
1.0	$8.38288175864 \times 10^{-10}$	$2.13968421631 \times 10^{-18}$	$8.38288173724192 \times 10^{-10}$

Table 6. Validation with previous result when $Bi_1 = Bi_2 = 10, Gr = 0 = \gamma = \eta, \alpha = 0.1$.

y	$\theta_1(y) - SRK4$ **Makinde & Aziz [34]**	$\theta_2(y) - SCCM$ **Present Result**	$\lvert \theta_1(y) - \theta_2(y) \rvert$
0	0.0004166959732611298	0.00041669643079810236	$4.57536972555978 \times 10^{-10}$
0.1	0.0007242078568868499	0.0007242204230993113	$1.256621246132922 \times 10^{-8}$
0.2	0.0008700589609001264	0.0008700665516844415	$7.59078431502528 \times 10^{-9}$
0.3	0.0009242312498706879	0.0009242382030372055	$6.953166517597101 \times 10^{-9}$
0.4	0.0009367330606715411	0.0009367393712081855	$6.310536644453524 \times 10^{-9}$
0.5	0.0009375671371315745	0.0009375727826577018	$5.645526127313893 \times 10^{-9}$
0.6	0.000936734339945098	0.0009367393712081855	$5.031263087466702 \times 10^{-9}$
0.7	0.0009242337880723514	0.0009242382030372052	$4.414964853774461 \times 10^{-9}$
0.8	0.0008700627611325915	0.0008700665516844411	$3.790551849616394 \times 10^{-9}$
0.9	0.0007242172747296661	0.0007242204230993107	$3.148369644634038 \times 10^{-9}$
1.0	0.0004166939431660092	0.0004166964307981019	$2.487632092722583 \times 10^{-9}$

Figure 2 represents the variation of third-grade material effect with velocity, temperature, entropy generation, and Bejan number, as shown in Figure 2a–d, respectively. The thickening effect is manifested in Figure 2a as the third-grade material parameter grows. Likewise in Figure 2b, thickening of the fluid is seen to make the non-Newtonian fluid colder. This is physically correct because the kinetic energy of the flow decreases as a result, the intermolecular bond is expected to increase. Therefore, the activation energy requirement will be much higher to induce chemical reaction. The result from Figure 2c reveals that as the third-grade material parameter rises, there is drop in the entropy generation at the cooled walls. This is reasonable, because a rise in the third-grade material parameter has been shown to reduce the flow velocity and temperature maximum in Figure 2a,b. Therefore, available energy for work is maximum with the thickened fluid. Finally, in Figure 2d the result of the heat irreversibility ratio credited Bejan is presented. If we suppose that the fluid velocity and density are both constant, then we have a case of slab with Be = 1. However, for a variable viscous third-grade flow, we have seen that velocity is a decreasing function of the third-grade parameter. As a result, the kinetic energy of the flow drops with increasing value of the parameter. Therefore, heat transfer irreversibility is expected to gain dominance over frictional irreversibility as shown in the plot.

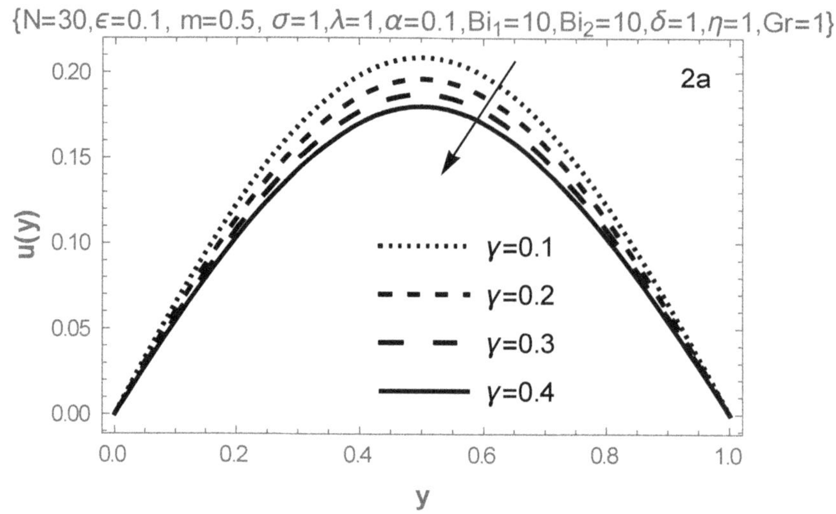

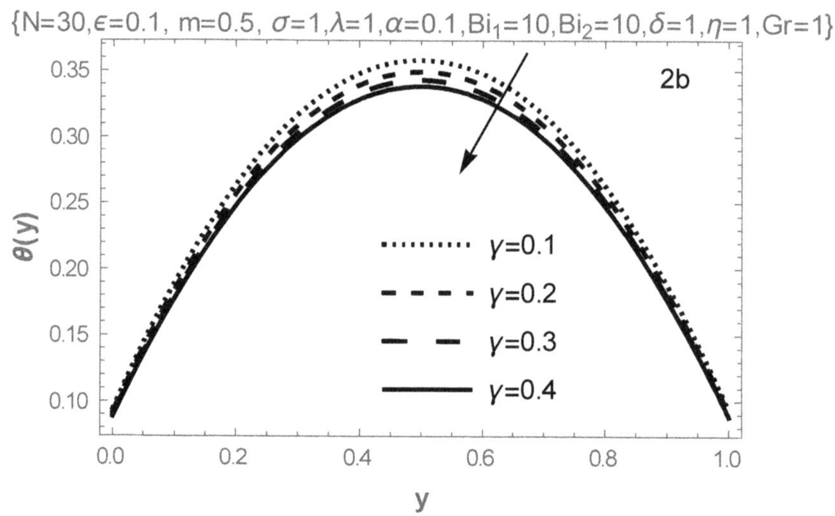

Figure 2. *Cont.*

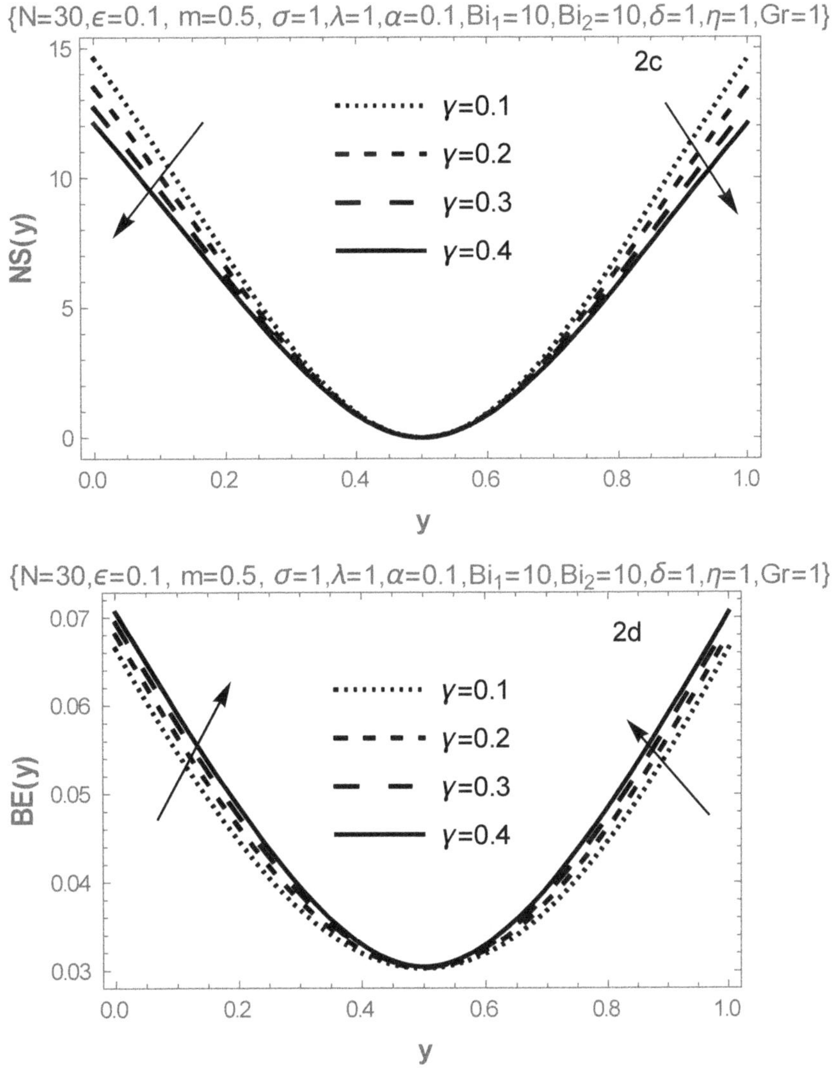

Figure 2. Variation of third-grade parameter.

Figure 3a–d represent the impact of buoyancy forces on the flow field on the velocity, temperature, entropy generation rate, and the Bejan number respectively. From Figure 3a, the velocity profile shows that when Gr = 1 implies a simple case when viscous force and buoyancy force are equal. Further increase in the Grashof number implies that the contribution of viscous force is weaker when compared with buoyancy, therefore, the flow velocity increases significantly in response to the nonlinear buoyancy forces. A similar behavior is seen in the profile for fluid temperature (see Figure 3b) resulting from increase in the kinetic energy of the fluid. Consequently in Figure 3c, the entropy profile is expected to increase due to increase in both flow and temperature maximum. In Figure 3d, rise in Grashof number implies increasing temperature as seen in Figure 3b. Therefore, increasing temperature translates to a decreased rate of heat transfer, this shows that irreversibility due to frictional interactions is more when compared with heat transfer irreversibility as revealed in Figure 3d. The graphical representations of the effect of the nonlinear convection component of the Grashof number on the flow velocity, temperature, entropy generation, and Bejan number respectively depicted in Figure 4a–d. The point when $\sigma = 0$ corresponds to the linearized buoyancy parameter when the nonlinear effect is neglected. As shown in the Figure 4a, this parameter has a significant effect on the fluid flow and cannot be neglected. Also, in Figure 4b, an increased value of the parameter is also seen to elevate the fluid temperature maximum. Similar effect is noticed in Figure 4c, the rate of entropy generation improves at both convective walls while in Figure 4d, the frictional irreversibility decreases within the flow channel, due to a decrease in heat transfer.

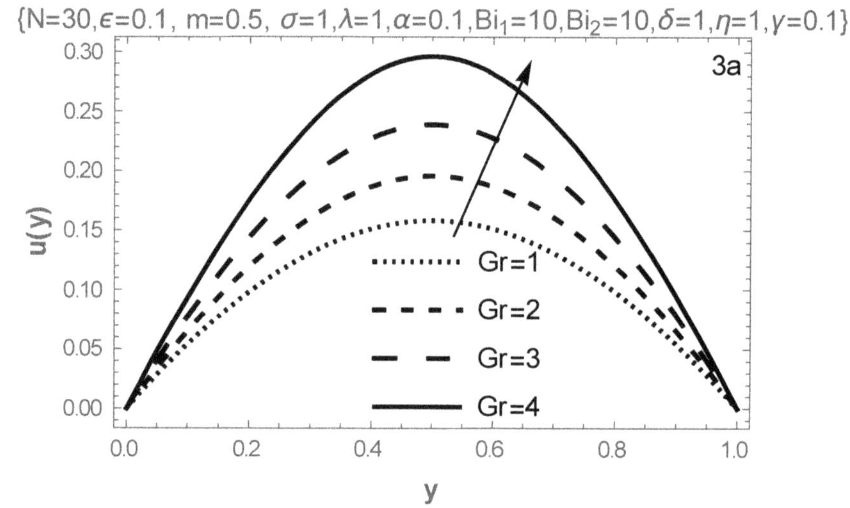

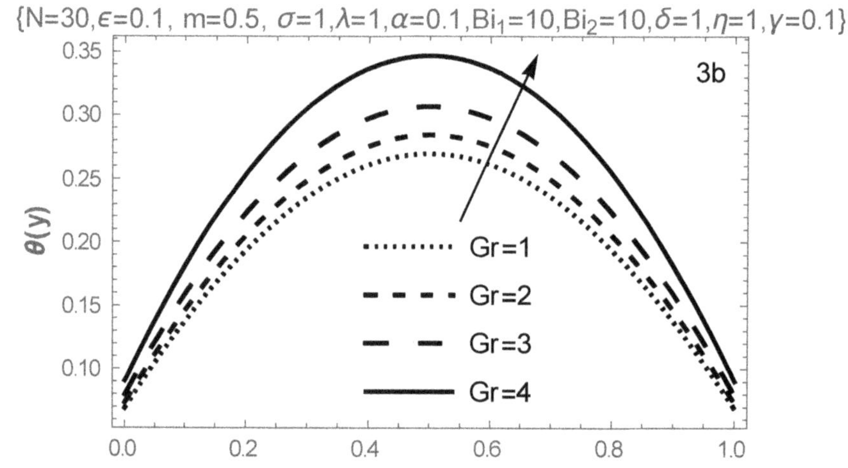

Figure 3. *Cont.*

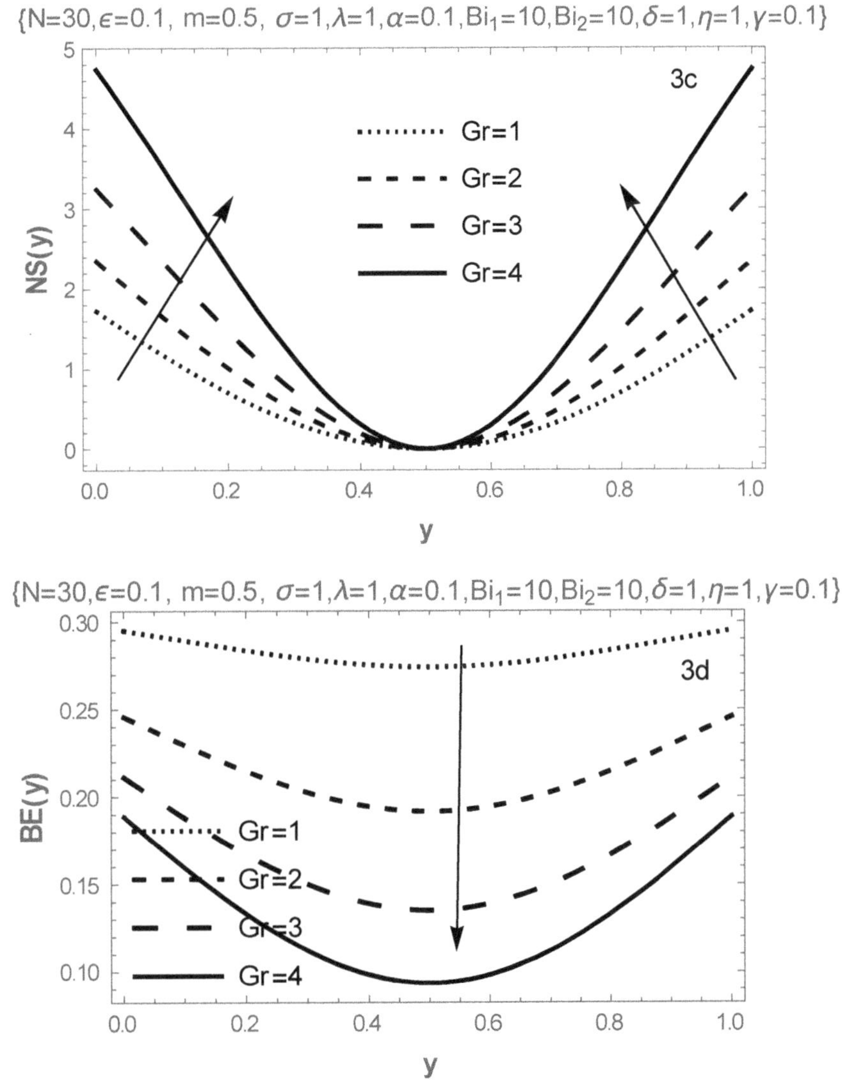

Figure 3. Variations in Grashof number.

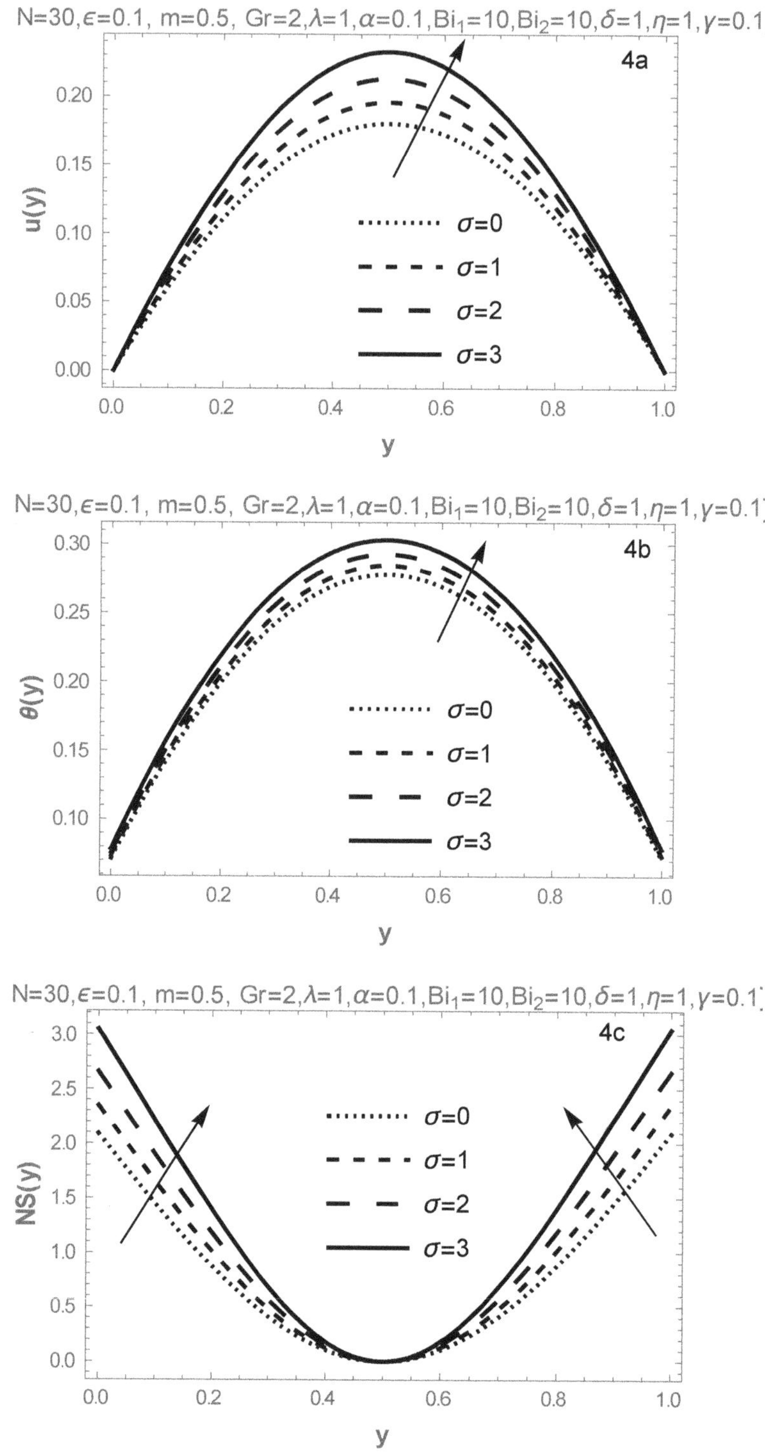

Figure 4. *Cont.*

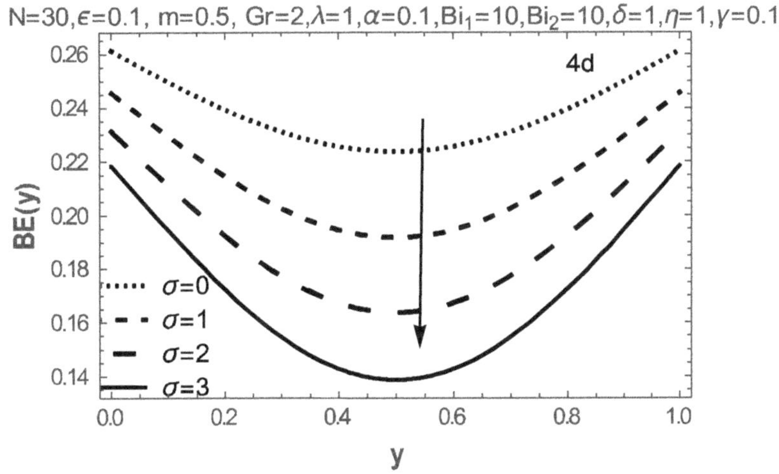

Figure 4. Variations in nonlinear Grashof number.

The influence of the reaction kinetics coefficient (λ) on the flow and thermal structure is presented in Figure 5a. It is evidently shown that the velocity distribution enhanced for larger λ. This is due to the thinning effect on the fluid viscosity and the heat transfer from the strongly exothermic reaction to the fluid. In Figure 5b, similar behavior is experienced with the fluid temperature, except at the cooled walls with increasing flow and temperature. As expected the entropy generation rate is to rise for increased values of λ as shown in Figure 5c, while larger λ significantly increase the Bejan profile. This in turn encourages heat transfer irreversibility over fluid friction irreversibility as seen in Figure 5d.

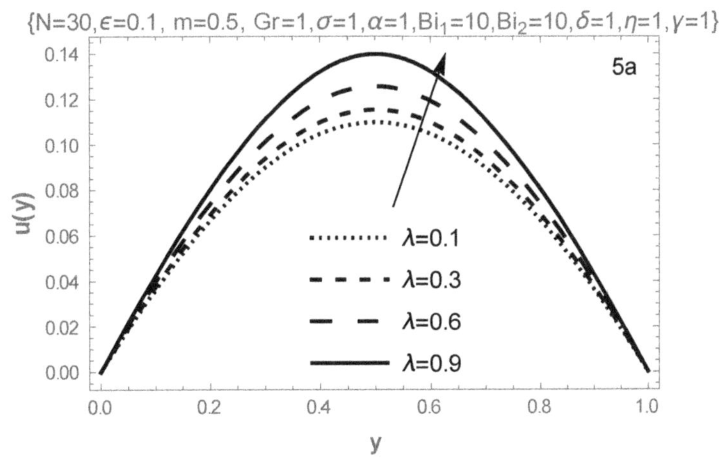

Figure 5. *Cont.*

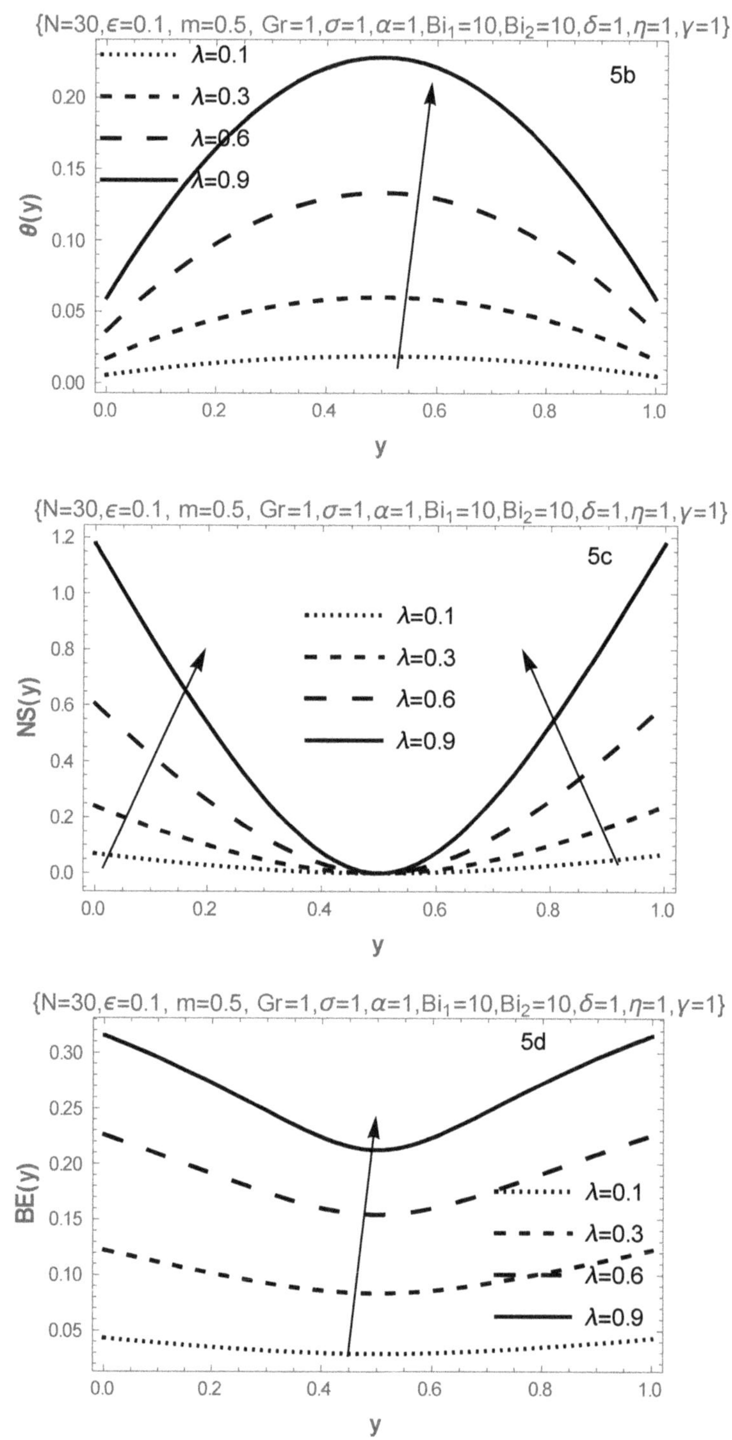

Figure 5. Variations in Frank-Kameneskii parameter.

Finally, Figure 6a–c shows the bifurcation plots for different reaction kinetics ($m = -2, 0, 0.5$). As observed from these plots, the blow-up point increases with increasing values of the exponent, thus thermal stability is enhanced.

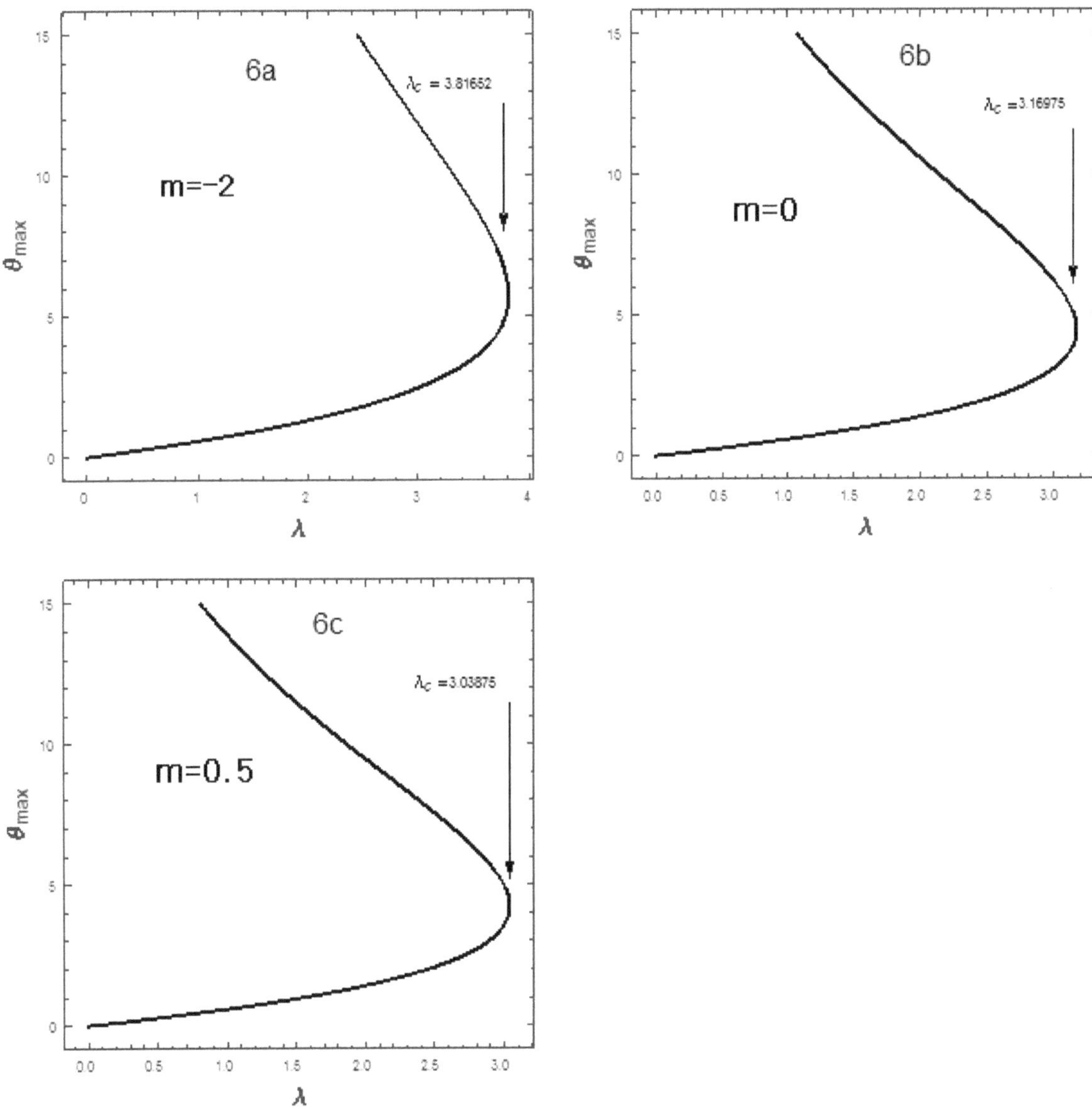

Figure 6. Bifurcation plots when $Gr = 7, \sigma = 0.1 = \varepsilon = \eta = \alpha, \delta = 0.2, \gamma = 0.3, Bi_{1,2} = 10$.

6. Concluding Remarks

In this study, a time-independent flow and heat transfer of reactive third-order fluid has been investigated numerically, taking dependence of some physical properties on temperature into consideration. The nonlinear equations were solved numerically by using a spectral Chebyshev weighted residual method and validated by shooting-Runge–Kutta approach. The main contributions to knowledge from the study are listed as follows:

- Flow, temperature, and entropy generation are enhanced with increasing values of the Grashof number, the quadratic component of buoyancy and Frank-Kameneskii parameter but reduces with increasing third-grade material parameter.
- Increasing values of third-grade parameter encourages the thermal stability of the flow, while increasing values of the linear and nonlinear buoyancy parameter destabilizes the flow.
- The Grashof number increase encourages the early occurrence of thermal runaway and exergy loss in the flow domain.

Future directions on this study are not limited to the developing flow, mass transfer, and elasto-hydrodynamics aspects of the work.

Author Contributions: Conceptualization, S.O.A.; Data curation, T.A.Y.; Formal analysis, S.O.A. and R.S.L.; Funding acquisition, R.S.L.; Investigation, T.A.Y.; Methodology, S.O.A.; Project administration, R.S.L.; Resources, T.A.Y.; Software, S.O.A.; Writing—original draft, S.O.A. All authors have read and agreed to the published version of the manuscript.

Funding: This research received no external funding.

Conflicts of Interest: The authors declare no conflict of interest.

Nomenclature

(x, y')	dimensional Cartesian coordinates, (m)
y	dimensional Cartesian coordinates
T_0	referenced temperature, (**K**)
T	dimensional fluid temperature (**K**)
θ	dimensionless fluid temperature
u'	the dimensional flow velocity (m/s)
u	the dimensionless flow velocity
E	activation energy (E/mols)
Q	reaction heat (J)
C_0	initial specie concentration (mol)
A	reaction rate constant
m	reaction exponent
ε	dimensionless activation energy
R	universal rate constant J/(**K**.mol)
μ_0	constant dynamic viscosity (Poise)
ρ	fluid density (Kg/m^3)
g	gravitational acceleration m/s^2
$\hbar$	Planck's constant (Js)
v	frequency of vibration N.s/m^2
P	fluid pressure (N/m^2)
k_0	referenced thermal conductivity J/(m**K**),
$(\bar{\alpha}, \bar{\eta})$	viscosity and thermal conductivity, respectively(1/**K**).
(α, η)	dimensionless variation parameters for viscosity and thermal conductivity, respectively.
Gr	modified Grashof number
σ	coefficient of the quadratic thermal expansion,
λ	Frank-Kameneskii parameter
δ	Viscous dissipation parameter
Bi	Biot number
γ	non-Newtonian material parameter
$h_{1,2}-$	coefficient of heat transfer 1/**K**

References

1. Salawu, S.O.; Fatunmbi, E.O.; Ayanshola, A.M. On the diffusion reaction of fourth-grade hydromagnetic fluid flow and thermal criticality in a plane couette medium. *Results Eng.* **2020**, *8*, 100169. [CrossRef]
2. Cui, J.; Munir, S.; Raies, S.F.; Farooq, U.; Razzaq, R. Non-similar aspects of heat generation in bioconvection from flat surface subjected to chemically reactive stagnation point flow of Oldroyd-B fluid. *Alex. Eng. J.* **2022**, *61*, 5397–5411. [CrossRef]

3. Salawu, S.O.; Oderinu, R.A.; Ohaegbue, A.D. Thermal runaway and thermodynamic second law of a reactive couple stress hydromagnetic fluid with variable properties and Navier slips. *Sci. Afr.* **2020**, *7*, e00261. [CrossRef]
4. Sadiq, M.A.; Hayat, T. Entropy optimized flow of Reiner-Rivlin nanofluid with chemical reaction subject to stretchable rotating disk. *Alex. Eng. J.* **2022**, *61*, 3501–3510. [CrossRef]
5. Okoya, S.S. Computational study of thermal influence in axial annular flow of a reactive third grade fluid with non-linear viscosity. *Alex. Eng. J.* **2019**, *58*, 401–411. [CrossRef]
6. Adesanya, S.O.; Falade, J.; Jangili, S.; Bég, O.A. Irreversibility analysis for reactive third-grade fluid flow and heat transfer with convective wall cooling. *Alex. Eng. J.* **2017**, *56*, 153–160. [CrossRef]
7. Salawu, S.O.; Kareem, R.A.; Shonola, S.A. Radiative thermal criticality and entropy generation of hydromagnetic reactive Powell–Eyring fluid in saturated porous media with variable conductivity. *Energy Rep.* **2019**, *5*, 480–488. [CrossRef]
8. Makinde, O.D. Hermite-Padé approximation approach to thermal criticality for a reactive third-grade liquid in a channel with isothermal walls. *Int. Commun. Heat Mass Transf.* **2007**, *34*, 870–877. [CrossRef]
9. Okoya, S.S. Disappearance of criticality for reactive third-grade fluid with Reynold's model viscosity in a flat channel. *Int. J. Non-Linear Mech.* **2011**, *46*, 1110–1115. [CrossRef]
10. Khan, S.A.; Khan, M.I.; Alzahrani, F. Melting heat transportation in chemical reactive flow of third grade nanofluid with irreversibility analysis. *Int. Commun. Heat Mass Transf.* **2021**, *129*, 105696. [CrossRef]
11. Makinde, O.D.; Chinyoka, T. Numerical study of unsteady hydromagnetic generalized Couette flow of a reactive third-grade fluid with asymmetric convective cooling. *Comput. Math. Appl.* **2011**, *61*, 1167–1179. [CrossRef]
12. Baranovskii, E.S.; Artemov, M.A. Steady flows of second-grade fluids in a channel. *Appl. Math.* **2017**, *13*, 342–353. [CrossRef]
13. Okoya, S.S. On the transition for a generalized Couette flow of a reactive third-grade fluid with viscous dissipation. *Int. Commun. Heat Mass Transf.* **2008**, *35*, 188–196. [CrossRef]
14. Hron, J.; LeRoux, C.; Malek, J.; Rajagopal, K.R. Flows of incompressible fluids subject to Navier's slip on the boundary. *Comput. Math. Appl.* **2008**, *56*, 2128–2143. [CrossRef]
15. Zehra, I.; Kousar, N.; UrRehman, K. Pressure dependent viscosity subject to Poiseuille and Couette flows via Tangent hyperbolic model. *Phys. A Stat. Mech. Its Appl.* **2019**, *527*, 121332. [CrossRef]
16. Manjunatha, G.; Rajashekhar, C.; Vaidya, H.; Prasad, K.V.; Vajravelu, K. Impact of heat and mass transfer on the peristaltic mechanism of Jeffery fluid in a non-uniform porous channel with variable viscosity and thermal conductivity. *J. Therm. Anal. Calorim.* **2020**, *139*, 1213–1228. [CrossRef]
17. Qasim, M.; Riaz, N.; Lu, D.; Afridi, M.I. Flow over a Needle Moving in a Stream of Dissipative Fluid Having Variable Viscosity and Thermal Conductivity. *Arab. J. Sci. Eng.* **2021**, *46*, 7295–7302. [CrossRef]
18. Saraswathy, M.; Prakash, D.; Muthtamilselvan, M.; Al Mdallal, Q.M. Arrhenius energy on asymmetric flow and heat transfer of micropolar fluids with variable properties: A sensitivity approach. *Alex. Eng. J.* **2022**, *61*, 12329–12352. [CrossRef]
19. Khan, A.A.; Zaib, F.; Zaman, A. Effects of entropy generation on Powell Eyring fluid in a porous channel. *J. Braz. Soc. Mech. Sci. Eng.* **2017**, *39*, 5027–5036. [CrossRef]
20. Singh, K.; Pandey, A.K.; Kumar, M. Entropy generation impact on flow of micropolar fluid via an inclined channel with non-uniform heat source and variable fluid properties. *Int. J. Appl. Comput. Math.* **2020**, *6*, 85. [CrossRef]
21. Shah, Z.; Kumam, P.; Deebani, W. Radiative MHD Casson Nanofluid Flow with Activation energy and chemical reaction over past nonlinearly stretching surface through Entropy generation. *Sci. Rep.* **2020**, *10*, 4402. [CrossRef] [PubMed]
22. Yusuf, T.A.; Kumar, R.N.; Prasannakumara, B.C.; Adesanya, S.O. Irreversibility analysis in micropolar fluid film along an incline porous substrate with slip effects. *Int. Commun. Heat Mass Transf.* **2021**, *126*, 105357. [CrossRef]
23. Agrawal, R.; Kaswan, P. Minimization of the entropy generation in MHD flow and heat transfer of nanofluid over a vertical cylinder under the influence of thermal radiation and slip condition. *Heat Transf.* **2022**, *51*, 1790–1808. [CrossRef]
24. Adesanya, S.O.; Ogunseye, H.A.; Lebelo, R.S.; Moloi, K.C.; Adeyemi, O.G. Second law analysis for nonlinear convective flow of a reactive couple stress fluid through a vertical channel. *Heliyon* **2018**, *4*, e00907. [CrossRef]
25. Xia, W.F.; Ahmad, S.; Khan, M.N.; Ahmad, H.; Rehman, A.; Baili, J.; Gia, T.N. Heat and mass transfer analysis of nonlinear mixed convective hybrid nanofluid flow with multiple slip boundary conditions. *Case Stud. Therm. Eng.* **2022**, *32*, 101893. [CrossRef]
26. Ibrahim, W.; Gadisa, G. Finite element solution of nonlinear convective flow of Oldroyd-B fluid with Cattaneo-Christov heat flux model over nonlinear stretching sheet with heat generation or absorption. *Propuls. Power Res.* **2020**, *9*, 304–315. [CrossRef]
27. Patil, P.M.; Shankar, H.F.; Hiremath, P.S.; Momoniat, E. Nonlinear mixed convective nanofluid flow about a rough sphere with the diffusion of liquid hydrogen. *Alex. Eng. J.* **2021**, *60*, 1043–1053. [CrossRef]
28. Bandara, S.; Carnegie, C.; Johnson, C.; Akindoju, F.; Williams, E.; Swaby, J.M.; Oki, A.; Carson, L.E. Interaction of heat generation in nonlinear mixed/forced convective flow of Williamson fluid flow subject to generalized Fourier's and Fick's concept. *J. Mater. Res. Technol.* **2020**, *9*, 11080–11086.
29. Yusuf, T.A.; Mabood, F.; Gbadeyan, J.A.; Adesanya, S.O. Nonlinear Convective for MHDOldroyd8-constant fluid in a channel with chemical reaction and convective boundary condition. *J. Therm. Sci. Eng. Appl.* **2020**, *12*, 1–19. [CrossRef]
30. IjazKhan, M.; Alzahrani, F.; Hobiny, A. Heat transport and nonlinear mixed convective nanomaterial slip flow of Walter-B fluid containing gyrotactic microorganisms. *Alex. Eng. J.* **2020**, *59*, 1761–1769. [CrossRef]
31. Ijaz, M.; Ayub, M. Nonlinear convective stratified flow of Maxwell nanofluid with activation energy. *Heliyon* **2019**, *5*, e01121. [CrossRef] [PubMed]

32. Srinivasacharya, D.; RamReddy, C.; Naveen, P. Double dispersion effect on nonlinear convective flow over an inclined plate in a micropolar fluid saturated non-Darcy porous medium. *Eng. Sci. Technol. Int. J.* **2018**, *21*, 984–995. [CrossRef]
33. Makinde, O.D. On thermal stability of a reactive third-grade fluid in a channel with convective cooling the walls. *Appl. Math. Comput.* **2009**, *213*, 170–176. [CrossRef]
34. Makinde, O.D.; Aziz, A. Second law analysis for a variable viscosity plane Poiseuille flow with asymmetric convective cooling. *Comput. Math. Appl.* **2010**, *60*, 3012–3019. [CrossRef]

mathematics

MDPI

Article

Dynamics of Heat Transfer Analysis of Convective-Radiative Fins with Variable Thermal Conductivity and Heat Generation: Differential Transformation Method

P. V. Ananth Subray [1,†], B. N. Hanumagowda [1], S. V. K. Varma [1], A. M. Zidan [2], Mohammed Kbiri Alaoui [2], C. S. K. Raju [3], Nehad Ali Shah [4,†] and Prem Junsawang [5,*]

1 School of Applied Sciences, REVA University, Bengaluru 560064, India
2 Department of Mathematics, College of Science, King Khalid University, P.O. Box 9004, Abha 61413, Saudi Arabia
3 Department of Mathematics, GITAM School of Science, Bangalore 562163, India
4 Department of Mathematics, Sejong University, Seoul 05006, Korea
5 Department of Statistics, Faculty of Science, Khon Kaen University, Khon Kaen 40002, Thailand
* Correspondence: prem@kku.ac.th
† These authors contributed equally to this work and are co-first authors.

Abstract: The study of convective heat transfer in differently shaped fins with radiation, internal heat generation and variable thermal conductivity was considered. The energy equation of the model was converted into the dimensionless form by adopting the proper variables, which was later solved using the differential transformation method. The impact of the parameters on the thermal performance, efficiency and heat transfer of the fins was analyzed graphically and also by performing thermal analysis on the fins. It was noticed that there was a significant effect on the thermal performance of the fins with different shapes, and also the heat transfer rate of the fin increased for improved values of the internal heat generation and radiation parameters. The exponential profile showed better results than other profiles, and the results obtained were supported by thermal analysis using ANSYS software.

Keywords: convection-radiation heat transfer; thermal analysis; differential transformation method; internal heat generation and variable heat conductivity

MSC: 76-10; 76R10

Citation: Ananth Subray, P.V.; Hanumagowda, B.N.; Varma, S.V.K.; Zidan, A.M.; Kbiri Alaoui, M.; Raju, C.S.K.; Shah, N.A.; Junsawang, P. Dynamics of Heat Transfer Analysis of Convective-Radiative Fins with Variable Thermal Conductivity and Heat Generation: Differential Transformation Method. *Mathematics* **2022**, *10*, 3814. https://doi.org/10.3390/math10203814

Academic Editor: Ramoshweu Solomon Lebelo

Received: 2 August 2022
Accepted: 11 October 2022
Published: 15 October 2022

1. Introduction

Heat enchantment has become an important factor that has captured the interest of many researchers. Increasing the heat transfer mainly depends on the heat transfer coefficient, the surface area available and the temperature difference between the surface and surrounding fluid. Fins are used as heat dissipators by increasing the surface area of the heated surface that is exposed to an ambient fluid. In particular, fins are electronic components, and diodes, transistors, etc., are made up of fins. Karus et al. [1] presented a general overview of fins. Using the above concepts, Gireesha and Sowmya [2] solved fin problems with heat distribution in an inclined fin. The study of horizontal fins with natural convection was considered by Popiel et al. [3]. In most cases, the electrical current generates internal heat that can be detected in electrical filaments or nuclear reactors exposed to the temperature. This is a nonlinear factor that does not allow an analytical solution. It can be solved using numerical or semi-analytical methods. A mathematical study of the fin with an internal heat source was studied by Minkler and Rouleau [4]. Recently, many researchers [5–7] have used a numerical approach to solve the fins of various shapes with an internal heat source. Sobamowo [8] investigated the effect of internal heat initiation and temperature-dependent heat conduction. Turkyilmazoglu [9] used variable heat conduction and heat distribution coefficients to obtain the rate of heat transfer through radial fins. Variable thermal properties in straight fins were reported by Ndlovu and Moitsheki [10]. Rohit et al. [11] studied temperature-relative heat transfer in a moving fin using the decomposition method. In recent years, thermal enhancement flow problems have been analyzed by a few authors [12,13]. The study of heat conduction between fins with a motion and in the presence of convection energy using the homotopy scheme was performed by Aziz and Khani [14]. Sowmya et al. [15] examined the heat performance in longitudinal fins with a heat source due to natural convection. A study of a porous medium and the radiation parameter was conducted by Hatami and Ganji [16] in a circular fin. Heat transfer and temperature distribution in circular convective radiative porous fins of different shapes were analyzed by Pasha [17]. Heat propagation in fins with radiation for different geometries was studied by Tarobi et al. [18]. A handful of researchers studied the effect of rectangular fins used in heat exchange systems and determined the dimensions of the fin to achieve better efficiency [19–21]. Shi et al. [22] studied the bio-convection flow of magneto-cross nanofluid containing gyrotactic microorganisms with activation energy.

Fins are widely used in industries to reduce the heat transfer rate of the appliances produced by them. As an example, Farhad et al. [23,24] studied the application of fins in air-conditioning and ice storage systems by arranging the fins in different combinations. Their study revealed that the length, shape and arrangement of the fins expedited heat transfer. Sabu et al. [25] studied the significance of nanoparticles' shape and thermo-hydrodynamic slip constraints on MHD alumina-water nanoliquid flows over a rotating heated disk. Jamal et al. [26] considered partially inclined baffles in a rectangular enclosure to study the turbulent and thermal behaviors of air using the finite volume method. They reported that the thermal performance of a heat exchanger and the reduction in pressure loss by adopting the designs that allow the maximum heat transfer rate with minimum energy coincide with the results of Demartini et al. [27]. Moreover, they found out that the heat transfer rate is directly related to the number of baffles present in the system. Meanwhile, Omid et al. [28] studied the performance of airflow in rectangular-shaped solar heaters with V-shaped ribs. Increasing the inclination of the ribs produces higher velocity and heat transfer. In this article, the DTM was used to solve the nonlinear energy equation describing the temperature distribution in fins with variable thermal conductivity, radiation and internal heat generation. DTM is a semi-analytical technique proposed by Zho [29] in 1986 to solve the initial value problems in electrical circuits to obtain precise nth derivative values. The solution for a system of differential equations by the DTM was explained by Fatma [30]. Two-dimensional DTM used to solve the differential equation was developed by Chen and Ho [31]. Ayaz [32] proved that DTM is better to solve a nonlinear problem than the Taylor series method. The DTM has been used to solve various problems in applied mathematics and physics such as systems of differential equations [33]. Fallo et al. [34] applied the 3D DTM for the first time to study heat transfer in a cylindrical spine fin with variable thermal properties. Chiba et al. [35] solved the one-dimensional phase change problem in a slab of finite thickness using the DTM. The finite Taylor series and the iteration operation described by the transformed equations derived from the original equation employing differential transformation operations can be utilized to assess the approximating solution. Several authors used the DTM concept to solve various types of equations [36–39].

This work aims to study the heat transfer of longitudinal fins with different geometries in the presence of a temperature-dependent heat source, thermal radiation and variable thermal conductivity by providing an analytical solution for the heat equation using the DTM approximation technique. A review of the above literature shows no attempt has been made to analyze the heat transfer for the above-considered profiles and effects using the DTM. Graphical comparison of heat transfer rate between the three profiles and the efficiency of the fins are discussed in this study, and also our study is supported by performing thermal analysis using ANSYS software.

2. Fundamental Operations of DTM

Let $\phi(r)$ be a function that is continuously differentiable in the domain D. Power series can be used for the representation of $\phi(r)$ and can be articulated in terms of the Taylor series [40] as follows:

$$\phi(r) = \sum_{e=0}^{\infty} \frac{\left(r - r_f\right)^e}{e!} \left[\frac{d^e y(r)}{dt^e}\right]_{r=r_f} \quad \forall\ r \in D \tag{1}$$

The Maclaurin series $\phi(r)$ is obtained by taking $r_i = 0$ in Equation (1) and can be expressed as:

$$\phi(r) = \sum_{e=0}^{\infty} \frac{(r)^e}{e!} \left[\frac{d^e \phi(r)}{dr^e}\right]_{r=0} \quad \forall\ r \in D \tag{2}$$

Franco [41] explained the use of differential transforms and expressed the function $\phi(r)$ as follows:

$$\varphi(e) = \sum_{e=0}^{\infty} \frac{H^e}{e!} \left[\frac{d^e \phi(r)}{dr^e}\right]_{r=0} \tag{3}$$

$\varphi(v)$ is the converted function, and $\phi(r)$ is the initial function. The differential function $\varphi(v)$ is restricted to $r \in [0\ ,\ H]$ where H is a permanent value and is assumed to be unity. The inverse differential transform $\varphi(v)$ can be expressed as:

$$\phi(v) = \sum_{e=0}^{\infty} \left(\frac{v}{H}\right)^e \varphi(e) \tag{4}$$

The functions and transformations used in our study are presented in Table 1. From this, it can be deduced that the differential transform is similar to the Taylor series. To get more accuracy, we consider a higher number of terms in the above series.

Table 1. Fundamental definitions of DTM.

Initial Function	Converted Function
$\phi(r) = \frac{dg(r)}{dx}$	$\varphi(v) = (v+1)G(v)$
$\phi(r) = \frac{d^2 g(r)}{dx^2}$	$\varphi(v) = (v+1)(v+2)G(v+1)$
$\phi(r) = 1$	$\varphi(v) = \delta(v)$
$\phi(r) = t$	$\varphi(v) = \delta(v-1)$
$\phi(r) = r^m$	$\varphi(v) = \delta(v-w) = \begin{cases} 1\ if\ v = w \\ 0\ if\ v \neq w \end{cases}$
$\phi(r) = g(r)h(r)$	$\varphi(v) = \sum_{w=0}^{v} H(v)G(v-w)$
$\phi(r) = e^{ar}$	$\varphi(v) = \frac{a^v}{v!}$

3. Mathematical Formulation

The fin length is L with the variable area $P(x)$ as shown in Figure 1. T_b is base temperature, T_a is ambient temperature, and the tip is presumed to be in convection. Constant heat h is maintained throughout the fin, while thermal conduction is temperature dependent and varies linearly. The energy equation is obtained considering the following assumptions:

- The temperature is a function of x and remains constant over time.
- The temperature variance due to fin thickness is neglected.
- The fin bed is kept at a steady temperature.
- Solid matrix and fluid are in a dynamic state of equilibrium.
- Fin is considered to be in a steady state.

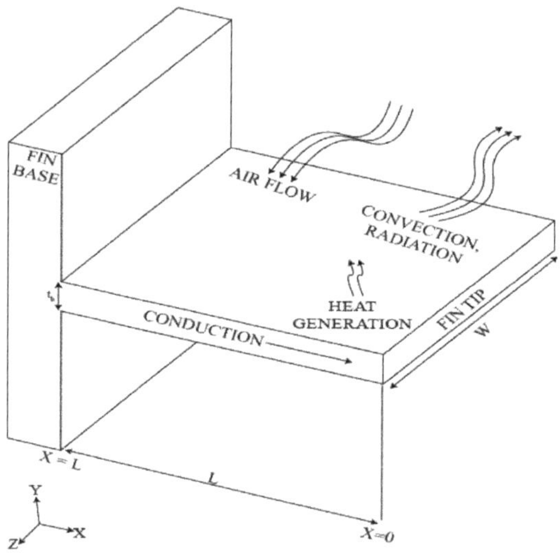

Figure 1. Representation of a rectangular fin.

The balanced energy equation under the above assumptions for the small element dx is:

$$\frac{d}{dx}\left[k(T) \times P(x) \times \frac{dT}{dx}\right] - \varepsilon\sigma\left(T^4 - T_a^4\right) - h(T - T_a) + q^* = 0 \tag{5}$$

The corresponding boundary constraints are defined as:

$$\frac{dT(0)}{dX} = 0$$

$$T(L) = T_b$$

Here, the variable heat conduction is stated as:

$$k(T) = k_a[1 + \zeta(T - T_a)] \tag{6}$$

where k_a is heat conduction at ambient temperature, and ζ is persistent. The fin is segregated into different profiles according to the difference in thickness along its length.

$$P(x) = b\Gamma(x) \tag{7}$$

where ω is the girth, and $\Gamma(x)$ is thickness along the length. Various geometries $\Gamma(x)$ can be considered as shown in Figure 2:

- For quadrilateral fin

$$\Gamma(x) = \Gamma_b \tag{8}$$

- For exponential fin

$$\Gamma(x) = \Gamma_b e^{a(x/L)} \tag{9}$$

- For convex fin

$$\Gamma(x) = \Gamma_b\left(\frac{x}{L}\right)^{0.5} \tag{10}$$

Dimensionless parameters are:

$$\theta = \frac{T}{T_b}, \theta_a = \frac{T_a}{T_b}\ X = \frac{x}{L}, N^2 = \left(\frac{hL^2}{k_b A_b}\right) Nr = \frac{\varepsilon \sigma L^2 T_b^3}{A_b k_a} G = \frac{L^2 q^*}{A_b k_a T_b} \tag{11}$$

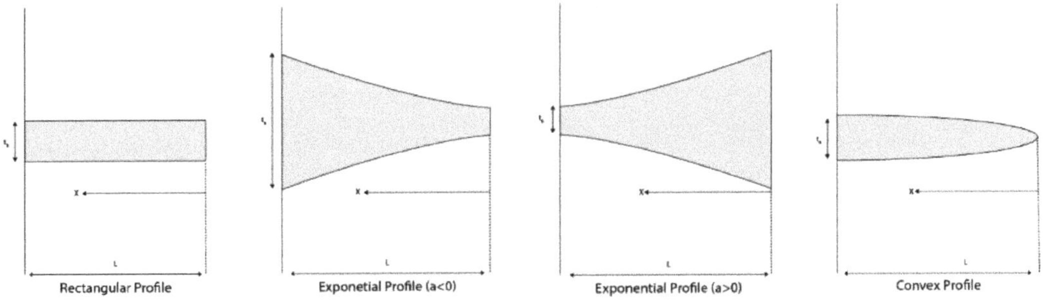

Figure 2. Schematic representation of fins with different profiles.

By applying equations in Equations (7)–(11) into Equation (5), we obtain:

- For rectangular profile

$$\beta\left(\frac{d\theta}{dX}\right)^2 + [1 + \beta(\theta - \theta_a)]\frac{d^2\theta}{dX^2} - Nr\left(\theta^4 - \theta_a^4\right) - N^2(\theta - \theta_a) + G = 0 \tag{12}$$

- For exponential profile

$$\begin{aligned} &e^{aX}[1 + \beta(\theta - \theta_a)]\tfrac{d^2\theta}{dX^2} + \mathrm{a}[1 + \beta(\theta - \theta_a)]e^{aX}\tfrac{d\theta}{dX} + e^{aX}\beta\left(\tfrac{d\theta}{dX}\right)^2 \\ &-Nr\left(\theta^4 - \theta_a^4\right) - N^2(\theta - \theta_a) + G = 0 \end{aligned} \tag{13}$$

- For convex profile

$$\beta\left(\frac{d\theta}{dy}\right)^2 + [1 + \beta(\theta - \theta_a)]\frac{d^2\theta}{dy^2} - Nr4y\left(\theta^4 - \theta_a^4\right) - N^2(\theta - \theta_a)4y + G4y = 0 \tag{14}$$

where $\beta = \zeta T_b$ and convective environment boundary conditions are:

$$\frac{d\theta(0)}{dX} = 0,\ \ \theta(1) = 1$$

4. Solution Method with DTM

Equations (12)–(14) are reduced to the Taylor series using the properties mentioned in Table 1. We obtain:

- For rectangular profile

$$\begin{aligned} &(e+1)(e+2)Q(e+2) + \beta\sum_{f=0}^{e} Q(f)(e-f+1)(e-f+2)Q(e-f+2) + \beta\sum_{f=0}^{e}(f+1)Q(f+1)(e-f+1)Q(e-f+1) \\ &-\beta\theta_a(e+1)(e+2)Q(e+2) - Nr\sum_{f=0}^{e}\sum_{k=0}^{e-f}\sum_{m-0}^{f-k} Q(f)Q(e-f)Q(f-k)Q(k-m) - N^2Q(e) + \left(Nr\theta_a^4 + N^2\theta_a + G\right)\delta(f) = 0 \end{aligned} \tag{15}$$

- For exponential profile

$$\begin{aligned} &\sum_{f=0}^{e}\tfrac{a^f}{d!}(e-f+1)(e-f+2)Q(e-f+2) + \beta\sum_{f=0}^{e}\tfrac{a^f}{f!}\sum_{s=0}^{e-f}Q(f)(e-f-s+1)(e-f-s+2)Q(e-f-s+2) - \beta\theta_a\sum_{d=0}^{e}\tfrac{a^f}{f!}(e-f+1)(e-f+2)Q(e-f+2) \\ &+a\sum_{f=0}^{e}\tfrac{a^f}{f!}(e-f+1)Q(e-f+1) + a\beta\sum_{f=0}^{e}\tfrac{a^f}{f!}\sum_{s=0}^{e-f}Q(f)(e-f-s+1)Q(e-f-s+1) - a\beta\theta_a\sum_{f=0}^{e}\tfrac{a^f}{f!}\sum_{s=0}^{e-f}(e-f-s+1)Q(e-f-s+1) \\ &\beta\sum_{f=0}^{e}\tfrac{a^f}{f!}\sum_{s=0}^{e-f}(f+1)Q(f+1)(f-e-s+1)Q(f-e-s+1) - Nr\sum_{f=0}^{e}\sum_{k=0}^{e-f}\sum_{m-0}^{f-k}Q(f)Q(e-f)Q(f-k)Q(k-m) - N^2Q(e) + \left(Nr\theta_a^4 + N^2\theta_a + G\right)\delta(f) = 0 \end{aligned} \tag{16}$$

- For convex profile

$$(e+1)(e+2)Q(e+2)+\beta\sum_{f=0}^{e}(f+1)Q(f+1)(e-f+1)Q(e-f+1)+\beta\sum_{f=0}^{e}Q(f)(e-f+1)(e-f+2)Q(e-f+2)$$
$$-\beta\theta_a(e+1)(e+2)Q(e+2)-4Nr\sum_{f=0}^{e}\sum_{k=0}^{f-e}\sum_{s-0}^{k-f}\sum_{m=0}^{s-k}\delta(f-1)Q(f-e)Q(k-f)Q(s-k)U(s-m)+4Nr\theta_a^4\sum_{f=0}^{e}\delta(f-1)Q(e-f) \quad (17)$$
$$-4N^2\sum_{f=0}^{e}\delta(f-1)Q(e-f)+4N^2\theta_a\sum_{f=0}^{e}\delta(f-1)Q(e-f)+4G\delta(e-1)=0$$

Boundary conditions can be reduced to:

$$Q(1)=0,\ \sum_{d=0}^{\infty}Q(f)=1 \quad (18)$$

Considering $Q(0)=a$ and using boundary constraints with the assistance of MATLAB software, the terms of the series can be obtained as follows:

- For rectangular profile

$$\begin{aligned}
Q[2] &= \frac{-G+aN^2+a^4Nr-N^2\theta_a-Nr\theta_a^4}{2(1+a\beta-\beta\theta_a)}\\
Q[3] &= 0\\
Q[4] &= \frac{N^2Q[2]-6\beta Q[2]^2}{12(1+a\beta-\beta\theta_a)}\\
Q[5] &= 0\\
Q[6] &= \frac{N^2Q[4]-30\beta Q[2]Q[4]}{30(1+a\beta-\beta\theta_a)}\\
Q[7] &= 0
\end{aligned} \quad (19)$$

and so forth

- For exponential profile

$$\begin{aligned}
Q[2] &= \frac{-G+aN^2+a^4Nr-N^2\theta_a-Nr\theta_a^4}{2(1+a\beta-\beta\theta_a)}\\
Q[3] &= \frac{-2aQ[2]-a\beta Q[2]-a^2\beta Q[2]+2a\beta\theta_a Q[2]}{3(1+a\beta-\beta\theta_a)}\\
Q[4] &= \frac{\begin{array}{l}-3a^2Q[2]+N^2Q[2]-2a\beta Q[2]-2a^2\beta Q[2]+3a^2\beta\theta_a Q[2]-4a\beta Q[2]^2-a^2\beta Q[2]^2-9aQ[3]-6a\beta Q[3]\\-3a^2\beta Q[3]+9a\beta\theta_a Q[3]\end{array}}{12(1+a\beta-\beta\theta_a)}
\end{aligned} \quad (20)$$

... and so forth

- For convex profile

$$\begin{aligned}
Q[2] &= 0\\
Q[3] &= -\frac{2\left(G-aN^2+aN^2\theta_a+aNr\theta_a^4\right)}{3(1+a\beta-\beta\theta_a)}\\
Q[4] &= 0\\
Q[5] &= 0\\
Q[6] &= \frac{4N^2Q[3]-4N^2\theta_aQ[3]-4Nr\theta_a^4Q[3]-15\beta Q[3]^2}{30(1+a\beta-\beta\theta_a)}\\
Q[7] &= 0\\
Q[8] &= 0
\end{aligned} \quad (21)$$

... and so forth

By substituting Equation (19) in Equation (4) for $H=1$

- For rectangular profile

$$\theta(X)=a+\frac{-G+aN^2+a^4Nr-N^2\theta_a-Nr\theta_a^4}{2(1+a\beta-\beta\theta_a)}X^2+\frac{N^2Q[2]-6\beta Q[2]^2}{12(1+a\beta-\beta\theta_a)}X^4+\dots \quad (22)$$

To obtain the values of a, we use Equation (18)

$$\theta(1) = a + \frac{-G + aN^2 + a^4Nr - N^2\theta_a - Nr\theta_a^4}{2(1 + a\beta - \beta\theta_a)} + \frac{N^2Q[2] - 6\beta Q[2]^2}{12(1 + a\beta - \beta\theta_a)} + \ldots \tag{23}$$

Solving Equation (23) using MATLAB software we obtain the exact value of a. The same procedure is repeated for the other profiles.

Fin Efficiency

The amount of heat transferred in a fin is determined with the help of the parameter called efficiency. It is a correlation between the actual heat shift in a fin to heat that would be transmitted if a complete fin is of the temperature of the fin bed. The non-dimensional equation for the efficiency of a rectangular profile is given by:

$$\eta = \frac{(1 + \beta(\theta - \theta_a))\left(\frac{d\theta}{dx}\right)_{x=1}}{Nr(\theta_b^4 - \theta_a^4) + N^2(\theta_b - \theta_a) - G} \tag{24}$$

5. Results

The current investigation presents the exploration of temperature differences associated with variable thermal conductivity, internal heat generation and radiation over the longitudinal fin of different profiles. The dimensionless energy equations of the fins are solved using the DTM. Results mainly referring to temperature field and thermal profiles are depicted graphically for three types of fins, namely rectangular, exponential and convex. The effects of Nr, G, Nc, θ_a and β on temperature fields are analyzed and discussed. Moreover, for all results reported here, the following values of variables are used unless otherwise indicated by the graphs or tables: $\beta = 0.5,\ N = 1,\ G = 0.1,\ Nr = 1\&\theta_a = 0.4$. Thermal analysis is performed and discussed using ANSYS software. The results of the present study are compared with the existing results of Languri et al. [42] and Arslanturk [43] (Table 2).

Table 2. Comparison of $\theta(X)$ obtained by different studies for rectangular fins by considering $\beta = 0,\ G = 0,\ Nr = 0,\ \theta_a = 0$ and $N = 0.5$.

X	HPM (Languri et al. [42])	ADM (Arslanturk [43])	VIM (Languri et al. [42])	DTM (Current Study)
		$\theta(X)$		
0	0.886819	0.886819	0.886819	0.886818
0.2	0.891257	0.891257	0.891257	0.8912567
0.4	0.904614	0.904615	0.904614	0.940614
0.6	0.927026	0.927026	0.927026	0.927027
0.8	0.958715	0.958716	0.958715	0.958715
1	1.000000	1.000000	1.000000	1.000000

The fluctuation in fin temperature due to variable heat conduction (β) is shown in Figure 3 for three different profiles. From this graph, it is noticed that the thermal gradient reduces gradually from the base to the tip of the fin for different values of β. The increment in β enhances the temperature field due to heat loss to the surrounding fluid from the fin surface. The results show that the fin-tip temperature for exponential profiles is greater than that of the other profiles.

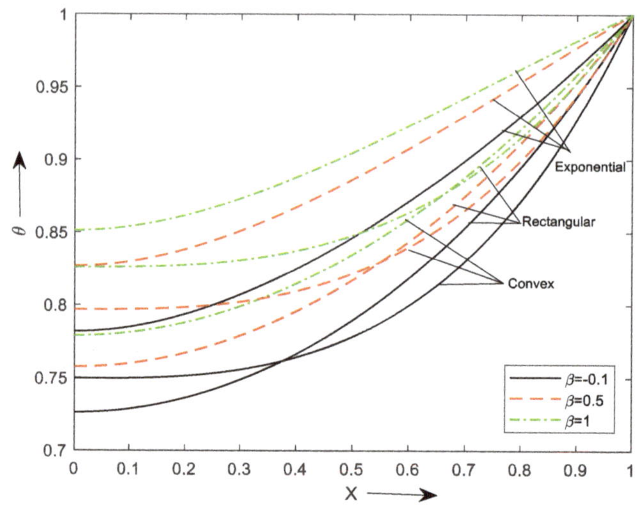

Figure 3. Temperature distribution of different profiles for diverse values of β.

Figure 4 shows the influence of the N on the thermal attribute of the fin. As the parameter N intensifies, the enriched heating pattern in the fin is noted which reduces the temperature rise. The contribution of this flow parameter is significant for enhancing the thermal transport of the fin. From the graph, we can notice that the exponential profile shows better performance which is followed by the rectangular and convex for various values on radiation parameters.

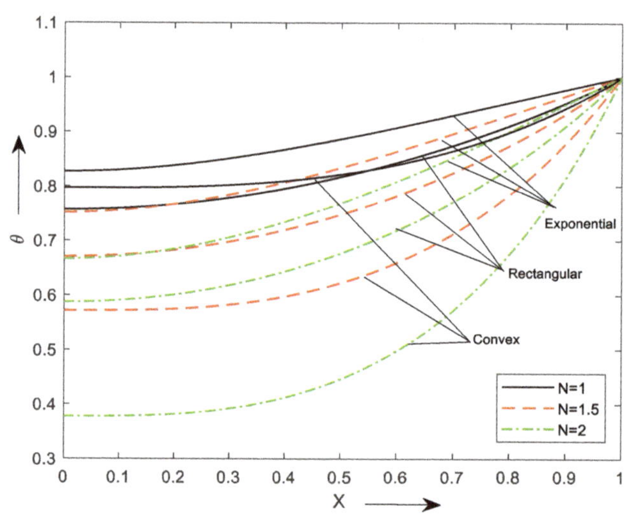

Figure 4. Temperature distribution of different profiles for diverse values of N.

The effect of the internal heat generation parameter (G) is depicted in Figure 5. For this, it is observed that the temperature of the fin can be enhanced with the values of G. Higher heat generation enhances fin temperature in steady-state conditions owing to the fact of larger dissipation of the heating environment due to the fin.

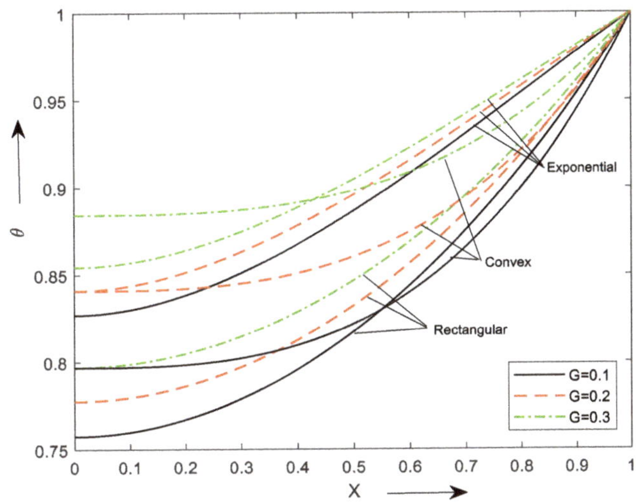

Figure 5. Temperature distribution of different profiles for diverse values of G.

The radiative parameter impact (Nr) is shown in Figure 6. With an increment in the radiation number, the thermal profile θ decreases steadily. The lower temperature inside the fin indicates a loss of ambient fluid temperature with radiative parameters.

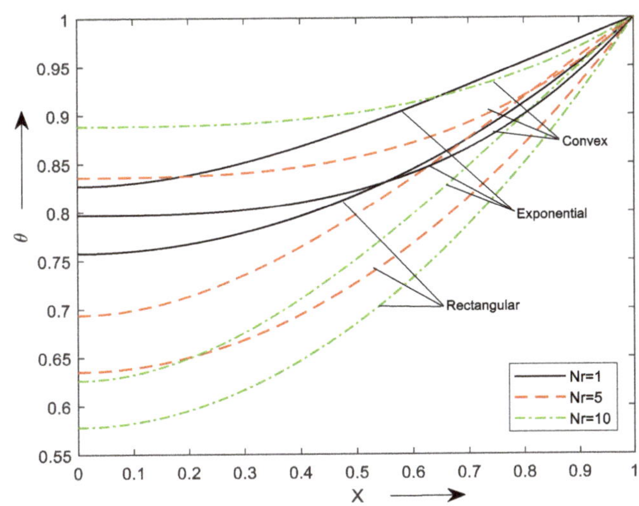

Figure 6. Behavior Nr on θ.

The Figure 7 shows variation in dimensionless ambient temperature (θ_a) on the temperature field. As θ_a increases, the temperature of the surrounding liquid increases, which affects the rate of heat transmission from fin to surface. This is noted with a rise in θ_a.

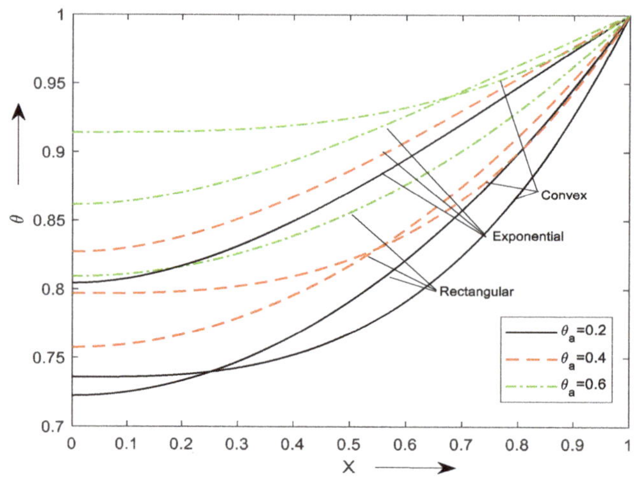

Figure 7. Behavior θ_a on θ.

The effect of parameters on temperature for rectangular, exponential and convex profiles is obtained on the same graph to understand the difference between each profile. Similar observations discussed above can be seen in the other two profiles. Meanwhile, the exponential-shaped fins exhibit better performance than the others. Heat transference at the fin base is an important study, which has many applications and is of the form

$$Q_b = \frac{d\theta(1)}{dX}$$

The effect of simultaneously varying Q_b with N for two different values of β and θ_a can be seen in Figure 8a,b, respectively, for all three profiles. From the graphs, it can be concluded that the value of Q_b is inversely related to the values of β and θ_a. Heat transfer is more at the base and then reduces to become constant at the fin tip. Which shows that the fin cools down earlier at the tip.

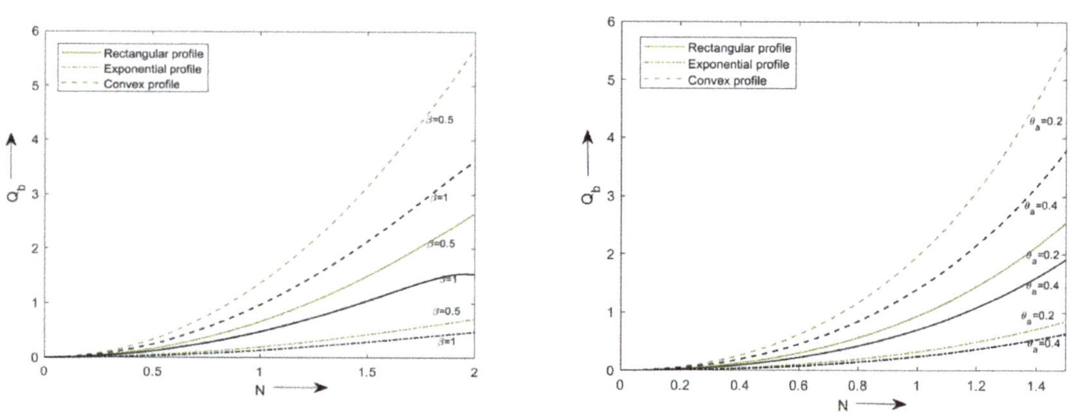

(a) (b)

Figure 8. (**a**) Variation of Q_b with N for several assigned values of β. (**b**) Variation of Q_b with N for several assigned values of θ_a.

6. Thermal Analysis

ANSYS is a tool that helps us understand the routine of a model from our study in a virtual environment. It uses governing equations to study the behaviors of the problem. To investigate the thermal behaviors, the following assumption is made on the fins:

- Aluminum alloy (AA6061) is considered a fin material as it is a good thermal and electrical conductor with heat conduction of 300 W/m K.
- Heat conduction is considered 1D and longitudinal.
- h is considered to be 39.9 W/m^{2}K above the fin surface.
- The fin base is kept at 550 K, and 283 is the ambient temperature.

Figure 9a–c illustrate longitudinal fin thermal propagation for rectangular, exponential and convex profiles, respectively. The maximum temperature observed was 550 K in all three profiles, and the fin tip temperature was 546 K, 530.28 K and 545 K, respectively. The temperature gradually decreased from the bed of the fin to the tip. Exponential fins have better results compared to other profiles. The results are drawn from the thermal analysis, which agrees well with our numerical results.

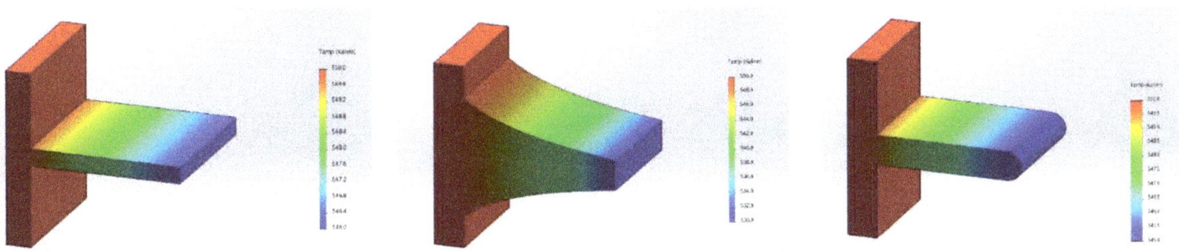

(a) (b) (c)

Figure 9. Temperature distribution of (**a**) rectangular; (**b**) exponential; (**c**) convex profile for aluminum alloy (AA6061).

The efficiency of the fin for several values of internal heat generation (G) versus the thermal expansion coefficient can be seen in Figure 10a. From the graph, we can depict that a smaller value of β efficiency is higher and decreases gradually. Moreover, as the value of heat production is increased, the efficiency is enhanced. This shows that by keeping the values of β smaller and values of G higher we can obtain efficient fins. A similar observation can be observed with the three different profiles considered in our study, but the exponential fin has fin efficiency in general.

Figure 10b shows the efficiency of the fin versus θ_a for different values of a radiative parameter. It can be observed that for a lower value of a radiative parameter and θ_a the efficiency is higher and reduces gradually as the values are increased. An exponential profile with a lower value of θ_a and Nr can be used to obtain the higher efficiency of the fin.

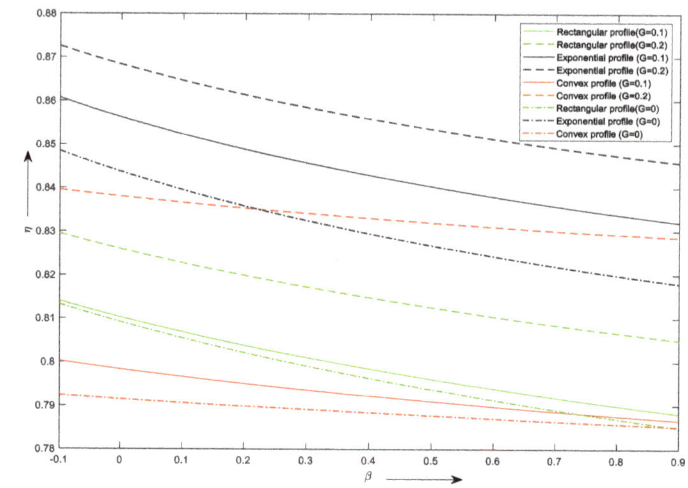

(a)

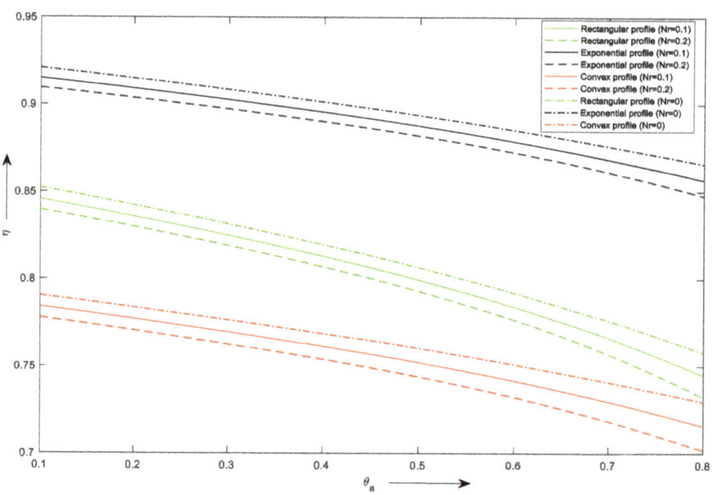

(b)

Figure 10. Influence of: (**a**) G and β on fin efficiency; (**b**) Nr and θ_a on fin efficiency.

7. Conclusions

The framework for temperature rate is presented in a longitudinal fin subject to internal heating, variable thermal conductivity and convective radiation. The Rosseland theory is used to determine the features of a radiative phenomenon. DTM approximations are followed for the simulation process. Graphical explanations are manifested for the consequence of parameters in the heat transfer of the fin. The key findings of this analysis are as follows:

- Upon enhancing the convection–conduction parameter, the thermal dispersal in the fin lowers.

- A strengthened heat transfer fine is observed for the radiative-conduction constant.
- The thermal rate of the fin improves with an augmented change in a heat-generating parameter.
- This scrutiny convinces us that DTM algorithms are efficient and convenient methods for nonlinear differential systems.
- Thermal radiation and natural convection have a significant influence on the cooling of a fin.
- In the steady state, fins dissipate heat to the environment because heat production within a fin surges the temperature of the fins.
- The temperature scatters of a fin for different profiles are calculated using the ANSYS software, considering aluminum alloy (AA6061) as the fin body material. The fin base has a higher temperature and reduces drastically toward the fin tip.

This work can be extended by considering the porous fins in the presence of a magnetic field and also by considering the porous fins with the nano and hybrid nanofluid with the effect of the shape factor.

Author Contributions: Conceptualization, M.K.A. and P.V.A.S.; Data curation, S.V.K.V.; Formal analysis, P.J.; Funding Hacquisition, P.J.; Methodology, A.M.Z.; Software, B.N.H.; Validation, C.S.K.R.; Writing—original draft, P.V.A.S.; Writing—review & editing, N.A.S. All authors have read and agreed to the published version of the manuscript.

Funding: This work received no external funds.

Institutional Review Board Statement: Not applicable.

Informed Consent Statement: Not applicable.

Data Availability Statement: All the data are present in this article.

Acknowledgments: The authors extend their appreciation to the Deanship of Scientific Research at King Khalid University, Abha 61413, Saudi Arabia, for funding this work through a research group program under grant number R.G.P.-2/21/43. This research received funding support from the NSRF via the Program Management Unit for Human Resources & Institutional Development, Research and Innovation, (grant number B05F650018).

Conflicts of Interest: The authors declare no conflict of interest.

Nomenclature

P	fin cross-section (m^2)
a	exponential parameter
h	heat transfer coefficient ($wm^{-1}k^{-1}$)
k	heat conduction ($wm^{-1}k^{-1}$)
Nr	radiative parameter
G	heat generation parameter
L	fin length (m)
N	convective parameter
T	temperature (k)
φ	transformed function
ϕ	original analytic function
a	fin base temperature
β	thermal expansion coefficient (K^{-1})
ζ	dimensional constant (K^{-1})
η	efficiency of the fin
U	transformed equation
θ	dimensionless temperature
a	ambient temperature
b	base of the fin

References

1. Kraus, A.D.; Aziz, A.; Welty, J.R. *Extended Surface Heat Transfer*; John Wiley: Hoboken, NJ, USA, 2002.
2. Gireesha, B.J.; Sowmya, G. Heat transfer analysis of an inclined porous fin using Differential Transform Method. *Int. J. Ambient Energy* **2022**, *43*, 3189–3195. [CrossRef]
3. Oleśkowicz-Popiel, C.; Blanch, R.O.; Wojtkowiak, J. Efficiency of the horizontal single pin fin subjected to free convection and radiation heat transfer. *Heat Transf. Eng.* **2007**, *28*, 299–309. [CrossRef]
4. Minkler, W.S.; Rouleau, W.T. The Effects of Internal Heat Generation on Heat Transfer in Thin Fins. *Nucl. Sci. Eng.* **1960**, *7*, 400–406. [CrossRef]
5. Venkitesh, V.; Mallick, A. Thermal analysis of a convective–conductive–radiative annular porous fin with variable thermal parameters and internal heat generation. *J. Anal. Calorim.* **2022**, *147*, 1519–1533. [CrossRef]
6. Majhi, T.; Kundu, B. New Approach for Determining Fin Performances of an Annular Disc Fin with Internal Heat Generation. In *Advances in Mechanical Engineering*; Springer: Singapore, 2020; pp. 1033–1043.
7. Das, R.; Kundu, B. Prediction of Heat-Generation and Electromagnetic Parameters from Temperature Response in Porous Fins. *J. Thermophys. Heat Transf.* **2021**, *35*, 761–769. [CrossRef]
8. Sobamowo, M.G. Analysis of convective longitudinal fin with temperature-dependent thermal conductivity and internal heat generation. *Alex. Eng. J.* **2017**, *56*, 1–11. [CrossRef]
9. Turkyilmazoglu, M. Exact heat-transfer solutions to radial fins of general profile. *J. Heat Trans.* **2016**, *30*, 89–93. [CrossRef]
10. Ndlovu, P.L.; Moitsheki, R.J. Steady state heat transfer analysis in a rectangular moving porous fin. *Propuls. Power Res.* **2020**, *9*, 188–196. [CrossRef]
11. Singla, R.K.; Das, R. Application of decomposition method and inverse prediction of parameters in a moving fin. *Energy Convers. Manag.* **2014**, *84*, 268–281. [CrossRef]
12. Shah, N.A.; Wakif, A.; El-Zahar, E.R.; Ahmad, S.; Yook, S.-J. Numerical simulation of a thermally enhanced EMHD flow of a heterogeneous micropolar mixture comprising (60%)-ethylene glycol (EG), (40%)-water (W), and copper oxide nanomaterials (CuO). *Case Stud. Therm. Eng.* **2022**, *35*, 102046. [CrossRef]
13. Sajjan, K.; Shah, N.A.; Ahammad, N.A.; Raju, C.S.K.; Kumar, M.D.; Weera, W. Nonlinear Boussinesq and Rosseland approximations on 3D flow in an interruption of Ternary nanoparticles with various shapes of densities and conductivity properties. *AIMS Math.* **2022**, *7*, 18416–18449. [CrossRef]
14. Aziz, A.; Khani, F. Convection-radiation from a continuously moving fin of variable thermal conductivity. *J. Frankl. Inst.* **2011**, *348*, 640–651. [CrossRef]
15. Sowmya, G.; Sarris, I.E.; Vishalakshi, C.s.; Kumar, R.S.V.; Prasannakumara, B.C. Analysis of transient thermal distribution in a convective–radiative moving rod using two-dimensional differential transform method with MULTIVARIATE pade approximant. *Symmetry* **2021**, *13*, 1793. [CrossRef]
16. Hatami, M.; Ganji, D.D. Thermal performance of circular convective-radiative porous fins with different section shapes and materials. *Energy Convers. Manag.* **2013**, *76*, 185–193. [CrossRef]
17. Pasha, A.v.; Jalili, P.; Ganji, D.D. Analysis of unsteady heat transfer of specific longitudinal fins with Temperature-dependent thermal coefficients by DTM. *Alex. Eng. J.* **2018**, *57*, 3509–3521. [CrossRef]
18. Torabi, M.; Aziz, A.; Zhang, K. A comparative study of longitudinal fins of rectangular, trapezoidal and concave parabolic profiles with multiple nonlinearities. *Energy* **2013**, *51*, 243–256. [CrossRef]
19. Zhang, H.; Sun, K. Conduction from longitudinal fin of rectangular profile with exponential vary heat transfer coefficient. *Adv. Mater. Res.* **2013**, *614–615*, 311–314. [CrossRef]
20. Buikis, A.; Pagodkina, I. Comparison of Analytical and Numerical Solutions for a Two-Dimensinal Longitudinal Fin of Rectangular Profile. *Latv. J. Phys. Technol. Sci.* **1996**, *33*, 177–187.
21. Kezzar, M.; Tabet, I.; Eid, M.R. A new analytical solution of longitudinal fin with variable heat generation and thermal conductivity using DRA. *Eur. Phys. J. Plus* **2020**, *135*, 120. [CrossRef]
22. Shi, Q.H.; Hamid, A.; Khan, M.I.; Naveen Kumar, R.; Punith Gowda, R.J.; Prasannakumara, B.C.; Shah, N.A.; Khan, S.U.; Chung, J.D. Numerical study of bio-convection flow of magneto-cross nanofluid containing gyrotactic microorganisms with activation energy. *Sci Rep.* **2021**, *11*, 16030. [CrossRef] [PubMed]
23. Afsharpanah, F.; Ajarostaghi, S.S.M.; Arıcı, M. Parametric study of phase change time reduction in a shell-and-tube ice storage system with anchor-type fin design. *Int. Commun. Heat Mass Transf.* **2022**, *137*, 106281. [CrossRef]
24. Afsharpanah, F.; Cheraghian, G.; Hamedani, F.A.; Shokri, E.; Soheil, S.; Ajarostaghi, M. Utilization of Carbon-Based Nanomaterials and Plate-Fin Networks in a Cold PCM Container with Application in Air Conditioning of Buildings. *Nanomaterials* **2022**, *12*, 1927. [CrossRef] [PubMed]
25. Sabu, A.S.; Wakif, A.; Areekara, S.; Mathew, A.; Shah, N.A. Significance of nanoparticles' shape and thermo-hydrodynamic slip constraints on MHD alumina-water nanoliquid flows over a rotating heated disk: The passive control approach. *Int. Commun. Heat and Mass Transf.* **2021**, *129*, 105711. [CrossRef]
26. Salhi, J.E.; Ajarostaghi, S.S.M.; Zarrouk, T.; Pour, M.S.; Salhi, N.; Salhi, M. Turbulence and thermo-flow behavior of air in a rectangular channel with partially inclined baffles. *Energy Sci. Eng.* **2022**, *10*, 3540–3558. [CrossRef]
27. Demartini, L.C.; Vielmo, H.A.; Möller, S.v. Numeric and Experimental Analysis of the Turbulent Flow through Channel With Baffle Plates. *J. Braz. Soc. Mech. Sci. Eng.* **2004**, *28*, 233–241. [CrossRef]

28. Kadijani, O.N.; Moghadam, H.K.; Ajarostaghi, S.S.M.; Asadi, A.; Pour, M.S. Hydrothermal performance of humid air flow in a rectangular solar air heater equipped with V-shaped ribs. *Energy Sci. Eng.* **2022**, *10*, 2276–2289. [CrossRef]
29. Zhou, J.K. *Differential Transformation Method and Its Application for Electrical Circuits*; Hauzhang University Press: Wuhan, China, 1986.
30. Ayaz, F. Solutions of the system of differential equations by differential transform method. *Appl. Math. Comput.* **2004**, *147*, 547–567. [CrossRef]
31. Chen, C.K.; Ho, S.H. Solving partial differential equations by two-dimensional differential transform method. *Appl. Math. Comput.* **1999**, *106*, 171–179.
32. Ayaz, F. On the two-dimensional differential transform method. *Appl. Math. Comput.* **2003**, *143*, 361–374. [CrossRef]
33. Kanth, A.S.V.R.; Aruna, K. Differential transform method for solving linear and non-linear systems of partial differential equations. *Phys. Lett. Sect. A Gen. At. Solid State Phys.* **2008**, *372*, 6896–6898. [CrossRef]
34. Fallo, N.; Moitsheki, R.J.; Makinde, O.D. Analysis of heat transfer in a cylindrical spine fin with variable thermal properties. *Defect Diffus. Forum* **2018**, *387*, 10–22. [CrossRef]
35. Chiba, R. A Series Solution for Heat Conduction Problem with Phase Change in a Finite Slab. *Abstr. Appl. Anal.* **2014**, 684293. [CrossRef]
36. Moradi, A.; Hayat, T.; Alsaedi, A. Convection-radiation thermal analysis of triangular porous fins with temperature-dependent thermal conductivity by DTM. *Energy Convers. Manag.* **2014**, *77*, 70–77. [CrossRef]
37. Ndlovu, P.L. Analytical study of transient heat transfer in a triangular moving porous fin with temperature dependant thermal properties. *Defect Diffus. Forum* **2019**, *393*, 31–46. Available online: www.scientific.net/DDF.393.31 (accessed on 1 August 2022). [CrossRef]
38. Abbasi, H.; Javed, A. Implementation of differential transform method (DTM) for large deformation analysis of cantilever beam. *IOP Conf. Ser. Mater. Sci. Eng.* **2020**, *899*, 012003. [CrossRef]
39. Aksoy, G. Application of differential transformation method for an annular fin with variable thermal conductivity. *Therm. Sci.* **2021**, 315. [CrossRef]
40. Joneidi, A.A.; Ganji, D.D.; Babaelahi, M. Differential Transformation Method to determine fin efficiency of convective straight fins with temperature dependent thermal conductivity. *Int. Commun. Heat Mass Transf.* **2009**, *36*, 757–762. [CrossRef]
41. Franco, A. An analytical method for the optimum thermal design of convective longitudinal fin arrays. *Heat Mass Transf.* **2009**, *45*, 1503–1517. [CrossRef]
42. Languri, E.M.; Ganji, D.D.; Jamshidi, N. Variational Iteration and Homotopy perturbation methods for fin efficiency of convective straight fins with temperature dependent thermal conductivity. In Proceedings of the 5th WSEAS International Conference on Fluid Mechanics (Fluids 08), Acapulco, Mexico, 25–27 January 2008; Volume 25.
43. Arslanturk, C. A decomposition method for fin efficiency of convective straight fins with temperature-dependent thermal conductivity. *Int. Commun. Heat Mass Transf.* **2005**, *32*, 831–841. [CrossRef]

MDPI
St. Alban-Anlage 66
4052 Basel
Switzerland
www.mdpi.com

Mathematics Editorial Office
E-mail: mathematics@mdpi.com
www.mdpi.com/journal/mathematics

www.ingramcontent.com/pod-product-compliance
Lightning Source LLC
Chambersburg PA
CBHW061603220926
51850CB00005B/409